Modern Methods of Plant Analysis

New Series Volume 1

Editors

H.F. Linskens, Nijmegen

J.F. Jackson, Adelaide

Cell Components

Edited by
H.F. Linskens and J.F. Jackson

Contributors

J.M. Anderson B. Andersson G.A. Berkowitz K. Cline
M. Gibbs R. Goldberg T. Hirokawa A.H.C. Huang
J.F. Jackson R.L. Jones B.A. Larkins Ch. Larsson
J.M. Miernyk G. Nagahashi C. Paech D.D. Sabnis
H. Takeda S.S. Thayer G. Wagner

With 96 Figures

Springer-Verlag
Berlin Heidelberg New York Tokyo

Professor Dr. HANS-FERDINAND LINSKENS
Botanisch Laboratorium
Faculteit der Wiskunde en Natuurwetenschappen
Katholieke Universiteit
Toernooiveld
NL-6525 ED Nijmegen
The Netherlands

Professor Dr. JOHN F. JACKSON
Department of Biochemistry
Waite Agricultural Research Institute
University of Adelaide
Glen Osmond, S.A. 5064
Australia

ISBN-13:978-3-642-82589-7 e-ISBN-13:978-3-642-82587-3
DOI: 10.1007/978-3-642-82587-3

2131/3130-543210

Introduction

Modern Methods of Plant Analysis

When the handbook *Modern Methods of Plant Analysis* was first introduced in 1954 the considerations were

1. the dependence of scientific progress in biology on the improvement of existing and the introduction of new methods;
2. the inavailability of many new analytical methods concealed in specialized journals not normally accessible to experimental plant biologists;
3. the fact that in the methods sections of papers the description of methods is frequently so compact, or even sometimes so incomplete, that experiments are difficult to reproduce.

These considerations still stand today.

The series was highly successful, seven volumes appearing between 1956 and 1964. Since today there is still a demand for the old series, the publisher has decided to resume publication of *Modern Methods of Plant Analysis*. It is hoped that the New Series will be as acceptable to those working in plant sciences and related fields as the early volumes undoubtedly were. It is difficult to single out the major reasons for success of any publication, but we believe that the methods published in the first series were up-to-date at the time and the descriptions as applied to plant material so complete in themselves that there was little need to consult other publications.

Editorial

The earlier series of *Modern Methods of Plant Analysis* was initiated by Michel V. Tracey, at that time in Rothamsted, later in Sydney, and by the late Karl Paech (1910–1955), at that time at Tübingen. The New Series will be edited by Paech's successor H. F. Linskens (Nijmegen, The Netherlands) and John F. Jackson (Adelaide, South Australia). Like the earlier editors, we are convinced "that there is a real need for a collection of reliable up-to-date methods for plant analysis in large areas of applied biology ranging from agriculture and horticultural experiment stations to pharmaceutical and technical institutes concerned with raw material of plant origin". The recent developments in the field of plant biotechnology and genetic engineering make it even more important for workers in the plant sciences to become acquainted with the more sophisticated methods, sometimes originating from biochemistry and biophysics, which have been developed in commercial firms, space science laboratories, nonuniversity research institutes and medical establishments.

Concept of the New Series

Many methods described in the biochemical, biophysical, and medical literature cannot be applied directly to plant material because of the special cell structure, surrounded by a tough cell wall, and the general lack of knowledge of the specific behavior of plant raw material during extraction procedures. Therefore all authors of this New Series have been chosen for their special experience in handling plant material, resulting in the adaptation of methods to problems of plant metabolism. Nevertheless each particular material from a plant species may require some modification of described methods and usual techniques. The methods are described critically, with indications as to their limitations. In general it will be possible to adapt the described methods to the specific needs of the users of this series, but nevertheless references have been made to the original papers and authors. While the editors have worked to plan in this New Series and made efforts to ensure that the aims and general layout of the contributions are within the general guidelines indicated above, we have tried not to interfere too much with the personal style of each author.

The First Volume – Cell Components and Organelles

The first volume in the New Series deals with *Cell Components and Organelles*. This has been planned so that many of the chapters contained in it will be useful as ground work for later volumes; the main point of the articles remains nevertheless plant analysis. Often with cell fractionation one must opt for either purity or maximum yield, as has been pointed out in the relevant chapters, relating as it does to methods which can be adopted for either analytical or preparative uses. In contrast to many other publications dealing with cell fractionation, we begin with two chapters on the cell wall, in the belief that it is this structure which sets most plant cells apart from their animal and bacterial counterparts. The tough, outer cell wall has made studies of plant cellular components extremely difficult in the past, and until the advent of protoplast formation, had rendered preparations for biochemical investigations extremely difficult. Perhaps the most significant breakthrough in plant cell fractionation in the past decade has been the use of enzyme breakdown of the cell-wall barrier, to produce protoplasts. This has allowed gentler techniques to be applied for functional organelle isolation. Protoplast formation is therefore dealt with in the third chapter, and is followed by a chapter giving practical guidance in the use of "markers" in cell fractionation. A treatment of the components and organelles then follows, from vacuoles to the various membranes of the cell, to the microtubules, chloroplasts, mitochondria, microbodies, protein and lipid bodies, and to the plant cell nucleus.

Finally, this New Series can be considered a continuation of the older one, brought up-to-date, so that the title *Modern Methods of Plant Analysis* remains justified. The continuity is demonstrated by the inclusion of a chapter by the son of the founder of the series, that by Dr. Christian Paech (Brookings, South Dakota).

The editors express their gratitude for the excellent cooperation with the publisher, with Dr. Dieter Czeschlik and especially with Ms. Linda Teppert.

Nijmegen and Adelaide, November 1985

H. F. LINSKENS
J. F. JACKSON

Contents

Cell-Wall Chemistry, Structure and Components
H. TAKEDA and T. HIROKAWA (With 16 Figures)

Protoplasts – for Compartmentation Studies
S. S. THAYER (With 2 Figures)

The Marker Concept in Cell Fractionation
G. NAGAHASHI (With 3 Figures)

Plasma Membranes
CH. LARSSON (With 10 Figures)

Vacuoles
G. WAGNER (With 1 Figure)

The Chloroplast Thylakoid Membrane – Isolation, Subfractionation and Purification of Its Supramolecular Complexes
B. ANDERSSON and J. M. ANDERSON (With 8 Figures)

The Isolation and Characterization of Nongreen Plastids
J. A. MIERNYK (With 9 Figures)

Mitochondria
J. F. Jackson (With 2 Figures)

Endoplasmic Reticulum
R. L. Jones (With 6 Figures)

Polyribosomes
B. A. LARKINS (With 10 Figures)

The Nucleus – Cytological Methods and Isolation for Biochemical Studies
J. F. JACKSON (With 3 Figures)

List of Contributors

ANDERSON, JAN M., CSIRO, Division of Plant Industry, P.O. Box 1600, Canberra ACT 2601, Australia

ANDERSSON, BERTIL, Department of Biochemistry, University of Lund, P.O. Box 124, S-221 00 Lund, Sweden

BERKOWITZ, GERALD A.,Horticulture and Forestry Department, Cook College, Rutgers University, New Brunswick, NJ 08903, USA

CLINE, KENNETH, Fruit Crops Department, University of Florida, Gainesville, FL 32611, USA

GIBBS, MARTIN, Institute for Photobiology of Cells and Organelles, Brandeis University, Waltham, MA 02254, USA

GOLDBERG, RENÉE, Laboratoire des Biomembranes et surfaces cellulaires végétales – Ecole Normale Supérieure, 24 rue Lhomond, F-75231 Paris Cedex 05, France

HIROKAWA, TOYOYASU, Department of Biology, Faculty of Science, Niigata University, Niigata 950-21, Japan

HUANG, ANTHONY H. C., Biology Department, University of South Carolina, Columbia, SC 29208, USA

JACKSON, JOHN F., Department of Biochemistry, Waite Agricultural Research Institute, University of Adelaide, Glen Osmond, S.A. 5064, Australia

JONES, RUSSELL L., Department of Botany, University of California, Berkeley, CA 94720, USA

LARKINS, BRIAN A., Department of Botany and Plant Pathology, Purdue University, West Lafayette, IN 47907, USA

LARSSON, CHRISTER, Department of Plant Physiology, University of Lund, P.O. Box 7007, S-220 07 Lund, Sweden

MIERNYK, JAN A., Biochemistry Department, 322A Chemistry Building, The University of Missouri, Columbia, MO 65211, USA

NAGAHASHI, GERALD, ARS, U.S. Department of Agriculture, Eastern Regional Research Center, 600 East Mermaid Lane, Philadelphia, PA 19118, USA

PAECH, CHRISTIAN, Department of Chemistry, Station Biochemistry, South Dakota State University, Brookings, SD 57007, USA

SABNIS, DINKAR D., Department of Plant Science, University of Aberdeen, Aberdeen, Scotland

TAKEDA, HIROSHI, Department of Biology, College of General Education, Niigata University, Niigata 950-21, Japan

THAYER, SUSAN S., Plant Growth Laboratory, University of California, Davis, CA 95616, USA

WAGNER, G., N-212F ASCN, University of Kentucky, Lexington, KY 40546-0091, USA

Cell-Wall Isolation, General Growth Aspects

R. Goldberg

1 Introduction

Cell walls have been described since the 17th century. This outer layer, surrounding the plasmalemma, was long considered as a passive, inert, plant skeleton, an intercellular cement which directed the shape of cells and consequently of the whole plant. Cell walls account for 20 to 60% of tissue dry matter. This insoluble material has been used by mankind since the beginning of time and its chemical composition has been thoroughly investigated throughout the 20th century. However, the wall's physiological functions were recognized only relatively recently. Cell walls may be considered from at least three different view points. (1) As a cell compartment between the external medium and the protoplasm, this barrier will affect absorption and exorption processes. (2) As a rigid envelope, its mechanical properties will limit the extension potential of the cell. (3) As the most external cellular layer, it is involved in recognition processes: for instance, some cell-wall polysaccharides might act as effectors in recognition systems between host and pathogens (Albersheim et al. 1980). Recently, knowledge of wall functions has greatly expanded and many experiments have been carried out on isolated cell walls. The purpose of this chapter is to answer two questions: first, how can pure isolated cell walls be obtained; second, what aspects can be studied with these isolated cell walls. We shall systematically consider the improved methods used for isolating cell walls and then the approaches adopted to investigate their composition, structure, and properties. Processes involved in cell-wall extension will be discussed separately.

2 Isolation Procedures

Isolation of plant cell walls firstly requires cell breakage and subsequent removal of all intracellular components (see Harris 1983). Cell-wall components should ideally be recovered in their totality, unmodified, and physiologically intact. All operations therefore have to be carried out at 3.5 °C to prevent enzymic degradation of cell-wall components. Only suspension cultures can be considered as a source of homogenous primary walls. In most cases, however, cell walls are isolated from organs which contain cells of several types and the preparations obtained are therefore heterogeneous.

2.1 Cell Breakage

Cell walls are generally isolated from fresh material but sometimes the tissue is previously frozen in liquid nitrogen and stored. The most commonly used media are aqueous (H_2O, buffers or hypertonic sucrose solutions which plasmolyze the cells) but since Kivilaan et al. (1959) glycerol has been used by a number of workers. Techniques chosen to break plant cells depend on the plant material. Suspension cultures are often broken by passage through a French pressure cell or by sonication. For sectioned organs (hypocotyls, epicotyls, coleoptiles, roots), mortar and pestle have generally been replaced by blenders with motors. Lignified tissues often are first ground in liquid nitrogen, lyophilized and the powder homogenized in a razor blender. With culture cells it is possible to avoid cell breakage and nevertheless to obtain sufficient quantities of wall components. These cells, indeed, are known to secrete into the medium mixtures of extracellular polysaccharides which are similar in composition to the components of the walls. With organs, Terry and Bonner (1980) have described a method enabling isolation without breakage, of the soluble (or weakly bound) components present in the periplasmic space. This is based on removal of components from the walls by low speed centrifugation (1000 *g*) of excised tissue segments. The solution centrifuged from the walls is devoid of contamination from cytoplasmic components, and the tissues remain able to grow and to respond to auxin.

2.2 Cell-Wall Recovery

The cell walls may be recovered from the homogenate by centrifugation at low speeds (250–1000 *g* for 5–15 min) followed by filtration through nylon mesh with pore size of 3–5 μm.

2.3 Removal of Contaminants

Starch granules, as well as heavy cytoplasmic organelles or plasmalemma fragments, may be retained with cell walls on nylon meshes. Cytoplasmic proteins may also be adsorbed on the walls during the isolation procedures. Enzymes are often used to remove starch and proteins. However, addition of enzymes presents certain dangers. Moscatelli et al. (1961) and Huber and Nevins (1977) reported that several commercial preparations of amylase from *Bacillus subtilis* were able to split β-1,3–1,4-glucans and then induce loss of wall components. Preparations of *Aspergillus niger* were also shown to contain β-1,3–1,4-glucanase activity (Bacic and Stone 1980). A prior purification of commercial preparations is therefore recommended. Furthermore, starch granules can also be removed by repeated washings with water (see Fig. 2). Use of protease can also alter the cell-wall composition: pronase degrades cell-wall proteins (Lamport 1965); moreover, commercial preparations often contain β-1,3-glucanase and several glycosidase activities (Davies and Domer 1977). Detergent solutions (0.1 to 2% Triton X or 0.5% SDS) or chloroform-methanol mixtures have often been used to wash the

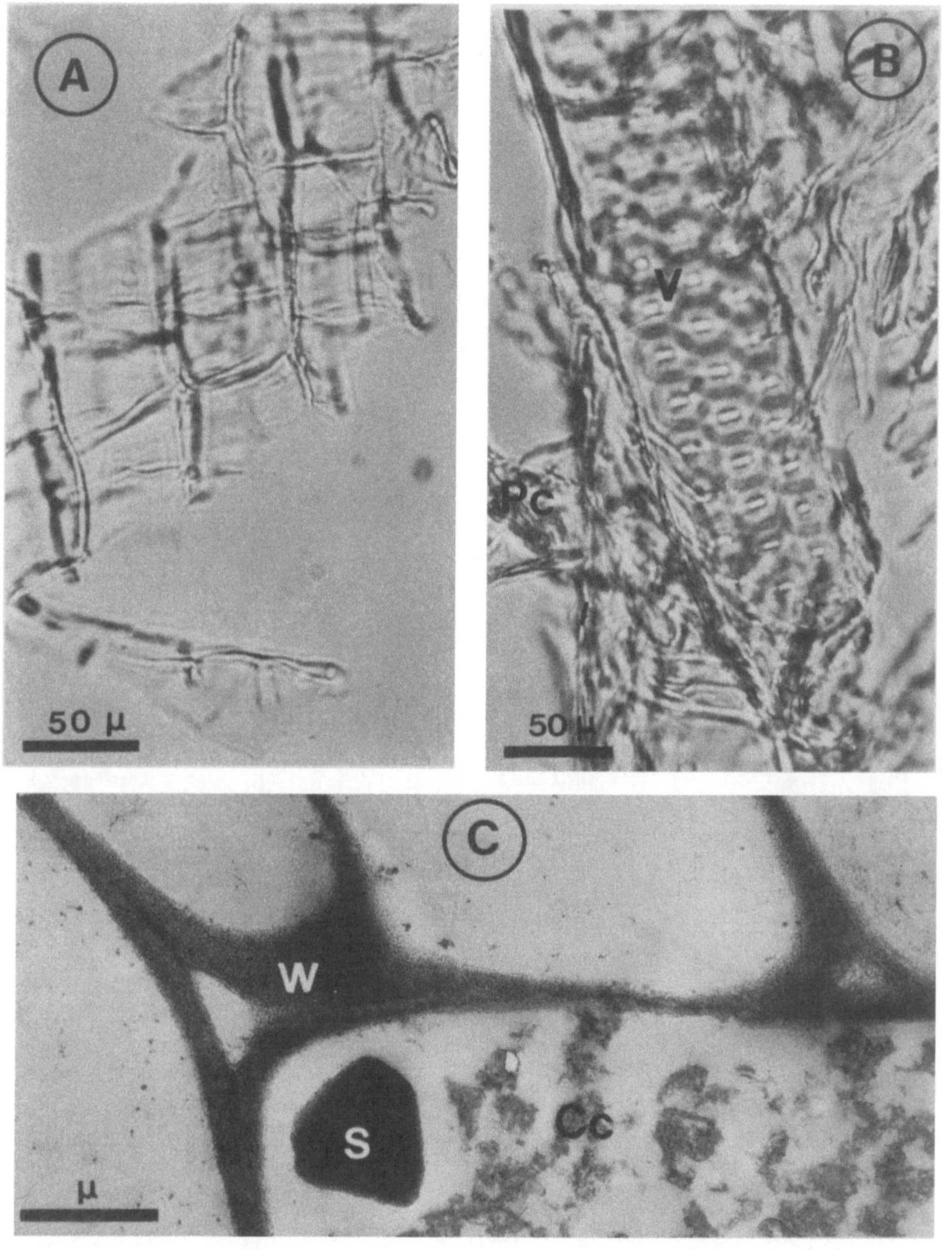

Fig. 1 A–C. Isolated cell-wall preparations. **A**, **B** light microscopic observations; **C** electron microscopic observation after PATAg reaction. **A** cell walls obtained from the upper part of mung bean hypocotyls according to the procedure represented in Fig. 3 B. **B** cell walls obtained from xylem tissues of *Populus* x *euramericana* stems; the tissues were ground in liquid nitrogen, lyophilized and homogenized with a Polytron; *V* vessel; *Pc* broken parenchyma cells; **C** cell walls isolated from mung bean hypocotyls, the Triton treatment was omitted and the walls insufficiently washed; starch (*S*) and cytoplasmic fragments (*Cc*) can therefore be noticed. (Courtesy of Prof. Roland)

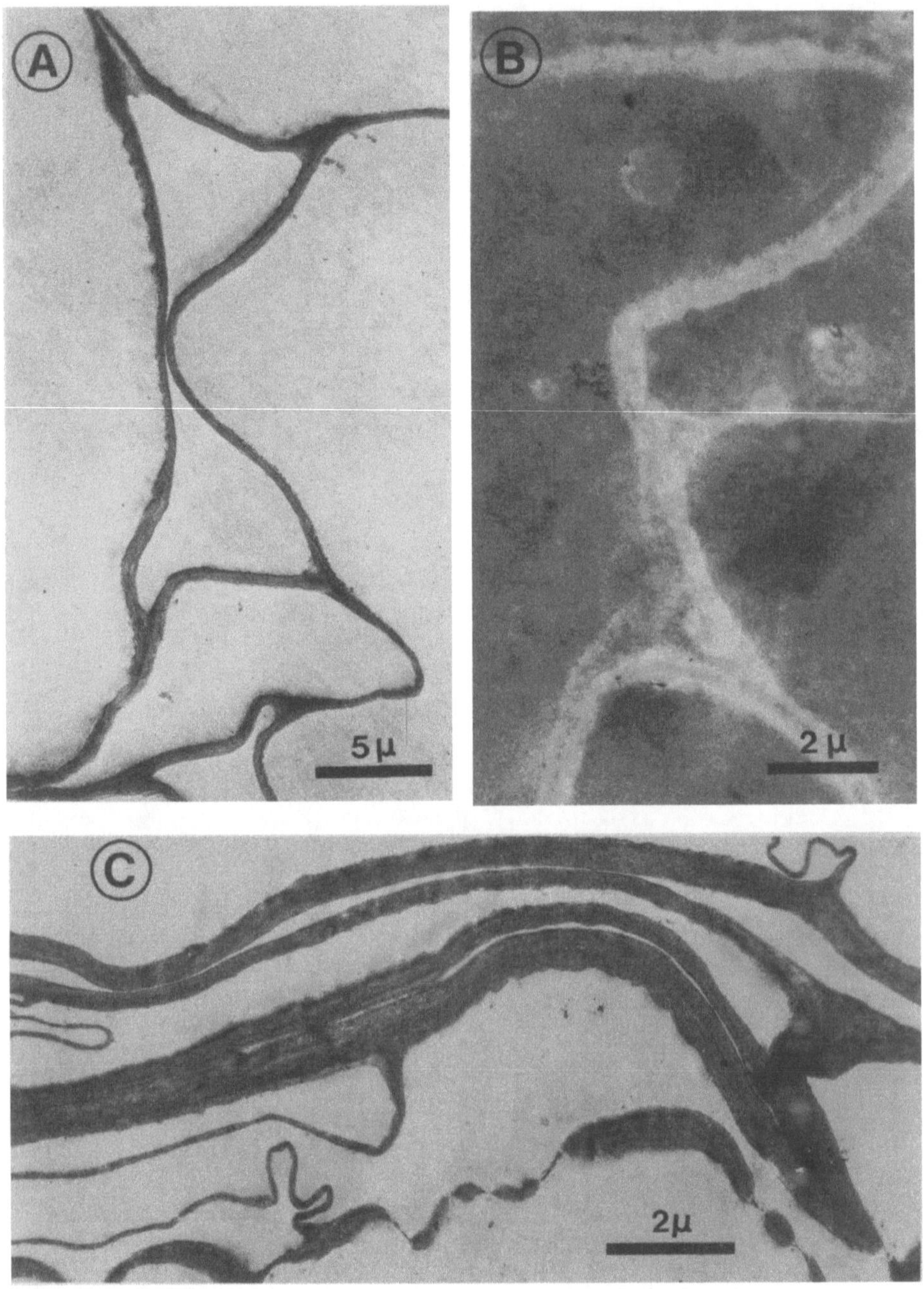

Fig. 2 A–C. Electron microscopic aspect of a pure fraction of isolated cell walls. **A**, **C** cell walls obtained from mung bean hypoçotyls according to the procedure represented in Fig. 3 B; PATAg staining; **A** young cells; **C** mature cells. (Courtesy of Prof. Roland). **B** cell walls isolated from *Acer pseudoplatanus* cultured cells, contrast $KMnO_4$ (Catesson et al. 1971)

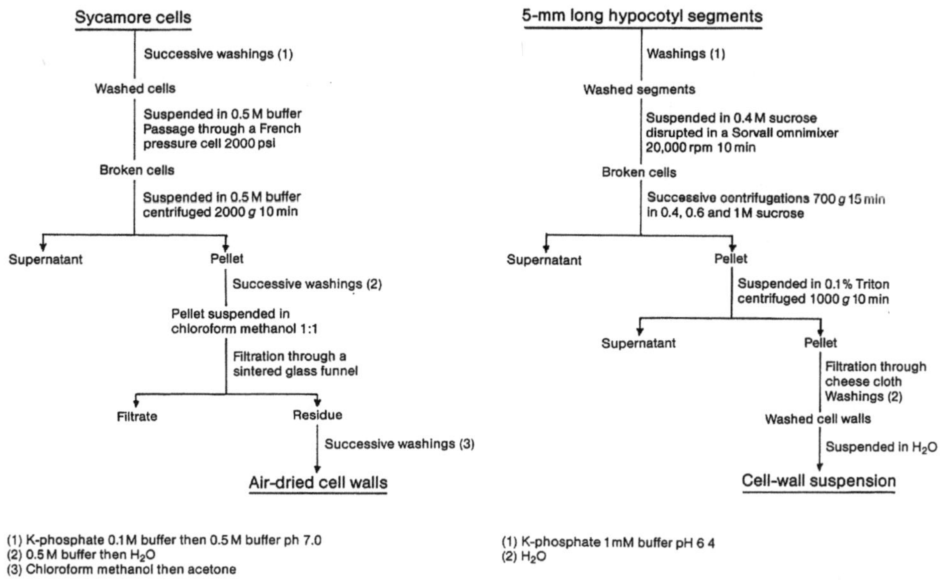

A B

Fig. 3. Flow charts for the isolation of cell walls from cell suspension cultures (**A**) and mung bean hypocotyl segments (**B**). **A** after Talmadge et al. (1973); **B** after Goldberg (1980)

cell walls in order to remove membranous contaminants. The nature of the investigations subsequently performed on the cell walls will determine the choice of experimental procedure. Study of the chemical composition of cell walls implies use of highly purified amylase or protease preparations and homogenization in a medium which does not solubilize any wall component. Investigations of the physiological functions of cell walls (ion exchange, cell extension, cell-wall assembly, recognition phenomena) may be carried out only if all cell-wall properties have been preserved: the charged molecules unmodified and the bound enzymes left in their regular microenvironment. Use of high ionic strength buffers must then be avoided during isolation operations. In any case, the stage of the isolated cell walls must be assessed microscopally. Light microscopy is often sufficient to detect the presence of starch granules (Fig. 1 A, B). Examination of ultrathin sections with transmission electron microscope (TEM) is necessary for detection of cytoplasmic contaminants (Fig. 1 C). The pictures reveal that primary cell walls obtained from young hypocotyl cells or from suspension cultures are not divided into small identical fragments. On the contrary, the shapes of the cells are still easily recognizable (Fig. 2 A, B): the cells have been broken at defined points, cleared out, and only the surrounding cell walls remain like a ghost network. When cell walls have been isolated from mature hypocotyl cells, the cellular shapes are much more difficult to recognize (Fig. 2 C). Chemical methods can also be used to assess the purity of cell-wall preparations. For instance, negative iodine test or lack of free glucose after an amylase treatment indicates the absence of starch in the preparations. Lack of lipids or lipoproteins in the cell-wall fragments indicates that they are free from plasmalemma.

Two complete flow-charts for obtaining isolated cell walls are depicted in Fig. 3. The method 3 B was used to isolate the cell walls of the Fig. 2. This relatively simple and quick procedure (about 2 h) avoids using saline solutions or enzymatic preparations which might modify the cell-wall composition; the electron microscopic observations reveal that the walls thus obtained are free of cytoplasmic contaminants. In conclusion, it appears that none of the routinely used procedures can be accepted without objection. The only rule is to adopt that which perturbs the cell wall least, depending on the type of investigation to be performed.

3 Composition and Ultrastructure of Plant Cell Walls

The composition of cell walls will be thoroughly investigated in the Chapter 2 of this Volume. Therefore, only standard methods, commonly used for investigating cell-wall chemical composition, will be here briefly summarized. In contrast, ultrastructural investigations will be more detailed.

3.1 Chemical Composition of Plant Cell Walls

3.1.1 Standard Extraction Procedures

The routine classification of cell-wall polysaccharides rests on their solubility properties. Usually, successive selective extractions are performed and four classes of biopolymers are obtained. Pectic substances can be extracted partly with boiling water, but principally with hot aqueous solutions of chelating agents for calcium (EDTA, EGTA, ammonium oxalate). Hemicelluloses are then solubilized with dilute alkali alone or with added borate and subdivided, according to their solubility at pH 6.0, into hemicelluloses A (insoluble) and B (soluble). The alkali-insoluble residue corresponds generally to cellulose which, unlike pectic substances or hemicelluloses, is a well-defined macromolecule. A routine flow-chart for the fractionation of cell-wall polysaccharides is given in Fig. 4. It is advisable to check for each stage of extraction: after EDTA treatment, for instance, the cell walls must be devoid of uronic acids. These can be estimated directly on the walls according to Ahmed and Labavitch (1977). However, the fractions extracted as pectic substances and hemicelluloses are actually very heterogeneous and embrace related polysaccharides, which must then be isolated.

3.1.2 Analysis of Polysaccharide Fractions

3.1.2.1 Chemical Methods

Conventional chemical methods are applied to isolate and investigate the different macromolecules (Aspinall 1980): gel filtration, ion exchange chromatography, high pressure liquid chromatography. The nature of the monomers is deter-

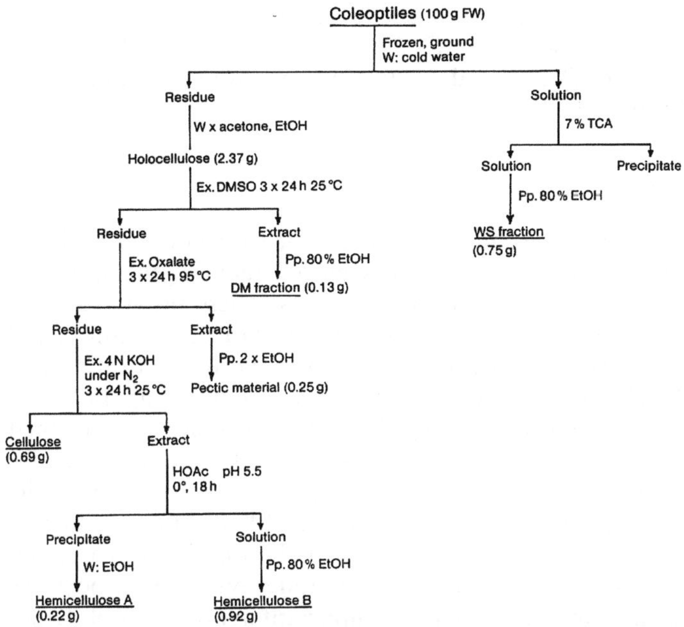

Fig. 4. Fractionation and yield of polymer fractions from oat coleoptiles. (After Wada and Ray 1978). *W* washing; *Pp* precipitation; *Ex* extraction; *Oxalate* 0,5% (NH_4) oxalate; *FW* fresh weight

mined after acid hydrolysis by GLC of suitably volatile derivatives. The structures of the molecules are assigned on the basis of methylation, isolation of methylated oligosaccharides after acidic or enzymatic hydrolysis and degradation involving periodate oxidation followed by reduction and mild hydrolysis of acyclic acetals. Methylated sugar derivatives are separated and characterized by GLC coupled with mass spectrometry (for details see MacNeil et al. 1982). However, chemical modifications (i.e., deesterification of uronic acids, loss of small molecules) can occur during these operations. Progress has been achieved with specific enzymes as splitting agents. Albersheim (1978) exhaustively analyzed the polysaccharides released by specific polysaccharide-degrading enzymes and presented a structural model of the walls of suspension-cultured sycamore cells. In this persuasive model, the polysaccharides are linked mainly with covalent bonds; only xyloglucans and cellulose are connected with hydrogen bonds. However, such a tightly related network has been criticized (Preston 1979) and other models have been proposed for dicotyledons (Monro et al. 1976) and monocotyledons (Darvill et al. 1978). The nature of the cell-wall glycoproteins was reviewed by Lamport (1980). Phenolic compounds, missing in most models, have yet been localized in primary walls (Fry 1979, 1983; Markwalder and Neukom 1976; Harris and Hartley 1976; Hartley and Harris 1978) and these may function by cross-linking polysaccharides. Other molecules such as lectins (Kauss and Bowles 1976) or polyamines (Dumortier et al. 1983; Vallée et al. 1983) should also be taken into account. In secondary cell walls, the nature of the association of lignin with hemicelluloses and cellulose is not yet well established. However, several recent analyses of lignin-carbohydrate complexes (LCC) showed that lignins could be associated with glucans or mannans (Yaku et al. 1979, 1981) or with xylans and galactoglucomannans (Eriksson et al. 1980).

3.1.2.2 Physical Methods

Cell-wall components are often crystalline, either in their native or recrystallized state. The structure and morphology of crystalline biopolymers is generally studied on purified samples by X-ray diffraction techniques. These provide a great deal of information for a given sample: molecular orientation, crystallinity, and even molecular structure. Such investigations have been performed on several polymers (Meier 1958; Jelsma and Kreger 1975; Marchessault at al. 1977). In 1954, Preston and Ripley introduced electron diffraction techniques performed directly within the electron microscope on smaller samples. To avoid decrystallization in vacuum, procedures have been developed to preserve the sample in its hydrated condition during the observation (Taylor et al. 1975). Electron diffraction data have been obtained by Chanzy et al. (1977, 1978) on several frozen hydrated cell wall components. Such characterization requires only minute portions from the wall of one cell and should become an important analytical tool. Other physical methods, such as infra-red spectroscopy or nuclear magnetic resonance spectroscopy, are now used for determination of polysaccharide structure. These have provided information on the configuration of glycosidic linkages.

3.2 Supramolecular Organization of Plant Cell Walls

Ultrastructural investigations of cell walls (isolated or in situ) have become routine methods of study. The general preparations and staining methods have been exhaustively described by Roland (1978). The most widely used method for TEM observation is fixation in glutaraldehyde-OSO_4 followed by embedding in Epon-araldite and staining by uranyl acetate-lead citrate. But the reaction of cell walls remains poor and the walls appear quite colorless (Fig. 2B). More specific methods have thus been developed for defined cell-wall components (see Roland 1978).

3.2.1 Morphological Observations

Observation of dispersed specimens spread directly onto grids can give information on fine morphology. The visualization is carried out either by shadow-casting with metal vaporization under vacuum (e.g., gold platinum, germanium, chromium) or by negative staining performed with an electron-opaque solution which outlines the profile of the structure (for technical protocols see Roland 1978).

3.2.2 Selective Staining of Polysaccharides

The cytochemical techniques do not strictly detect polysaccharides, but they allow the visualization of specific chemical functions present in the constituent sugars.

3.2.2.1 Visualization of Esterified Carboxyl Groups

Albersheim (1965) and Albersheim et al. (1960) developed a cytochemical test for electron microscopy from a method previously used to quantify esterified polyuronides. Esterified pectic groups are first reacted with basic hydroxylamine and turned into pectic hydroxamic acids which are then treated with ferric ions to form an insoluble complex.

3.2.2.2 Detection of Acidic Functions

Acidic groups such as carboxyl of uronides can be visualized by such positively charged colloidal metallic particles as colloidal iron or cationized ferritin. This technique can be applied not only on cell walls in situ but also to polysaccharides extracted from the walls.

3.2.2.3 Periodic Oxidation of Glycol Groups

Glycol groups are first oxidized to aldehydes with periodic acid. Condensation of aldehyde groups is then performed with thiocarbohydrazide (Seligman et al. 1965) and the final visualization obtained with silver proteinate (Thierry 1967). This method concerns only compounds carrying free vic-glycols (constitutive sugars bound between carbon 1 and carbon 4) and branches reduce the number of available vic-glycols. As 1,4-glycosidic bonds are very abundant in cell walls, these generally present very strong reactions (Fig. 2 A, B). However, this technique does not provide any information on the organization of the cell-wall components. It is therefore useful to selectively extract compounds present in the walls before performing the detection technique. Such substractive localization often unmasks subunits from compact associations. Extractions can be performed either chemically (and then coupled to the analysis of the solubilized substances, Reis 1980, 1981) or enzymatically (Roland and Vian 1981). This last technique seems more specific but requires highly purified enzyme preparations. The differences between chemical and enzymatic extractions are illustrated in Fig. 5. Such substractive localizations have led to considerable knowledge on the three-dimensional architecture of cell walls. Roland et al. (1982) have thus shown the highly ordered structure, "twisted plywood", of elongating cell walls. This ordered construction, labile and transient in primary cell walls, appears on the contrary to be consolidated and stable in secondary walls (Parameswaran and Liese 1982; Roland and Mosiniak 1982). Aspects of cell-wall architecture in an epidermal cell and in a vessel after methylamine extraction are illustrated in Fig. 6 A and B.

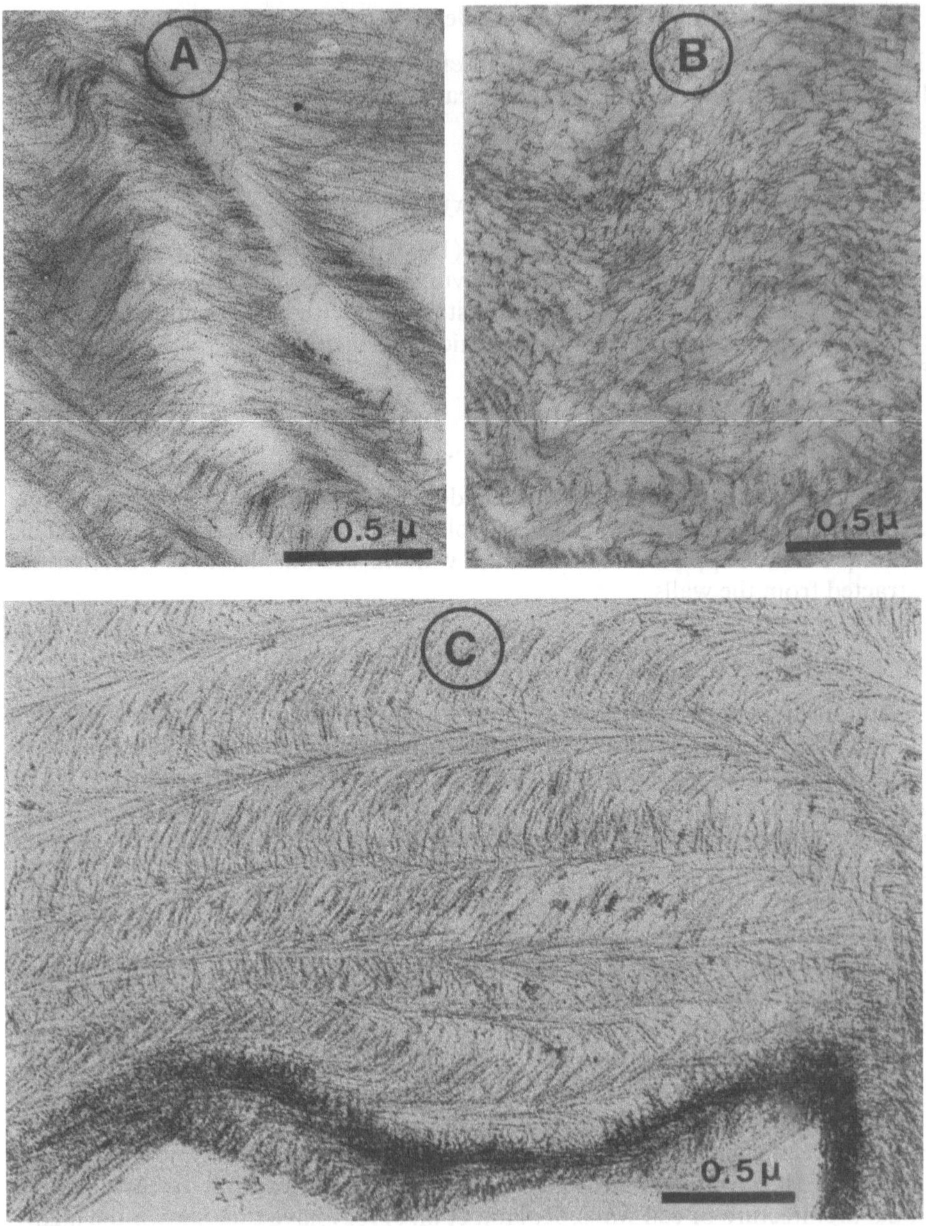

Fig. 5 A–C. Detection of vic-glycol groups with PATAg reaction after partial removal of matrix material. **A**, **B** isolated cell walls depectinized by boiling water (2 × 1 h, 100 °C). **A** cell walls obtained from the hook of the mung bean hypocotyl; **B** cell walls obtained 1 cm below. **C** thin section of mung bean hypocotyl tissues previously treated by a purified endopolygalacturonase. (**A**, **B** courtesy of Prof. Roland; **C** Roland and Vian 1981)

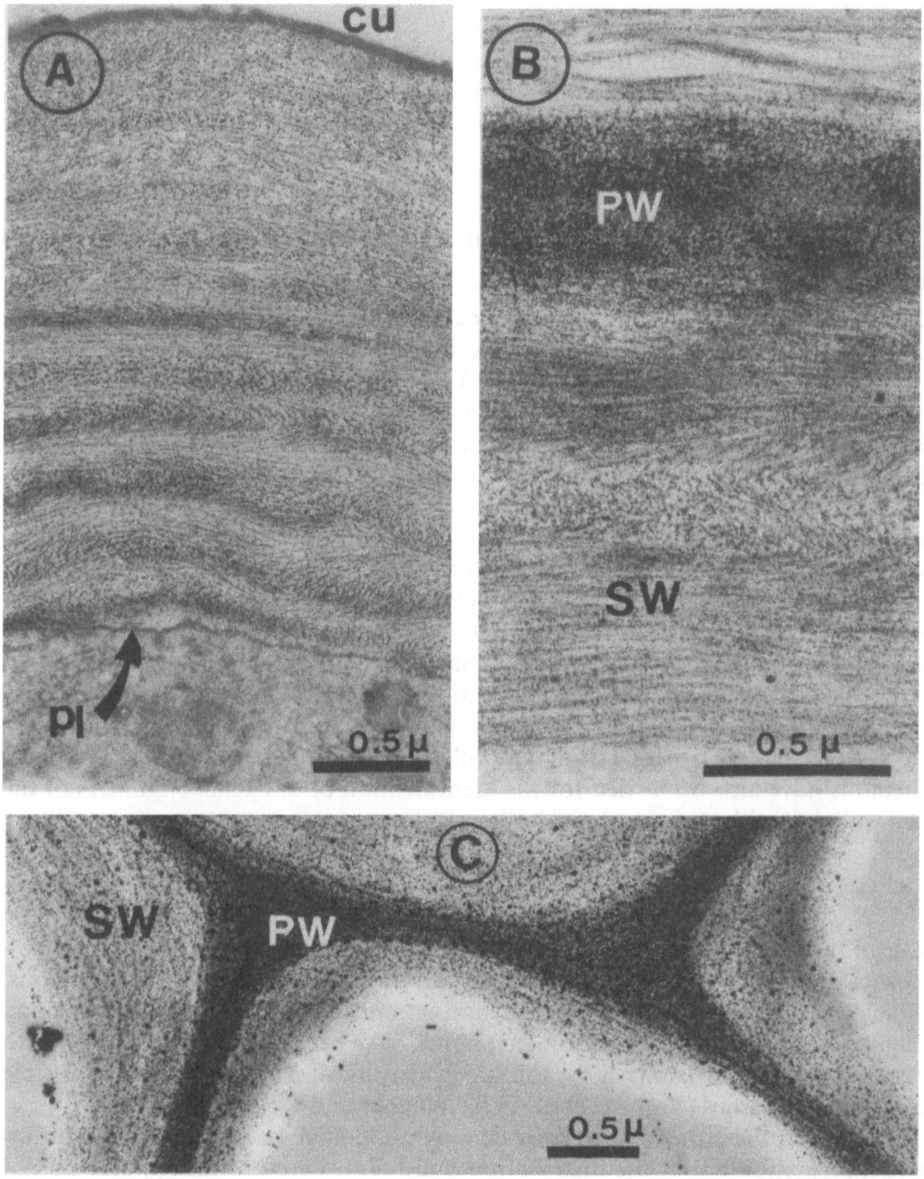

Fig. 6 A, B. Detection of vic-glycol groups with PATAg reaction after methylamine treatment. **A** mung bean hypocotyl epidermal cells; *Cu* cuticle; *Pl* plasmalemma. (Courtesy of Dr. Reis); **B** cross-section of *Robinia pseudoacacia* wood; the primary wall (*PW*) gives a strong reaction, several successive layers can be noticed in the secondary wall (*SW*). **C** detection of lignin by chlorine water-ethanolamine-silver nitrate; longitudinal section of *Ulmus campestris* wood (**B**, **C**, courtesy of Dr. Czaninski)

3.2.3 Visualization of Lignin

Lignin can be specifically detected at the ultrastructural level by successive chlorine water-alcoholic ethanolamine-silver nitrate treatments (Robarts 1967; Czaninski 1979; Chafe and Chauret 1974; Hoffmann and Parameswaran 1976). Parallel layers can thus be visualized in secondary walls (Fig. 6C).

3.2.4 Identification of Wall Components by Means of Affinity Methods

Lectins can bind specifically with defined hexose molecules and therefore are now considered as promising tools for the localization of given monomers inside the cell walls. The lectins, linked to their specific substrate, are then visualized by fixation on the sugar moiety of glycosylated ferritin. This technique was used for the detection of fucose-containing polysaccharides (Rougier et al. 1979) and for the localization of mannans in yeast cell walls. However, technical difficulties remain in the sizes of the lectins, which do not allow good penetration inside the walls.

Enzymes were also used to localize macromolecules at the electron microscope level. Enzymes, initially labeled with gold and made opaque for electrons, can reveal their corresponding substrate when they are applied on tissue thin sections. This method, developed by Bendayan (1981) for the detection of nucleic acids, was recently adapted by Vian et al. (1983) for the visualization of xylans in cell walls by means of a xylanase-gold complex. Such investigations require preliminary purification of the enzyme preparation. The specificity of the enzyme activity and of the labeling must be checked. However, this new technique may open up a new field of investigation in wall topochemistry. Immunocytochemical methods have also been tried for the identification of antigenic oligomers of the cell wall, but with major technical difficulties.

3.2.5 Detection and Estimation of Cations

Use of the standard fixatives glutaraldehyde and osmium allows loss of most soluble cations. However, cell walls are known to bind preferentially calcium ions, which can be detected with potassium antimonate (Tandler et al. 1970; Wick and Hepler 1982). Freeze substitution techniques have been used in conjunction with electron probe microanalysis for the subcellular localization of various diffusible ions in plant tissues (Van Zyl et al. 1976; Wagner and Rossbacher 1980).

However, these procedures provide only qualitative or at best semi-quantitative information about the distribution and relative concentration, and do not discriminate between free and bound forms (Roux and Scoloum 1982). Bound forms can be estimated in isolated cell walls by flame spectrophotometry after solubilization in 0.1 N HCl.

3.2.6 Ultracryotomy

Routine techniques for microscopy involve embedding in plastic, which limits access of reagents in cytochemical reactions and successive chemical treatments which can lead to the denaturation and inactivation of cellular structures. With cryotomy, freezing is used both as a fixative and hardening agent for the specimen to minimize chemical treatment and to avoid plastic embedding. Wall subunits can then be visualized sharply by using negative staining. Moreover, by using glycolytic enzymes on ultrathin sections, the progressive unmasking of fibrillar subunits can be followed. Characterization and localization of glucidic components of cell walls can then also be performed by using their lectin-binding properties (Vian 1981). However, the technical difficulties of these methods, especially the routine production of meaningful ultrathin frozen sections, should not be overlooked.

In conclusion, a better understanding of the structure of the cell wall may depend upon joint structural and quantitative studies; these have already been attempted (Reis 1980, 1981). Such joint investigations might reduce the gap that still remains between biochemistry and cytochemistry.

4 Properties of Plant Cell Walls

4.1 Exchange Properties of Plant Cell Walls

Cell walls carry numerous charged radicals as amine or carboxyl groups which will be ionized according to the pH and the ionic strength of the solution bathing the cell walls. Cell walls will behave as ion exchangers, and plasma membranes will not be exposed to the mass ionic conditions of the external medium. They are actually in contact with a modified medium, the composition of which is affected by the interactions between the outer solution and the pecto-cellulosic cell walls (Sentenac and Grignon 1981). Experiments with yeast have shown the effects of the surface ionic conditions on transport kinetics (Theuvenet and Borst-Pauwels 1976; Roomans and Borst-Pauwels 1979). Two approaches are generally encountered. On the one hand, theoretical models are developed. In most of them the ionic behavior of cell walls has been described by the well-known models of Donnan (Pitman 1965; Dainty et al. 1960) or Gouy-Chapman (Shone 1966; Dainty and Hope 1961). Demarty et al. (1978) have extended the classical Donnan theory, taking into account the activity coefficients to describe ion–ion and ion–water interactions. In Sentenac and Grignon's model (1981), the ionic environment of cell wall results from simultaneous electrostatic interactions between free ions and specific association equilibria, including acido-basic one.

By the alternative approach, experiments are performed with isolated cell walls in order to corroborate the theoretical models. Sentenac's model, for instance, was supported by estimations of K^+, Ca^{2+}, Mg^{2+} concentrations in isolated cell walls in different incubation baths. Potentiometric titrations have been performed on isolated cell walls (Schönherr and Bukovac 1973; Morvan et al. 1979), using various bases in order to estimate the exchange capacities for different conterions. Cell walls behave as ion exchange resins of the weak acid type. These biophysical investigations have shown that the micro-layer of the plasmalemma concentrates divalent cations, with a high selectivity for Ca^{2+}, while, owing to the surface repulsion of anions, it is relatively poor in NO_3^-, P_1^- or So_4^{2-}. The accumulation of Ca^{2+} modulates in turn the local fluxes of cations and anions. The apoplasm is not a mass compartment, but consists of very different microenvironments, products of three factors: the external medium, the cell wall, and the plasmalemma.

4.2 Enzymatic Properties

4.2.1 Cytochemical Investigations

Methods used for cytochemical localization of enzymes in plant cells were reviewed in 1978 by Sexton and Hall. In order to localize specific enzymes at their normal sites within intact cells, the conditions in the incubation medium are manipulated so that an insoluble, electron-dense product is produced by the enzymatic activity. At least two critical steps in enzyme cytochemistry should be stressed. First, tissue preparation should not modify enzyme activity and cellular fine structure: second, diffusion of enzyme or reaction products may have significant effect on the validity of the localization observed. Only few enzymes, essentially phosphatases and peroxidases, have been localized in the cell walls by histochemical techniques (Fig. 7).

4.2.1.1 Cell-Wall Phosphatase Activities

Phosphatase activities have long been recognized in cell walls (Hall and Butt 1969; Gahan and Mac Lean 1969; Halperin 1969; Poux 1970; Catesson et al. 1971). In most cases, adaptation of Gomori's technique, based on lead capture, has been used. In spite of criticism of lead-based methods (Washitani and Sato 1976), their use has produced a great deal of valuable information about the distribution of phosphatases in plants (see Poux 1973). Similar localizations have been observed by Esau and Charvat (1975) with azodye methods. Moreover, Catesson et al. (1971) observed phosphatase activities in isolated cell walls, but also in cell walls of whole, unbroken cells, which excludes artifacts resulting from diffusion processes.

4.2.1.2 Cell-Wall Peroxidase Activities

The original Graham and Karnovski method (1966), based on oxidation of DAB, has been widely and successfully used for light and electron microscopic cytochemical localization of plant peroxidases. Recently, two new, noncarcinogenic and sensitive hydrogen donors have been employed for histochemical studies of

Fig. 7 A–C. Cell-wall enzymatic activities. **A, C** phosphatase activity detected by Gomori's technique; **A** *Acer pseudoplatanus* cultured cells; **C** isolated cell walls from the same (Catesson et al. 1971). **B** peroxidase activity detected with DAB and H_2O_2 at pH 9.0 on a longitudinal section of *Triticum vulgare* stem (Courtesy of Dr. Czaninski). *CW* cell wall; *PC* parenchyma cell; *PW* primary cell; *SW* secondary wall; *V* vacuole; *VE* vessel

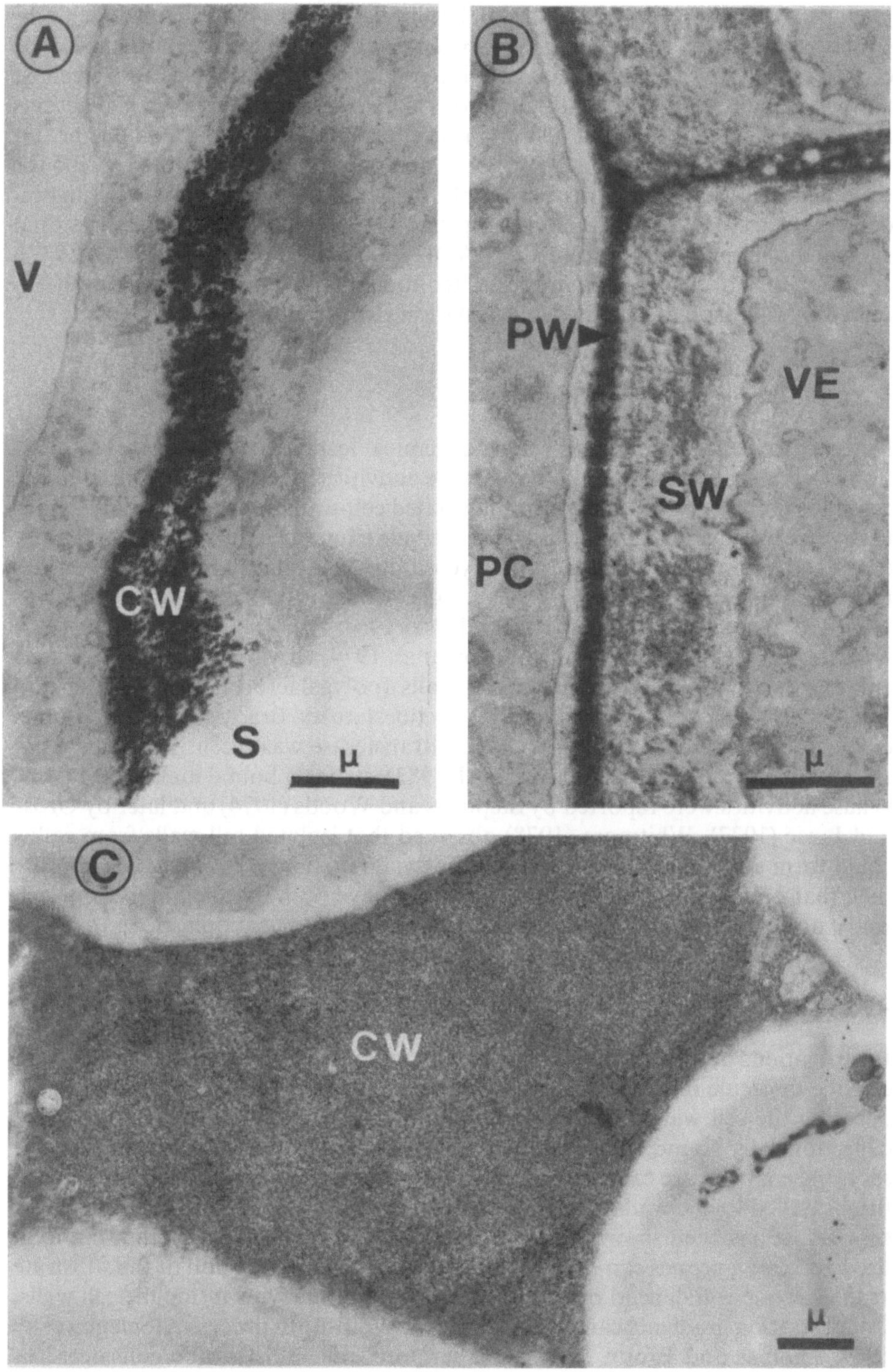
A
V
CW
S
μ
B
PW
VE
SW
PC
μ
C
CW
μ

plant peroxidases (Imberty et al. 1984). Preliminary results indicate that both substrates (tetramethylbenzidine and p-phenylenediamine-pyrocatechol) should be of interest to visualize peroxidase activities on ultrathin sections. However, most peroxidase substrates react with numerous isoperoxidases, and positive reactions are observed in cell walls but also in intracellular organelles. Syringaldazine, introduced by Harkin and Obst (1973), seems to be a substrate highly specific for the cell-wall isoperoxidases involved in the oxidative polymerization of lignin precursors (Catesson et al. 1978; Goldberg et al. 1983). H_2O_2 necessary for this final step might be generated in situ from NADH (Gross 1977; Gross and Janse 1977). A malate dehydrogenase activity restricted to lignifying cell walls was indeed observed by Catesson et al. (1978) on carnation stem sections.

4.2.2 Biochemical Investigations

Detection of cell-wall activities via histochemical techniques provides semi-quantitative results at best and only few enzyme activities have been recognized in the cell walls in this way. On the contrary, isolated cell-wall preparations allow quantitative assays and determination of the enzyme properties. In addition to peroxidases and phosphatases, numerous glycosidases and glucanases have been detected and investigated on isolated cell walls (see for instance Heyn 1969; Nevins 1970; Goldberg 1977; Chang and Bandurski 1964; Fan and Mac Lachlan 1967; Keegstra and Albersheim 1970; Johnson et al. 1974, Koyama et al. 1981). Pectic enzymes have also been found in cell walls (polygalacturonases: Wallner and Walker 1975; Pressey and Avants 1977; pectinesterases: Brady 1976; Mac Donnel et al. 1945; Versteeg et al. 1978). Glycosyltransferase was recently isolated from *Acer pseudoplatanus* cell walls (Nari et al. 1983); strongly bound malate deshydrogenase activities were reported by Stephens and Woods (1974) and later by Gross and Janse (1977). Whitmore (1978) observed that isolated cell walls from callus could form a lignin–carbohydrate complex via peroxidasic oxidation. It appears then that only hydrolase or oxidase activities have been detected in isolated cell walls; no synthetase activity has ever been reported in the apoplasm.

4.2.2.1 Properties of Immobilized and Solubilized Cell-Wall Enzymes

Investigations of the properties of cell-wall enzymes are generally performed with enzyme molecule in free solutions. When the molecules can be solubilized by incubating the cell walls in salt solutions of high ionic strength, the enzymes are called ionically bound; when a subsequent pectinase-cellulase treatment is necessary, the enzymes are called covalently bound. In some cases, the enzymes are so firmly linked to the walls that it is impossible to free them from their insoluble support. It has been shown both theoretically and experimentally that immobilized and free enzymes can behave differently. The concentration of the substrate at the enzyme will depend on the localization of the enzyme inside the cell walls. Concentration gradients can result simply from diffusion processes (Selegny et al. 1969; Thomas and Broun 1972). Furthermore, cell walls can be considered as polyelectrolites and therefore the local concentration of ions is significantly different from their concentration in bulk solution. Near the enzymes the local pH will then be modified. Goldstein et al. (1964) have predicted that the kinetic behavior of free and bound enzymes can differ owing to the changes in substrate concentration induced by electrostatic effects. As expected by this polyelectrolyte theory, Crasnier et al. (1980) have observed differences in the kinetic parameters of bound and free cell-wall phosphates. Association of the enzyme with a polyanion modified kinetic properties and sensitivity to ions. Furthermore, cell-wall enzymes are located in a defined microenvironment whose composition may modulate their activity. Cell-wall enzymatic properties inferred from investigations upon solubilized enzymes may therefore be inexact; enzyme activities should better be estimated in situ on isolated cell walls, but these measurements require homogenous cell-wall preparations.

4.2.2.2 Biological Functions

The presence of enzyme activities at the cell surface will modify the organic molecules translocated via the apoplasm or entering the cells. In 1965, Hawker and Hatch had already reported that cell-wall invertases were important parts of the translocation machinery. Later Bowen and Hunter (1972) demonstrated that in the stem of sugar cane the invertase located on the parenchyma cell walls hydrolyzes the sucrose of the medium before this sugar enters the cell. Similarly, Thomas and Webb (1979) reported that in leaves of *Cucurbita pepo* the transport of galactosides is facilitated by the prior hydrolysis in the periplasm of galactosyl sucrose oligosaccharides. Cell-wall phosphatases may also participate in regulation of cell metabolism, since Lugon et al. (1974) observed that they hydrolyzed the nucleotides present in the medium, the nucleosides thus formed being then absorbed by the cells. Differentiation of the cell walls can also be controlled by cell-wall enzymes: two steps of the lignification process are known to be mediated by wall enzymes. The oxidative polymerization of cinnamyl alcools is catalyzed by peroxidases (Griesbach 1977; Gross 1977) and the NADH used for the generation of H_2O_2 is formed via a wall-bound malate dehydrogenase (Elstner and Heupel 1976; Gross and Janse 1977; Gross et al. 1977). The role of peroxidases in lignification might explain the enhancement of cell-wall peroxidase levels observed in disease resistance (Vance et al. 1980; Hammerschmidt and Kuc 1980; Fleuriet and Deloire 1982). Some cell-wall peroxidases may also function as auxin oxidases (Palmieri et al. 1978; Kokkinakis and Brooks 1979). Another oxidasic activity localized in cell walls is polyamine oxidase, found in leaves of oat seedlings by Kaur-Sawhney et al. (1981), and which could be involved in polyamine metabolism.

Since Kivilaan's first report of the autolysis process (1959), in vitro hydrolysis of isolated cell walls in the absence of any added organic molecule has often been observed and investigated (Katz and Ordin 1967; Kivilaan et al. 1971; Huber and Nevins 1979). Cell-wall hydrolases were then thought to regulate the turnover of cell-wall constituents and to be involved in cell-loosening processes (this point will be discussed in Sect. 5 devoted to growth aspects). Cell-wall softening which occurs during fruit maturation was also related to the levels of cell-wall pectic enzymes such as pectinemethylesterases and polygalacturonases (Brady 1976; Pressey and Avants 1977; Versteeg et al. 1978).

In conclusion; the presence of enzymes in the apoplasm is now well established. Biochemical investigations are necessary to understand their possible functions and regulation. However, parallel histochemical approaches allow a histological localization of activities and observation of unbroken cells, whereas isolated cell-wall suspensions generally result from heterogeneous tissues. Moreover, histochemistry provides tools for studying specialized cells occurring at relatively low densities in very heterogeneous tissues. Cytochemical and biochemical methods of analysis are thus, once more, highly complementary.

4.3 Mechanical Properties

Mechanical properties of walls can be estimated on deproteinized material. Subjected to a constant applied stress, walls behave as a viscoelastic material. Removal of the stress enables distinction between elastic (reversible) and plastic (irreversible) extension processes. According to Cleland (1982), wall extensibility can be defined as the ability of cell walls to undergo irreversible extension when under constant stress. This cell-wall characteristic is thought, since Heyn (1931), to control the rate of cell elongation. Methods used for measuring cell-wall extensibility have been discussed by Cleland (1971, 1982). The most widely used technique is the Instron method (Cleland 1967; Masuda 1969; Olson et al. 1965; Penny et al. 1972; Courtney and Morré 1980a; Van Volkenburgh et al. 1983). With this instrument, segments are fastened between two clamps and the bottom clamp is lowered at a constant rate; the load which develops across the section is then measured by an electronic load transducer attached to the upper immobile clamp. Mechanisms modifying cell-wall extensibility will be discussed in the next chapter.

5 Growth Aspects

Any increment of plant-cell size implies an increase of the wall area. Fine measures of elongation are now routinely performed with transducers (Evans 1976; Prat 1978; Penny et al. 1972, 1973). A complete system is described in Fig. 8.

The biological sample (S) is set in an incubation chamber with a continuous flow of incubation medium (about 10 ml h^{-1}. The properly named auxanometer is composed of a displacement transducer (DT, Philips PR 9310) set on a microscope stand. The sample displacement induces a current, the strength of which is proportional to the extent of the displacement; the converter (C, Philips PR 9870) receives this electric signal. The transducer position is regulated by the micrometric screw of the support; when the displacement exceeds the auxanometer scale (i.e., 250 µm), reset to zero is automatically performed by the motor (M). A multipoint galvanometric recorder (R) unables six simultaneous experiments. In routine experiments, the elongation curves are converted by a digitizing surface (DS) and replotted by a microcomputer (MC). Averages and successive derivates are calculated and the data stored on a disc (D). The curves are redrawn at the chosen scale by a digital plotter (DP). The final precision is then about 2 µm. In short-term experiments, very fine measurements can be obtained by a direct microcomputer analysis of the signal given by the converter; the effective accuracy corresponds then to the theoretical accuracy of the displacement transducer itself, i.e. 0.1 µm.

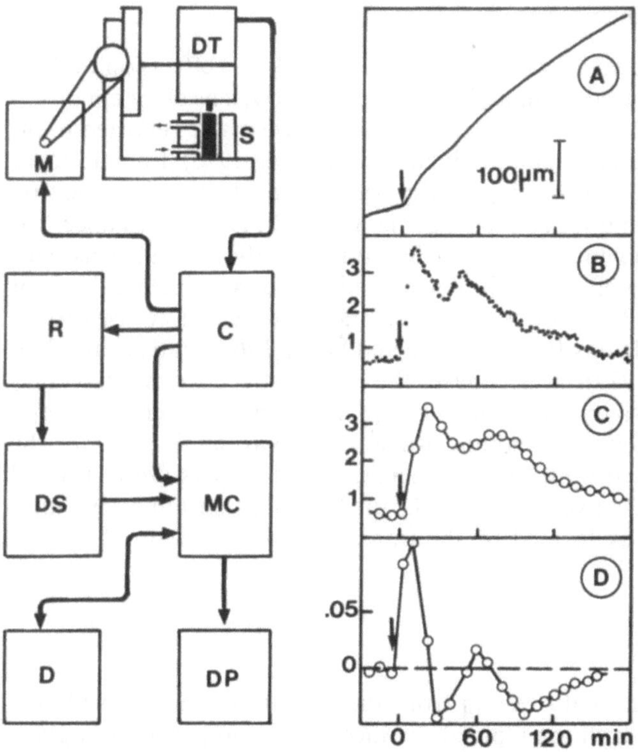

Fig. 8 A–D. Diagram of a complete auxanometric system and measurements performed with mung bean hypocotyl segments. *S* biological sample; *DT* displacement transducer; *C* converter; *M* motor; *R* multipoint galvanometric recorder; *DS* digitizing surface; *MC* microcomputer; *D* disk; *DP* digital plotter. **A, B, C, D** figures correspond to an experiment performed with abraded mung bean hypocotyl segments; *arrow* indicates a pH change from 7.7 to 4.7 in the incubation chamber. **A** elongation in µm, experimental recording; **B** elongation rate in µm min^{-1}, calculated from **A** experimental data with 2-min interval measurements; **C** elongation rate calculated from an average of eight experimental elongation curves with 10-min interval measurements. **D** acceleration in µm min^{-2} derived from curve **C**. (Courtesy of R. Prat)

Growth involves at least three processes: cell-wall loosening resulting from modification of the bonds which hold the polysaccharide network, stretching of the wall under turgor pressure and deposition of new wall material. Cell-wall stretching can be considered as materializing the cell-wall loosening, i.e., the increment of the wall extensibility.

5.1 Cell-Wall Loosening

Wall loosening can be estimated either on isolated walls or on live tissues. Routine tests (creep test, stress relaxation, instron) have been described and compared by Cleland (1982).

5.1.1 Wall-Loosening-Inducing Agents

Auxins have long been known as wall-loosening factors. Auxin effects on wall extensibility have been measured on monocot coleoptiles as well as dicot epi or hypocotyls (see Cleland 1982). In some cases, gibberellins were also reported to increase wall extensibility (Adams et al. 1975; Nakamura et al. 1975). Enhancement of wall extensibility by H^+ ions has repeatedly been demonstrated since the first reports of Rayle and Cleland (1970) or Hager et al. (1971) (see for instance Yamamoto et al. 1974; Cleland and Rayle 1975; Sakurai et al. 1977, 1979). This acid-induced cell-wall-loosening effect can be considered as the experimental basis of the "acid growth theory". Exogenously applied glucanases were also used to increase wall extensibility (Nevins 1975; Masuda and Wada 1967; Wada et al. 1968; Yamamoto and Nevins 1979). β-glucanase, the most active enzyme, also presented transglucosylase activity. These observations support the "chemical creep" hypothesis (Albersheim 1976; Cleland and Rayle 1972; Nari et al. 1983) for the wall-loosening process; according to this schema, wall polymers are restrained from moving by cross-links; glycosylase breaks the bonds, extension occurs until restrained by new cross-link formed by the transglycosylase activity of the enzyme.

5.1.2 Nature of the Broken Bonds

Increase of cell-wall extensibility results from the rupture of bonds inside the polysaccharide network which maintained the mechanical properties at a given state. The nature of these bonds has generally been investigated by analyzing either the molecules released from the wall during growth reactions or the changes occurring in the cell-wall composition. Different results were reported for mono and dicots perhaps due to differences in cell-wall composition. Monocot cell walls are known to be rich in β-1,3 glucans or mixed β-1,3 β-1,4 glucans and auxin-induced growth is accompanied by marked release of glucose (Loescher and Nevins 1972; Nevins 1975; Yamamoto et al. 1980). Moreover analysis of the cell-wall composition has shown a decrease in the molecular weight of the noncellulosic glucans and a positive correlation between auxin response and the amount of noncellulosic glucose present in the cell walls (Sakurai et al. 1979; Zarra and Masuda 1979). These results parallel the autolytic activities of isolated cell walls which release into the medium glucose and β-1,3 oligoglucans (Katz and Ordin 1967; Lee et al. 1967; Kivilaan et al. 1971; Huber and Nevins 1979). Moreover, autolysis, auxin-induced elongation and β-glucosidase are all inhibited by the same antibiotic, nojirimycin (Nevins 1975; Zarra and Nicolas 1979; Labrador and Nicolas 1982). However, the involvement of β-glycosyl linkages in the maintenance of cell-wall rigidity is still controversial, as Courtney and Morré (1980b) could not detect any significant changes in the glucans in auxin-treated coleoptile walls; but if transglycosylation occurs, no glucose release will be noticed in the medium. The presence of β-1,3 glucans in dicotyledonous walls was claimed by Buchala and Franz (1974), Heininger and Franz (1980), and Goldberg and Prat (1982), but rejected by Albersheim (1978) and Cleland (1982). Labavitch and Ray (1974) reported the release from pea epicotyls of xyloglucans during the development of auxin- and H^+-induced growth response. These observations were in good agreement with Albersheim's representation of the cell-wall chemical structure. Furthermore, Masuda (1982) showed that IAA and H^+ treatments induce in both oat coleoptile and azuki bean epicotyl segments a decrease in the molecular weight of the xyloglucans. However, Nishitani and Masuda (1980) had failed in their attempt to obtain xyloglucan release during extension and investigators found other sugars, i.e., arabinose (Darvill et al. 1978), to be solubilized from the walls. Particular attention is now often given to the wall pectic fraction. Enhancement of uronide metabolism was reported to occur during growth responses (Nishitani and Masuda 1980; Terry et al. 1981). Autolytic freeing of acidic polysaccharides from isolated cell walls was observed at acidic pH (Bates and Ray 1979; Steele and Nevins 1979). Lastly, changes in the oriented structure of acidic wall polysaccharides during extension have been demonstrated by infrared analysis of cell walls from monocots known for their low pectin content and also from dicots (Morikawa et al. 1978; Hayashi et al. 1980). The well-established stiffening effect of calcium argues also for the possible involvement of pectic substances in the regulation of wall extensibility, although some observations indicate that calcium bridges would not be the load-bearing bonds in the walls (Cleland and Rayle 1978; Tepfer and Cleland 1979; Courtney and Morré 1980b). The release of uronic acids during growth response could be caused either by the disruption of hydrogen bonds which, according to Preston (1979), join the polyuronides to xylan molecules, or from enzymic cleavage of the covalent bonds which, in the Albersheim model (Mac Neil et al. 1979), link the xyloglucans to pectic polymers.

5.2 Deposition of Wall Material

It appears evident that wall synthesis is required to maintain the structure of the walls and auxin is known to enhance glucan synthetase activity. However, initiation of growth can be obtained without wall synthesis. Since the first observations of Penny et al. (1972), the kinetics of auxin-induced elongation have often been shown to be biphasic (Vanderhoef and Stahl 1975; Vanderhoef et al. 1976; Bouchet et al. 1983). The first auxin response, according to the "acid growth theory", can be considered to result from disruption from acid labile bonds inside the polysaccharide matrix; the second phase (Vanderhoef 1980; Vanderhoef and Dute 1981) would correspond to a supply of wall material. Hydrolases might be involved in this late phase as Mac Lachlan (1976) reported that these enzymes do regulate the biosynthesis of polysaccharides that takes place via an insertion mechanism. The exact sites of synthesis of the various wall subunits remain controversial. It is generally admitted that cellulose synthesis occurs at the surface of the plasmalemma, while pectic and hemicellulose components are synthesized by the Golgi apparatus and derived secretory vesicles (see Mollenhauer and Morré 1966). However, in view of the complexity of the cell-wall models, the question arises as to the chronology of construction of such precise sequences. Moreover, events occurring in the periplasm are difficult to analyze and cannot be studied by biochemical methods. Cytochemistry and ultracryotomy have provided valuable information (Vian 1982). Both approaches revealed that an ordered disposition appears when the fibrils come in contact with the preexisting wall. It has been shown that some wall fractions, and mainly those extracted by alkali, have the ability to aggregate in fibrillar elements in vitro (Reis 1978; Hills et al. 1975; Roland et al. 1977). In vivo arrangements might then result in part from self-assembly processes regulated by the physichochemical conditions encountered in the periplasm.

5.3 Growth Direction

Microfibril orientation has often been proposed to determine the preferred direction of extension of a cell wall (Frey-Wissling 1976; Richmond et al. 1980). There have been some reports of hormone-induced changed in the orientation of microtubules (Shibaoka 1974; Sawhney and Srivastava 1975; Steen and Chadwick 1981; Lang et al. 1982). However, the exact involvement of microtubules in wall morphogenesis is still a matter of discussion (Robinson 1977), while determination of orientation of cellulose microfibrils deposition in elongating cells is complicated by the polylamellate structure of epidermal and some internal walls. Recently, using ethrel and colchicine, Vian et al. (1982) demonstrated that induced changes in the direction of growth are accompanied by changes in the cell-wall texture, but chronological observations did not indicate that the changes in texture occur prior to changes in the growth axis (Reis et al. 1982).

In conclusion, from the reports discussed in this chapter, the cell walls appear as living cellular barriers which on the one hand modulate transport and exchange processes and are involved in recognition phenomenon; on the other hand, in addition to these properties resulting from their external location, the cell walls are also partly able to regulate their own growth.

Acknowledgement. I wish to thank Dr. S. Brown for his linguistic help.

References

Adams PA, Montague MJ, Tepfer M, Rayle DL, Ikuma M, Kaufman PB (1975) Effect of gibberellic acid on the plasticity and eleasticity of Avena stem segments. Plant Physiol 56:757–760

Ahmed AER, Labavitch JM (1977) A simplified method for accurate determination of cell wall uronide content. J Food Chem 1:361–365

Albersheim P (1965) A cytoplasmic component stained by hydroxylamine and iron. Protoplasma 60:131–135

Albersheim P (1976) The primary cell wall. In: Bonner J, Varner JE (eds) Plant Biochemistry, 3rd edn. Academic Press, New York, pp 225–274

Albersheim P (1978) Concerning the structure and biosynthesis of the primary cells walls of plants. In: Manners DJ (ed) Biochemistry of carbohydrates II. Int Rev Biochem Vol 16. Baltimore MD, Univ Park Press, pp 127–150

Albersheim P, Mühlethaler K, Frey-Wissling A (1960) Stained pectin as seen in the electron microscope. J Biophys Biochem Cytol 8:501–506

Albersheim P, Mc Neil M, Darvill AG, Valent B, Hahn MG, Robertsen B, Aman P, Franzen LE, Desjardins A, Ross LM, Spellman M (1980) Recognition between plant cell and microbes is regulated by complex carbohydrates. 2nd Int Congr Cell Biol, Berlin. Eur J Biol 22:230

Aspinall GO (1980) Chemistry of cell wall polysaccharides in the biochemistry of plants. J. Preiss (ed). Academic Press, New York, pp 473–500

Bacic A, Stone B (1980) A (1→3) and (1→4)-linked β-D-glucan in the endosperm cell walls of wheat. Carbohydr Res 82:372–377

Bates GM, Ray PM (1979) pH dependent release of polymers from isolated cell walls. Plant Physiol 63:5–21

Bendayan M (1981) Electron microscopical localization of nucleic acids by means of nuclease-gold complexes. Histochem J 13:699–710

Bouchet MM, Prat J, Goldberg R (1983) Kinetics of indole acetic acid-induced growth in hypocotyl sections of Vigna radiata. Physiol Plant 57:95–100

Bowen JE, Hunter JE (1972) Sugar transport in immature internodal tissue of sugarcane II. Mechanism of sucrose transport. Plant Physiol 49:789–793

Brady CJ (1976) The pectinesterase of the pulp of the banana fruit. Aust J Plant Physiol 3:163–172

Buchala AJ, Franz G (1974) A hemicellulosic β-D glucan from the hypocotyls of Phaseolus aureus. Phytochemistry 13:1887–1889

Catesson AM, Goldberg R, Winny MC (1971) Etude d'activités phosphatasiques acides dans les cellules d'Acer pseudoplatanus cultivées en suspension. C R Acad Sci Paris 272:2078–2081

Catesson AM, Czaninski Y, Monties B (1978) Caracte'res histochimiques des peroxidases pariétales dans les cellules en cours de lignification. C R Acad Sci Paris 268:1787–1790

Chafe SC, Chauret G (1974) Cell-wall structure in the xylem parenchyma of trembling aspen. Protoplasma 80:129–147

Chang CW, Bandurski RS (1964) Exocellular enzymes of corn roots. Plant Physiol 39:60–64

Chanzy H, Guizard C, Vuong R (1977) Electron diffraction on frozen hydrated polysaccharides. J Microsc 111:143–150

Chanzy H, Imada K, Vuong R (1978) Electron diffraction from the primary wall of cotton fibers. Protoplasma 94:299–306

Cleland RE (1967) Extensibility of isolated cell walls: measurements and changes during cell elongation. Planta 74:197–209

Cleland RE (1971) Mechanical behaviour of isolated Avena coleoptile walls subjected to constant stress. Plant Physiol 47:805–811

Cleland RE (1982) Wall extensibility: hormones and wall extrusion. In: Tanner W, Loewus FA (eds) Plant Carbohydrates II. Encyclopedia of plant physiology. Springer, Berlin Heidelberg New York, pp 255–276

Cleland RE, Rayle DL (1972) Absence of auxin-induced stored growth in Avena coleoptiles and its implication concerning the mechanism of wall extrusion. Planta 106:61–71

Cleland RE, Rayle DL (1975) Hydrogen ion entry as a controlling factor in the acid-growth response of green pea stem sections. Plant Physiol 55:547–549

Cleland RE, Rayle DL (1978) Auxin, H^+-excretion and cell elongation. Bot Mag Tokyo Spec Issue 1:125–139

Courtney JS, Morré DL (1980a) Studies on the role of wall extensibility in the control of cell expansion. Bot Gaz 141:56–62

Courtney JS, Morré DJ (1980b) Studies on the chemical bases of auxin-induced cell wall loosening. Bot Gaz 141:63–68

Crasnier M, Noat G, Ricard J (1980) Purification and molecular properties of acid phosphatase from Sycamore cell walls. Plant Cell Environ 3:217–224

Czaninski Y (1979) Cytochimie ultrastructurale des parois du xylème secondaire. Biol Cell 35:97–102

Dainty J, Hope AB (1961) The electric double layer and the Donnan equilibrium in relation to plant cell walls. Aust J Biol Sci 14:541–551

Dainty J, Hope AB, Denby C (1960) Ionic relations of cells of *Chara australis* II. The indiffusible anions of the cell wall. Aust J Biol Sci 13:267–276

Darvill AG, Smith CJ, Hall MA (1978) Cell wall structure and elongation growth in *Zea mays* coleoptile tissue. New Phytol 80:503–516

Davis TE, Domer JE (1977) Glycohydrolase contamination of commercial enzymes frequently used in the preparation of fungal cell walls. Anal Biochem 80:593–600

Demarty M, Morvan C, Thellier M (1978) Exchange properties of isolated cell walls of *Lemna minor* L. Plant Physiol 62:477–481

Dumortier FM, Flores HE, Shekhawat NS, Galston AW (1983) Gradients of polyamines and their biosynthetic enzymes in coleoptiles and roots of corn. Plant Physiol 72:915–918

Elstner EF, Heupel A (1976) Formation of hydrogen peroxide by isolated cell walls from horseradish (*Armoracia lapathifolia*). Planta 130:175–180

Eriksson Ö, Goring DAI, Lindgren BO (1980) Structural studies on the chemical bonds between lignins and carbohydrates in spruce wood. Wood Sci Technol 14:267–279

Esau K, Charvat I (1975) An ultrastructural study of acid phosphatase localization in cells of *Phaseolus vulgaris* phloem by the use of the azodye method. Tissue Cell 7:619–630

Evans ML (1976) A new sensitive root auxanometer. Preliminary studies of the interaction of auxin and acid pH in the regulation of intact root elongation. Plant Physiol 58:599–601

Fan DF, Mac Lachlan GA (1967) Studies on the regulation of cellulase activity and growth in excised pea epicotyl sections. Can J Bot 45:1837–1844

Fleuriet A, Deloire A (1982) Aspects histochimiques et biochimiques de la cicatrisation des fruits de tomate blessés. Z Pflanzenphysiol 107:259–268

Frey-Wissling A (1976) The plant cell wall. Borntraeger, Berlin

Fry SC (1979) Phenolic components of the primary cell wall and their possible role in the hormonal regulation of growth. Planta 146:343–351

Fry SC (1983) Feruloylated pectins from the primary cell wall: their structure and possible functions. Planta 157:111–123

Gahan AB, Mc Lean J (1969) Subcellular localization and possible functions of acid β-glycerophosphatases and naphtol esterases in plant cells. Planta 89:126–135

Goldberg R (1977) On possible connections between auxin induced growth and cell wall glucanase activities. Plant Sci Lett 8:233–242

Goldberg R (1980) Cell wall polysaccharidase activities and growth processes: a possible relationship. Physiol Plant 50:261–264

Goldberg R, Prat R (1982) Involvement of cell wall characteristics in growth processes along the Mung bean hypocotyl. Plant Cell Physiol 23:1145–1154

Goldberg R, Catesson AM, Czaninski Y (1983) Some properties of syringaldazine oxidase, a peroxidase specifically involved in the lignification process. Z Pflanzenphysiol 110:267–279

Goldstein L, Levin Y, Katchalsky E (1964) A water insoluble polyanionic derivative of trypsin. II. Effect of polyelectrolyte carrier on the kinetic behaviour of the bound trypsin. Biochemistry 3:1913–1919

Graham RC, Karnovski MJ (1966) The early stages of absorption on injected horseradish peroxidase in the proximal tubules of mouse kidney: ultrastructural cytochemistry by a new technique. J Histochem Cytochem 14:291–302

Griesbach H (1977) Biochemistry of lignification. Naturwissenschaften 64:619–625

Gross GG (1977) Biosynthesis of lignin and related monomers. Recent Adv Phytochem 11:141–184

Gross GG, Janse C (1977) Formation of NADH and hydrogen peroxide by cell wall-associated enzymes from Forsythia xylem. Z Pflanzenphysiol 84:447–452

Gross GG, Janse C, Elstner EF (1977) Involvement of malate, monophenols, and the superoxide radical in hydrogen peroxide formation by isolated cell walls from horseradish (*Armoracia lapathifolia Gilib*) Planta 136:271–276

Hager AH, Menzel H, Krauss A (1971) Versuche und Hypothese zur Primäwirkung des Auxins beim Streckungswachstum. Planta 100:47–75
Hall JL, Butt VS (1969) Adenosine triphosphatase activity in cell wall preparations and excised roots of barley. J Exp Bot 20:751–762
Halperin W (1969) Ultrastructural localization of acid phosphatase in cultured cells of *Daucus carota*. Planta 88:91–102
Hammerschmidt R, Kuc J (1980) Enhanced peroxidase activity and lignification in the induced systemic protection of cucumber. Phytopathology 70:689
Harkin JM, Obst JR (1973) Lignification in trees: indicative of exclusive peroxidase participation. Science 180:296–298
Harris PJ (1983) Cell walls. In: Hall JL, Moore AL (eds) Isolation of membranes and organelles from plant cells. Academic Press, New York, pp 25–53
Harris PJ, Hartley RD (1976) Detection of bound ferulic of the gramineae by ultraviolet fluorescence microscopy. Nature (Lond) 259:508–510
Hartley RD, Harris PJ (1978) Degradibility and phenolic components of cell walls of wheat in relation to susceptibility to *Puccinia struiformis*. Ann Appl Biol 88:153–158
Hayashi R, Morikawa H, Nakajima N, Senda M (1980) Oriented structure of pectic polysaccharides in pea epidermal cell walls. Plant Cell Physiol 21:999–1005
Hawker JS, Hatch MD (1965) Mechanism of sugar storage by mature stem of sugar cane. Physiol Plant 18:444–453
Heininger U, Franz G (1980) The role of NDP-glucose pyrophosphorylase in mung bean seedlings in relation to cell wall biosynthesis. Plant Sci Lett 17:443–450
Heyn ANJ (1931) Der Mechanism der Zellstreckung. Rec Trav Bot Neerl 28:113–244
Heyn ANJ (1969) Glucanase activity in coleoptiles of Avena. Arch Biochem Biophys 132:442–449
Hills GJ, Phillips JM, Gay MR, Roberts K (1975) Self assembly of plant cell wall in vitro. J Mol Biol 96:431–441
Hoffmann P, Parameswaran N (1976) On the ultrastructural localization of hemicelluloses within delignified tracheids of spruce. Holzforschung 30:62–70
Huber DJ, Nevins DJ (1977) Preparation and properties of a β-D-glucanase for the specific hydrolysis of β-D-glucans. Plant Physiol 60:300–304
Huber DJ, Nevins DJ (1979) Autolysis of the cell wall β-D-glucan in corn coleoptiles. Plant Cell Physiol 20:201–212
Imberty A, Goldberg R, Catesson AM (1984) Tetramethylbenzidine and p-phenylenediamine-pyrocatechol for peroxidase histochemistry and biochemistry: two new, non carcinogenic chromogens for investigating lignification process. Plant Sci Lett (in press)
Jelsma J, Kreger DR (1975) Ultrastructural observations on $(1 \rightarrow 3)$-β-D-glucan from fungal cell walls. Carbohydr Res 43:200–203
Johnson KD, Daniels D, Dowler MJ, Rayle DL (1974) Activation of Avena coleoptile cell wall glycosidases by hydrogen ions and auxin. Plant Physiol 53:224–228
Katz M, Ordin L (1967) A cell wall polysaccharide-hydrolyzing enzyme system in *Avena sativa* coleoptiles. Biochim Biophys Acta 141:126–134
Kaur-Sawhney R, Flores ME, Galston AW (1981) Polyamine oxidase in oat leaves: a cell wall localized enzyme. Plant Physiol 68:494–498
Kauss H, Bowles DJ (1976) Some properties of carbohydrate-binding proteins (lectins) solubilized from cell walls of Phaseolus aureus. Planta 130:169–174
Keegstra K, Albersheim P (1970) The involvement of glycosydases in the cell wall metabolism of suspension cultures Acer pseudoplatanus cells. Plant Physiol 45:675–678
Kivilaan A, Beaman TC, Bandurski RS (1959) A partial chemical characterization of Maize coleoptile cell walls prepared with the aid of a continually renewable filter. Nature 184:81–82
Kivilaan A, Bandurski RS, Schulze A (1971) A partial characterization of an autolytically solubilized cell wall glucan. Plant Physiol 48:389–393
Kokkinakis DM, Brooks JL (1979) Hydrogen peroxide-mediated oxidation of indole-3-acetic acid by tomato peroxidase and molecular oxygen. Plant Physiol 64:220–223
Koyama T, Hayashi T, Kato Y, Matsuda K (1981) Degradation of xyloglucan by wall

bound enzymes from soybean tissue. I. Occurrence of xyloglucandegrading enzymes in soybean cell walls. Plant Cell Physiol 22:1191–1198
Labavitch JM, Ray PM (1974) Relationship between promotion of xyloglucan metabolism and induction of elongation by indolacetic acid. Plant Physiol 54:499–502
Labrador E, Nicolas G (1982) Autolytic activities of the cell wall in rice coleoptiles. Effects of nojirimycin. Physiol Plant 55:345–350
Lamport DTA (1965) The protein component of primary cell walls. Adv Bot Res 2:151–218
Lamport DTA (1980) Structure and function of plant glycoproteins. In: Preiss J (ed) The biochemistry of plants. Academic Press, New York, pp 501–543
Lang JM, Eisinger WR, Green PB (1982) Effects of ethylene on the orientation of microtubules and cellulose microfibrils of pea epicotyl cells with polylamellate cell walls. Protoplasma 110:5–14
Lee S, Kivilaan A, Bandurski RS (1967) In vitro autolysis of plant cell walls. Plant Physiol 42:968–972
Loescher W, Nevins DJ (1972) Auxin-induced changes in Avena coleoptile cell wall composition. Plant Physiol 50:556–563
Lugon M, Goldberg R, Guern J (1974) Evidence for the intervention of cell wall phosphatases in AMP utilisation by *Acer pseudoplatanus* cells. Plant Sci Lett 3:165–171
Mac Donnel LR, Jansen EF, Lineweaver F (1945) The properties of orange pectinesterase. Arch Biochem 6:389–401
Mac Lachlan GA (1976) A potential role for endo-cellulase in cellulose biosynthesis. Appl Polymer Symp 28:645–658
Mac Neil M, Darvill AG, Albersheim P (1979) The structural polymers of the primary cell walls of dicots. Fortschr Chem Org Naturst 37:191–248
Mac Neil M, Darvill AG, Aman P, Franzen LE, Albersheim P (1982) Structural analysis of complex carbohydrates using high performance liquid chromatography, gaz chromatography and mass spectrometry. In: Ginsburg V (ed) Methods in enzymology, vol 83. Academic Press, New York, pp 3–45
Marchessault RH, Deslandes Y, Ogawa K, Sundararajan PR (1977) X-ray diffraction data for $\beta(1 \rightarrow 3)$D-flucan. Can J Chem 55:300–303
Markwalder HU, Neukom H (1976) Diferulic acid as a possible crosslink in hemicelluloses from wheat germ. Phytochemistry 15:836–837
Masuda Y (1969) Auxin-induced cell expansion in relation to cell wall extensibility. Plant Cell Physiol 10:1–9
Masuda Y (1982) Auxin-induced changes in hemicellulosic polysaccharides. 11th Int Conf Plant Growth Substances, 12–16 July 1982. Aberystwyth, Wales, (abstracts) p 10
Masuda Y, Wada S (1967) Effect of β-1,3-glucanase on the elongation growth of oat coleoptile. Bot Mag Tokyo 80:100–102
Meier H (1958) The structure of the cell walls and cell wall mannans from Ivory nuts and from dates. Biochim Biophys Acta 28:229–240
Mollenhauer HH, Morré DJ (1966) Golgi apparatus and plant secretion. Annu Rev Plant Physiol 17:27–46
Monro JA, Penny D, Raymond WB (1976) The organization and growth of primary cell walls of lupin hypocotyls. Phytochemistry 15:1193–1198
Morikawa H, Kitamura S, Senda M (1978) Effect of auxin on changes in the oriented structure of wall polysaccharides in response to mechanical extension in oat coleoptile cell walls. Plant Cell Physiol 19:1553–1155
Morvan C, Demarty M, Thellier M (1979) Titration of isolated cell walls of *Lemna minor* L. Plant Physiol 63:1117–1122
Moscatelli EA, Ham EA, Rickes EL (1961) Enzymatic properties of a β-glucanase from Bacillus subtilis. J Biol Chem 236:2858–2862
Nakamura T, Sekine S, Arai K, Takahashi V (1975) Effects o gibberellic acid and IAA on stress-relaxation properties of pea hook cell wall. Plant Cell Physiol 16:127–138
Nari J, Noat G, Ricard J, Franchini E, Moustacas AM (1983) Catalytic properties and tentative function of a cell wall β-glucosyltransferase from soybean cells cultured in vitro. Plant Sci Lett 28:313–320

Nevins DJ (1970) Relation of xylosidases to bean hypocotyl growth. Plant Physiol 46:458–462

Nevins DJ (1975) The in vitro stimulation of IAA-induced modification of Avena cell wall polysaccharides by an exo-glucanase. Plant Cell Physiol 16:495–503

Nishitani K, Masuda Y (1980) Modifications of cell wall polysaccharides during auxin-induced growth in azuki bean hypocotyl segments. Plant Cell Physiol 21:169–181

Olson AC, Bonner J, Morré DJ (1965) Force extension analysis of Avena coleoptile cell walls. Planta 66:127–133

Palmieri S, Odoardi M, Soressi GP, Salamini F (1978) Indoleacetic acid oxidase activity in two high-peroxidase tomato mutants. Physiol Plant 42:85–90

Parameswaran N, Liese W (1982) Ultrastructural localization of wall components in wood cells. Holz Roh- und Werkstoff 40:145–155

Penny P, Penny D, Marshall DC, Heyes JK (1972) Early responses of excised stem segments to auxins. J Exp Bot 23:23–26

Penny D, Penny P, Marshall DC (1973) High resolution measurements of plant growth. Can J Bot 52:959–969

Pitman MG (1965) The location of the Donnan free space in disk of beetroot tissue. Aust J Biol Sci 18:547–553

Poux N (1970) Localisation d'activités enzymatiques dans le méristème radiculaire de *Cucumis sativus* L. III. Activité phosphatasique acide. J Microsc 9:407–434

Poux N (1973) Localisation d'activités enzymatiques en microscopie électronique, application aux cellules végétales. Ann Univ ARERS 11:81–94

Prat R (1978) Gradient of growth, spontaneous changes in growth rate and response to auxin in excised hypocotyl segments of *Phaseolus aureus*. Plant Physiol 62:75–79

Pressey R, Avants JK (1977) Occurrence and properties of polygalacturonase in Avena and other plants. Plant Physiol 60:548–553

Preston RD (1979) Polysaccharide conformation and cell wall function. Annu Rev Plant Physiol 30:55–78

Preston RD, Ripley GW (1954) Electron diffraction diagrams of cellulose microfibrils in Valonia. Nature 174:76

Rayle DL, Cleland RE (1970) Enhancement of wall loosening and elongation by acid solutions. Plant Physiol 46:250–253

Reis D (1978) Precisions cytochimiques sur l'assemblage in vitro des hemicelluloses de l'hypocotyle de Soja (*Phaseolus aureus* Roxb.) Ann Sci Nat Bot Biol Veg 19:163–193

Reis D (1980–1981) Cytochimie ultrastructurale des parois en croissance par extractions ménagées. Effets comparés du diméthylsulfoxide et de la méthylamine sur le démasquage de la texture. Ann Sci Nat Bot Biol Veg 2,3:121–136

Reis D, Mosiniak M, Vian B, Roland JC (1982) Cell wall and cell shape changes in texture correlated with an ethylene-induced swelling. Ann Sci Nat Bot Biol Veg 4:115–133

Richmond PA, Metraux JP, Taiz L (1980) Cell expansion patterns and directionality of wall mechanical propertics in Nitella. Plant Physiol 65:211–217

Robards AW (1967) The xylem fibres of Salix fragilis. J Ro Micr Soc 87:329–352

Robinson DG (1977) Plant cell wall synthesis. Adv Bot Res 5:89–151

Roland JC 1978) General preparation and staining of thin sections. In: Hall JL (ed) Electron microscopy and cytochemistry of plant cells. Elsevier, Amsterdam, pp 1–62

Roland JC, Mosiniak M (1982) On the twisting pattern, texture and layering of the secondary cell walls of lime wood. Proposal of an unifying model. IAWA (Int Assoc Wood Anat) Bull 4:15–26

Roland JC, Vian B (1981) Use of a purified endopolygalactoronase for a topochemical study of elongating cell walls at the ultrastructural level. J Cell Sci 48:333–343

Roland JC, Vian B, Reis D (1977) Further observations of cell wall morphogenesis and polysaccharide arrangements during plant growth. Protoplasma 91:125–141

Roland JC, Reis D, Mosiniak M, Vian B (1982) Cell wall texture along the growth gradient of the mung bean hypocotyl: ordered assembly and dissipative processes. J Cell Sci 56:303–318

Roomans GM, Borst-Pauwels GWFH (1979) Interactions of cations with phosphate uptake by *Saccharomyces cerevisiae*. Effect of surface potential. Biochem J 178:521–527

Rougier M, Kieda C, Monsigny M (1979) Use of lectin to detect the sugar components of maize root cap slime. J Histochem Cytochem 27:878–881

Roux SJ, Scholum RD (1982) Role of calcium in mediating cellular functions important for growth and development in higher plants. In: Calcium and cell function, vol 3. Academic Press, New York, pp 409–453

Sakurai N, Nevins DJ, Masuda Y (1977) Auxin- and hydrogen ions-induced cell wall loosening and cell extension in Avena coleoptile segments. Plant Cell Physiol 18:371–380

Sakurai N, Nishitani K, Masuda Y (1979) Auxin-induced changes in the molecular weight of hemicellulosic polysaccharides of the Avena coleoptile cell wall. Plant Cell Physiol 20:1349–1357

Sawhney VK, Srivastava LM (1975) Wall fibris and microtubules in normal and gibberellic acid-induced growth of lettuce hypocotyl cells. Can J Bot 53:824–835

Schönherr J, Bukovac MJ (1973) Ion exchange properties of isolated Tomato fruit cuticular membrane: exchange capacity, nature of fixed charges and cation selectivity. Planta 109:73–93

Selegny E, Broun G, Thomas D (1969) Enzymes en phase insoluble. Variation de l'activité enzymatique en fonction de la concentration en substrat. Effets dits „régulateurs“ des membranes enzymatiques. C R Acad Sci Paris 269:1330–1333

Seligman AM, Hanker JS, Wasserkrug H, Dmochowski H, Katzoff L (1965) Histochemical demonstration of some oxidized macromolecules with thiocarbohydrazide (TCH) or thiosemicarbozide (TSC) and osmium tetroxide. J Histochem Cytochem 13:629–639

Sentenac H, Grignon C (1981) A model for predicting ionic equilibrium concentrations in cell walls. Plant Physiol 68:415–419

Sexton R, Hall JL (1978) Enzyme cytochemistry. In: Hall JL (ed) Electron microscopy and cytochemistry of plant cells. Elsevier/North Holland, Amsterdam, pp 63–148

Shibaoka H (1974) Involvement of wall microtubules in gibberellin promotion and kinetin inhibition of stem elongation. Plant Cell Physiol 15:255–263

Shone MGT (1966) The initial uptake of ions by barley roots. J Exp Bot 17:89–95

Steele GW, Nevins DJ (1979) Autolysis of dicot cell wall. Plant Physiol 63:1–51

Steen DA, Chadwick AV (1981) Ethylene effects in pea stem tissue. Evidence of microtubule mediation. Plant Physiol 67:460–466

Stephens GJ, Woods RKS (1974) Release of enzymes from cell walls by an endopectate trans-eliminase. Nature 251:358

Talmadge KM, Keegstra K, Bauer WD, Albersheim P (1973) The structure of plant cell walls I. The macromolecular components of the walls of suspension-cultured sycamore cells with a detailed analysis of the pectic polysaccharides. Plant Physiol 51:158–173

Tandler CJ, Libanati CM, Sanchis CA (1970) The intracellular localization of inorganic cations with potassium pyroantimonate. Electron microscope and microprobe analysis. J Cell Biol 45:355–366

Taylor KA, Chanzy H, Marchessault RH (1975) Electron diffraction for hydrated crystalline biopolymers: nigeran. J Mol Biol 92:165

Tepfer M, Cleland RE (1979) A comparison of acid-induced cell wall loosening in Valomia ventricosa and in oat coleoptiles. Plant Physiol 63:898–902

Terry ME, Bonner BA (1980) An examination of centrifugation as a method of extracting an extracellular solution from peas, and its use for the study of indoleacetic acid-induced growth. Plant Physiol 66:321–35

Terry ME, Jones RL, Bonner BA (1981) Soluble cell wall polysaccharides released from pea stems by centrifugation I. Effects of auxin. Plant Physiol 68:531–537

Theuvenet APR, Borst-Pauwels GWFH (1976) The influence of surface charge on the kinetics of ion-translocation across biological membranes. J Theor Biol 57:313–329

Thiery JP (1967) Mise en évidence des polysaccharides sur coupes fines en microscopie électronique. J Microsc 6:987–1018

Thomas D, Broun G (1972) Monoenzymatic model membranes: diffusion-reaction kinetics and phenomena. Biochimie 54:229–244
Thomas B, Webb JA (1979) Intracellular distribution of α-galactosidase in leaves of Cucurbita pepo. Can J Bot 57:1904–1911
Vallée JC, Vansuyt G, Negrel J, Perdrizet E, Prévost J (1983) Mise en évidence d'amines liées à des structures cellulaires chez *Nicotiana tabacum* et *Lycopersicum esculentum*. Physiol Plant 57:143–148
Vance CP, Kirk TK, Sherwood RT (1980) Lignification as a mechanism of desease resistance. Annu Rev Phytopathol 18:259–288
Vanderhoef LN (1980) Auxin-regulated cell enlargment: is there action at the level of gene expression? In: Leaver CJ (ed) Genome organization and expression in plants. Plenum, New York, pp 159–173
Vanderhoef LN, Dute RL (1981) Auxin-regulated wall loosening and sustained growth in elongation. Plant Physiol 67:146–149
Vanderhoef LN, Stahl CA (1975) Separation of two responses to auxin by means of cytokinin inhibition. Proc Natl Acad Sci USA 72:1822–1825
Vanderhoef LN, Stahl CA, Williams CA, Brinkmann KA, Greenfield JC (1976) Additional evidence for separable responses to auxin in soybean hypocotyl. Plant Physiol 58:817–819
Van Vokenburgh E, Hunt S, Davies WJ (1983) A simple instrument for measuring cell-wall extensibility. Ann Bot 51:669–672
Van Zyl J, Forrest QG, Hocking C, Pallaghy CK (1976) Freeze substitution of plant and animal tissue for the localization of water soluble compounds by electron probe microanalysis. Micron 7:213–224
Versteeg C, Rombouts FM, Pilnik W (1978) Purification and some characteristics of the two pectinesterase isoenzymes from orange. Lebensm Wiss Technol 11:267–274
Vian B (1981) La technique des coupes au froid appliquée au matériel végétal. Bull Soc Bo Fr 128:89–112
Vian B (1982) Organized microfibril assembly in higher plant cells. In: Brown Jr RM (ed) Cellulose and other natural polymar systems. Plenum, New York, pp 23–43
Vian B, Mosiniak M, Reis D, Roland JC (1982) Dissipative process and experimental retardation of the twisting in the growing cell wall. Effect of ethylene-generating agent and colchicine: a morphogenetic revaluation. Biol Cell 46:301–310
Vian B, Brillouet JM, Satiat-Jeunemaitre B (1983) Structural visualization of xylans in cell walls of hardwood by means of xylanase-gold complex. Biol Cell 49:179–182
Wada S, Ray P (1978) Matrix polysaccharides of oat coleoptile cell walls. Phytochemistry 17:923–931
Wada S, Tanimoto E, Masuda Y (1968) Cell elongation and metabolic turn-over of the cell wall as affected by auxin and cell wall degrading enzymes. Plant Cell Physiol 9:369–376
Wagner G, Rossbacher R (1980) X-ray microanalysis and chlorotetracycline staining of calcium vesicles in the green algua Mongeotia. Planta 149:298–305
Wallner SJ, Walker JE (1975) Glycosidases in cell wall-degrading extracts of ripening tomato fruits. Plant Physiol 55:94–98
Washitani I, Sato S (1976) Reliability of lead salt precipitation method of acid phosphatase localization in plant cells. Protoplasma 89:157–170
Whitmore FW (1978) Lignin-carbohydrate complex formed in isolated cell walls of callus. Phytochemistry 17:421–425
Wick SM, Hepler PK (1982) Selective localization of intracellular Ca^{2+} with potassium antimonate. J Histochem Cytochem 30:1190–1204
Yaku F, Tsuji S, Koshijima T (1979) Lignin carbohydrate complex III. Formation of micelles in the aqueous solution of acidic lignin carbohydrate complex. Holzforschung 33:54–59
Yaku F, Tanaka R, Koshijima T (1981) Lignin carbohydrate complex IV. Lignin as side chain of the carbohydrate in Björkman LCC. Holzforschung 35:177–181
Yamamoto R, Nevins DJ (1979) A transglucosylase from *Sclerotinia libertiana*. Plant Physiol 64:193–196

Yamamoto R, Makai K, Masuda Y (1974) Auxin and hydrogen ion actions on light-grown pea epicotyl segments III. Effect of auxin and hydrogen ion on stress-relaxation properties. Plant Cell Physiol 15:1027–1038

Yamamoto R, Sakurai N, Shibata K, Masuda Y (1980) Effects of auxin on the structure of hemicelluloses of Avena coleoptiles. Plant Cell Physiol 21:373–381

Zarra I, Masuda Y (1979) Growth and cell wall changes in rice coleoptiles growing under different conditions II. Auxin-induced growth in coleoptile segments. Plant Cell Physiol 20:1125–1133

Zarra I, Nicolas G (1979) Effect of cytokinins and najirimycin on auxin-induced growth in rice coleoptiles. Plant Cell Physiol 20:679–682

Cell-Wall Chemistry, Structure and Components

H. TAKEDA and T. HIROKAWA

1 Introduction

Plant cell walls give shape or rigidity as well as protection and support to plant cells and tissue. The cell walls are constructed not only of a single substance, but consist of many substances which are distinguished chemically and physically.

From the chemical viewpoint the most abundant components in plant cell walls are polysaccharides. Even in cell-wall polysaccharides there are homoglycans such as cellulose, mannan, xylan, and chitin and numerous heteroglycans which are composed of two or more different sugars. The polysaccharides are composed of various kinds of sugar and their derivatives with various linkages to give diverse chemical structure (see the references of Aspinall 1981; Percival and McDowell 1981). Glycoproteins play an important role in the construction of the cell wall (see the references of Lamport 1965; Lamport and Catt 1981). In some plants, aromatic compounds such as lignin are deposited for hardening the cell wall (see the reference of Higuchi 1981).

Morphological or physical characteristics of plant cell walls are the layered structure which encloses the cell. The layers are discerned as primary and secondary walls, and further, an intercellular layer (middle lamella) is present in the boundary of the adjacent cells as a cementing substance. All cell-wall substances are insoluble matter surrounding the cell itself. Some of them have a crystalline and fibrous structure, and others are deposited as matrix (filler) substances among the microfibrillar framework (see the references of Roelofsen 1965; Colvin 1981; Robinson 1981). These cell-wall materials are composed of different constituents and different linkages (see Katō 1981). These macromolecular components are interconnected with each other to make a functional wall structure (Keegstra et al. 1973). Further, the chemical structures of cell-wall substances differ from plant to plant and from tissue to tissue. This diversity is in striking contrast to the fact that the constituents of the other organelles resemble each other regardless of species. Thus, plant cell walls must be analyzed with methods adapted to suit the various kinds of plant material.

In this chapter methods of histochemical analysis of cell walls will be described as used on a monolayered thallus, *Prasiola*, and methods of chemical analysis will be described using the unicellular alga *Chlorella*, with special reference to their growth.

2 Histochemical Analysis of Cell Walls

For the study of plant cell walls it is important to use plant material which has thick and simple cell walls. Analytical methods of histochemical study are shown by using *Prasiola japonica* as an example (Takeda et al. 1967). *Plasiola japonica* is a freshwater green alga, with a thallus 3–6-cm high. The vegetative regions of the thalli are one-layered, as shown in Fig. 1. Each cell has cell walls of moderate thickness. A schematical figure drawn by following the results of histochemical analyses is shown in Fig. 2. The surfaces of both sides of the frond are covered with thin membrane A, and monolayered cells sandwiched between the membranes are each enveloped by two clearly distinguished layers, B (outside) and C (inside). Further, the boundary regions between adjacent cells and also between

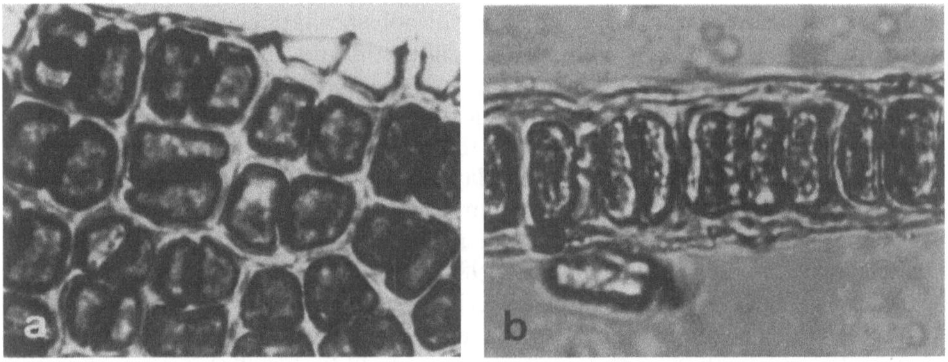

Fig. 1 a, b. Photomicrograph of the thallus of *Prasiola japonica.* **a** surface view; **b** cross-section

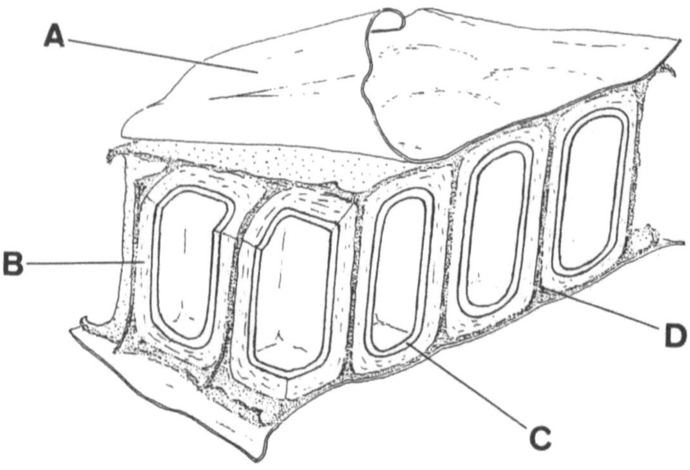

Fig. 2. Schematic representation of the structure of *Prasiola japonica* cell wall (*A–D* see text)

cells and the thin membrane A are clearly distinct as layer D. The cell-wall structure of *Prasiola japonica* can be made clear by the histochemical methods which will be described below.

2.1 Specific Stainings

Ruthenium red [1] is a specific reagent used to stain the acidic part of cell wall which is the pectic substance, while zinc chloride-iodine [2] is used for the cellulose part. After staining with these reagents, the fronds are observed microscopically. With ruthenium red a part corresponding to layer D is stained red and layer B pink. On the other hand, layer C is stained deep blue with zinc chloride-iodine.

2.2 Staining with Fluorescent Brightener

Cellulose binds with the fluorescent brightener, Calcofluor White ST (American Cyanamid Co., Wayne, New Jersey, USA) and emits fluorescence. By using a fluorescence microscope the brightener-bound cellulose can be detected. Calcofluor White ST binds specifically with β-1,4-hexapyranosan (Maeda and Ishida 1967), and is thus used for the examination of plant cell walls (Nagata and Takebe 1970). The *Prasiola* cell wall is also stained with this brightener.

2.3 Anisotropy Test

Plant cell walls are generally constructed of crystalline fibril and matrix gel. The former is discerned as having positive anisotropy when a polarizing microscope is used. In the *Prasiola* cell wall layers A and C show positive anisotropy.

2.4 Selective Dissolution

2.4.1 Alkali Treatment

A *Prasiola* frond is treated in 4% NaOH solution at room temperature for 10 min. The frond is degraded and the cells are separated from each other, as shown in Fig. 3. In this sample the boundary region, which was stained red with ruthenium red, disappeared. This indicates that the dissolved part is the layer D.

---

1 Ruthenium red solution: Ruthenium red is dissolved (0.01%) in 0.1% ammonium chloride (pH 9.0, adjusted with ammonium hydroxide)

2 Zinc chloride-iodine solution: $ZnCl_2$ 10 g is dissolved in 7 ml water. To this viscous solution 2 ml I_2-KI solution (KI 1.0 g, I_2 0.1 g, water 10 ml) is added. Supernatant solution after centrifugation is used

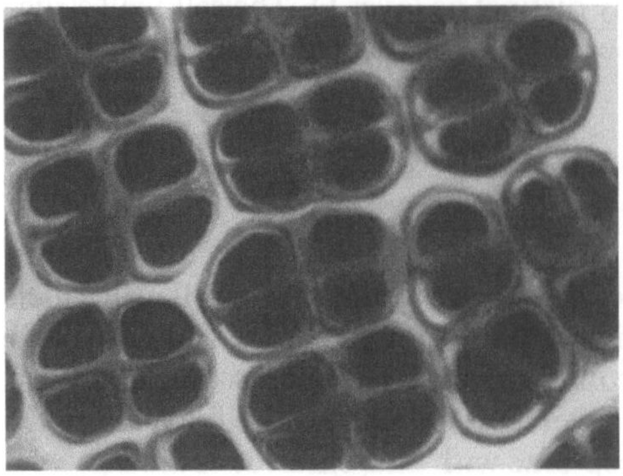

Fig. 3. Surface view of alkali-treated thallus of *Prasiola*

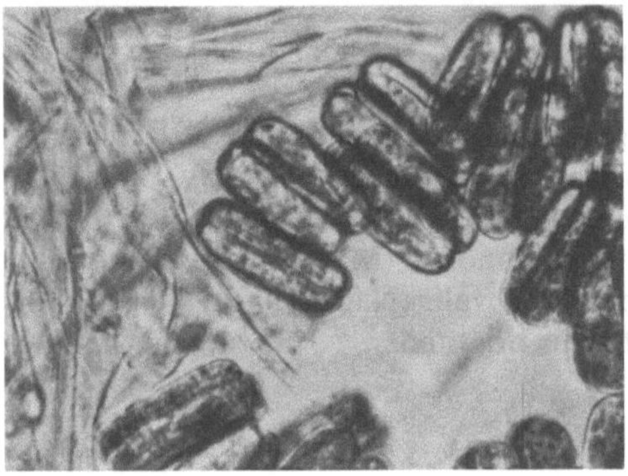

Fig. 4. Disintegration of *Prasiola* cell wall by autolysis

2.4.2 Cuprammonium Solution [3] (Schweitzer's Reagent) Treatment

A *Prasiola* frond is soaked in cuprammonium solution at room temperature for 20 h. Under the microscope it can be seen that layer C has disappeared with the cell content leaving the net-like structure of the cell wall. Cuprammonium solution is known as a specific solvent for cellulose. The cell wall remaining after this treatment is not stained with zinc chloride-iodine. From this result it is clear that by cuprammonium treatment layer C is dissolved, and layers B and D remain, leaving the net-like shape.

2.4.3 Enzymic Digestion

a) Autolytic Digestion. A *Prasiola* frond is soaked in water with toluene as an antiseptic, and kept at room temperature. After 4–5 days the *Prasiola* frond is degraded to pieces, and the soaking fluid becomes viscous. Dispersed cells and thin membrane A are observed microscopically as shown in Fig. 4. This thin membrane shows positive anisotropy. The cells isolated from the frond keep their cell

3 Cuprammonium solution: Aqueous solution of $CuSO_4$ is added to NaOH solution. Formed precipitate of cupric hydroxide is collected by centrifugation, and dissolved in minimal amount of 28% ammonia water

Table 1. Enzymes for cell-wall degradation

Commercial name	Main enzymes	Origin	Manufacturer
Cellulase Onozuka R-10	Cellulase	*Trichoderma viride*	Kinki Yakult (Japan)
Meicelase	Cellulase	*Trichoderma viride*	Meiji Seika (Japan)
Cellulysin	Cellulase	*Trichoderma viride*	Calbiochem (USA)
Driselase	Cellulase Hemicellulase Polygalacturonase	*Irpex lacteus*	Kyowa Hakko (Japan)
Macerozyme R-10	Hemicellulase Polygalacturonase	*Rhizopus arrizus*	Kinki Yakult (Japan)
Pectinase	Polygalacturonase	*Aspergillus niger*	Sigma (USA)
Pectolyase Y-23	Polygalacturonase Pectinlyase	*Aspergillus japonicus*	Kikkoman Shoyu (Japan)
Zymolyase	β-1,3-glucanase	*Arthrobacter luteus*	Kirin Brewery (Japan)

content, and the surface coat of the cells is stained with zinc chloride-iodine solution. This dissolution is an autolytic phenomenon. When the frond was heated previously, this phenomenon was not observed, suggesting that the layer B is dissolved by some endogenous cell-wall-dissolving enzyme. In some algae cell walls degrading enzymes – autolysines – are reported (Schlösser 1981).

b) Cell-Wall Digestion by Specific Enzymes. The cell-wall digestion test is usually carried out using specific cell-wall-degrading enzymes. The enzyme is in some cases extracted from the mid-gut gland of a particular snail, *Helix pomatia*. Recently, several enzymes from microbes have been sold commercially. Some of them are listed in Table 1. The *Prasiola* frond which has previously been heat-treated to inactivate the autolytic activity is incubated with cellulase (Meicelase) from *Trichoderma viride*. Figure 5 is the photomicrograph of the result, i.e., the innermost layer C together with the cell contents is lost, leaving the net-like framework of the cell wall. This is stained pink with ruthenium red but not with zinc chloride-iodine. This figure is similar to the result after dissolution by cuprammonium solution. The frond treated with Macerozyme R-10 is broken down, leaving

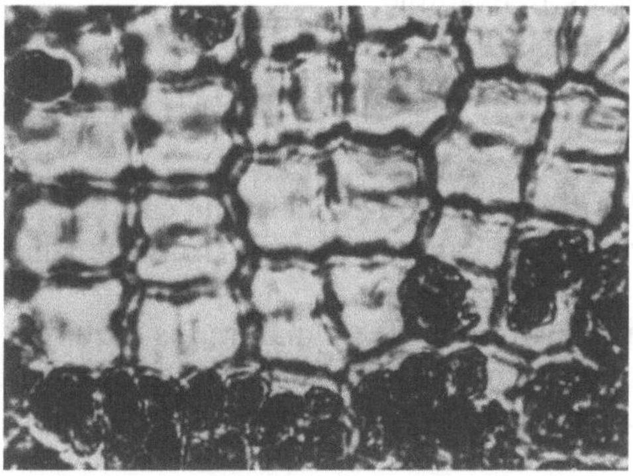

Fig. 5. *Prasiola* frond acted upon by *Trichoderma*-cellulase

Table 2. Some characteristics in the cell-wall layers of *Prasiola japonica*

Treatment	Layer of cell wall			
	A	**B**	**C**	**D**
Ruthenium red	−	+	−	+++
$ZnCl_2$-I_2	−	−	+++	−
Anisotropy	+	−	+	−
Autolysis	**no** [a]	sol [b]	no	sol
Meicelase	no	no	sol	no
Macerozyme R-10	no	sol	no	sol
Pectinase	no	no	no	sol
4% NaOH (cold)	res [c]	res	res	sol
(warm)	sol	res	res	sol
Product of acid hydrolysis	Xylose and glucose	Mannose and xylose	Glucose	Uronic acid

[a] No reaction
[b] Solubilized
[c] Resistant
[d] A xylomannan, studied by Takeda et al. (1968)

layer C and the thin membrane A. The result after treatment with Pectinase (Sigma) from *Aspergillus niger* give the same appearance as that after alkali treatment (Fig. 3). Table 2 summarizes the results mentioned above. Further, each cell-wall layer is isolated and hydrolyzed (described later). Constituent sugars are also listed in Table 2.

The structure of the *Prasiola japonica* cell wall presented in Fig. 2 results from many experiments and much observation, and is considered to be reliable.

3 Quantitative Analysis of Cell Walls

In the growing plant cell, the cell wall changes in quantity as well as in quality. Here the analytical methods used on the plant cell walls are described with special relation to cell growth by using *Chlorella* as an example.

3.1 Plant Materials

The *Chlorella* for the analyses should be the monoclone. The clone can be collected outside and isolated in the laboratory, or can be obtained from the algal collection center.

For the chemical analysis a considerable amount of cell material is required, so *Chlorella* is grown in a liquid culture. Contamination with any other strain or other microbes can cause serious error. Therefore, a pure culture is an absolute requisite.

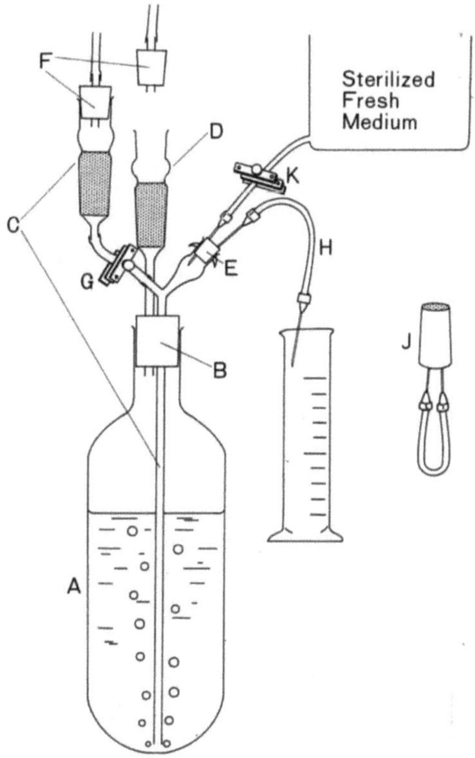

Fig. 6. Culture apparatus for microalgae. Culture vessel *A* is a flat and oblong glass flask, and is closed by a silicon stopper *B* with tubes *C* and *D*. In the bulbous parts of tubes *C* and *D* silicon sponge plugs are inserted, and the side arm of tube *C* is closed with a silicon stopper *E*. After autoclaving, the apparatus is kept in a thermo-bath, and algal cells are aseptically inoculated by a syringe through the stopper *E*. CO_2-enriched air is supplied from air-tubing *F* through the tube *C*, and the vessel *A* is illuminated. For sampling the algal suspension, the screw cock *G* is closed and air is sent through the tube *D*, then the culture is taken out through the silicon tube *H* connected to the tube *C* passing through the stopper *E*. The tube *H* has previously been sterilized after injection needles were put through the silicon plug as shown in *J*. After taking out the sample the tube *H* is withdrawn, and the air-tubing *F* is disconnected from the tube *D*, then the cock *G* is re-opened. Fresh medium is supplied by opening the screw cock *K*. These procedures are repeated by hand at every sampling time

3.1.1 Pure Culture

A photoautotrophic organism, *Chlorella* is cultured in inorganic culture medium[4] with aeration of CO_2-enriched air, and under illumination.

4 Composition of inorganic culture medium for *Chlorella*. Major medium: KNO_3 1.25 g, KH_2PO_4 1.25 g and $MgSO_4$ 1.25 g are dissolved in 1 l water. Arnon A_5 solution: H_3BO_3 2.86 g, $MnCl_2 \cdot 4H_2O$ 1.81 g, $ZnSO_4 \cdot 7H_2O$ 0.222 g, $(NH_4)_6Mo_7O_{24} \cdot 4H_2O$ 0.018 g and $CuSO_4 \cdot 5H_2O$ 0.079 g are dissolved in 1 l water. Iron solution: $FeSO_4 \cdot 7H_2O$ 27.8 g is dissolved in 1 l N/10 H_2SO_4. Calcium solution: $CaCl_2 \cdot 2H_2O$ 50 g is dissolved in 1 l water
Culture medium is prepared by adding 1 ml each of Arnon A_5 solution, iron solution and calcium solution to 1 l of the major medium

Chlorella is inoculated into the inorganic culture medium in a culture flask which has previously been autoclaved, and is cultured with 1% CO_2-enriched air bubbling through it, and under illumination.

When *Chlorella* has grown, and an experimental sample is required, or when the cell density has increased too much, some of the *Chlorella* suspension is taken out and an equal volume of fresh medium is added. The apparatus convenient for the sampling and dilution is shown in Fig. 6. Material obtained from such a culture gives precise analytical data. For the studies on the qualitative as well as quantitative changes of the cell wall in the course of cell growth synchronous culture is used.

3.1.2 Synchronous Culture

If *Chlorella* is cultured in alternating light and darkness at intervals (light 14–16 h, dark 10–8 h), the growth phases of all cells synchronize. Culture in the exclusive synchronous culture apparatus (Senger et al. 1972) is most desirable, but a simple apparatus as shown in Fig. 6 is also available, i.e., at intervals quantities of *Chlorella* culture are taken out. What remains in the flask is diluted by the addition of fresh medium, so that the volume of culture in the flask is always approximately the same, and so that cell density does not become too great and thereby prevent illumination from reaching all the cells.

3.1.3 Harvesting of Cells

Chlorella suspension is taken quantitatively from the culture apparatus. A part of this suspension is used for the cell measurements. The greater part of the suspension is quantitatively taken and centrifuged. The resulting pellet is washed twice with water. This is the analytical sample.

3.2 Measurement of Cell Growth

The growth of *Chlorella* is measured periodically, at least whenever a sample is taken.

a) Cell Volume. Total cell volume in the unit cell suspension is measured. The most common method is the measurement by hematocrit. For the measurement of cell volume of microalgae such as *Chlorella* a special hematocrit tube shown in Fig. 7 is used. Ten ml *Chlorella* suspension is pipetted into the hematocrit tube and centrifuged. The condition of centrifugation (rpm × min) must be determined previously. The measured value is expressed as ml (µl) packed cell volume (PCV). Further, there is another method to measure optically the turbidity of cell suspension. In this case the relation between PCV and turbidity should be checked previously.

b) Number of Cells. The number of microalgae cells is calculated as the number per liter of culture by using a hemacytometer. If one is available, the use of an electronic counting machine such as the Coulter Counter is convenient.

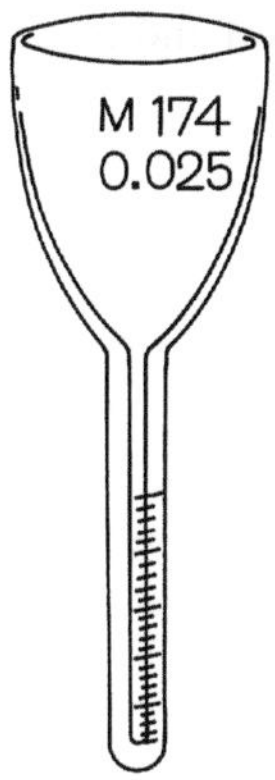

Fig. 7. Hematocrit tube for measuring cell volume of microalgae. *M 174* serial number; *0.025* maximal packed volume (0.025 ml)

The measurements of cell volume and cell number are important for ascertaining the stage of growth. From these values not only the degree of synchronization but also average cell volume and average surface area of a cell can be measured.

3.3 Preparation and Fractionation of Cell Walls

3.3.1 Disruption of Cells

One of the most characteristic features of the cell wall is its insolubility in water. Therefore, mechanical disruption is desired as the first step. There are many methods of disruption. For *Chlorella* a French pressure cell is the most useful tool. By passing them through the press twice, more than 99% of the cells are quickly disrupted. A sonicator and a cell homogenizer which disrupts the cells by the vigorous shaking with glass beads can be used. For the disruption of tissue cells or cell aggregate a Waring blendor or a Potter's glass homogenizer is useful. Methods and tools for cell disruption should be chosen depending on the plant materials.

3.3.2 Separation and Purification of Cell Walls

To a quantity of cell homogenate a detergent, sodium lauryl sulfate (SLS), is added to a concentration of 0.1 to 0.5%, and then stirred. Then this homogenate is centrifuged at 10,000 *g* for 30 min to remove soluble components. The precipitate is suspended in 0.2% SLS and centrifuged. Lipoproteinous component of the cell membrane is solubilized and removed. The precipitate is then washed with 1 M NaCl to remove nucleoprotein (Punnett and Derrenbacker 1966). The remaining insoluble part is washed with water repeatedly to remove SLS and NaCl, and then boiled under reflux in 80% (v/v) ethanol twice and acetone-ether (1 : 1) successively to remove lipids. The insoluble part is washed with water, and treated with pure α-amylase and nuclease to remove contaminated starch and nucleic acid. Solid residue is washed with water, and lyophilized to give a whitish powder of cell wall (whole cell wall). When only the carbohydrate is analyzed, the cell-wall

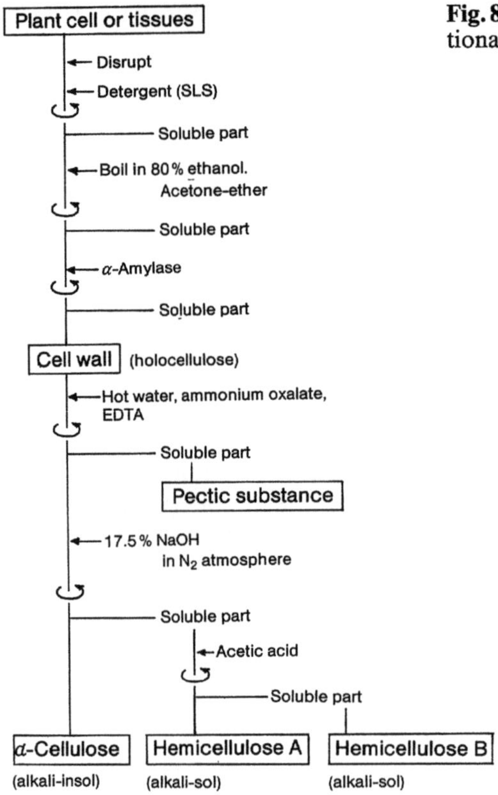

Fig. 8. Procedure for the preparation and fractionation of cell wall

sample is treated with protease to remove the peptides which might affect the measurement of carbohydrates.

3.3.3 Fractionation of Cell Walls

The cell wall is in some cases analyzed directly, but in most cases is analyzed after being fractionated according to its solubility (Rogers and Perkins 1968). However, all plant cell walls do not always have the same solubility, and in some cases it is not necessary to fractionate in detail. Cell walls are first of all fractionated by selective dissolution in alkaline solutions of various concentrations. Dimethyl sulfoxide is an effective solvent for the extraction of acetylated polysaccharides (Bouveng and Lindberg 1965). The carbohydrate moiety of the cell wall is separated into three main fractions, the pectic substance, hemicellulose, and α-cellulose. One of the most representative methods for the preparation and fractionation of plant cell walls is shown in Fig. 8. Chitin is known to occur as a cell-wall component of lower plants (fungi and green algae). Its isolation requires the use of more rigorous methods to remove contaminating substances (Bemiller 1965). Cell-wall peptides or proteins are localized in the hemicellulose and α-cellulose fractions. The hemicellulose is further fractionated into various components (Blake and Richards 1971).

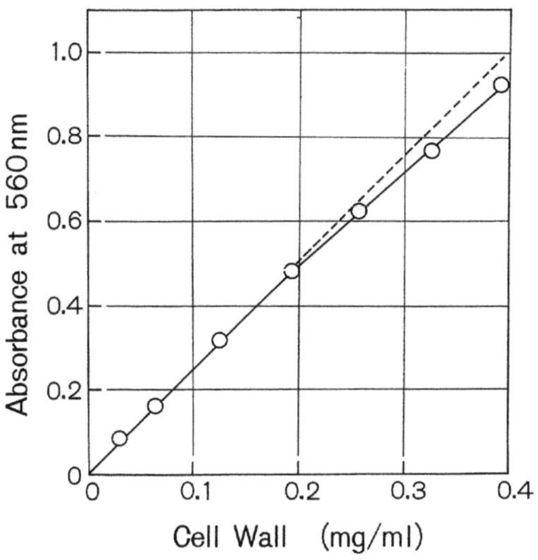

Fig. 9. Relation between turbidity and amount of cell wall

3.4 Quantitative Analysis of Whole Cell Walls

Isolated cell wall is measured quantitatively according to the methods described below.

3.4.1 Gravimetry

This is a fundamental method to be applied in any case whatever the chemical components of the cell wall. Isolated cell wall is placed on a stainless steel planchet, dried under an infrared lamp, and weighed in a microbalance. This method is simple and accurate, but requires rather a large amount of sample (about 5 mg).

3.4.2 Turbidimetry

There is a correlation between the amount of cell wall and the turbidity of the cell-wall suspension. Turbidity is measured by using a turbidimeter or spectrophotometer. A calibration curve is shown in Fig. 9. For the measurement described above, a sonicator can effectively be used to disperse the cell wall.

3.4.3 Colorimetry

a) Phenol-Sulfuric Acid Method (Total Sugar). This method (Dubois et al. 1951) is used for cell-wall samples in which the main component is neutral sugar or uronic acid. Into 1 ml suspension of cell-wall sample 1 ml of 5% phenol solution is added and mixed. To this mixture 5 ml of concentrated sulfuric acid (96%) is added so that the stream hits the liquid surface directly to produce good mixing and even heat distribution. A sample of cell-wall-containing sugar is colorized depending on the quantity. The color is measured photometrically at 490 nm. A lin-

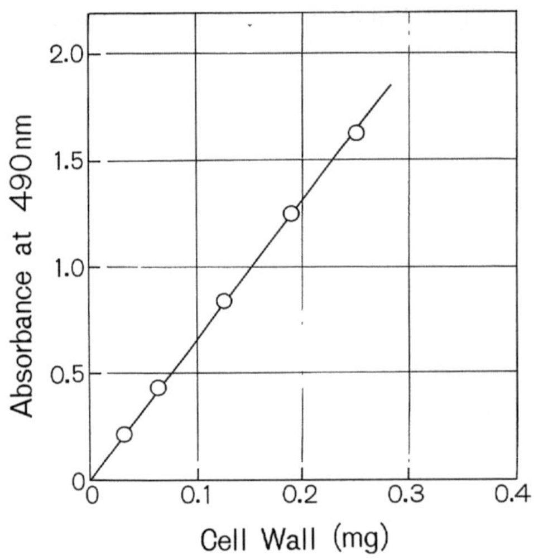

Fig. 10. Calibration curve of cell-wall carbohydrates measured by the phenol-sulfuric acid method

ear relationship is observed between the cell-wall amount and absorbance at 490 nm, as shown in Fig. 10. The phenol-sulfuric acid method is sensitive, but it cannot be used for cell wall composed mainly of amino sugar.

b) Ninhydrin Method (Total Amino Compound). To 1.5 ml of the sample solution whose pH is adjusted to 5 ± 0.2, 1.2 ml of potassium cyanide-ninhydrin solution [5] is added and heated at 100 °C for 15 min to give a blue-violet color. After cooling the test solution is added 3 ml of 60% (v/v) ethanol, and the absorbance is measured at 570 nm.

c) Elson-Morgan Method (Total Amino Sugar). To 2 ml of a sample solution containing hexosamine (5–50 μg) in a test tube with a ground glass stopper is added 1 ml of a 4% by volume solution of acetylacetone in 1.25 N sodium carbonate, and the tube is heated for 1 h at 90 °C. Then 8 ml of ethanol and 1 ml of a solution of 1.6 g of *N,N*-dimethylaminobenzaldehyde in a mixture of 30 ml of ethanol and 30 ml of conc. hydrochloric acid are added. A red color with an absorption maximum at 530 nm appears and is stable for at least 1 h (Dische 1962b).

d) Various Color Reactions of Carbohydrates. The carbohydrates of cell constituents including hexoses, pentoses, and hexuronic acids are photometrically measured according to the methods of respective color reactions (Dische 1962a).

e) Hydroxamic Acid-Ferric Ion Reaction (Pectic Substance). For measurement of the amount of pectic substance in the cell wall this method (McCready and Reeve 1955; Reeve 1959; Gee et al. 1959) is applicable. Methyl ester of the carboxyl group in pectin reacts with hydroxylamine in aqueous alkali at room temper-

5 KCN-ninhydrin solution: 1. KCN solution: 0.01 M KCN solution 5 ml is mixed with 250 ml Methyl Cellosolve. 2. ninhydrin soution: ninhydrin 2.5 g is dissolved in 50 ml Methyl Cellosolve. Solutions 1 and 2 are mixed

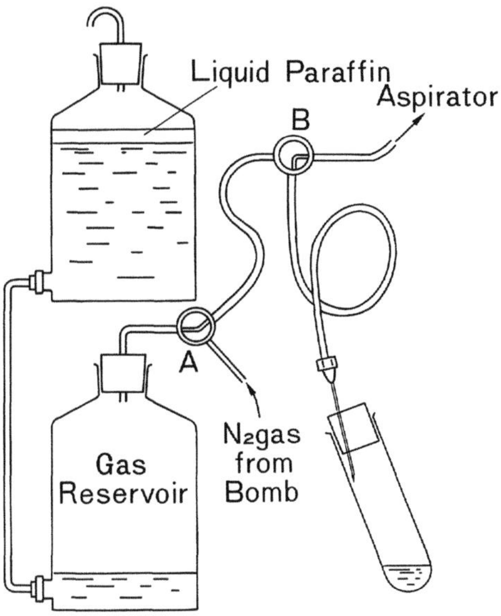

Fig. 11. Simple device for gas exchange. The gas reservoir is previously filled with N_2 gas by the operation of cock *A*. By turning the cock *B* the N_2 supply is cut off and suction commenced

ature to form pectin hydroxamate. This reacts with ferric chloride in an acid solution to form the insoluble red-colored complex, ferric hydroxamic acid. The complex is measured photometrically at 515–530 nm, using the opal glass method (Shibata 1959). In view of the characteristics of this reaction, the total amount of pectin should be measured after being methyl-esterified artificially.

f) Quantitative Analyses of Cell Wall During the Growth of Chlorella. The amount of *Chlorella* cell wall was measured using synchronized cells of *Chlorella ellipsoidea* Gerneck (IAM C-27)[6] (Takeda and Hirokawa 1978, 1979).

The cell wall of the C-27 strain was fractionated into two fractions by treating in 0.4 M NaOH at 37 °C for 20 h. In order to prevent the oxidation by alkali, air in the reaction tube is replaced by nitrogen gas, using the device shown in Fig. 11. The alkali-soluble fraction is mostly composed of neutral sugars, and the alkali-insoluble fraction of amino sugar. The amount of the former is measured by the phenol-sulfuric acid method, and that of the latter by gravimetry, the ninhydrin method and the Elson-Morgan method, during synchronous growth. As shown in Fig. 12 two fractions of C-27 cell wall showed the characteristic quantitative changes during the cell cycle. The alkali-soluble cell wall increased rapidly during the period of cell expansion. On the other hand, measured by the three different methods the alkali-insoluble cell wall gave the same result each time, namely that the amount did not change during the period of cell expansion, and increased rapidly at the time of cell multiplication.

Cell wall covers the cell surface. So, the cell-wall amount was recalculated in terms of per unit cell surface area. Surface area can be computed from the data

6 Stock culture strain at the algal collection of the Institute of Applied Microbiology (IAM), University of Tokyo

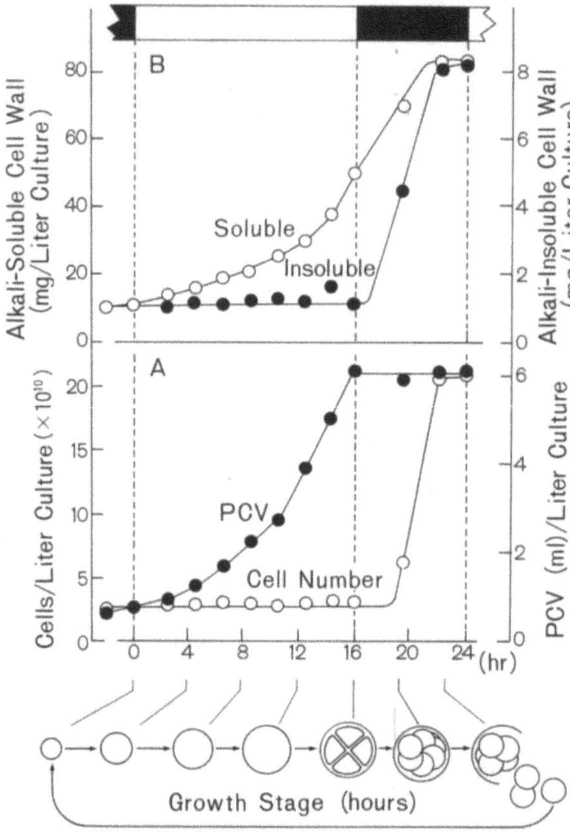

Fig. 12. A, B. Changes in the amounts of cell-wall constituents during the cell cycle of *Chlorella ellipsoidea* C-27. **A** Changes in cell number and packed cell volume. **B** Changes in the amounts of two cell-wall constituents. Schematic figures represent the growth of *Chlorella* cell

of PCV and number of cells. Cell wall amounts of both alkali-soluble and -insoluble cell walls per cell surface area changed respectively through the cell cycle as shown in Fig. 13. From this figure it is speculated that the quantity of alkali-soluble cell wall per cell surface area was almost constant during the period of cell expansion, indicating that alkali-soluble cell wall increased in proportion to the cell surface area. This suggests that alkali-soluble cell wall keeps a constant thickness during the growth. On the other hand, alkali-insoluble cell wall per cell surface area decreased gradually during the period of cell expansion. This decrease would mean that the alkali-insoluble cell wall becomes thinner just as in the case of an inflating balloon. By measuring both the cell-wall amount and cell growth throughout the cell cycle the interesting phenomena described above were found.

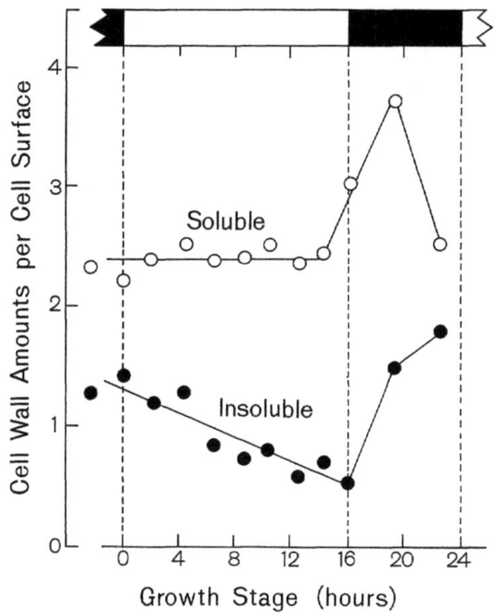

Fig. 13. Relative changes in amounts of the two cell-wall constituents per cell surface area during the course of the cell cycle

4 Qualitative Analysis of Cell-Wall Materials

4.1 Acid Hydrolysis

Acid hydrolysis is the most common method of obtaining the monomeric units of cell walls, but the conditions are different, depending on the materials (Bell 1955; Hirst and Jones 1955; Jermyn 1955; Adams 1965; Mankarios et al. 1979; Kennedy and White 1983). The most common hydrolytic conditions are:

1. heat in 0.01 M mineral acid at 80–90 °C for 15–25 h – for furanoside pentoglycans.
2. heat in 2.5% H_2SO_4 at 100 °C for 1 h – for pectic substances.
3. heat in 0.5 M H_2SO_4 at 100 °C for 1–3 h – for hemicellulosic polysaccharides.
4. heat in 1 M HCl at 100 °C for 1–3 h – for hemicellulosic polysaccharides.
5. heat in 2 M trifluoroacetic acid at 100 °C for 3–6 h – for hemicellulosic and pectic substances.
6. solubilize in 72% H_2SO_4 at room temperature for 4 h, then dilute to 1–4% H_2SO_4 with water and heat at 100 °C for 2–4 h – for cellulose and other β-hexoglycans.
7. heat in 3–6 M HCl at 110 °C for 3–18 h – for N-acetyl hexosamine-containing polysaccharides and also proteins.

Acid Hydrolysis of Cell-Wall Materials of Chlorella. Chlorella cell walls fractionated into alkali-soluble and -insoluble parts were hydrolyzed respectively as described below.

a) Alkali-Soluble Fraction. An exactly weighed cell-wall sample (about 2 mg) was placed in a small test tube, and 1 ml 0.5 M H_2SO_4 was added. After putting in a silicon rubber cork, the air contents were replaced by a nitrogen gas, using the device shown in Fig. 11. Under the reduced pressure the needle was withdrawn, and the test tube was heated at 100 °C in a thermo-bath for 2.5 h. The sample was then cooled and carefully neutralized with $BaCO_3$ powder. Solid substances ($BaSO_4$ and excess $BaCO_3$) were removed by centrifugation or filtration. Soluble fractions were collected and concentrated using an evaporator.

b) Alkali-Insoluble Fraction. Cell-wall material in this fraction is usually not hydrolyzed in diluted acids such as 0.5 M H_2SO_4 or 1 M HCl, and more rigorous conditions are required. However, in a strong acid solution some of the constituents will be degraded, so the hydrolytic conditions should be chosen carefully depending on the sample. For the alkali-insoluble cell wall of C-27 *Chlorella*, to liberate neutral sugar, hydrolysis was performed using 72% H_2SO_4. The alkali-insoluble cell wall (about 2 mg) was soaked in 0.5 ml 72% H_2SO_4 at room temperature for 2 h, then diluted with water to 4% H_2SO_4 and heated for 4 h at 100 °C. The procedure after hydrolysis was the same as in the case of the alkali-soluble fraction.

For the analyses of amino groups in the alkali-insoluble fraction 6 M HCl hydrolysis was adopted. The alkali-insoluble cell wall of C-27 *Chlorella* was hydrolyzed in 6 M HCl at 110 °C for 18 h. This is the condition used for the hydrolysis of the peptide bond of protein. In the hydrolysis the nitrogen gas filling and evacuation are important. Hydrolyzate is evaporated to produce concentrated sample as well as to remove HCl, using a rotary evaporator. For the removal of HCl the trap flask is kept alkaline, and N_2 gas is blown over the surface of the evaporating sample through a capillary tube using the apparatus shown in Fig. 11.

4.2 Enzymic Hydrolysis

As described before with reference to the *Prasiola* cell wall, the use of specific enzymes is convenient for ascertaining the chemical nature of cell walls; but enzymes effective to hydrolyze all cell walls are not always available, at least at the present time. The enzymes listed in Table 1 can be used effectively.

5 Chromatographic Analysis of Cell-Wall Constituents

Chromatography is one of the most easy and suitable methods to separate and to identify the constituents of biological materials. The data of chromatographies are collected in the handbook of Żweig and Sherma (1972). Here, the techniques for thin-layer and liquid chromatographies are described.

5.1 Thin-Layer Chromatography

5.1.1 Neutral Sugars and Uronic Acids

Silica gel plate for thin-layer chromatography is excellent for separation and colorization, but it cannot directly be used for neutral sugars because of their weak polarity. Lato et al. (1969) and Ghebregzabher et al. (1976), however, found a way of using a silica gel plate for neutral sugars by impregnating it with sodium acetate, monosodium phosphate or disodium phosphate. For example, a silica gel plate is sprayed with 0.2 M monosodium phosphate and kept at room temperature for 10 h. By heating at 110 °C for 1 h the plate is activated. Cell-wall hydrolyzates are spotted on the plate, and developed with a solvent, for example, ethyl acetate : acetic acid : methanol : water (60 : 15 : 15 : 10, v/v). Sugars are colorized using any of the following reagents [7]: diphenylamine-aniline-phosphoric acid, aniline-hydrogen phthalate, aniline-trichloroacetic acid and p-anisidine-phosphoric acid.

p-Anisidine-phosphoric acid reagent is suited for quantitative analysis because brown spots appear and are stable for at least 2 h. The other solvents are suited for the qualitative analyses because the visible spots have different colors depending on hexose, pentose and uronic acid. Coloring reagent is sprayed evenly, and heated at 100 °–110 °C for 3–10 min.

Further, the silver nitrate method can be used for the detection of reducing sugar with its high sensitivity. Silver nitrate reagent [8] is sprayed on to the plate, and dried in dim light. By spraying alkaline-ethanol solution [8] reducing sugars appear as black spots on the white background. The plate is fixed by dipping in aqueous sodium hyposulfite solution, and a stable chromatogram is obtained.

5.1.2 Amino Acids and Amino Sugars

Amino acids and amino sugars can be separated directly on a thin-layer silica gel plate. If plenty of amino acids are contained in the sample two-dimensional chromatography is recommended. For example, the solvent systems are: (1) n-butanol : acetic acid : water (3 : 1 : 1, v/v), (2) phenol : water (75 : 25, w/w, containing 20 mg NaCN per 100 g mixture). In some cases two-dimensional separation by high voltage thin-layer electrophoresis [9] (1) and thin-layer chromatography (2) can be carried out.

---

7 Diphenylamine-aniline-phosphoric acid: 2 g diphenylamine and 2 g aniline are dissolved in 100 ml acetone, and 10 ml of 85% phosphoric acid is added. Aniline-hydrogen phthalate: 1 g aniline and 1.6 g phthalic acid are dissolved in 100 ml water saturated butanol. Aniline-trichloroacetic acid: 2 g aniline and 2 g trichloroacetic acid are dissolved in 100 ml ethyl acetate. p-Anisidine-phosphoric acid: 0.5 g p-anisidine is dissolved in 4 ml hot 85% phosphoric acid, and 90 ml water-saturated butanol is added

8 Silver nitrate reagent: 0.1 ml saturated aqueous silver nitrate solution is diluted with 20 ml acetone, and the precipitate formed is dissolved by addition of a minimal amount of water. Alkaline-ethanol: 2 g sodium hydroxide is dissolved in a minimal amount of water, and 100 ml ethanol is added

9 Thin-layer electrophoresis: formic acid : acetic acid : water (26 : 12 : 1000, v/v, pH 1.9), 100 V/cm, 30 min

Amino compounds on the plate are colorized with ninhydrin reagent (0.2% ninhydrin in water-saturated butanol) at 100 °–105 °C. Proline and hydroxyproline give a yellow color, and the other amino compounds blue-violet. Most amino compounds are identified from their Rf values. For identifying amino sugars in amino compounds the Elson-Morgan and silver nitrate methods are adopted. For the Elson-Morgan test, the plate is sprayed with acetyl acetone reagent [10] and heated at 105 °C for 5 min, then Ehrlich reagent [11] is sprayed on and heated at 90 °C for 5 min. Free amino sugars give a red color and N-acetyl amino sugar gives violet, but amino acids do not colorize.

For the qualitative analyses of amino sugars (glucosamine or galactosamine), not only the cochromatography but also the ninhydrin oxidation method [12] (Stoffyn and Jeanloz 1954) is applied.

5.1.3 Thin-Layer Chromatographic Analyses of the Constituents of *Chlorella* Cell Walls

On the cell wall of *Chlorella* Becker and Shefner (1964), Loos and Meindl (1982) and Takeda and Hirokawa (1984) reported the data analyzed by thin-layer chromatography. These data, together with the report of Northcote et al. (1958), are quite different from each other. The components of *Chlorella* cell wall were shown to be very diverse in cases of different species or strains.

For thin-layer chromatography both hand-made and ready-coated plates can be used, but the latter are better suited for quantitative analyses because the coating of silica gel is certain to be of a uniform thickness. Colorized spots on the plate are measured qualitatively as well as quantitatively, using a scanning densitometer.

5.2 Liquid Chromatography

The analytical sample applied on the column is separated by the appropriate eluting solvent. This is liquid chromatography. By combining with analytical apparatus the chromatograph is converted to an autoanalyzer. The completed apparatus is an amino acid-autoanalyzer.

5.2.1 Amino Acids and Amino Sugars

Amino acids and amino sugars in the cell-wall hydrolyzate can be analyzed qualitatively as well as quantitatively by using an amino acid autoanalyzer. Figure 14

10 Acetyl acetone reagent: 0.1 ml acetyl acetone is dissolved in 10 ml n-butanol. To this solution 0.5 ml of 50% aqueous KOH-ethanol mixture (1:4) is added just before use

11 Ehrlich reagent: 0.2 g p-dimethylaminobenzaldehyde is dissolved in 6 ml ethanol. To this solution 6 ml 35% HCl is added

12 Ninhydrin oxidation: Cell-wall sample 1–2 mg is hydrolyzed in 6 M HCl. After removal of HCl, 0.2 ml of 2% ninhydrin solution in 4% pyridine is added and heated at 100 °C for 1 h. After cooling the sample is examined by thin-layer chromatography. Glucosamine gives arabinose and galactosamine lyxose

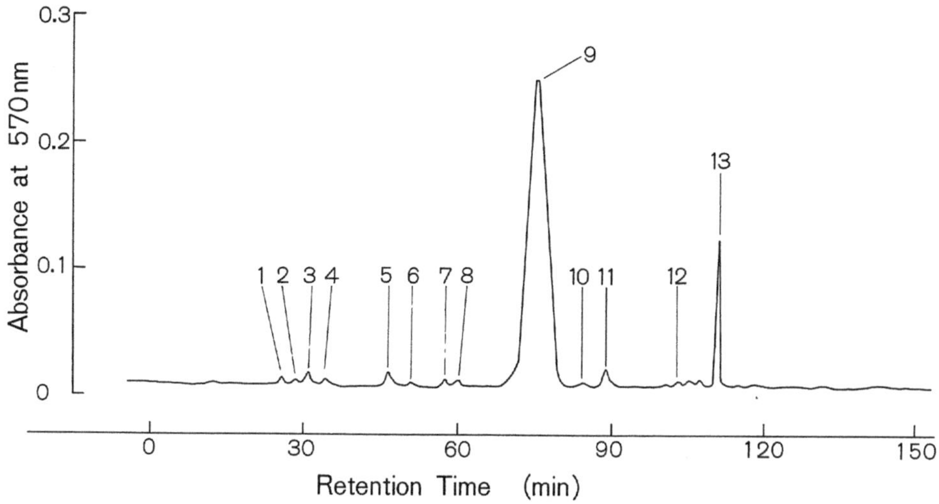

Fig. 14. Autoanalysis of the amino group in the alkali-insoluble cell wall of *Chlorella ellipsoidea* C-27. *1* aspartic acid; *2* threonine; *3* serine; *4* glutamic acid; *5* glycine; *6* alanine; *7* cystine; *8* valine; *9* glucosamine; *10* phenylalanine; *11 unknown*; *12* lysine; *13* ammonia

shows the chromatogram of acid hydrolyzate of alkali-insoluble cell wall of *Chlorella ellipsoidea* C-27.

5.2.2 Neutral Sugars

Neutral sugars form anionic complexes with borate ion and are able to be resolved by chromatography using anion exchange resin (Khym and Zill 1952). Autoanalyzing systems for sugar are illustrated (Walborg et al. 1975), and are sold commercially. However, a sugar autoanalyzer can be made in the laboratory (Hirokawa and Takeda 1980). As shown in Fig. 15 the column (3 mm × 450 mm) filled with anion exchange resin, Aminex A-25 (17.5 ± 2 µm) is equilibrated with 0.1 M boric acid-NaOH (pH 7.0) at 52 °C. The hydrolyzate is applied through the sample injector (SI), and eluted with a gradient buffer mixture of 0.1 M boric acid-NaOH (pH 7.0) and 0.15 M borax. The eluent is quantitatively mixed in a reaction mixer (RM) with 1% phenol and 88% (w/w, sp.gr. 1.802) H_2SO_4 (flow rate of three liquids are: sample : phenol : H_2SO_4 = 0.23 : 0.6 : 1.9 ml min^{-1}), and the mixture is heated for 2.5 min at 115 °C by passing through the reaction coil (RC) in the heating bath (HB). The flow of color is measured photometrically at 490 nm. Figure 16 is the recorded profile of the hydrolyzate of alkali-soluble cell wall of *Chlorella ellipsoidea* C-27 (Takeda and Hirokawa 1984). Liquid chromatography-autoanalysis is one of the most accurate methods for the separation and quantitative analysis of sugar.

The old concept that the plant cell wall is a rigid and inert structure constructed as the end-product of a plant has now given way to the new idea that the cell walls of a growing plant are undergoing turn-over (reviewed by Labavitch 1981, Takeda and Hirokawa 1982, 1983). This change of idea has affected the

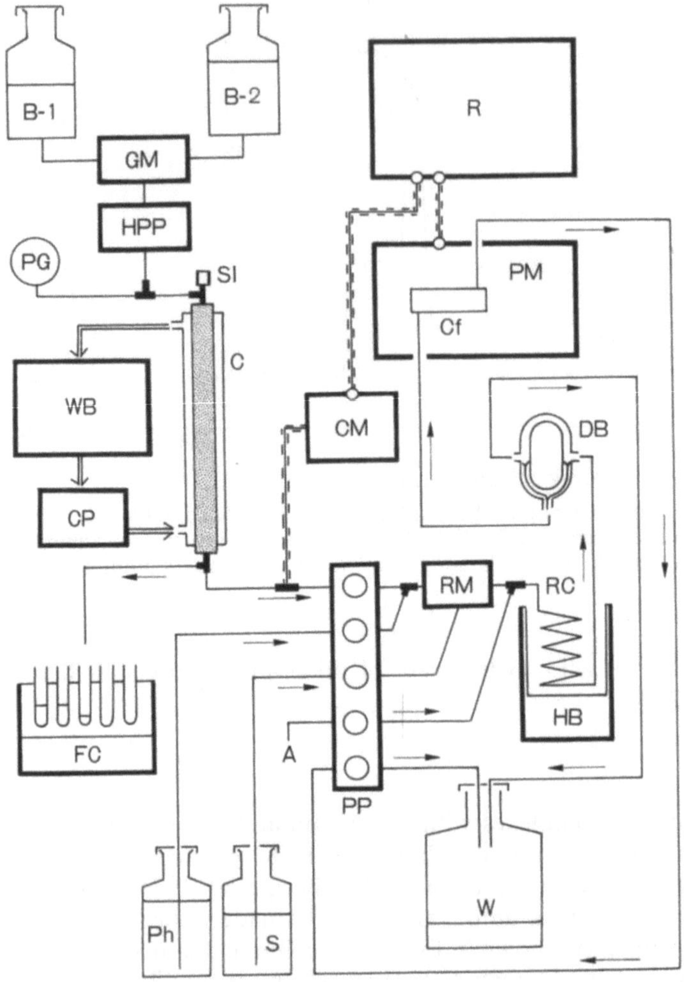

Fig. 15. Flow diagram of the modified sugar autoanalyzer. *B-1* elution buffer 1; *B-2* elution buffer 2; *GM* gradient mixer; *HPP* high pressure pump; *PG* pressure gauge; *SI* sample injector; *C* column; *WB* water bath; *CP* circulation pump; *FC* fraction collector; *Ph* phenol reagent; *S* sulfuric acid reagent; *A* air-inlet; *RM* reaction mixer; *RC* reaction coil; *HB* heating bath; *DB* debubbler; *Cf* flow cell; *CM* conductivity meter for gradient monitoring; *R* recorder; *PM* photometer; *PP* proportioning pump; *W* waste bottle

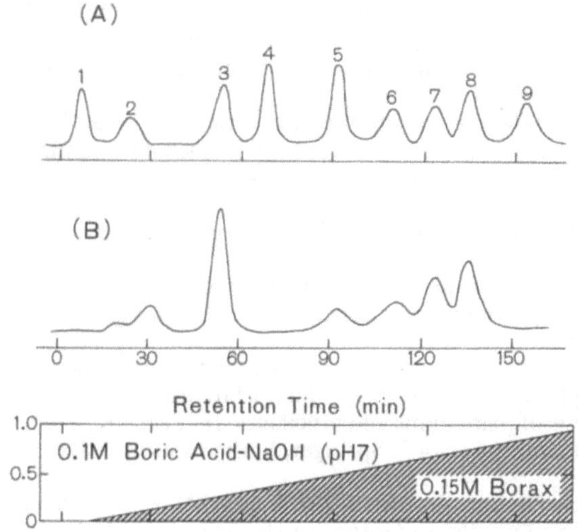

Fig. 16 A, B. Chromatogram of neutral sugars in the alkali-soluble cell wall of *Chlorella ellipsoidea* C-27. **A** Standard sugars: *1* sucrose; *2* maltose; *3* rhamnose; *4* ribose; *5* mannose; *6* arabinose; *7* galactose; *8* xylose; *9* glucose. **B** Hydrolyzate. Two peaks at the retention times 20 and 30 min are *unknown*. *Bottom figure* represents the changes in the mixing ratio (v/v) of the two elution buffers

methods of cell-wall research. Cell walls which undergo morphological, chemical, and physical changes must be studied more dynamically. The methods of enzymic study, tracer study, and electron microscopic study, as well as the methods of structural study of cell-wall polysaccharides and glycoproteins required, can be found in the appropriate literature in each case.

References

Adams GA (1965) Complete acid hydrolysis. In: Whistler RL (ed) Methods in carbohydrate chemistry, vol 5. Academic Press, New York, pp 269–276

Aspinall GO (1981) Constitution of plant cell wall polysaccharides. In: Tanner W, Loewus FA (eds) Plant carbohydrates II. Encyclopedia of plant physiology, new series, vol 13 B. Springer, Berlin Heidelberg New York, pp 3–8

Becker M, Shefner AM (1964) Thin-layer and paper chromatographic analyses of the carbohydrates in the cell wall of *Chlorella pyrenoidosa* 7-11-05. Nature 202:803–804

Bell DJ (1955) Mono- and oligosaccharides and acidic monosaccharide derivatives. In: Peach K, Tracey MV (eds) Modern Methods of Plant Analysis, vol 2. Springer, Berlin Göttingen Heidelberg, pp 1–54

Bemiller JN (1965) Chitin. In: Whistler RL (ed) Methods in carbohydrate chemistry, vol 5. Academic Press, New York, pp 103–106

Blake JD, Richards GN (1971) An examination of some methods for fractionation of plant hemicelluloses. Carbohyd Res 17:253–268

Bouveng HO, Lindberg B (1965) Native acetylated wood polysaccharide. Extraction with dimethyl sulfoxide. In: Whistler RL (ed) Methods in carbohydrate chemistry, vol 5. Academic Press, New York, pp 147–150

Colvin JR (1981) Ultrastructure of the plant cell wall: Biophysical viewpoint. In: Tanner W, Loewus FA (eds) Plant Carbohydrates II. Encyclopedia of plant physiology, new series, vol 13 B. Springer, Berlin Heidelberg New York, pp 9–24

Dische Z (1962 a) Color reactions of carbohydrates. In: Whistler RL, Wolfrom ML (eds) Methods in carbohydrate chemistry, vol 1. Academic Press, New York, pp 475–514

Dische Z (1962b) Color reactions of hexosamines. In: Whistler RL, Wolfrom ML (eds) Methods in carbohydrate chemistry, vol 1. Academic Press, New York, pp 507–512
Dubois M, Gilles K, Hamilton JK, Rebers PA, Smith F (1951) A colorimetric method for the determination of sugars. Nature 168:167
Gee M, Reeve RM, McCready RM (1959) Measurement of plant pectic substances. Reaction of hydroxylamine with pectinic acids. Chemical studies and histochemical estimation of the degree of esterification of pectic substances in fruit. Agric Food Chem 7:34–38
Ghebregzabher M, Rufini S, Monaldi B, Lato M (1976) Thin-layer chromatography of carbohydrates. J Chromatogr 127:133–162
Higuchi T (1981) Biosynthesis of lignin. In: Tanner W, Loewus FA (eds) Plant carbohydrates II. Encyclopedia of plant physiology, new series, vol 13B. Springer, Berlin Heidelberg New York, pp 194–224
Hirokawa T, Takeda H (1980) Automatic analysing system for the liquid chromatography of carbohydrates. J Gen Educ Dept Niigata Univ 10:97–106 (in Japanese)
Hirst EL, Jones JKN (1955) The analysis of plant gums and mucilages. In: Peach K, Tracey MV (eds) Modern methods of plant analysis, vol 2. Springer, Berlin Göttingen Heidelberg, pp 275–294
Jermyn MA (1955) Cellulose and hemicellulose. In: Peach K, Tracey MV (eds) Modern methods of plant analysis, vol 2. Springer, Berlin Göttingen Heidelberg, pp 197–225
Katō K (1981) Ultrastructure of the plant cell wall: Biochemical viewpoint. In: Tanner W, Loewus FA (eds) Plant carbohydrates II. Encyclopedia of plant physiology, new series, vol 13B. Springer, Berlin Heidelberg New York, pp 29–46
Keegstra K, Talmadge KW, Bauer WD, Albersheim P (1973) The structure of plant cell walls III. A model of the walls of suspension-cultured sycamore cells based on the interconnections of the macromolecular components. Plant Physiol 51:188–196
Kennedy JF, White CA (1983) Analytical methods for identification and determination of macromolecular structure. In: Bioactive carbohydrates. Ellis Horwood, Chichester, pp 66–87
Khym JX, Zill LP (1952) The separation of sugars by ion exchange. J Am Chem Soc 74:2090–2094
Labavitch JM (1981) Cell wall turnover in plant development. Annu Rev Plant Physiol 32:385–406
Lamport DTA (1965) The protein component of primary cell walls. In: Preston RD (ed) Advances in botanical research, vol 2. Academic Press, London, pp 151–218
Lamport DTA, Catt JW (1981) Glycoproteins and enzymes of the cell wall. In: Tanner W, Loewus FA (eds) Plant carbohydrates II. Encyclopedia of plant physiology, new series, vol 13B. Springer, Berlin Heidelberg New York, pp 133–165
Lato M, Brunelli B, Ciuffini G (1969) Thin-layer chromatography of carbohydrates on silica gel impregnated with sodium acetate, monosodium phosphate and disodium phosphate. J Chromatogr 39:407–417
Loos E, Meindl D (1982) Composition of the cell wall of *Chlorella fusca*. Planta 156:270–273
Maeda H, Ishida N (1967) Specificity of binding of hexapyranosyl polysaccharides with fluorescent brightener. J Biochem (Tokyo) 62:276–278
Mankarios AT, Jones CFG, Jarvis MC, Threlfall DR, Friend J (1979) Hydrolysis of plant polysaccharides and GLC analysis of their constituent neutral sugars. Phytochemistry 18:419–422
McCready RM, Reeve RM (1955) Test for pectin based on reaction of hydroxamic acids with ferric ion. Agr Food Chem 3:260–262
Nagata T, Takebe I (1970) Cell wall regeneration and cell division in isolated tobacco mesophyll protoplasts. Planta 92:301–308
Northcote DH, Goulding KJ, Horne RW (1958) The chemical composition and structure of the cell wall of *Chlorella pyrenoidosa*. Biochem J 70:391–397
Percival E, McDowell RH (1981) Algal walls – Composition and biosynthesis. In: Tanner W, Loewus FA (eds) Plant carbohydrates II. Encyclopedia of plant physiology, new series, vol 13B. Springer, Berlin Heidelberg New York, pp 277–316

Punnett T, Derrenbacker EC (1966) The amino acid composition of algal cell walls. J Gen Microbiol 44:105–114

Reeve RM (1959) Histological and histochemical changes in developing and ripening peaches II. The cell walls and pectins. Am J Bot 46:241–248

Roelofsen PA (1965) Ultrastructure of the wall in growing cells and its relation to the direction of the growth. In: Preston RD (ed) Advances in botanical research, vol 2. Academic Press, London, pp 69–149

Robinson DG (1981) The assembly of polysaccharide fibrils. In: Tanner W, Loewus FA (eds) Plant carbohydrates II. Encyclopedia of plant physiology, new series, vol 13 B. Springer, Berlin Heidelberg New York, pp 25–28

Rogers HJ, Perkins HR (1968) Cell walls and membranes. E & FN Spon, London

Schlösser UG (1981) Algal wall-degrading enzymes – Autolysines. In: Tanner W, Loewus FA (eds) Plant carbohydrates II. Encyclopedia of plant physiology, new series, vol 13 B. Springer, Berlin Heidelberg New York, pp 333–351

Senger H, Pfau J, Werthmüller K (1972) Continuous automatic cultivation of homocontinuous and synchronized microalgae. In: Prescott DM (ed) Methods in cell physiology, vol 5. Academic Press, New York, pp 301–323

Shibata K (1959) Spectrophotometry of translucent biological materials – Opal glass transmission method. In: Glick D (ed) Methods of biochemical analysis, vol 7. Inerscience, New York, pp 77–109

Stoffyn PJ, Jeanloz RW (1954) Identification of amino sugars by paper chromatography. Arch Biochem Biophys 52:373–379

Takeda H, Hirokawa T (1978) Studies on the cell wall of *Chlorella* I. Quantitative changes in cell wall polysaccharides during the cell cycle of *Chlorella ellipsoidea*. Plant Cell Physiol 19:591–598

Takeda H, Hirokawa T (1979) Studies on the cell wall of *Chlorella* II. Mode of increase of glucosamine in the cell wall during the synchronous growth of *Chlorella ellipsoidea*. Plant Cell Physiol 20:989–991

Takeda H, Hirokawa T (1982) Studies on the cell wall of *Chlorella* III. Incorporation of photosynthetically fixed carbon into cell walls of synchronously growing cells of *Chlorella ellipsoidea*. Plant Cell Physiol 23:1033–1040

Takeda H, Hirokawa T (1983) Studies on the cell wall of *Chlorella* IV. Incorporation of glucose-carbon into cell walls of synchronously growing cells of *Chlorella ellipsoidea*. Plant Cell Physiol 24:1157–1161

Takeda H, Hirokawa T (1984) Studies on the cell wall of *Chlorella* V. Comparison of the cell wall chemical compositions in the strains of *Chlorella ellipsoidea*. Plant Cell Physiol 25:287–295

Takeda H, Nisizawa K, Miwa T (1967) Histochemical and chemical studies on the cell wall of *Prasiola japonica*. Bot Mag 80:109–117

Takeda H, Nisizawa K, Miwa T (1968) A xylomannan from the cell wall of *Prasiola japonica* Yatabe. Sci Rep Tokyo Kyoiku Daigaku Sect B 13:183–198

Walborg Jr EF, Kondo LE, Robinson JM (1975) Automated determination of saccharides using ion-exchange chromatography of their borate complexes. In: Wood WA (ed) Methods in enzymology, vol 41. Academic Press, New York, pp 10–21

Zweig G, Sherma J (eds) (1972) CRC handbook series in chromatography, section A: general data and principles, vol 1, 2. CRC Press, Boca Raton

Protoplasts – for Compartmentation Studies

S. S. Thayer

1 Introduction

Compartmentation of plant constituents is an important means of regulating plant metabolism (Dennis and Miernyk 1982). Perhaps the most significant breakthrough in the last 10 years in this field has been the use of protoplasts as a source for cell fractionation experiments and organelle isolations. Although protoplasts were first isolated using cell-wall-degrading enzymes in 1960 (Cocking 1960), it was not until the mid 1970's that functional organelles were isolated from protoplasts (Galun 1981; Nishimura et al. 1976; Rathnam and Edwards 1976; Wagner and Siegelman 1975). Since that time, protoplasts have been used in hundreds of studies of plant metabolism.

The focus of this chapter is on the use of protoplasts for compartmentation studies. The advantages of the use of protoplasts will be discussed as well as the adverse effects of isolation on protoplast metabolism. In addition, several methods will be presented for protoplast lysis and separation of multiple components (subcellular fractions) of the protoplasts. The isolation and purification of single components of lysed protoplasts and the appropriate use of markers will be discussed in subsequent chapters.

2 Advantages of the Use of Protoplasts for Compartmentation Studies

The primary barrier to localization studies in plants has been the presence of a rigid cell wall. Previously, vigorous homogenization techniques had to be used in order to release organelles or subcellular fractions from plant tissues, causing their lysis, and fractionation of certain tissues such as the leaves of the economically important cereals (grasses) was virtually impossible. The removal of the cell wall by the use of cellulolytic and pectinolytic enzymes permitting gentle lysis has opened up vast new possibilities in localization studies and other areas (Wagner et al. 1978). The primary advantage of using protoplasts is gentle cell lysis which enables the investigator to recover a high percentage of intact organelles. Because fewer organelles are disrupted during lysis of protoplasts than in other homogenization techniques, the purity of the resultant organelles is higher (Wagner and Siegelman 1975).

The use of protoplasts also permits calculation of various activities on a per cell basis. This has advantages in many areas of localization and physiology. If protoplasts from different tissues of the same organ can be prepared, the specific activity of markers or other constituents in each tissue can be compared on a cellular basis, as was done in Sorghum (Kojima et al. 1979). This type of information cannot be obtained from crude homogenates.

Also, the ability to prepare and fractionate protoplasts has permitted tissue localization experiments and subcellular localization of constituents within those protoplasts. Examples of this are (1) the study of C_4 photosynthesis by preparation of mesophyll protoplasts and bundle-sheath strands or protoplasts (Moore et al. 1984), (2) the localization of pathways of synthesis and degradation of compounds such as the cyanogenic glucoside dhurrin, which is synthesized and stored in epidermal cells and degraded by enzymes localized in the mesophyll (Kojima et al. 1979; Wurtele et al. 1982), (3) the study of leaf guard cells by release of all other epidermal and mesophyll cells by wall-degrading enzymes (Robinson et al. 1983).

In addition to enzyme or metabolite localization studies, protoplasts have been useful in uptake studies and those on subcellular metabolite movement. At various intervals, the intracellular location of the compound taken up or the metabolite studied may be determined by rapid fractionation of protoplasts (Sect. 5.1.2).

3 Protoplast Isolation and Its Effect on Cellular Metabolism

If one wishes to use protoplasts for localization studies, it is essential to preserve the physiological functioning and structural integrity of the cells during isolation. The status of the protoplasts must be carefully monitored and compared to that of the parent tissue in order to assure correct evaluation of compartmentation experiments. Several aspects of the protoplast isolation technique are deleterious to the function of intact protoplasts as well as the sub-cellular components isolated from them. One should be aware of them and minimize their effects whenever possible.

3.1 Isolation Procedures

The technique of protoplast isolation has been recently reviewed (Ruesinik 1980). An extensive table is presented including enzymes used for cell-wall removal, osmotica, pH, incubation conditions and references for the successful release of protoplasts from numerous plant and tissue sources, callus and suspension cultures (Ruesinik 1980). Also discussed are protoplast stability, viability, and methods of purifying protoplasts following wall removal. Therefore, this area will not be covered.

Researchers have been limited in the past by the inability to prepare protoplasts from certain tissues. A newly developed enzyme Onozuka RS cellulase (Yakult Honsha Co., Ltd., 8-21 Shingikancho, Nishinomiya, 662, Japan) is produced by a mutant strain of *Trichoderma viride*. The RS preparation has twice the filter-paper-decomposing activity and five times the xylanase activity of Onozuka R10 cellulase (also marketed by Calbiochem as Cellulysin). Onozuka RS is capable of dissolving walls of a wider range of plant tissues including those of the bundle sheath of several C_4 plants (Moore et al. 1984) and tobacco suspension culture cells (Nagata et al. 1981) which are not readily degraded by R10. In general, Onozuka RS degrades walls faster than Onozuka R10 and the resultant protoplasts are more stable (Moore et al. 1984). The new CEL and CELF enzymes by Worthington (see Sect. 3.4) may also have different specificities than R10 since they are produced from *T. reesei* instead of *T. viride*.

3.2 Effect of Isolation pH

The pH of the protoplast isolation medium can effect various enzymes activities within the protoplasts. Protoplasts isolated from primary barley leaves at pH 5.5 had about one-third the catalase activity of protoplasts isolated at pH 6.6 [29 and 81% of activity of whole leaf control respectively (Netzley 1983)]. Phenylalanine ammonia lyase and malate dehydrogenase activities were not affected by the pH of the isolation medium (Netzley 1983). The decreased activities of these enzymes was not due to the buffer used or the presence of cellulase (Netzley 1983). The tissue was most sensitive to pH during the initial phase of incubation [during plasmolysis and early wall removal (Netzley 1983)].

3.3 Effect of Plasmolysis

Plant cells would lyse upon removal of their walls if a suitable osmoticum were not present. Plasmolysis of the tissue, though necessary, is a violent event. In an electron micrograph study of plasmolysis (Prat 1972) it was shown that the strands of cytoplasm connecting cells through plasmodesmota are stretched and broken during plasmolysis. Therefore, the cells are no longer part of a symplast, and their function may be affected. For example, isolated mesophyll protoplasts photosynthesize and form starch and sucrose, yet they are unable to export soluble metabolites (Edwards et al. 1976), and the subcellular localization of these products may be affected (Boller and Alibert 1983). The protoplast surface area decreases during plasmolysis [a fact which must be taken into account in uptake studies (Morris and Thain 1980)]. The cells respond to plasmolysis as they would to water stress (Morris et al. 1981). Ferritin or virus particles added to the medium during plasmolysis, or to the isolated protoplasts, can be taken up by pinocytosis (Cocking 1970). Therefore other materials, such as digestive enzymes, may also be taken up into the protoplasts. For this reason it may be preferable to pre-plasmolyze tissue before the addition of the wall-degrading enzymes. Plasmolysis also affects solute uptake, photosynthetic CO_2 fixation and O_2 evolution in leaf tissue

(Morris et al. 1981). Some osmotica may not be as damaging as others (Ruesinik 1980) and several should be tried when developing a new protoplast isolation system. The lowest concentration of osmoticum which permits isolation of stable protoplasts should be used to prevent excessive osmotic stress.

3.4 Effect of Enzyme Contaminants

Many interested in localization of various activities and components have noted a loss of activity during protoplast preparation. These losses have been due to the presence of salts, protease, lipase, nuclease, and peroxidase activities in the crude cellulase and pectinase products commercially available. One approach in dealing with this problem has been to purify these enzymes. Satisfactory results have been obtained by (1) desalting by (a) gel filtration (Ruesinik 1980), (b) dialysis (Lin et al. 1977; Schilde-Rentschler 1977), or (c) use of activated charcoal (Ohlrogge et al. 1980), (2) inhibition of proteases (Boller and Kende 1979), (3) isolation of active fractions following separation by column chromatography (Ishii and Mogi 1983; Slabas et al. 1980), and (4) extraction with acidified acetone, fractional precipitation with ethanol, gel-filtration, charcoal adsorption (Schenk and Hildebrandt 1969).

Alternatively, two new cellulases (CEL and CELF) and a new pectinase from Worthington (Freehold, NH 07728, USA) now available are low in salt or salt-free, and low in protease, nuclease, and lipase activities. Because these enzymes are highly purified, lower concentrations may be used (0.5–1.5% cellulase and 0.1–0.5% pectinase). Several preliminary findings indicate that much improved recovery of certain activities is possible with the CEL enzyme (Davies 1984). Lin (1983) has used the CEL enzyme for the preparation of soybean leaf protoplasts.

The plasma membrane, being in direct contact with the wall-degrading enzymes, is especially vulnerable to adverse effects of contaminants. The loss of Con A-binding sites (La Fayette et al. 1983) and an increase in membrane permeability (Taylor and Hall 1976) have been observed. Because of the intercellular connections broken during plasmolysis and the interaction of the plasma membrane and cell wall [e.g., in wall synthesis (Wagner et al. 1978)] one must be cautious in the study of plasma membrane function in protoplast preparations.

Intracellular processes may also be affected by enzymatic wall removal. Ethylene production in tobacco leaf discs is induced within 30–60 min by Cellulysin (Anderson et al. 1982) possibly mimicking a wound response to wall removal. The tonoplast ATPase is inactive unless protoplasts are prepared using desalted enzymes (Lin et al. 1977). Nitrate reductase (NR) activity is absent in protoplasts of corn prepared in 2% Onozuka R10 cellulase and 0.1% Sigma pectinase (Rhizopus) for 4 h. In vitro NR activity is inhibited in the presence of the cellulase/pectinase preparation as well. Nitrite reductase and glutamine synthetase are unaffected (Davies 1984). NR activity *is* present in corn protoplasts prepared using the Worthington CEl enzyme (Davies 1984).

More general effects of reduced stability and viability of protoplasts prepared using unpurified enzyme preparations have been observed by many workers (Lu et al. 1983; Quail 1979; Slabas et al. 1980).

4 Protoplast Lysis

High proportions of intact organelles can be released from gently lysed protoplasts. As with all protoplast techniques, there is considerable variation in the effectiveness of these procedures with protoplasts of different plants and/or tissues. In all cases, lysis should be monitored microscopically to assure protoplasts are all lysed and to prevent damage to organelles. The most frequently used method of protoplast disruption is to force the protoplasts through nylon net (10–20 μm mesh). The mesh size depends on the size of the protoplasts and the size of organelles being isolated. Lysis is accomplished by either pushing the protoplasts through the net which is attached over the opening of a syringe [without needle (Rathnam and Edwards 1976)] or by centrifuging the protoplasts through the net (Hampp 1980; Hampp and Goller 1983; La Fayette et al. 1983) and Fig. 2a, b). The net can be attached to the syringe opening by a small piece of polypropylene tubing. Passage of protoplasts through a small piece of Miracloth (Calbiochem, CA) inserted at the base of a 0.7 × 32 mm needle on a syringe causes lysis (Nishimura and Beevers 1978). Shear force of pushing the protoplasts through a long (10 × 0.2 mm) needle (Martinoia et al. 1981) releases intact vacuoles and chloroplasts, Osmotically induced lysis coupled with an increase in pH has been used primarily in the release of intact vacuoles (Chap. 4, this vol.) which are usually not preserved in other methods.

5 Protoplast Fractionation

Protoplast fractionation may be carried out in a variety of ways, depending on the goal of the compartmentation study. Unfortunately, space does not permit discussion of many useful methods including that of partition of components in an aqueous polymer two-phase system which separates particles on the basis of charge and hydrophobicity of the membrane surface (Larsson 1983). In this chapter only methods which separate multiple components of the protoplast will be examined. Those methods specific for the isolation of individual components such as vacuoles or chloroplasts will be covered in other chapters.

5.1 Density Gradient Fractionation

Most of the fractionation techniques are based on density gradient centrifugation which separates organelles/components according to their density. The most complete, widely used separation system is that of the linear sucrose density gradient. The history and theoretical basis of this technique has been described (DeDuve 1971; Price 1983). The advantages of this method are (1) the separation of the major organelles (ER, chloroplasts, mitochondria, cytoplasm/vacuolar contents, microbodies) from each other and (2) the ability to account for all the

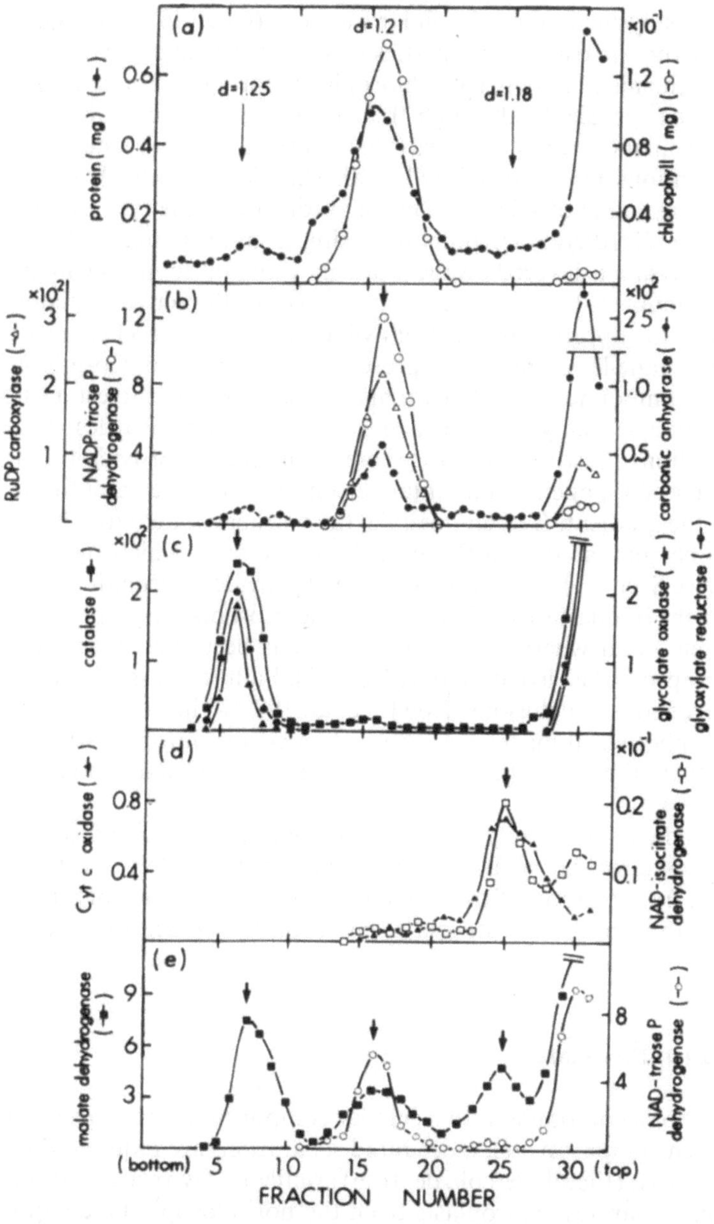

Fig. 1. Localization of enzyme activities in the separated fractions after sucrose density gradient centrifugation. Spinach leaf protoplasts were mechanically lysed (through Miracloth) and 0.5 ml of the lysate (containing approximately 2 mg chl) was layered on top of a 15 ml preformed linear sucrose gradient (35–60% w/v) dissolved in 0.02 M Tricine-NaOH buffer pH 7.5. The gradient was spun at 24,000 rpm for 3 h at 4 °C using a Beckman-Spinco SW 25-3 rotor. After centrifugation 0.5 ml fractions were collected and used for the assays. All enzyme activities but carbonic anhydrase are expressed as μmol substrates utilized or products formed/min^{-1} per 0.5 ml fraction. (Nishimura et al. 1976)

activity in the protoplast lysate in the subcellular fractions (Quail 1979). A major disadvantage comes from the osmotic properties of sucrose which can cause deterioration in the biochemical activities of chloroplasts during prolonged (3-h) centrifugation in sucrose gradients although the chloroplasts still contain their full complement of stromal (soluble) enzymes (Nishimura et al. 1976). Other disadvantages and precautions are discussed by Nagahashi (Chap. 17, this Vol.).

Figure 1 shows an example of a linear sucrose gradient and the type of data that one might obtain. Gradients are prepared using a gradient maker (Price 1983). The protoplast lysate is layered on the gradient, centrifuged in a swinging bucket rotor and fractionated. Fractions are then assayed for the activity or compound of interest and for appropriate markers which aid the researcher in the identification of the organelles/components present in each fraction. Co-migration of the activity or compound with a known marker has been used to establish the intracellular location of many enzymes and natural products (Bowles et al. 1979). Several precautions must be taken for these techniques to provide useful localization data. First of all, one must maintain a complete balance sheet for all activities assayed in the gradient. If at all possible, the markers or activities being studied should also be compared with those of the tissue from which the protoplasts originated (see Sect. 6.1). This comparison will provide information as to whether the constituent in questions is localized in the protoplast (symplast) or apoplast (cell wall, etc.) and whether the constituent has been inactivated, degraded, or lost during protoplast isolation (see Sect. 3.2). Failure to assay all fractions may lead to erroneous conclusions (Bowles et al. 1979). The correct use of biochemical and morphological markers is critical to a correct interpretation of data (Bowles et al. 1979; Quail 1979). A thorough discussion of markers is presented in other chapters.

Discontinuous gradients are rarely used for fractionation of the entire protoplast lysate but are used for the isolation of particular organelles (e.g., vacuoles or chloroplasts) or in fractionation of the different components of one organelle such as the envelope, thylakoids, and stroma of chloroplasts (Douce and Joyard 1979).

5.2 Rapid Fractionation Procedures

For studies on uptake of a compound into a cellular component or on movement of metabolites between compartments, more rapid separations are required than can be obtained by sucrose (Ficoll, Percoll, dextran) gradients. Two such methods will be discussed: one separates chloroplasts from the nonchloroplastic components of the cell ("cytoplasm") in 20 s (Robinson and Walker 1979) and the second separates plastids, mitochondria, and cytoplasm from protoplasts within 60 s (Hampp 1980; Hampp and Goller 1983).

The first method which separates chloroplasts and cytoplasm (shown in Fig. 2a) involves rupture of the protoplasts through nylon mesh and passage of the intact chloroplasts through silicone oil (Type AR 150, viscosity 150 cS, density 1.04 g ml^{-1}, Wacker Chemie, Munich, Germany). The "cytoplasm" remains in the supernatant. The pellet normally contained 95–97% of the chlorophyll, 85–

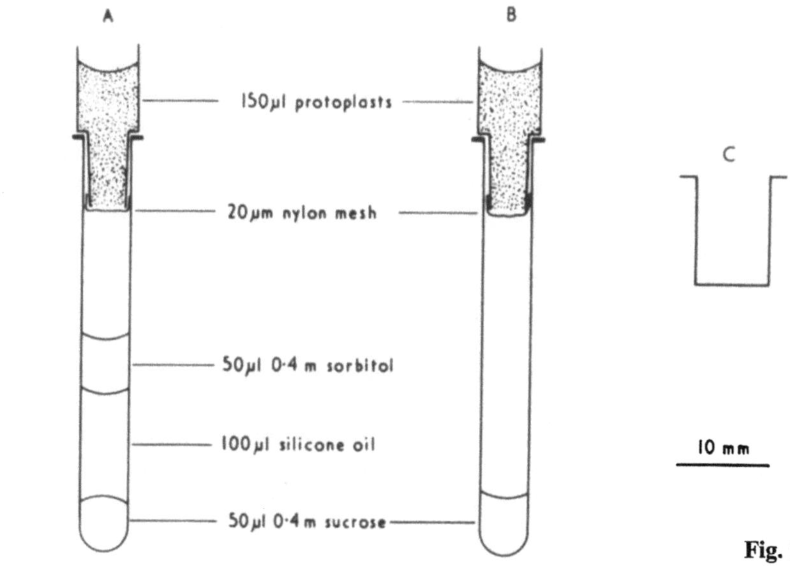

Fig. 2a

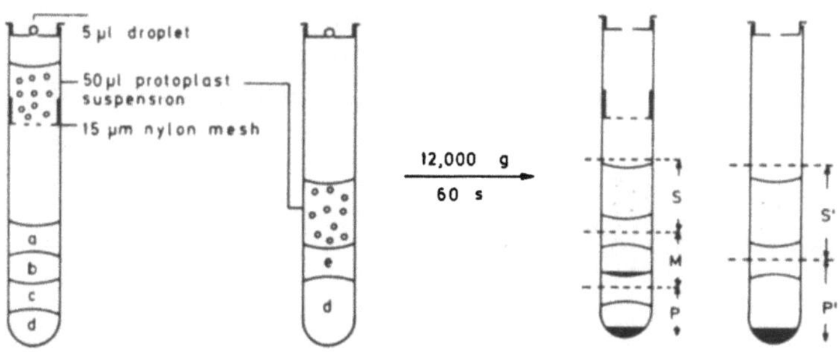

Fig. 2b

Fig. 2 a. Diagrammatic representation of the centrifuge tubes used for rapid fractionation of protoplasts. The adapters are made by cutting the ends from 1-ml disposable syringes, and these are inserted into plastic centrifuge tubes (0.4 ml capacity) with a piece of nylon mesh between the adapter and the tube. The protoplasts are added into the adapter and held in place by the mesh. If the silicone oil layer is present (**A**), the chloroplasts travel through the oil and are collected in the lower layer (sucrose) whereas the remainder of the extract is retained in the upper layer (sorbitol). In the absence of silicone oil (**B**) the whole extract is collected. For the termination of activity, the lower sucrose layer can be replaced with a 1 N $HClO_4$ solution. A cross-section of the metal holders (made from 0.65 mm thick shim steel) used to hold the centrifuge tubes with adapters is shown in (**C**). (Robinson and Walker 1979). **b** Schematic presentation of the procedure of homogenization and fractionation of isolated intact protoplasts. Protoplast aliquots were pipeted onto a 15 µm nylon mesh (left gradient) or directly onto layer (**e**) and the tubes closed with punctured caps, containing 5 µl of a Tris medium or of 0.3 M HCl. By centrifugation (60 s, 12,000 *g*) protoplasts were either ruptured by passing through the net and the resulting homogenate fractionated on a density gradient, or filtered through a layer of silicon oil (*e*). After the end of the run, the tubes were rapidly frozen in degassed liquid N_2 and the fractions were recovered by slicing through the microtubes at the positions indicated. *P*, *M*, *S* pellet, middle fraction, supernatant of ruptured protoplasts; *P'*, *S'* pellet and supernatant of intact protoplasts; *a*, *c*, *e* silicon oils of densities 1.030 Type AR, 1.070 Type CR, and 1.028 Type AR respectively; *b* 0.4M sucrose ± inhibitors; *d* 0.6M sucrose or 15% (v/v) $HClO_4$. (Hampp and Goller 1983)

90% of the two chloroplast stromal enzymes NADP-glyceraldehyde-3-phosphate dehydrogenase and ribulose bisphosphate carboxylase. Levels of mitochondrial and peroxisomal enzymes in the pellet did not exceed 15% and contamination of the pellet by soluble cytoplasm was less than 5%. Addition of bovine serum albumin (0.05–0.2%) did not reduce "cytoplasmic" contamination of the pellet and presence of EDTA (1–2 mM) significantly altered the fractionation results. For assays the upper layer (silicone oil supernatant) was removed and then the tubes were sliced through the middle of the silicone oil layer and the pellet collected using a syringe (Robinson and Walker 1979; Wirtz et al. 1980).

The second microgradient system capable of fractionating mitochondria, plastids, and cytoplasm within 60 s is shown in Fig. 2b (Hampp 1980; Hampp and Goller 1983). As in the previous system protoplasts are ruptured by centrifugation through nylon mesh and fractionated through silicone oil. Metabolic reactions may be immediately quenched in each fraction, allowing determination of metabolites therein. More than 70% of the plastids are recovered intact in the pellet (d) with less than 1% contamination by mitochondrial or cytoplasmic enzymes. About 50% of the mitochondria collect in layer (b), which also contains about 18% of the plastid markers, 15% of peroxisomal markers and about 2% of soluble cytoplasmic markers. Organelle membranes apparently penetrate the silicone layers before any effect of the acidification (by the HCl droplet to quench reactions) in the supernatant fraction occurs (Hampp 1980; Hampp and Goller 1983; Wirtz et al. 1980). In both methods a Beckman 152 microfuge was used.

6 Methods to Relate Protoplast Activity to That of Intact Tissue

These methods have been developed for leaf tissue, but the use of markers other than chlorophyll could be used to extend these methods to nongreen tissues as well. In leaf tissue, the comparison of the chlorophyll content of leaf segments or discs (with a known weight or leaf area) and that of a known number of mesophyll protoplasts can be used to *estimate* the number of mesophyll cells per g fresh weight or cm^2 (Morris and Thain 1980; Netzley 1983). For this calculation to be valid, one must assume that the *average* chlorophyll content, on a cell basis, is the same in both systems (Morris and Thain 1980). In leaves which have more than one chlorophyll-containing tissue (i.e., spongy and palisade parenchyma or mesophyll and bundle-sheath cells in C_4 plants) the assumption may not be valid. One would have to determine the relative proportions of the two tissue types which are released as protoplasts to use this method. Also, in grass leaves especially, the chlorophyll content/10^6 cells may vary along the length of the leaf (Netzley 1983). In spite of these precautions, the comparison of chlorophyll content provides a useful means of relating protoplast activities to intact leaves.

Another method useful when protoplasts of multiple tissues are isolated from one organ is to estimate the relative volumes of each tissue type present in the intact leaf [using sections of fixed material (Kojima et al. 1979) or counting microscopically the number of cells, e.g., epidermal and mesophyll per cm^2 leaf area]. In this way, relative cell numbers and volumes can be used to compare protoplast activity to intact leaves (Kojima et al. 1979).

7 Concluding Remarks

Our understanding of the role of different compartments in cell metabolism has been deepened by studies using protoplasts. Yet, we must understand that protoplasts may not be equivalent in structure or function to those cells in situ in the originating tissue. Care must be taken to preserve the in situ functioning of the cells during isolation if useful results are to be obtained using protoplasts.

References

Anderson JD, Mattoo AK, Lieberman M (1982) Induction of ethylene biosynthesis in tobacco leaf discs by cell wall digesting enzymes. Biochem Biophys Res Commun 107:588–596

Boller T, Alibert G (1983) Photosynthesis in protoplasts from *Melilotus alba*: Distribution of products between vacuole and cytosol. Z Pflanzenphysiol 110(2):231–238

Boller T, Kende H (1979) Hydrolytic enzymes in the central vacuole of plant cells. Plant Physiol 63:1123–1132

Bowles DJ, Quail PH, Morre DJ, Hartman GC (1979) Use of markers in plant cell fractionation. In: Reid E (ed) Plant organelles: Methodological surveys B. Biochemistry, vol 9. Ellis Horwood, Chichester, pp 207–224

Cocking EC (1960) A method for the isolation of plant protoplasts and vacuoles. Nature 187:962–963

Cocking EC (1970) Virus uptake, cell wall regeneration, and virus multiplication in isolated plant protoplasts. Int Rev Cytol 28:89–124

Davies HM, Faciotti D (1984) Effects of mesophyll protoplast isolation on enzymes of nitrogen metabolism. Plant Physiol [Suppl] 75:99

DeDuve C (1971) Tissue fractionation: past and present. J Cell Biol 50:20D–55D

Dennis DT, Miernyk JA (1982) Compartmentation of non-photosynthetic carbohydrate metabolism. Annu Rev Plant Physiol 33:27–50

Douce R, Joyard J (1979) Structure and function of the plastid envelope. Adv Bot Res 7: 1–17

Edwards GE, Huber SC, Guiterrez M (1976) Photosynthetic properties of plant protoplasts. In: Peberdy JF, Rose AH, Rogers HJ, Cocking EC (eds) Microbial and Plant Protoplasts. Academic Press, London, pp 299–322

Galun E (1981) Plant protoplasts as physiological tools. Annu Rev Plant Physiol 32:237–266

Hampp R (1980) Rapid separation of the plastid, mitochondrial, and cytoplasmic fractions from intact leaf protoplasts of *Avena*. Planta 150:291–298

Hampp R, Goller M (1983) Compartmentation of labeled fixation products in intact mesophyll protoplasts from *Avena*. Planta 159:314–321

Ishii S, Mogi Y (1983) Identification of enzymes that are effective for isolating protoplasts from grass leaves. Plant Physiol 72:641–644

Kojima M, Poulton JE, Thayer SS, Conn EE (1979) Tissue distributions of dhurrin and of enzymes involved in its metabolism in leaves of *Sorghum bicolor*. Plant Physiol 63:1022–1028

LaFayette PR, Breidenbach RW, Travis RL (1983) Radiolabeled lectin binding by SDS-solubilized glycoproteins from soybean root plasma membranes. Plant Physiol [Suppl] 72(1):15

Larsson C (1983) Partition in aqueous polymer two-phase systems: a rapid method for separation of membrane particles according to their surface properties. In: Hall JL, Moore AL (ed) Isolation of membranes and organelles from plant cells. Academic Press, London, pp 277–309

Lin W (1983) Isolation of mesophyll protoplasts from mature leaves of soybeans. Plant Physiol 73:1067–1069

Lin W, Wagner GJ, Siegelman HW, Hind G (1977) Membrane bound ATPase of intact vacuoles and tonoplasts isolated from mature plant tissue. Biochim Biophys Acta 465:110–117

Lu DY, Davey MR, Cocking EC (1983) A comparison of the cultural behavior of protoplasts from leaves, cotyledons and roots of *Medicago sativa*. Plant Sci Lett 31:87–99

Martinoia E, Heck U, Wiemken A (1981) Vacuoles as storage compartments for nitrate in barley leaves. Nature 289:292–294

Moore BD, Ku MSB, Edwards GE (1984) Isolation of leaf bundle sheath protoplasts from C_4 dicot species and intracellular localization of selected enzymes. Plant Sci Lett 35: 127–138

Morris P, Thain JF (1980) Comparative studies of leaf tissue and isolated mesophyll protoplasts I. O_2 exchange and CO_2 fixation. J Exp Bot 31:83–95

Morris P, Thain JR (1980) Comparative studies of leaf tissue and isolated mesophyll protoplasts II. Ion relations. J Exp Bot 31:97–104

Morris P, Linstead P, Thain JF (1981) Comparative studies of leaf tissue and isolated protoplasts III. Effects of wall-degrading enzymes and osmotic stress. J Exp Bot 32: 801–811

Nagata T, Okada K, Takebe I, Matsui C (1981) Delivery of tobacco mosaic virus RNA into plant protoplasts by reverse-phase evaporation vesicles (liposomes). Mol Gen Genet 184:161–165

Netzley D (1983) Tissue localization of phenolic metabolism in barley primary leaves. Thesis Miami Univ, Oxford, OH 45056

Nishimura M, Beevers H (1978) Isolation of intact plastids from protoplasts from castor bean endosperm. Plant Physiol 62:40–43

Nishimura M, Graham D, Akazzawa T (1976) Isolation of intact chloroplasts and other cell organelles from spinach leaf protoplasts. Plant Physiol 58:309–314

Ohlrogge JB, Garcia-Martinez JL, Adams D, Rappaport L (1980) Uptake and subcellular compartmentation of gibberellin A_1 applied to leaves of barley and cowpea. Plant Physiol 66:422–427

Quail PH (1979) Plant cell fractionation. Annu Rev Plant Physiol 30:425–484

Prat R (1972) Contribution a l'etude des protoplasts vegetaux I. Effect du traitement d'isolement sur la structure cellulaire. J Microsc (Paris) 14:85–114

Price CA (1983) General principles of cell fractionation. In: Hall JL, Moore AL (ed) Isolation of membranes and organelles from plant cells. Academic Press, London, pp 1–24

Rathnam CKM, Edwards GE (1976) Protoplasts as a tool for isolating functional chloroplasts from leaves. Plant Cell Physiol 17:177–186

Robinson SP, Walker DA (1979) Rapid separation of the chloroplast and cytoplasmic fractions from intact leaf protoplasts. Arch Biochem Biophys 196(2):319–323

Robinson NL, Zeiger E, Preiss J (1983) Regulation of ADP glucose synthesis in guard cells of *Commelina communis*. Plant Physiol 73:862–864

Ruesinik A (1980) Protoplasts of plant cells. Methods Enzymol 69:69–84

Schenk RU, Hildebrandt AC (1969) Production of protoplasts from plant cells using purified commercial cellulase. Crop Sci 9:629–631

Schilde-Rentschler L (1977) Role of the cell wall in the ability of tobacco protoplasts to form callus. Planta 135:177–181

Slabas AR, Powell AJ, Lloyd CW (1980) An improved procedure for the isolation and purification of protoplasts from carrot suspension culture. Planta 147:283–286

Taylor ARD, Hall JL (1976) Some physiological properties of protoplasts isolated from maize and tobacco tissues. J Exp Bot 27:383–391

Wagner GJ, Siegelman HW (1975) Large-scale isolation of intact vacuoles and isolation of chloroplasts from protoplasts of mature plant tissues. Science 190:1298–1299

Wagner GJ, Butcher IV HC, Siegelman HW (1978) The plant protoplast. Bioscience 28:95–101

Wirtz W, Stitt M, Heldt HW (1980) Enzymic determinations of metabolites in the subcellular compartments of spinach. Plant Physiol 66:187–193

Wurtele ES, Thayer SS, Conn EE (1982) Subcellular localization of a UDP-glucose: aldehyde cyanohydrin β-glucosyl transferase in epidermal plastids of *Sorghum* leaf blades. Plant Physiol 70:1732–1737

The Marker Concept in Cell Fractionation

G. NAGAHASHI

1 Introduction

The intent of this chapter is to provide a practical foundation for anyone interested in performing subcellular fractionation of plant tissue. It is not its purpose to discuss the validity of various markers, since this topic has recently been thoroughly reviewed (Quail 1979). Specific marker assay procedures are given in other chapters of this book and are also available elsewhere (Hall and Moore 1983). Catalogs of "commonly accepted" markers have been published (Bowles et al. 1979; Quail 1979) and should be referred to by the reader.

Historical perspectives and recent reviews have been written on animal (DeDuve 1971, 1975; Morré et al. 1979; Schneider and Hogeboom 1951) and plant cell fractionation (Bowles et al. 1979; Price 1983; Quail 1979) and they should be read before attempting to isolate subcellular components.

2 The Marker Concept

To study biochemical properties and functions of subcellular components, one approach is to disrupt cells and separate various organelles and membranes. The isolated components are identified by their associated markers. A marker is any identification tag which can be used to distinguish one subcellular component from all others after cells are disrupted. It can be a constituent (naturally occurring biochemical entity) of a given component, or it can be exogenously imposed with unique specificity.

2.1 Basic Concepts

An ideal marker is unique to one subcellular component and it is homogeneously dispersed throughout the component population (i.e., the component population is homogeneous in biochemical composition). The ideal marker is likely to be an exception rather than rule (Morré et al. 1979; Quail 1979), since the most commonly used markers appear to have a primary location in one cellular component and a secondary location elsewhere (Morré et al. 1979). For example, NADH cytochrome C reductase activity is associated with both mitochondria and ER, and ATPases have several subcellular locations (Leonard et al. 1973). Other markers are associated with membranes and also have a soluble counterpart. When markers have more than one location but can be distinguished by selective inhibition or removal of the soluble counterpart, they can still be confidently used to identify particular subcellular components. Since many marker enzymes do not have a single location, predominant localization within a single component is now an accepted criteria for markers (Morré et al. 1979).

Markers may be associated with distinct membrane domains or regions, as has been shown for the Golgi apparatus (Morré et al. 1979, 1983) and, hence, only mark a certain part of an organelle. Regional differences in plasma membranes are also possible (Morré et al. 1979) and it is likely that members of the endomembrane system will exhibit regional domains (Quail 1979). The use of multiple markers will hopefully given a more accurate representation of the distribution of any individual subcellular component.

In general, any marker can be used in one of two modes (Quail 1979). A positive marker is used to show enrichment or purification of a particular component. Negative markers are used to assess the degree of contamination of a particular component by other membranes and organelles.

2.2 Types of Marker

In two recent reviews (Bowles et al. 1979; Quail 1979) markers were classified either as morphological or biochemical. A third category (cytochemical) is added here for clarity. Both morphological and cytochemical markers are monitored and quantitated with the electron microscope.

2.2.1 Morphological

The inherent morphology of some organelles can be used to identify them in subcellular fractions. Except for ER, this technique cannot be used to identify membranous components which vesiculate or fragment during cell disruption. Although morphology provides a specific way to monitor certain organelles, there are several drawbacks when marking in this manner.

The amount of material in thin sections as represented in a micrograph is several orders of magnitude less than that used in biochemical assays (Quail 1979). Quantitation of morphological components (morphometric analysis) is a long and tedious process which requires analysis of a statistically significant number of unbiased electron micrographs. Pelleted fractions are most difficult to analyze because of the stratification observed. Thin sections from various areas of the pellet must be analyzed before the actual contents can be determined. Isolated fractions are best prepared for EM by filtration on Millipore filters (Quail 1979), which eliminates stratification and facilitates their analysis.

The size of the organelles under examination is also a critical factor when evaluating micrographs. For example, if one mitochondrion is consistently seen for every nine chloroplasts in a thin section, one could erroneously conclude that the chloroplast fraction is 90% pure. Taking into account that a mitochondrion can be as small as one tenth size of a chloroplast, the plastid fraction may actually be 50% pure. With the difficulties aside, morphological markers should be used in conjunction with biochemical markers whenever possible.

2.2.2 Cytochemical

Because the specificity of a cytochemical procedure inherently depends upon the presence of a unique constituent or constituents in a subcellular component, these markers are biochemical in nature. Quantitation of cytochemical markers usually requires examination by electron microscopy and for this reason they are not included in the biochemical category.

Various procedures have been used to selectively stain or selectively enhance staining of ER (Hepler 1981), PM (Quail 1979), and plastid membranes (Hurkman et al. 1979) in intact plant cells. The staining procedure for ER has not yet been applied to isolated fractions, as is the case for PM and plastid membranes (Hurkman et al. 1979, Quail 1979). The question of specificity and the assumptions made when the PM stain is applied to isolated membranes have been discussed (Quail 1979). The same assumptions are made for any cytochemical staining procedure. A minimum control for cytochemical staining procedures requires that intact tissue and isolated fractions be stained at the same time under identical conditions. Cytochemical markers are quantitated by morphometric analysis of electron micrographs and incur the same problems described for morphological markers.

Other cytochemical procedures are not used as markers per se but are used to confirm the subcellular localization of biochemical markers. These procedures involve the subcellular localization of enzymes in intact tissue at the ultrastructural level. Electron-dense reaction products generated directly or indirectly by the enzyme of interest are detected via electron microscopy. The major assumption is that the reaction product remains near or at the site of the enzyme. Cytochemical localization of an enzyme in intact tissue can lend credence to the assignment of its subcellular location by cell fractionation procedures.

2.2.3 Biochemical

Biochemical markers are usually intrinsic constituents of a given organelle or membrane. The most common are enzymatic, and enrichment in marker activity is taken to mean enrichment of the associated membrane or organelle. Nonenzymatic constituents such as chlorophyll (chloroplasts and thylakoids) and cardiolipin (inner mitochondrial membrane) are also used as markers. A third type of biochemical marker is one that is extrinsically imposed on membranes such as surface labeling of the plasma membrane with a radioisotope. Biochemical markers are considerably easier to quantitate than morphological or cytochemical markers. A representative sample of any isolated fraction is easy to obtain and quantitation of any biochemical marker is less subjective than quantitative morphometry.

3 Preservation of Marker Enzyme Activity During Cell Disruption

3.1 Choice of Material

Unless chloroplasts are the object of study, etiolated or nongreen tissues are frequently used for subcellular fractionation since they are low in phenolic compounds as well as chloroplast pigments. Good cell-free systems are characterized by lack of browning and by low absorbance at 260 nm (mainly due to phenolics; Loomis 1974). In addition to phenolics and polyphenol oxidase (PPO) activity, some plant tissues are high in acyl hydrolase activity (Moreau 1985a, b), while others have considerable proteolytic activity (Alpi and Beveers 1981; Caldwell and Haug 1980; Gardner et al. 1971). These hydrolases can contribute to marker enzyme degradation. The extent of these hydrolytic activities in nongreen compared to green tissue is not clear.

The presence of phenolics and PPO activity is readily determined by homogenizing the tissue in basic homogenization medium without additives (see Sect. 3.2). If the crude homogenate turns brown during isolation or after short-term storage, phenolics and PPO activity are likely. There are no quick ways to determine the presence of acyl hydrolase or protease activity. Crude homogenates or crude fractions isolated by differential centrifugation should be assayed for markers immediately after fractionation. The crude fractions should then be stored on ice and marker enzyme activity monitored over a period of time. Loss of marker activity after several hours of storage can indicate a potential acyl hydrolase or protease problem. Several approaches can be used to minimize degradation of membrane constituents and these will be discussed shortly.

3.2 Homogenization Procedure

Historically, plant subcellular fractionation is based on the quantitative approach and techniques employed by animal cell biochemists. Fractionation of plant cells is inherently more difficult because of the presence of a cell wall, the presence of secondary metabolites (especially phenolic compounds), and the presence of hydrolytic activity compartmented in vacuoles and plastids. Conventional approaches to fractionation require fairly harsh homogenization procedures to break open cells, and consequently secondary metabolites and hydrolytic enzyme activity are released when subcellular compartments are ruptured.

Homogenization techniques include razor blade chopping (by hand or mechanically driven chopper), mortar and pestle, ground glass homogenizers (Dounce), Potter-Elvehjem homogenizer, Virtis, Polytron, and Waring blendor. The latter three should not be used on plant tissues that are known to contain high levels of phenolics and PPO activity. Whirling blade mechanisms actually whip air into the homogenate (Loomis 1974) and will provide the oxygen necessary for PPO activity. Mechanically driven razor blade choppers also produce considerable frothing and personal experience with potato leaves has shown an enhanced browning of the homogenate when this technique is employed.

The most commonly used cell disruption techniques are razor blade chopping and grinding by mortar and pestle since they produce low to medium shear forces and consequently least damage to intact organelles. These procedures probably produce a low yield of organelles when compared to the total number of organelles present in intact tissue. The report by Jacobsen (1968) indicated that grinding by mortar and pestle produced a low yield (17%) of proplastids in the crude homogenate. The total number of proplastids in the intact etiolated corn leaves was determined two different ways, and all isolated fractions were related back to the total number present before homogenization. The major loss of proplastids was presumably due to unbroken cells which were removed by filtering through Miracloth. Another approach to estimating cell breakage compared the phospholipid content of unground roots to that of the crude homogenate. This report (Fisher and Hodges 1969) indicated that 42% of the membranes were released by mortar and pestle grinding. Even with fairly harsh homogenization procedures, the apparent yield of organelles and membranes from intact tissue appears to be low.

The homogenization procedure may have a direct effect on the distribution of marker enzymes separated by differential centrifugation. With low shear force (razor blade chopping), as much as 60% of PM and ER marker activity (Lord et al. 1972; Nagahashi and Beevers 1978) are pelleted with the crude mitochondrial fraction. With medium shear (mortar and pestle), PM is evenly distributed between the crude mitochondrial and crude microsomal fraction (Leonard and Vanderwoude 1976). Higher shear techniques such as the polytron probably generates smaller microsomal vesicles and consequently, as much as 70 to 80% of the plasma membranes will remain in the post-mitochondrial supernatant (Koehler et al. 1976). The disadvantage of high shear homogenization is that lower yields of intact organelles are probable, which means that the microsomal fraction will have considerable contamination of plastid membranes, mitoplasts, inner and outer mitochondrial membranes, and possibly nuclear membranes.

3.3 Use of Additives in the Homogenization Medium

In many plant tissues, phenolics are mainly responsible for loss of marker enzyme activity. Phenolics can complex with proteins via H-bonds (Loomis and Battaile 1966) and they can also be oxidized by PPO in the presence of oxygen to form quinones (Mayer and Harel 1979). Quinones can readily oxidize functional groups of proteins (Loomis and Battaile 1966) and/or nonenzymatically polymerize (Maillard reaction) to form dark pigments which covalently bond to proteins (Loomis 1974). Phenolics which are released from disrupted cells can be removed by adsorbants (Table 1). The removal of phenolics by insoluble polyvinylpyrollidone is most practical, since bound phenolics can be removed by low speed centrifugation and the remaining supernatant can be fractionated and assayed without concern for interference from the additive. Oxidation of phenolics is prevented by homogenizing in the presence of reducing reagents and/or inhibitors of PPO activity (Table 1).

Table 1. Additives which may be used in the homogenization medium to sustain (protect) marker enzyme activities

Mode of action	Additive or protective agent
Protease inhibition	Phenylmethanesulfonyl fluoride (Alpi and Beevers 1981; Caldwell and Haug 1980; Gardner et al. 1971; Scherer 1981), leupeptin (Alpi and Beevers 1981), glycerol (Alpi and Beevers 1981), iodoacetamide (Alpi and Beveers 1981), p-Cl-mercuribenzoate (Alpi and Beevers 1981)
Lipolytic acyl hydrolase and phospholipase inhibition	Nupercaine (dibucaine) (Bishop and Oertle 1983; Scherer and Morré 1978), BSA (Galliard 1974; Loomis 1974), EDTA (Moore and Proudlove 1983; Philipp et al. 1976), EGTA (Galliard 1974; Miflin 1974), sulfhydryls (Moore and Proudlove 1983), choline and ethanolamine (Scherer and Morré 1978), sodium fluoride (Philipp et al. 1976)
Polyphenolic oxidase inhibition and prevention of browning	Sodium mercaptobenzothiazole (Anderson 1968; Loomis 1974), sodium ascorbate (Anderson 1968; Moore and Proudlove 1983), sodium metabisulfite (Anderson 1968; Loomis 1974; Moore and Proudlove 1983), polyvinylpyrrolidone (Loomis 1974; Loomis and Battaile 1966), BSA (Loomis 1974), thiols and reducing agents (Anderson 1968)
Adsorbs free fatty acids	BSA (Anderson 1968; Galliard 1974; Loomis 1974)
Adsorbs phenolic compounds and tannins	Soluble polyvinylpyrrolidone (Loomis 1974), BSA (Anderson 1968; Loomis 1974), Insoluble polyvinylpyrrolidone (Loomis 1974; Loomis and Battaile 1966)

In addition to phenolics and PPO activity, the hydrolytic activity of lipolytic acyl hydrolases and proteases can present a major problem during membrane isolation. The degradative effect of these enzymes may be somewhat controlled by pH since both types of hydrolases are usually active under acidic conditions (Alpi and Beevers 1981; Galliard 1974). Homogenization should be performed in the cold (0 °–4 °C) with an osmoticum (0.25–0.5 M sucrose, sorbitol, or mannitol) buffered (0.03 to 0.15 M hydrogen ion buffers such as HEPES, TES, MES [1], and Tricine) between pH 7.0 to 8.0. Other additives can be added to the basic homogenization medium to inhibit acyl hydrolase and protease activity (Table 1). Additives should be used with discretion, since it must be determined whether they have a direct effect on marker enzyme activity, have an effect on the assay procedure itself, or interact with one another and consequently interfere with certain enzyme assays (Koundal et al. 1983).

The use of additives can also be counter-productive in another manner. For example, the presence of metabisulfite inhibited PPO activity in tobacco leaves but also produced optimal peptidase activity (Loomis 1974). Our experience with

1 HEPES = N-2-hydroxyethyl piperazine-N′-2-ethane sulfonic acid
MES = (2-[N-morpholino]) ethane sulfonic acid
TES = (N-tris [hydroxymethyl]) methyl-2-aminoethane sulfonic acid
Tricine = (N-tris [hydroxymethyl]) methyl glycine

potato leaves has indicated that sulfhydryls (DTT or β-mercaptoethanol) can retard browning; however, much higher levels of acyl hydrolase activity are maintained (Moreau 1985a).

3.4 Gel Filtration to Remove Soluble Hydrolytic Activity

Regardless of the composition of the homogenization medium, it is necessary to separate membranes and organelles from soluble hydrolases and secondary metabolites as rapidly as possible. Fast separation is usually achieved by differential centrifugation or by centrifugation of unpelleted homogenates directly on sucrose gradients (see Sect. 4.3.2). Recently, several workers (Boller and Kende 1979; Jones 1980; Scherer 1981; Van der Wilden et al. 1980) have used gel filtration as the initial separation step. The crude homogenate is layered on a Sepharose 4B (Boller and Kende 1979; Jones 1980; Van der Wilden et al. 1980) or Sepharose 2B-CL column (Scherer 1981). Apparently, most of the subcellular components pass in the void volume, while soluble hydrolytic enzymes are retarded by the column. This technique has the advantage of removing undesirable soluble enzymes without pelleting organelles; however, very little quantitative data is available showing percent recovery of markers after gel filtration (Van der Wilden et al. 1980) and short separation time may require small columns with small sample size.

Even with rapid separation of membranes from the supernatant, there is no guarantee that soluble enzymes will not adhere to isolated membranes or be trapped inside isolated membrane vesicles. Protoplast preparations may provide a way to isolate various subcellular compartments intact (especially vacuoles and plastids). Protoplasts can be lysed by gentle techniques and the subsequent release of hydrolases and phenolics can be minimized. The use of protoplasts to study subcellular compartments and the problems associated with this approach are covered elsewhere in this book.

4 Methods Used to Separate Markers

4.1 General Approaches to Cell Fractionation

Separation of organelles and membranes is based on differences in size, density, or surface properties. Purification based on differences in surface properties is relatively new and includes such techniques as phase partition (Larsson and Anderson 1979; Yoshida et al. 1983), free flow electrophoresis (Dubacq and Kader 1978; Hannig and Heidrich 1974; Morré et al. 1983), and affinity column chromatography (Schroeder et al. 1982). The application of these techniques to plant tissue needs further evaluation.

The most common approach to subcellular fractionation has been a combination of differential centrifugation (size separation) and density gradient centrifugation. Differential centrifugation is the oldest form of subcellular fractionation and will be discussed first.

4.2 Differential Centrifugation

Low speed centrifugation (250 to 3000 g) is typically used to pellet most of the nuclei (Price 1979), plastids (Jacobson 1968), cell wall fragments (Nagahashi et al. 1985), and unbroken cells. The supernatant is centrifuged at higher forces (6000 to 20,000 g) to sediment most of the mitochondria (Nagahashi and Hiraike 1982) and microbodies which have a similar size. The post-mitochondrial supernatant is centrifuged at high speeds (40,000 to 120,000 g) to pellet microsomes. Microsomal fractions are heterogeneous mixtures consisting largely of ER, PM, Golgi membranes (Koehler et al. 1976; Leonard et al. 1973; Nagahashi and Beevers 1978; Nagahashi and Kane 1982; Philipp et al. 1976), and probably tonoplast. The post-microsomal supernatant contains the "soluble" enzymes [considerable microsomes are still in suspension in 80,000 g supernatants (Koehler et al. 1976) but can be pelleted at 120,000 g (Nagahashi and Hiraike 1982)].

Recently, differential centrifugation has been used primarily to collect large organelles or to obtain a microsomal fraction to use for further purification. In either case, optimum centrifugation conditions should be determined to achieve best separation with minimum cross-contamination. A recent report (Nagahashi and Hiraike 1982) has determined the optimum centrifugal force and centrifugation time for separating crude microsomes from crude mitochondria. Best separation was achieved when lower centrifugal forces (6000 to 8000 *g* for 20 min) were used to pellet mitochondria. These optimum conditions may only apply to tissues homogenized by mortar and pestle and not to tissue homogenized by other techniques.

4.2.1 Preparative vs. Analytical Cell Fractionation

Historically, the goal of preparative fractionation was to isolate and purify a morphologically identifiable component for the subsequent determination of its biochemical and/or physical properties. The major problems with this approach were the lack of criteria used to determine purity (quantitative reliability) and the erroneous assignment of biochemical properties to an incompletely "purified" component (DeDuve 1971).

Analytical cell fractionation focuses on the actual objects of the analyses (biochemical constituent or enzyme activity) and the determination of these entities among all fractions separated from the crude homogenate. Quantitative recovery of markers in all fractions is the characteristic trait of the analytical approach and was insisted upon by Claude (DeDuve 1971) and championed by Schneider and Hogeboom (1951; DeDuve 1971). This approach has been recommended for plant cell fractionation studies (Quail 1979).

4.2.2 Need for Quantitation

To determine the subcellular location of an enzyme, the distribution of its activity and whatever markers being tested should ideally be related back to the total crude homogenate. The reported data should include the total activity and specific activity of all markers, as well as the activity of the enzyme of interest in all isolated fractions. These complete balance sheets are in line with the analytical fractionation approach and are necessary for several reasons. If a substantial increase in total enzyme activity is recovered compared to the original homogenate, the increase could be due to removal of an endogenous inhibitor during subsequent fractionation or the presence of an activator which preferentially associates with a particular fraction. A substantial loss in total activity could be due to the concentration of an endogenous inhibitor in one or more isolated fractions. Recombination of fractions (mixing experiments) can be used to determine which fraction(s) contain the inhibitor or activator, and can provide further insight as to how the activities associated with subcellular compartments are related to the activity of the whole cell.

Assuming complete recovery of the starting activities, the distribution observed can help pare down the possible subcellular sites of the enzyme of interest. If most of the activity is found in the post-microsomal supernatant, the small amount associated with particulate fractions must seriously be considered as an adsorption phenomena (Schneider and Hogeboom 1951).

If most of the enzyme activity is associated with a subcellular component, quantitation of all isolated fractions can be used to help choose a specific crude fraction to further purify (Nagahashi and Beevers 1978). In some cases, a combined crude membrane fraction (Leonard and Vanderwoude 1976; Nagahashi and Baker 1984; Nagahashi and Kane 1982) will be chosen or an unpelleted crude fraction (Lord et al. 1972) will be used for further purification and analysis. In all of these cases, the choice of a crude fraction to purify further was based on initial results from differential centrifugation experiments.

4.2.3 Problems with Complete Quantitation and Interpretation of Data

If the homogenization medium contains additives, they may interfere with biochemical assays performed on original crude homogenates and supernatant fractions. For example, additives such as EDTA in the crude homogenate could interfere with Mg^{2+}-ATPase activity due to the potential chelation of Mg^{2+} in the reaction mix. Other additives such as PVP may give a positive reaction to the Lowry procedure (Pertoft and Laurent 1977). The presence of BSA in the crude homogenate may also lead to falsely elevated protein estimates, since BSA can bind phenolics. In this case, the background protein (BSA content) will not be accurately accounted for and, hence, the apparent specific activities of the crude homogenate or supernatant fractions may be lower as a result of this artifact. The effects of additives on marker enzyme activity as well as assay procedures can be minimized by diluting the crude homogenate in homogenization medium minus the additives. Interference can likewise be minimized in isolated pellets by resuspension in the absence of additives. In some cases, it is possible to dialyze out the additives before performing enzyme assays. In tissues such as potato leaves or tubers, dialysis cannot be performed fast enough, since marker activity is lost rapidly even at 0 °–4 °C.

In cases where a large percentage of the total crude homogenate activity is soluble, the small percentage associated with subcellular components should not automatically be assumed as artifactual adsorption. A membrane-bound cellulase represented 5–10% of the total cellulase activity isolated from kidney bean abscission zones (Koehler et al. 1976), and 5–8% of the total triose phosphate isomerase activity from various tissues was associated with plastids (Miflin 1974). Enzymes like triose phosphate isomerase that have a cytosolic form as well as a soluble form inside an organelle can confound the distribution pattern of activity in isolated fractions. Enzyme profiles in this case will depend upon the amount of damage done to originally intact organelles.

If the small percentage of total activity associated with particulate fractions is actually due to a soluble contaminant, addition of BSA to the homogenization medium can be effective in preventing the nonspecific adsorption of this activity (Dalling et al. 1972). This treatment is not always effective, and washing the isolated pellets may be necessary to remove soluble contaminants (Elias and Givan 1978). The soluble contaminants removed by washing should be combined with the post-microsomal supernatant to insure accurate bookkeeping.

The pH of the homogenization medium may increase or decrease the adsorption of "soluble" enzymes to crude particulate fractions (Jaynes et al. 1972). At pH 5.0, most of the β-glucosidase activity from sweetclover leaves and bean hypocotyls (Jaynes et al. 1972) was strongly associated with a crude mitochondrial fraction (17,000 *g* for 10 min). At pH 8.5, most of the activity was found in the post-mitochondrial supernatant (Jaynes et al. 1972). Homogenization pH is normally not considered in this light, but one should be aware that acidic or basic conditions may greatly influence the distribution of soluble enzymes during differential centrifugation.

4.3 Linear Density Gradient Centrifugation

Crude fractions separated by differential centrifugation can be further purified in density gradients. Linear gradients are recommended over step or discontinuous gradients because better resolution (as indicated by higher specific activities) can be obtained (Quail 1979).

4.3.1 Density Gradient Material

The most commonly used materials for density gradient centrifugation (Ford et al. 1983) can be grouped under the following categories: (1) sugars and related polymers (sucrose, sorbitol, mannitol, dextran, and ficoll); (2) salt solutions (cesium chloride); (3) colloidal silica suspensions (Ludox, percoll); and (4) iodinated compounds (urografin, renografin, metrizamide, nycodenz). All of these gradient materials have certain properties which make them nonideal. The osmotic effects of sucrose (Price 1982, 1983), anomalous effects of ficoll (Leonard and Vanderwoude 1976), interaction of metrizamide with proteins and enzymes (Ford et al. 1983), and interference of Percoll with protein analysis (Pertoft and Laurent 1977; Yakmyshyn et al. 1982) and enzyme assays (Yakmyshyn et al. 1982) are examples of these properties.

In spite of its osmotic effects, sucrose is the most widely used gradient material because it is inexpensive, transparent, highly soluble and does not interfere with most marker assays. Although results from sucrose density gradient analysis will be discussed at length, the types of problem encountered, the need for quantitation of markers, and problems in data interpretation will apply to density gradient separation regardless of gradient material used.

4.3.2 Pelleted vs. Unpelleted Overlays

Pelleted fractions isolated by differential centrifugation are further separated by centrifuging in density gradients. Two objections to the use of pelleted overlays have been raised. It has been argued that pelleting is potentially destructive (Quail 1979) and furthermore may enhance nonspecific adsorption of membranes to each other (Ray 1977). The evidence which has led to these objections is unclear. Large intact organelles such as nuclei and chloroplasts may be fragile and break during pelleting and resuspending (Tautvydas 1971), but this is not always the case (Elias and Givan 1978). In contrast, pelleting of microsomes may be nondestructive, since they represent a population of small-sized vesicles derived from broken membraneous structures. Pelleted fractions have the advantage that they can be washed to remove soluble enzymes and additives from the homogenization medium. Resuspended pellets can be applied to density gradients either as an overlay or underlay (flotation method).

If pelleting causes adherence problems, crude particulate fractions can be collected on sucrose cushions to avoid pelleting (Koehler et al. 1976; Ray 1977) or alternatively, unpelleted fractions can be overlaid directly on density gradients (Koehler et al. 1976). This latter method has been used by Lord et al. (1972, 1973) and has been recommended as the method of choice for density gradient centrifugation (Quail 1979). It should be pointed out that the decision by Lord et al. (1972) to use an unfractionated overlay (270 g supernatant) resulted from their initial differential centrifugation experiments. Unfractionated homogenates will provide the greatest amount of soluble enzyme contamination in density gradients and, furthermore, only small volumes can be overlaid on gradients centrifuged in commonly used swinging bucket rotors (SW 25.1, SW 27, SW 28, SW 28.1). The SW 25.2 rotor does have a large capacity and can take up to 30 ml of homogenate over a 30 ml gradient (Fig. 1).

4.3.3 Soluble Enzyme Contamination in Gradients

Soluble enzymes can contaminate sucrose gradients in three different ways. They may nonspecifically adhere to membranes (Dalling et al. 1972; Mandala et al. 1982) and be carried into density gradients. The degree to which the binding is pH-dependent is not known; however, this form of contamination can lead to an erroneous distribution of activity in the fractionated gradient. Plastids are a ma-

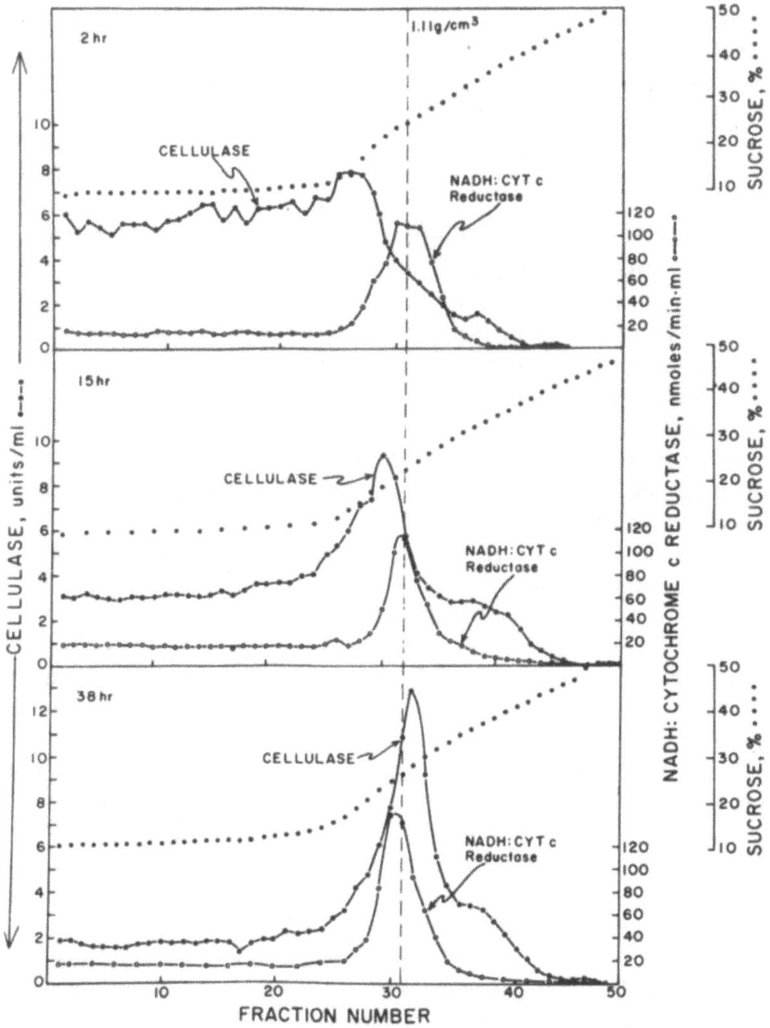

Fig. 1. Sedimentation velocity gradient of peak I (soluble cellulase). An 80,000 g supernatant (30 ml) from the homogenate of bean abscission zones was layered over a linear 15 to 50% (w/w) sucrose density gradient and centrifuged for 2, 15, or 38 h. The three gradients were aligned at a common density of 1.11 g cm^{-3} and plotted as shown. Fractions of 1.2 ml were collected. (Koehler et al. 1976)

jor site of adsorption by soluble enzymes (Dalling et al. 1972; Elias and Givan 1978), and BSA has been used to prevent this binding (Dalling et al. 1972). BSA will not reduce the adsorption of all soluble enzymes and in this situation, repeated washing of crude organelle pellets prior to density gradient centrifugation is necessary to remove soluble enzyme contamination (Elias and Givan 1978).

Secondly, soluble enzymes can enter gradients during prolonged centrifugation (15 h or longer) even when they are not adhering to a membrane or organelle

(Koehler et al. 1976; Lord et al. 1972; Nagahashi and Baker 1984; Nagahashi et al. 1984). This contamination is exacerbated when unfractionated homogenates are used; however, unwashed pellets also carry over considerable soluble enzyme activity. After prolonged centrifugation, soluble enzymes will contaminate light density membranes (Koehler et al. 1976; Lord et al. 1972; Nagahashi and Baker 1984) and will appear to be localized with ER or possibly tonoplast and Golgi membranes.

To determine if enzyme activity near the top of density gradients is actually associated with a light density membrane or is a soluble contaminant, identical linear gradients should be centrifuged for various time periods (Fig. 1) (Nagahashi and Baker 1984; Nagahashi et al. 1985a, b). If associated with a membrane, the distribution of enzyme activity will not change once the membrane is at equilibrium density. If soluble, the activity will continue to move further into the gradient during extended centrifugation time. Alternatively, differently shaped linear gradients can be centrifuged until membranes are isopycnic. The gradient shape will directly effect the density of a soluble enzyme (Nagahashi et al. 1985b) but will not effect the equilibrium density of any given membrane.

To minimize this soluble enzyme contamination, Lord et al. (1973) layer a 5 ml step of 20% sucrose on top of a preformed linear sucrose gradient (32–60%). The crude homogenate (5 ml of a 270 g supernatant) is layered on top of the step which retards the movement of soluble enzymes into the gradient during short term centrifugation (4 h). However, "soluble" cytidyl transferase moved further into the gradient with prolonged centrifugation (24 h). This particular gradient design also produces zone-narrowing in the region of the ER, since a normal linear gradient (Lord et al. 1973) shows a much broader peak of ER-associated activity. Unfortunately, for tissues with considerable Golgi bodies and well-developed vacuoles (unlike castor bean endosperm), the gradient-induced zone narrowing (Price 1982) will stockpile tonoplast and Golgi membranes in the region of ER.

Thirdly, soluble enzymes can be entrapped in membrane vesicles during cell disruption. This form of contamination will be greatest in the crude microsomal fraction and will be carried into sucrose gradients to the density of the membrane vesicles which contain the enzyme. Sodium carbonate has been used to remove entrapped soluble enzymes (Fujiki et al. 1982) and this treatment converts closed vesicles to open sheets. Entrapped proteins and peripheral membrane proteins are released in soluble form. It has not been reported whether this process can be reversed to form resealed vesicles from open sheets.

Isolated membranes from density gradients could be treated with sodium carbonate to determine if enzyme activity associated with vesicles is a soluble form inside or a peripheral membrane protein. If the activity remains with the membranes after treatment, it is likely to be an integral protein.

4.3.4 Equilibrium Density Centrifugation (Isopycnic Conditions)

Maximum separation of subcellular components in density gradients can only be achieved when membranes are at their equilibrium density. Although little attention has been given to this fact, a few reports (Koehler et al. 1976; Leonard and

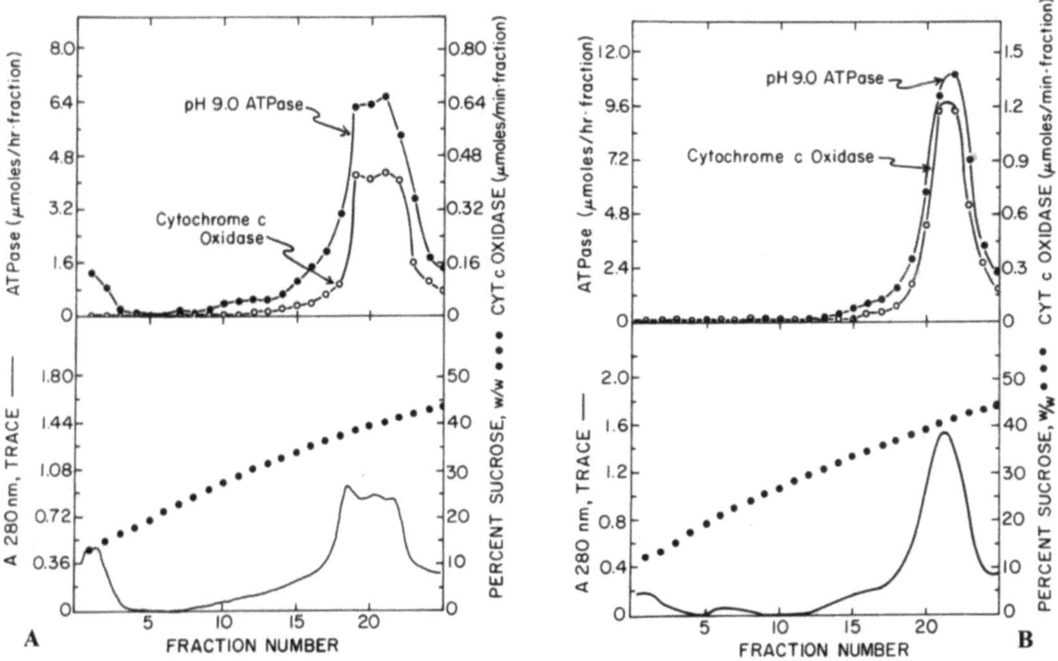

Fig. 2. A Distribution of pH 9.0 ATPase and Cyt c oxidase when a 13,000 g fraction from corn roots was centrifuged for 1.5 h in a linear sucrose gradient. (Leonard and Vanderwoude 1976) **B** Distribution of pH 9.0 ATPase and Cyt c oxidase obtained when a 13,000 g fraction from corn roots was centrifuged for 15 h in a linear sucrose gradient. (Leonard and Vanderwoude 1976)

Vanderwoude 1976; Nagahashi and Baker 1984) have indiciated that membranes and organelles do not reach isopycnic conditions (as judged by marker enzyme distribution) after centrifugation for 1.5 to 2 h at high speeds (80,000 to 85,000 *g*). For the mitochondrial (cyt c oxidase, pH 9.0 ATPase), plasma membranes (pH 6.5 K^+-ATPase), and ER (antimycin A insensitive NADH cyt c reductase) markers (Nagahashi and Baker 1984), short term centrifugation (Figs. 1, 2A, 3A) resulted in broad peaks of activity with corresponding broad peaks of absorbance at 280 nm (Figs. 2A, 3A). These data are interpreted to indicate that not all members of a given component population have reached isopycnic conditions. Afer 15 h of centrifugation, the markers are much sharper in resolution (Figs. 1, 2B, 3B). Further centrifugation up to 38 h (Fig. 1) or 40 h (Nagahashi and Baker 1984) did not change the resolution of the ER marker, regardless of whether pelleted overlays or unpelleted overlays were used. Consistent with these results is a recent report on the subcellular fractionation of guinea pig enterocytes (Ford et al. 1983). Mitochondria, lysosomes, and plasma membranes all showed much tighter banding patterns in sucrose gradients centrifuged for 16 h (100,000 *g*) compared to 2 h (30,000 *g*). This report (Ford et al. 1983) also showed differences in membrane densities when short- and long-term centrifugation were performed with metrizamide and nycodenz gradients.

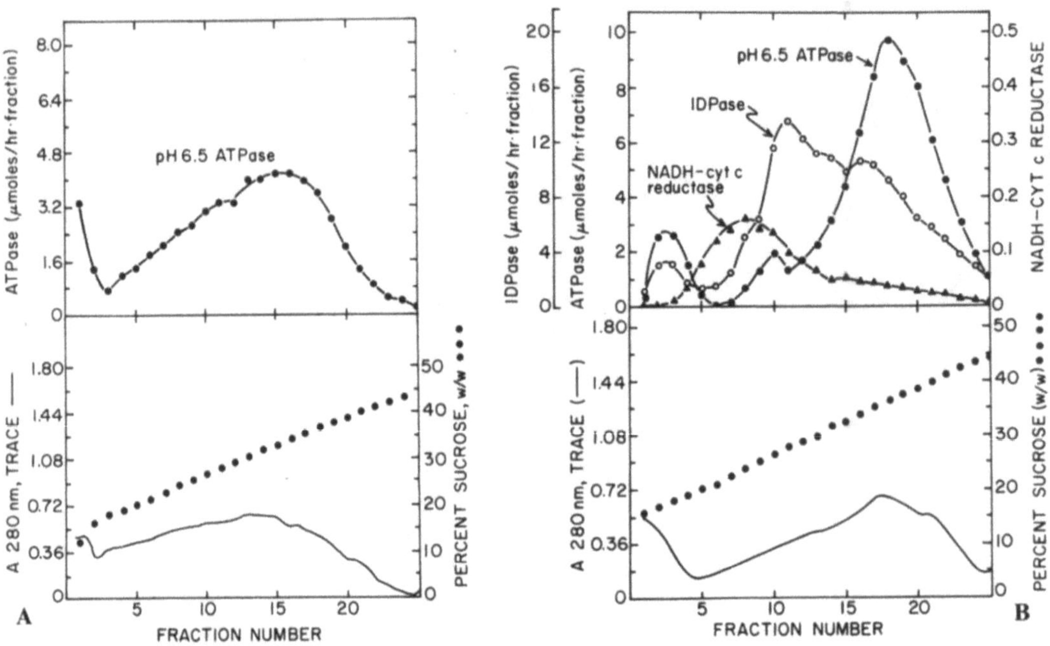

Fig. 3. A Distribution of pH 6.5 ATPase obtained when a 13,000 to 80,000 g fraction from corn roots was centrifuged for 1.5 h in a linear sucrose gradient. (Leonard and Vanderwoude 1976). **B** Distribution of pH 6.5 ATPase, latent IDPase, and NADH cyt c reductase obtained when a 13,000 to 80,000 g fraction from corn roots was centrifuged for 15 h in a linear sucrose gradient. (Leonard and Vanderwoude 1976)

It is not known whether centrifugation for 6 to 8 h will be the minimum time necessary to reach equilibrium conditions. With the common type of swinging bucket rotors, centrifugation for 6 to 8 h presents an inconvenient time period since homogenization, centrifugation, fractionation, and enzyme assays cannot be performed during the same work day. Preparation and analysis of gradients centrifuged for 2 h or less can be done in 1 day and, hence short-term centrifugation is used by many workers. Overnight centrifugation (15 h) is recommended, since maximum separation of markers is insured and a complete day for marker assays is available.

4.3.5 Other Factors Which Influence Marker Enzyme Profiles Across a Gradient

The level of membrane protein will affect the detection of enzyme activity especially if the assay procedure is rather insensitive. At low protein levels in sucrose gradients, some peaks of enzyme activity are not readily detectable, but become noticeable when the protein level in the gradient is increased (Nagahashi et al. 1984). In these cases, marker assays may need to be modified (larger aliquots assayed or longer incubation periods) before activity can be detected.

Another factor which can influence marker enzyme activity is the presence of additives in the gradient. EDTA and Mg^{2+} can alter marker enzyme and protein profiles when added to sucrose gradients. When ribosomes are retained ($+Mg^{2+}$), RER markers have a density range of 1.15–1.18 g/cm^{-3} in sucrose gradients (Quail 1979). The density of ER markers when ribosomes are removed (+EDTA) is much lighter (1.10–1.12 g/cm^{-3}). Other additives such as buffers and sulfhydryls may help stabilize marker activity especially if prolonged centrifugation is used. Higher levels of marker activity may be sustained with protective reagents and may result in more accurate recoveries in gradient fractions.

Finally, the resolution of markers in density gradients can be diminished by the collection of large fractions, as well as by inappropriate centrifugation conditions (Sect. 4.3.4). Best resolution has been achieved when small fractions (0.6 ml to 1.5 ml) are collected during fractionation of large gradients (38 to 60 ml) (Koehler et al. 1976; Leonard and Vanderwoude 1976; Lord et al. 1972; Nagahashi and Baker 1984; Nagahashi and Kane 1982).

4.3.6 Need for Quantitation and Lack of Negative Marker Activity

The total activity and specific activity of markers before and after density gradient centrifugation should be determined. Quantitation of all gradient fractions including the material which pellets through the gradient should be determined (Quail 1979) and related back to the crude overlay. Incomplete recovery of marker activity after density gradient fractionation can be due to any of the reasons discussed earlier (Sect. 3.3 and 4.2.2).

Markers used in the negative mode should be monitored to determine the degree of contamination of a "purified" fraction. Lack of negative marker activity is usually interpreted to mean no contamination by membranes carrying the negative marker. This is only one possibility, since any of the following is also possible: (1) the negative marker may be unstable and activity is lost during gradient centrifugation; (2) the negative marker chosen may not be ubiquitous; (3) low protein levels in the gradient may prevent detection of the negative marker; (4) the enzyme assay is performed incorrectly. In these cases, membranes or organelles contaminating the "purified" fraction are actually present but cannot be detected. The use of multiple markers combined with morphological examination of the isolated fraction will provide a clearer picture of the degree of contamination in "purified" fractions.

5 Concluding Remarks

The scope of this article has been to present basic concepts about markers and suggest a basic approach to plant cell fractionation. The value of differential centrifugation experiments cannot be overemphasized. Membrane fractions separated by simple centrifugation experiments can be used to work out marker assay procedures, determine the stability of marker enzymes, and determine if PPO and hydrolytic activity are present. Even with relatively poor resolving power, analytical fractionation by differential centrifugation can provide quantitative information about the distribution of biochemical properties over a number of fractions which together represent the whole tissue.

To study biochemical functions through cell fractionation is no easy task, and should be performed with as much rigor as technically possible. The major problems which occur during cell fractionation and the potential erroneous conclusions drawn from experiments without appropriate controls have been discussed. The approach proposed and the problems of data analysis discussed should be taken neither as gospel nor the opinions of others, but only as the views of this author.

References

Alpi A, Beevers H (1981) Proteinases and enzyme stability in crude extracts of castor bean endosperm. Plant Physiol 67:499–502

Anderson JW (1968) Extraction of enzymes and subcellular organelles from plant tissues. Phytochemistry 7:1973–1988

Bishop DG, Oertle E (1983) Inhibition of potato tuber lipoxygenase and lipolytic acyl hydrolase activities by nupercaine. Plant Sci Lett 31:49–53

Boller T, Kende H (1979) Hydrolytic enzymes in the central vacuole of plant cells. Plant Physiol 63:1123–1132

Bowles DJ, Quail PH, Morré DJ, Hartmann GC (1979) Use of markers in plant-cell fractionation. In: Reid E (ed) Plant organelles, vol 9. Ellis Horwood, Chichester, pp 207–224

Caldwell CR, Haug A (1980) Kinetic characterization of barley root plasma membrane-bound Ca^{2+}- and Mg^{2+}-dependent adenosine triphosphatase activities. Physiol Plant 50:183–193

Dalling MJ, Tolbert NE, Hageman RH (1972) Intracellular location of nitrate reductase and nitrite reductase I. Spinach and tobacco leaves. Biochim Biophys Acta 283:505–512

DeDuve C (1971) Tissue fractionation: past and present. J Cell Biol 50:20D–55D

DeDuve C (1975) Exploring cells with a centrifuge. Science 189:186–194

Dubacq J-P, Kader J-C (1978) Free flow electrophoresis of chloroplasts. Plant Physiol 61:465–468

Elias BA, Givan CV (1978) Density gradient and differential centrifugation methods for chloroplast purification and enzyme localization in leaf tissue. The case of citrate synthase in *Pisum sativum* L. Planta 142:317–320

Fisher J, Hodges TK (1969) Monovalent ion stimulated adenosine triphosphatase from oat roots. Plant Physiol 44:385–395

Ford T, Rickwood D, Graham J (1983) Buoyant densities of macromolecules, macromolecular complexes, and cell organelles in nycodenz gradients. Anal Biochem 128:232–239

Fujiki Y, Hubbard AL, Fowlers S, Lazarow PB (1982) Isolation of intracellular membranes by means of sodium carbonate treatment: application of endoplasmic reticulum. J Cell Biol 93:97–102

Galliard T (1974) Technique for overcoming problems of lipolytic enzymes and lipoxygenases in the preparation of plant organelles. In: Fleischer S, Packer L (eds) Methods Enzymol 31:520–528

Gardner G, Pike CS, Rice HV, Briggs WR (1971) Disaggregation of phytochrome in vitro – a consequence of proteolysis. Plant Physiol 48:686–693

Hall JL, Moore AL (eds) (1983) Isolation of membranes and organelles from plant cells. Academic Press, London

Hannig K, Heidrich H-G (1974) The use of continuous preparative free-flow electrophoresis for dissociating cell fractions and isolation of membranous components. In: Fleischer S, Packer L (eds) Methods Enzymol 31:746–766

Hepler PK (1981) The structure of the endoplasmic reticulum revealed by osmium tetraoxide – potassium ferricyanide staining. Eur J Cell Biol 26:102–110

Hurkman W, Morré DJ, Bracker CE, Mollenhauer HH (1979) Identification of etioplast membranes in fractions from soybean hypocotyls. Plant Physiol 64:398–403

Jacobson AB (1968) A procedure for isolation of proplastids from etiolated maize leaves. J Cell Biol 38:238–244

Jaynes TA, Haskins FA, Gorz HJ, Kleinhofs A (1972) Solubility of β-glucosidase in homogenates of sweetclover leaves and bean hypocotyls. Plant Physiol 49:277–279

Jones RL (1980) The isolation of endoplasmic reticulum from barley aleurone layers. Planta 150:58–69

Koehler DE, Leonard RT, Vanderwoude WJ, Linkins AE, Lewis LN (1976) Association of latent cellulase activity with plasma membranes from kidney bean abscission zones. Plant Physiol 58:324–330

Koundal KR, Sawhney SK, Sinha SK (1983) Oxidation of 2-mercaptoethanol in the presence of tris buffer. Phytochemistry 22:2183–2184

Larsson C, Anderson B (1979) Two-phase methods for chloroplasts, chloroplast elements, and mitochondria. In: Reid E (ed) Plant organelles. Ellis Horwood, Chichester, pp 35–46

Leonard RT, Vanderwoude WJ (1976) Isolation of plasma membranes from corn roots by sucrose density gradient centrifugation. An anomalous effect of ficoll. Plant Physiol 57:105–114

Leonard RT, Hansen D, Hodges TK (1973) Membrane-bond adenosine triphosphatase activities of oat roots. Plant Physiol 51:749–754

Loomis WD (1974) Overcoming problems of phenolics and quinones in the isolation of plant enzymes and organelles. In: Fleischer S, Packer L (eds) Methods Enzymol 31:528–554

Loomis WD, Battaile J (1966) Phenolic compounds and the isolation of plant enzymes. Phytochemistry 5:423–438

Lord JM, Kagawa T, Beevers H (1972) Intracellular distribution of enzymes of the cytidine diphosphate choline pathway in castor bean endosperm. Proc Natl Acad Sci USA 69:2429–2432

Lord JM, Kagawa T, Moore TS, Beevers H (1973) Endoplasmic reticulum as the site of lecithin formation in castor bean endosperm. J Cell Biol 57:659–667

Mandala S, Mettler IJ, Taiz L (1982) Localization of the proton pump of corn coleoptile microsomal membranes by density gradient centrifugation. Plant Physiol 70:1743–1747

Mayer AM, Harel E (1979) Polyphenol oxidases in plants. Phytochemistry 18:193–215

Miflin BJ (1974) The location of nitrite reductase and other enzymes related to amino acid biosynthesis in the plastids of roots and leaves. Plant Physiol 54:550–555

Moore AL, Proudlove MO (1983) Mitochondria and sub-mitochondrial particles. In: Hall JL, Moore AL (eds) Isolation of membranes and organelles from plant cells. Academic Press, London, pp 153–184

Moreau RA (1985 a) Membrane-degrading enzymes in the leaves of *Solanum tuberosum*. Phytochemistry 24:411–414

Moreau RA (1984 b) Membrane-degrading enzymes in the tubers of *Solanum tuberosum*. J Agric Food Chem 33:36–39

Morré DJ, Cline GB, Coleman R, Evans WH, Glaumann H, Headon DR, Reid E, Siebert G, Widnell CC (1979) Markers for membranous cell components. Eur J Cell Biol 20:195–199

Morré DJ, Morré DM, Heidrich-H-G (1983) Subfractionation of rat liver Golgi apparatus by free-flow electrophoresis. Eur J Cell Biol 31:263–274

Nagahashi G, Baker AF (1984) β-glucosidase activity in corn root homogenates: problems in subcellular fractionation. Plant Physiol 76:861–864
Nagahashi J, Beevers L (1978) Subcellular localization of glycosyl transferases involved in glycoprotein biosynthesis in the cotyledons of *Pisum sativum* L. Plant Physiol 61:451–459
Nagahashi J, Hiraike K (1982) Effects of centrifugal force and centrifugation time on the sedimentation of plant organelles. Plant Physiol 69:546–548
Nagahashi J, Kane AP (1982) Triton-stimulated nucleoside diphosphatase activity: subcellular localization in corn root homogenates. Protoplasma 112:167–173
Nagahashi G, Seibles TS, Jones SB, Rao J (1985a) Purification of cell wall fragments by sucrose gradient centrifugation. Protoplasma (in press).
Nagahashi G, Seibles TS, Tu S-I (1985b) The pH dependent distribution of β-glucosidase activity in isolated particulate fractions. Plant Sci Lett 38:173–178
Philipp E-I, Franke WW, Keenan TW, Stadler J, Jarasch E-D (1976) Characterization of nuclear membranes and endoplasmic reticulum isolated from plant tissue. J Cell Biol 68:11–29
Pertoft H, Laurent TC (1977) Isopycnic separation of cells and cell organelles by centrifugation in modified colloidal silica gradients. In: Catsimpoolas N (ed) Methods of cell separation, vol 1. Plenum, New York, pp 25–65
Price CA (1979) Isolation of plant nuclei. In: Reid E (ed) Plant Organelles. Ellis Horwood, Chichester, pp 200–206
Price CA (1982) Centrifugation in density gradients. Academic Press, London
Price CA (1983) General principles of cell fractionation. In: Hall JL, Moore AL (eds) Isolation of membranes and organelles from plant cells. Academic Press, New York, pp 1–24
Quail PH (1979) Plant cell fractionation. Annu Rev Plant Physiol 30:425–484
Ray PM (1977) Auxin-binding sites of maize coleoptiles are localized on membranes of the endoplasmic reticulum. Plant Physiol 59:594–599
Scherer GFE (1981) Auxin-stimulated ATPase in membrane fractions from pumpkin hypocotyls (*Cucurbita maxima* L.) Planta 151:434–438
Scherer GFE, Morré DJ (1978) Action and inhibition of endogenous phospholipases during isolation of plant membranes. Plant Physiol 62:933–937
Schneider WC, Hogeboom GH (1951) Cytochemical studies of mammalian tissues: the isolation of cell components by differential centrifugation: A review. Cancer Res 11:1–20
Schroeder F, Fontaine RN, Kinder DA (1982) LM fibroblast plasma membrane subfractionation by affinity chromatography on con A-sepharose. Biochim Biophys Acta 690:231–242
Tautvydas KJ (1971) Mass isolation of pea nuclei. Plant Physiol 47:499–503
Van der Wilden W, Gilkes NR, Chrispeels MJ (1980) The endoplasmic reticulum of mung bean cotyledons. Plant Physiol 66:390–394
Yakmyshyn LM, Walker K, Thomson ABR (1982) Use of percoll in the isolation and purification of rabbit small intestinal brush border membranes. Biochim Biophys Acta 690:269–281
Yoshida S, Uemura M, Niki T, Sakai A, Gusta LV (1983) Partition of membrane particles in aqueous two-polymer phase system and its practical use for purification of plasma membranes from plants. Plant Physiol 72:105–114

Plasma Membranes

CH. LARSSON

1 Introduction

The plasma membrane is the outer, permeability barrier of the plant cell. This position imposes several important functions on the plasma membrane, such as transport of ions and photosynthetic products, hormone binding and responses, and cell-wall synthesis. In addition, the plasma membrane is likely to have an important role in resistance toward pathogens, in frost hardiness, and in blue light photomorphogenesis. Some of these events may well be studied with protoplasts, but preparations of isolated plasma membranes will certainly have increasing importance in the unveiling of the functions of the plasma membrane.

A prerequisite for studies on the plasma membrane is the isolation of sufficient quantities of high purity, a demand which is difficult to meet with different types of density gradient centrifugation, at present the most commonly used method for plasma membrane isolation (reviewed by Hall 1983). This is particularly true when green tissue is used as the starting material, since chloroplast fragments are not easily removed from plasma membranes by separation methods based on differences in density. Furthermore, for many types of study, e.g., transport or hormone binding, it is invaluable to have preparations of sealed plasma membrane vesicles with a defined sidedness, a demand probably also not met by preparations obtained by density gradient centrifugation, as revealed by a recent study (Randall and Ruesink 1983), indicating that such preparations contain a mixture of right-side-out and inside-out vesicles, with a relatively high proportion of unsealed material.

As an alternative to gradient centrifugation, partition in aqueous dextran-polyethylene glycol two-phase systems was introduced in the purification of plasma membranes (Widell and Larsson 1981; Lundborg et al. 1981). By this method membrane particles are separated according to their surface properties (Albertsson 1971) rather than according to their size and density. Thus, preparations of membrane vesicles with uniform surface properties and thereby uniform sidedness can be obtained. The power of the method in this respect has earlier been demonstrated by the preparation of inside-out and right-side-out vesicles from both chloroplast thylakoids (Andersson and Åkerlund 1978) and erythrocytes (Steck 1974). The plasma membrane vesicles obtained from plant material using phase partition are right-side-out and mainly sealed (Larsson et al. 1984).

Phase partition is easily scaled up, and the plasma membrane preparations are very pure; e.g., chlorophyll-free preparations are readily obtained from leaves of both mono- and dicotyledons (Kjellbom and Larsson 1984). The method is also rapid; plasma membranes can be obtained within 2 to 4 h from homogenization.

Thus, phase partition meets the demands discussed above; namely to produce relatively large quantities of pure plasma membranes, and moreover, sealed plasma membrane vesicles with a defined sidedness. Below, the method is presented in detail both from practical and theoretical viewpoints, and the properties of the plasma membrane preparations are discussed.

2 Theory of Phase Partition

2.1 The Phase System

When aqueous solutions of Dextran T 500 and polyethylene glycol 3350 (earlier referred to as polyethylene glycol 4000) are mixed a two-phase system may be obtained (Albertsson 1971). This occurs if the final concentrations of polymers fall within the two-phase region indicated in the phase diagram (Fig. 1). This phenomenon is not unique for the Dextran/polyethylene glycol couple, but holds for several couples of water-soluble polymers (Albertsson 1971).

Phase systems containing Dextran and polyethylene glycol have a polyethylene glycol-rich upper phase and a Dextran-rich lower phase. The composition of each phase can be determined from the phase diagram (Fig. 1). Since both phases contain mainly water, aqueous polymer two-phase systems provide a suitable environment for biological material, unlike phase systems made up of water and organic solvents. Furthermore, sucrose, buffers and salts may be added to the aqueous phase systems to produce the tonicity, pH, and ionic strength required for maintaining optimal activity.

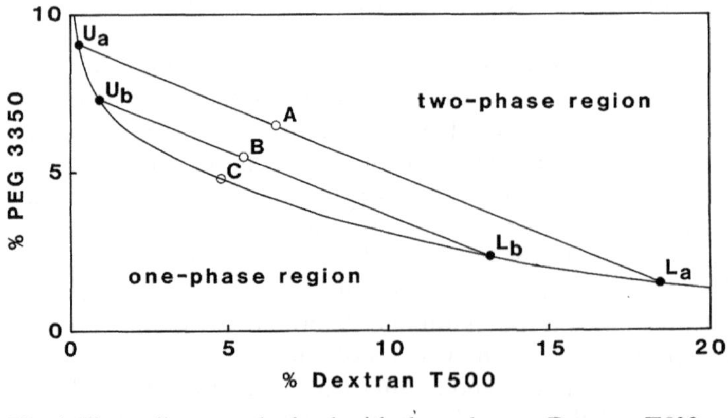

Fig. 1. Phase diagram obtained with the polymers Dextran T500 and polyethylene glycol 3350, and H_2O at 4 °C. At polymer concentrations above the curved line a two-phase system is obtained. The compositions of the upper and lower phases of phase systems *A* and *B* are given by the points U_a, U_b, L_a, and L_b respectively. *C* is the so-called critical point. Percent composition is w/w. Note that the phase diagram is very temperature-dependent

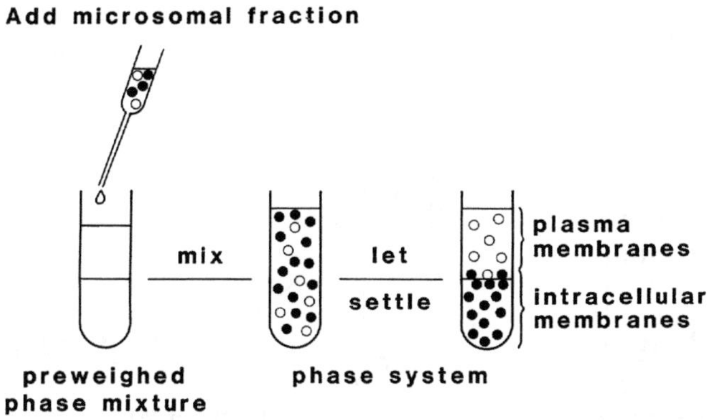

Fig. 2. Partition of membrane vesicles in a two-phase system. The microsomal fraction, suspended in the medium used in the phase system, is added to a preweighed phase mixture to give a phase system of the desired final composition. The phase system is mixed and left to settle (phase settling may be accelerated by centrifugation). Plasma membranes partition in the polyethylene glycol-rich upper phase, whereas intracellular membranes partition at the interface and in the Dextran-rich lower phase

2.2 Partition of Membrane Particles

When a crude membrane preparation, such as a microsomal fraction, is added to the phase system, the different particles will distribute between the upper phase and the interface plus lower phase, as illustrated in Fig. 2. The distribution of the different membrane particles will be determined by their respective surface properties.

To achieve optimal separation between different membranes in a crude preparation one must be able to direct the partition of the membranes. Since the plasma membrane is known to have a very high affinity for the upper phase compared to intracellular membranes (Larsson and Andersson 1979), conditions are sought where the plasma membranes end up mainly in the upper phase and the intracellular membranes mainly at the interface or in the lower phase. This is done by changing polymer and/or salt concentrations in a systematic way.

2.3 Effects of Polymer Concentrations

Close to the critical point of the phase system (Fig. 1), all membrane particles are almost completely distributed to the upper phase. When the polymer concentrations are increased, the membranes start to collect at the interface (Fig. 3). This response to the increasing polymer concentrations is most pronounced for mitochondrial membranes and for the endoplasmic reticulum, whereas chloroplast thylakoids, Golgi and particularly plasma membranes are less sensitive. The reasons for these differences are probably related to hydrophobic/hydrophilic properties of the membrane surfaces (Albertsson 1971); it can be assumed that the up-

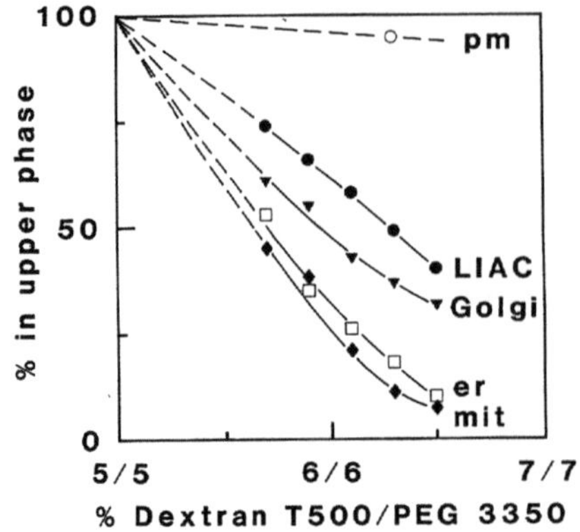

Fig. 3

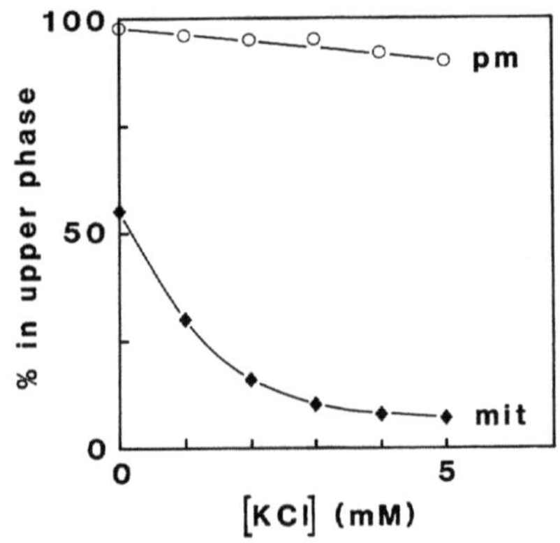

Fig. 4

Fig. 3. Effects of increasing polymer concentrations on the partition of different membrane vesicles in a microsomal fraction from wheat root. Markers used were: Cytochrome c oxidase (◆) for mitochondria, antimycin A-resistant NADH-cytochrome c reductase (□) for the endoplasmic reticulum, IDPase (▼) for Golgi, and light-reducible cytochrome (●), LIAC, which is present in both plasma membranes and endoplasmic reticulum. The partition of plasma membranes (○) has been calculated from the peak of plasma membranes in the counter-current distribution diagram in Fig. 6; the partition of all membranes has been extrapolated back to the critical point (Fig. 1) where all material is in the upper phase. Note that the plasma membranes has an extreme affinity for the upper phase even at relatively high polymer concentrations, in marked contrast to intracellular membranes. Apart from polymers, the phase systems contained 0.25 M sucrose and 5 mM potassium phosphate, pH 7.8. Temperature, 5 °C. (Data from Lundborg et al. 1981)

Fig. 4. Effects of increasing Cl^- concentration on the partition of plasma membranes and mitochondria in a phase system composed of 5.7% Dextran T500, 5.7% polyethylene glycol 3350, 5 mM potassium phosphate, pH 7.8. Temperature, 4 °C. The "positive" upper phase created by HPO_4^{2-} is titrated away by Cl^-; this markedly affects the partition of mitochondria, but not that of plasma membranes. Other intracellular membranes behave similarly to mitochondria. Plasma membranes and mitochondria, respectively, were purified from cauliflower inflorescences and repartitioned in a series of phase systems; the partition was determined by the absorbance at 280 nm (○) and cytochrome c oxidase activity (◆) respectively

per phase becomes increasingly hydrophobic when the polymer concentrations, and thereby the polyethylene glycol concentration in the upper phase, are increased (Fig. 1).

2.4 Effects of Salts

The addition of a salt to the phase system normally creates an electrostatic potential difference between the two phases (Johansson 1970; Albertsson 1971). This happens if the ions have different affinity for the phases. For example, HPO_4^{2-} has a stronger affinity for the lower phase than K^+ or Na^+, and a "positive" upper phase is obtained with phosphate buffer. An electrostatic potential difference of opposite sign is obtained with, e.g., KCl or NaCl. Even if all biological membranes are negatively charged at neutral pH, their surface net charge density differs. Therefore, separation may be obtained by varying the salt composition (Fig. 4). Useful ions which give a "positive" upper phase are phosphate^{2-}, sulfate, citrate, and maleate (di- or trivalent anions), whereas acetate and chloride (monovalent anions), for example, give "negative" upper phases.

2.5 Multistep Procedures

An adequate separation of plasma membranes from intracellular membranes is usually not obtained in a single step as illustrated in Fig. 2. Therefore, multistep procedures have to be used to improve separation further. For preparative work batch procedures of a few steps may be sufficient (Fig. 5).

In Table 1 calculations are made for the yield and purification of plasma membranes by the batch procedure outlined in Fig. 5. The degree of purification is, of course, increased the more dissimilar the plasma membranes and the intracellular membranes are in their partition behavior (compare cases I and II), and if the number of steps is increased. Extracting the lower phases with a second upper phase (giving $U_{3'}$) increases the yield, but also increases contamination somewhat. Advantages with the batch procedure are that it can be carried out relatively rapidly (about 10 min per step), and with a minimum of equipment; and it can easily be scaled up.

Another useful multistep procedure is counter-current distribution employing an automatic apparatus (Albertsson 1971; Albertsson et al. 1982). Counter-current distribution is in principle an extension of the batch procedure in Fig. 5, using 60 or even 120 lower phases and feeding in fresh upper phase until all lower phases have received an upper phase. Counter-current distribution is, with the equipment used until now, a rather lenghty procedure (about 9 h for 60 steps), but due to its superior separatory power most useful for analytical purposes (Fig. 6). Note, however, that a new apparatus has recently been developed, which considerably reduces the time for counter-current distribution (Åkerlund 1984). Automatic apparatuses for counter-current distribution are obtainable from the Central Workshop, Chemical Center, University of Lund, S-221 00 Lund, Sweden.

Table 1. Yield of plasma membranes (pm) and intracellular membranes (icm) in the batch procedure outlined in Fig. 5

Step	Phase	I[a]		II	
		% pm	% icm	% pm	% icm
2	U_2	52	3	66	1
	$U_{2'}$	29	5	24	1
3	U_3	37	1	53	0.1
	$U_{3'}$	32	1	30	0.1

[a] In I it is assumed that 80% of the plasma membranes partition in the upper phase, but only 20% of the intracellular membranes. In II the relation is 90% to 10%

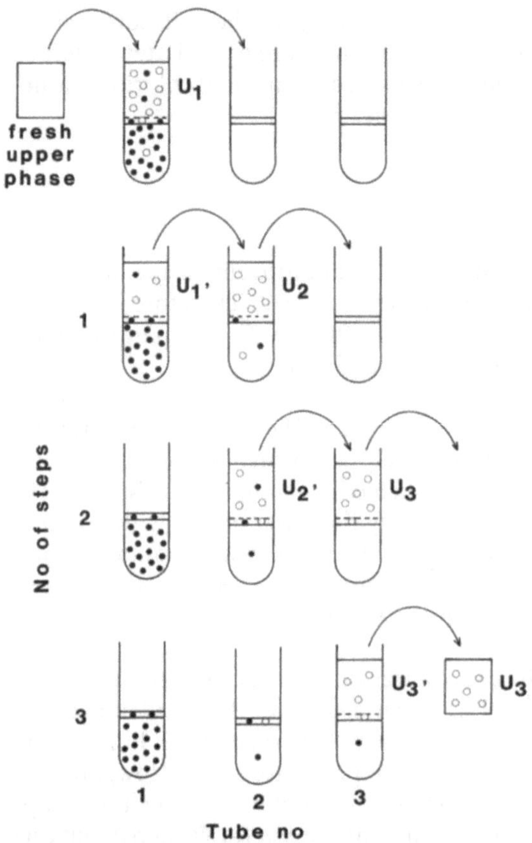

Fig. 5. Separation of plasma membranes (○) and intracellular membranes (●) by a batch procedure of three steps (*1, 2, 3*). The microsomal fraction suspended in the medium used in the phase system is added to a preweighed phase mixture to give a phase system of the desired final composition. The phase system is well mixed (by, e.g., 20 to 30 inversions of the tube) and allowed to settle. Phase settling is facilitated by centrifugation in a swinging bucket centrifuge for a few minutes. Then, 90% (indicated by the *dashed line*) of the upper phase (U_1) is removed, and repartitioned twice with fresh lower phase (tubes no 2 and 3) to increase the purity (giving U_3). To increase the yield of plasma membranes, the lower phases may be reextracted, in sequence, with fresh upper phase (giving $U_{3'}$). The upper phases (U_3 and $U_{3'}$) are pooled, diluted severalfold, and the plasma membranes collected by ultracentrifugation. Fresh upper and lower phases are obtained from a bulk phase system prepared separately (see Sect. 3.2). In the example it is assumed that 90% of the plasma membranes partition in the upper phase, while 90% of the intracellular membranes partition at the interface or in the lower phase (cf. Table 1). In addition to fresh lower phase, tubes no 2 and 3 also contain a small volume of upper phase corresponding to the 10% left in tube no 1. This is to keep a constant upper phase volume throughout the procedure

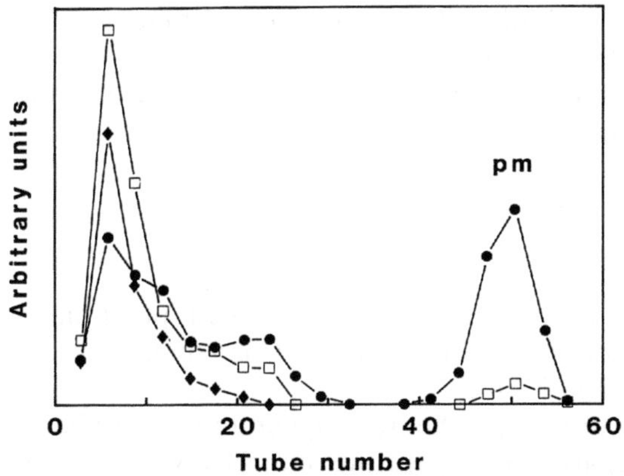

Fig. 6. Counter-current distribution of a microsomal fraction from wheat root. The automatic apparatus of Albertsson (1971) was used and 57 steps made. The phase system contained 6.3% Dextran T500, 6.3% polyethylene glycol 3350, 0.25 M sucrose, 5 mM potassium phosphate, pH 7.8. Temperature, 5 °C. Markers used were: cytochrome c oxidase (◆), antimycin A-resistant NADH-cytochrome c reductase (□), and LIAC (●). Note that LIAC (light-reducible cytochrome) gave two peaks; the *left-hand peak* probably represents endoplasmic reticulum, whereas the *peak to the far right* represents plasma membranes. A small peak of cytochrome c reductase is associated with the plasma membranes (discussed in Sect. 4.5). (Data from Lundborg et al. 1981)

3 Experimentals

3.1 Chemicals

Dextran T500 (average mw 500,000) is obtained from Pharmacia Fine Chemicals, Uppsala, Sweden. Polyethylene glycol 3350 (average mw 3350) is purchased from Union Carbide, New York, USA. Note that Union Carbide has recently changed the designation of polyethylene glycol 4000 to 3350, to reflect more closely the average molecular weight of the polymer.

The following stock solutions are prepared:

20% (w/w) Dextran T500: Dextran powder contains some percent water, and therefore the exact concentration of the solution should be determined. Layer 220 g of Dextran on 780 g of water. Heat on a water bath with gentle stirring until all Dextran is dissolved. Dilute about 5 g of the solution to 25 ml with water and measure the optical rotation with a polarimeter at 589 nm. The specific rotation is +199 degree ml g^{-1} dm^{-1}, and thus the concentration (% w/w) is given by the following equation:

$$\frac{\text{Optical rotation}}{199} \cdot \frac{25\ (\text{ml})}{\times\ (\text{g})} \cdot 100 .$$

Adjust to 20% (w/w) concentration with water.

40% (w/w) Polyethylene glycol 3350: Make 400 g up to 1000 g with water. The polymer stock solutions should be stored in the cold or frozen.

Sucrose, salt, and buffer solutions: Prepare these 4–20 times stronger than the final concentrations in the phase systems.

3.2 Preparation Procedure

The exact preparation procedure for plasma membranes varies with plant material. As an example, the preparation of plasma membranes from spinach leaves has been chosen (Kjellbom and Larsson 1984):

A bulk phase system (50–300 g), to provide the fresh upper and lower phases in the batch procedure (Fig. 5) is prepared in a separating funnel by weighing and pipeting from stock solutions (Table 2). After temperature equilibration in the cold room the phase system is well mixed and allowed to settle, usually overnight. The upper and the lower phase are collected and stored separately in the cold or frozen. A phase mixture is prepared similarly (Table 2) in a centrifugation tube of suitable size.

Spinach leaves (125 g) are homogenized (kitchen homogenizer) for 3 × 20 s in 275 ml of 0.5 M sucrose, 50 mM HEPES-KOH, pH 7.5, 5 mM ascorbic acid, 4 mM cysteine and 0.6% polyvinylpolypyrrolidone (to remove phenolic compounds). The homogenate is filtered through a 240 μm nylon cloth, and the bulk of chloroplasts and mitochondria pelleted at 10,000 *g* for 10 min. A microsomal pellet is obtained from the supernatant by centrifugation at 50,000 *g* for 30 min. This pellet is suspended to a total volume of 10 ml in 0.33 M sucrose, 3 mM KCl, 5 mM potassium phosphate, pH 7.8,; and 9.0 g (containing about 50 mg protein) of the suspension is added to the 27.0 g phase mixture to give a 36.0 g phase system with a final composition of 6.2% (w/w) Dextran T500, 6.2% (w/w) polyethylene glycol 3350, 0.33 M sucrose, 3 mM KCl, 5 mM potassium phosphate, pH 7.8. The phase system is thoroughly mixed by 20–30 inversions of the tube and phase-settling accelerated by centrifugation in a swinging bucket centrifuge at 1500 *g* for 3 min. Further processing is according to the batch procedure in Fig. 5. The final upper phases (U_3 and $U_{3'}$) are diluted at least twofold with a suitable medium and the plasma membranes collected by centrifugation at 100,000 *g* for

Table 2. A 300-g phase system and a 27-g phase mixture are prepared from stock solutions

	Phase system	Phase mixture
20% (w/w) Dextran T500	93.00 g	11.16 g
40% (w/w) Polyethylene glycol 3350	46.50 g	5.58 g
Sucrose (solid)	33.89 g	3.05 g
0.2 M potassium phosphate, pH 7.8	7.50 ml	0.675 ml
2 M KCl	0.45 ml	0.041 ml
Add H_2O to a final weight of	300.00 g	27.00 g

30 min. The pellet is white and no chlorophyll is detected by standard procedures. The yield is about 4 mg protein.

The medium used for homogenization may be varied in several ways without changing the final result very dramatically. We use an isotonic or slightly hypertonic medium (0.25–0.5 M sucrose or sorbitol) to keep organelles as intact as possible. In this way fewer of the mitochondria and chloroplasts end up in the microsomal fraction. The choice of buffer at this stage is not critical, but should of course be one which preserves enzyme activities. The medium may include excess EDTA or excess Mg^{2+}, BSA or other protective agents like protease inhibitors.

By contrast, the composition of the phase system is very critical. With both Dextran and polyethylene glycol batch to batch variation occurs. Therefore, published polymer concentrations can not be used directly, but rather the optimal concentrations must be determined for each batch of polymer. A good strategy is to start with an experiment using published conditions. If the material partitions too much in the upper phase (e.g., a green upper phase is obtained with leaf material) the polymer concentrations are too low. On the other hand, if the material (including the plasma membranes!) partions too much at the interface and in the lower phase, the polymer concentrations are too high. Optimal polymer concentrations are determined by partitioning the material in a series of phase systems with increasing polymer concentrations (cf. Fig. 3), either above or below the polymer concentrations used in the pilot experiment. When the optimal polymer concentrations are known, the phase system can be optimized with respect to chloride concentration (cf. Fig. 4). Since the partition of plasma membranes, in contrast to intracellular membranes, is relatively insensitive to both increased polymer and chloride concentrations (Figs. 3 and 4) it is often wise to exploit both properties for optimal separation. For these experiments phase systems with a total weight of 4.00 g (3.50 g phase mixture plus 0.50 g sample) are suitable. After mixing and phase settling (20 to 30 min at unit gravity, and constant temperature), 90% of the upper phase is transferred to another test tube, and both phases diluted to the same final volume (the same height). To determine the partition of different membranes, appropriate assays are performed on aliquots from the diluted phases. Protein may be determined according to Bearden (1978).

Although the composition of the phase system is very critical, some variations are allowed. However, usually the conditions have to be optimized again after each change. Sorbitol may be used instead of sucrose. To avoid potassium and phosphate (e.g., to be able to measure K^+-ATPase activity in samples directly withdrawn from the upper phase) H_2SO_4-Tris or maleate-Tris may be used as buffers, since SO_4^{2-} and maleate both give "positive" upper phases. Chloride may be added as HCl-Tris. Note that the concentrations of SO_4^{2-}, maleate, and chloride are the essential ones (and hence written first), not the concentration of Tris. Low concentrations of EDTA (0.1 mM), BSA (0.1%) and protease inhibitors may be included without affecting the partition of the material. Note finally, that a constant temperature should be used throughout, since the partition is temperature-sensitive.

4 Purity of the Preparations

During the last years, phase partition has successfully been applied for the preparation of plasma membranes from several plant species and tissues, like corn coleoptiles, corn shoots and barley shoots (Widell and Larsson 1981), wheat roots (Lundborg et al. 1981), oat roots and shoots (Widell et al. 1982), cauliflower inflorescences (Widell and Larsson 1983), needles of Scots pine (Hellergren et al. 1983), crown tissue of Orchard grass (Yoshida et al. 1983) and winter rye (Uemura and Yoshida 1983), and spinach and barley leaves (Kjellbom and Larsson 1984). These preparations probably have a purity higher than 90%, and often close to 100%, although the lack of absolute markers for the plasma membrane and for most intracellular membranes (Quail 1979) makes it difficult to give an exact figure. However, several of the proposed plasma membrane markers, as well as markers for contaminants, have now been used with plasma membranes isolated by phase partition, and the results are discussed below.

4.1 Specific Staining

Thin sections for electron microscopy may be stained selectively for plasma membranes by silicotungstic or phosphotungstic acid at low pH (Roland 1978). According to this criterion the plasma membrane preparations contain hardly any other material than smooth, closed vesicles of plasma membrane (0.1 to 1.0 μm in diameter) since all, or almost all, vesicles are well stained. This is the case with preparations from such diverse material as oat shoots and roots (Widell et al. 1982), corn coleoptiles (Widell and Larsson 1983), Scots pine (Hellergren et al. 1983), Orchard grass (Yoshida et al. 1983), winter rye (Uemura and Yoshida 1983), and barley leaves (Fig. 7). In our hands this has been the most absolute marker for the plasma membrane, and we have not observed any dense staining of intracellular membranes. A prerequisite for the validity of the method is, of course, that in sections of intact tissue the plasma membrane and only the plasma membrane should be well stained.

4.2 K^+-Stimulated, Mg^{2+}-Dependent ATPase

ATPase activity, dependent on Mg^{2+}, further stimulated by K^+, and with a pH optimum at 6.0–6.5 was introduced by Hodges and Leonard (1974) as a marker for the plasma membrane. Usually, only the K^+-stimulated increment (K^+-ATPase) is used as a marker, since several different Mg^{2+}-ATPases are present in plant cells. The K^+-ATPase activity was enriched more than tenfold in plasma membranes from oat roots, and about 50% of the total activity in the microsomal fraction was recovered in the final plasma membrane preparation (Widell et al. 1982). Similar results were obtained with plasma membranes from needles of Scots pine (Hellergren et al. 1983).

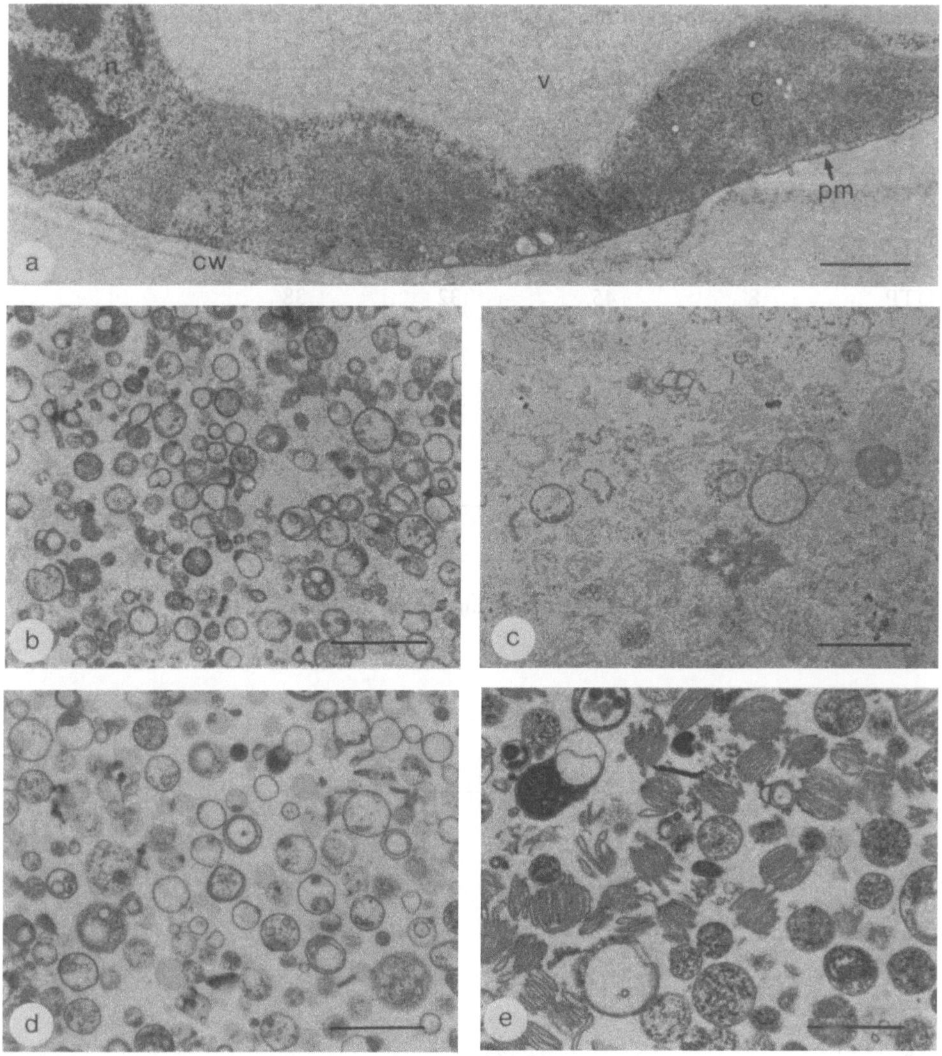

Fig. 7. Silicotungstic acid staining (Roland 1978) of: **a** section of a barley leaf; **b** the plasma membrane fraction; **c** the microsomal fraction used as the starting material. Plates **d** and **e** show the plasma membrane and microsomal fractions respectively, stained with uranyl acetate for comparison. Note the specificity of the silicotungstic acid stain for the plasma membrane. *c* chloroplast; *cw* cell wall; *n* nucleus; *pm* plasma membrane; *v* vacuole. Bar = 1 μm. (From Kjellbom and Larsson 1984)

With pure preparations of plasma membrane the ATPase is more conveniently assayed as the total K^+, Mg^{2+}-ATPase activity, since the K^+-stimulation with some species is very low compared to the basal Mg^{2+}-ATPase activity, and therefore difficult to determine (e.g., Yoshida et al. 1983); although a low K^+-stimulation may also be due to suboptimal assay conditions (Sommarin et al. 1985). The K^+, Mg^{2+}-ATPase shows a relatively high specificity for ATP, which,

Table 3. Substrate specificity of the K^+-stimulated, Mg^{2+}-dependent ATPase of plasma membranes

Substrate	% K^+, Mg^{2+}-ATPase activity			
	Oat[a]	Wheat[a]	Scots pine[b]	Orchard grass[c]
ATP	100	100	100	100
CTP	1	53	5	–
GTP	2	27	1	–
ITP	2	8	36	36
UTP	8	45	32	38
ADP	4	2	5	8
CDP	0	39	8	30
GDP	0	12	17	–
IDP	3	13	2	17
UDP	3	43	18	–
AMP	0	0	0	0
PNPP[d]	0	1	6	1

[a] Sommarin et al. (1985)
[b] Hellergren et al. (1983)
[c] Yoshida et al. (1983)
[d] p-Nitrophenyl phosphate, a substrate of unspecific acid phosphatases

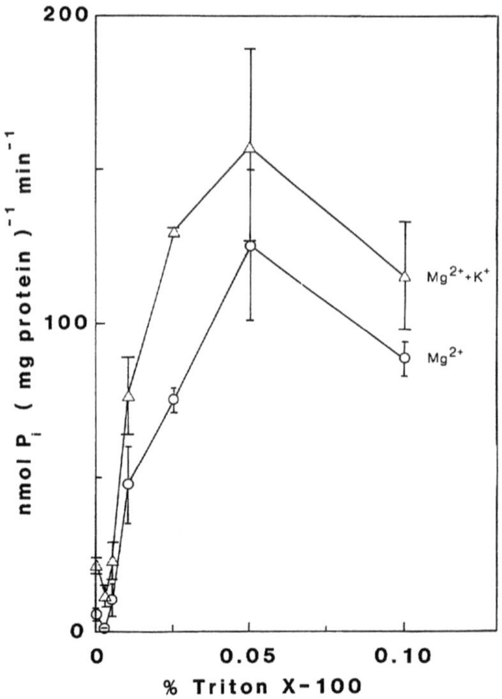

Fig. 8. Latency of K^+-stimulated, Mg^{2+}-dependent ATPase activity of isolated plasma membrane vesicles from cauliflower inflorescences. The activity was either assayed with Mg^{2+} only (○) or with Mg^{2+} + K^+ (△). Each value represents the mean (and maximum variation) of two independent experiments. (Larsson et al. 1984)

however, differs among species (Table 3). A very high specificity was obtained with oat plasma membranes, results being very similar to those obtained by Hodges and Leonard (1974) with the K^+-ATPase of oat plasma membranes prepared by sucrose gradient centrifugation, and by Vara and Serrano (1982) with the partially purified enzyme.

Addition of detergent (e.g., Triton X-100) at low concentrations stimulates the activity severalfold (Fig. 8), indicating that the plasma membrane vesicles are sealed with the ATP-binding site on the inner side. Since the ATP-binding site is generally assumed to be on the cytoplasmic side of the plasma membrane, the observed latency suggests that the vesicles are right-side-out and sealed (sealed being defined as not permeable to ATP).

The ATPase shows the expected reaction pattern toward inhibitors, i.e., it is inhibited by presumed inhibitors of plasma membrane ATPase, like vanadate and stilbestrol, but is not affected by azide or oligomycin, inhibitors of mitochondrial ATPase (Yoshida et al. 1983; Sommarin et al. 1985), or by ammonium molybdate, an inhibitor of unspecific acid phosphatases (Sommarin et al. 1985).

4.3 Glucan Synthetase II

Several different glucan synthetases are probably present in plants, since there exist several different types of polyglucans. One of these glucan synthetases is glucan synthetase II, which may be used as a marker for the plasma membrane (Ray 1979). The recovery of glucan synthetase II activity in the plasma membrane fraction may be relatively high (Table 4), but still often represents an underestimation of the recovery of plasma membranes. This is due to the fact that the assay is not specific for plasma membrane-bound glucan synthetase (Quail 1979; Ray 1979), so that also glucan synthetases bound to intracellular membranes contribute to the measured activity. The specificity of the marker may be increased by assaying

Table 4. Recovery of protein and markers in membrane fractions from barley and spinach leaves. (Data from Kjellbom and Larsson 1984)

	Protein		GS II		Cyt c ox		Cyt c red		Chl	
	mg	%	act[a]	%	act	%	act	%	mg	%
Barley:										
Microsomes	58	100	12.5	100	7,400	100	470	100	6.7	100
Plasma m.	1.3	2.2	4.7	38	42	0.6	6	1	nd[b]	0
Intrac. m.	28	48	5.4	43	5,100	69	280	60	3.1	46
Spinach:										
Microsomes	49	100	63	100	19,000	100	610	100	4.0	100
Plasma m.	4.0	8	37	59	92	0.5	29	5	nd	0
Intrac. m.	22	45	11	17	14,200	75	360	59	2.8	70

[a] Total activity in nmol min^{-1}
[b] Not detectable by standard procedures

at the pH optimum of the plasma membrane-bound activity, since this optimum may be rather narrow (Larsson et al. 1984). The plasma membrane-bound activity shows latency on addition of detergent similar to the K^+, Mg^{2+}-ATPase, suggesting that the binding site for UDP-glucose is on the cytoplasmic side (Larsson et al. 1984). This latency is not observed with all species, however, probably due to inhibition of the activity by the detergent used. Thus it should be possible to increase the specificity of the assay further by including a suitable detergent in the assay medium.

4.4 Light-Reducible b-Cytochrome

Poff and Butler (1975) described a blue light-induced absorbance change in fungi, reflecting the reversible reduction of a b-type cytochrome. The presence of a similar b-type cytochrome in higher plants was later established by Brain et al. (1977), and the cytochrome was suggested to be located in the plasma membrane, and to have a role in blue light photomorphogenesis (Senger and Briggs 1983). This localization has later been supported by several studies (Leong and Briggs 1981; Leong et al. 1981; Widell and Larsson 1981, 1983; Lundborg et al. 1981; Widell et al. 1982, 1983; Kjellbom and Larsson 1984), and the light-induced absorbance change (LIAC) may therefore be used as a marker for the plasma membrane, and is enriched severalfold in plasma membrane preparations from different material. However, there are also light-reducible b-cytochrome in intracellular membranes,

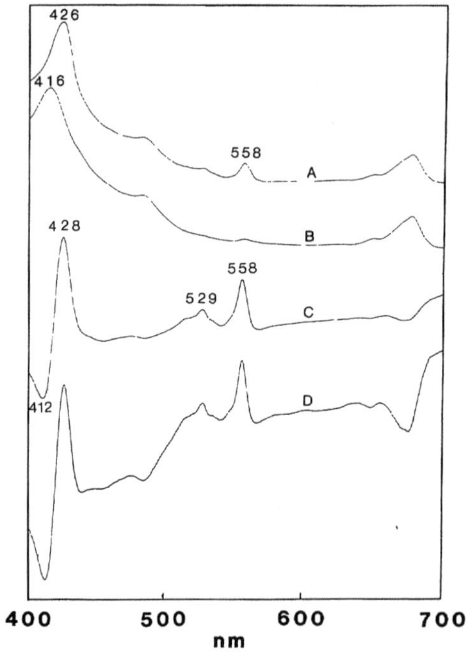

Fig. 9. Low temperature (77 K) spectra of the plasma membrane preparation from spinach. *A* dithionite reduced; *B* air oxidized; *C* dithionite reduced minus oxidized; *D* blue light-reduced minus oxidized. Light-reducible cytochrome (LIAC) is assayed as the difference between the maximum and minimum absorbance in the Soret region of the light minus dark difference spectrum [i.e., $\Delta(A^{428}\text{-}A^{412})$ in D], at room temperature. A small chlorophyll peak is seen at about 680 nm, although chlorophyll is not detected by standard procedures. (Data from Kjellbom and Larsson 1984)

probably mainly in the endoplasmic reticulum (Widell and Larsson 1983; Widell et al. 1983). Thus, LIAC is not an absolute marker for the plasma membrane.

Spectra of isolated plasma membranes indicate the existence of one cytochrome with an α-band around 558 nm (Fig. 9), typical for b-cytochromes. Action spectra of the light reduction shows a maximum at 450 nm and indicate that a flavin is responsible for the primary absorption of light (Widell and Larsson 1983).

4.5 Markers for Contaminants

Mitochondrial membranes (cytochrome c oxidase) are present in the plasma membrane preparations, but to a very low extent. Calculations from the data in Table 5 show that only 0.6% or less of the membranes in the plasma membrane fractions are inner mitochondrial membranes. This is the case even when plasma membranes are prepared from a "mitochondrial" pellet (9 kp, Table 5). The relatively low specific absorbance change (LIAC) in the plasma membrane fraction obtained from the 9000 g pellet (9 kp) is therefore not due to contamination by mitochondria, but may rather be due to the presence of cytoskeleton, since a lot of fibrous material was observed in electron micrographs of this fraction (C. Larsson and S. Widell, unpublished). Thus, also low speed pellets may be used as starting material for plasma membranes.

Cytochrome c reductase activity is found in all plasma membrane preparations (Fig. 6, Table 4, Widell and Larsson 1983). Recent experiments show that the plasma membrane contains cytochrome P-450/420, as judged from CO-difference spectra (C. Schelin, P. Kjellbom, S. Widell and C. Larsson, unpublished). The presence of cytochrome P-450/420 in the plasma membrane would explain the presence of cytochrome c reductase activity in the plasma membrane preparations. Furthermore, it is probable that the cytochrome P-450/420, which is known to be associated with a flavoprotein, is identical with the plasma membrane-bound light-reducible b-cytochrome. This would be consistent with the presence of light-reducible b-cytochrome also in the endoplasmic reticulum (see

Table 5. Total protein, and specific activities of LIAC and cytochrome c oxidase in plasma membrane preparations obtained from different centrifuge fractions. Material, cauliflower inflorescences. (Data from Widell and Larsson 1983)

Fraction	Protein mg	LIAC $\Delta(\Delta A)\ 10^3\ mg^{-1}$	Cyt c oxidase $nmol\ min^{-1} mg^{-1}$
pm (9 kp[a])	1.70	6.8	12
pm (21 kp)	1.44	13.2	6
pm (50 kp)	1.76	15.6	9
Mitochondria	4.44	2.1	1,800

[a] Pellets obtained after differential centrifugation: 9 kp, 9,000 *g* for 20 min; 21 kp, 21,000 *g* for 20 min, 50 kp, 50,000 *g* for 45 min

Sect. 4.4), which is known to contain relatively high activities of cytochrome P-450/420. Since the bulk of cytochrome c reductase partitions in the lower phase (Figs. 3 and 6, Table 4) the contamination of the isolated plasma membranes by endoplasmic reticulum is probably of the same order as for mitochondria, i.e., below 1%.

IDPase activity, a marker for Golgi, is negligible in most plasma membrane preparations (Lundborg et al. 1981, Table 3). However, some activity is detected in plasma membranes from wheat and Orchard grass (Table 3). Whether these activities are due to the K^+, Mg^{2+}-ATPase, or to Golgi contamination is uncertain.

A possible contaminant is the tonoplast membrane. NO_3^- sensitive ATPase activity may be used as a marker for this membrane (O'Neill et al. 1983). This activity is low or absent in the isolated plasma membranes (Sommarin et al. 1985).

A mostly overlooked problem is the contamination by cytoplasm. When the plasma membrane is fragmented on homogenization and resealed into small vesicles, these are bound to trap some cytoplasm, as shown by the presence of cytoplasmic malic enzyme in plasma membrane preparations (Pupillo and Del Grosso 1981). A way to remove this contamination is to open up the isolated plasma membrane vesicles by suspension in a hypotonic medium, or by sonication, and then re-pellet the membranes.

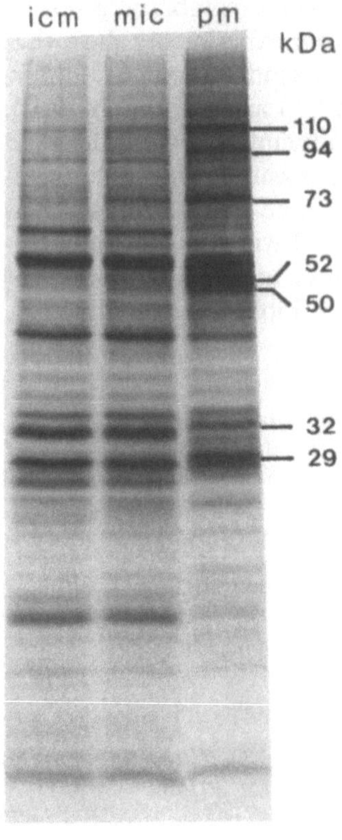

Fig. 10. SDS-polyacrylamide gelelectrophoresis of the plasma membrane (*pm*) fraction from spinach leaves. The polypeptide patterns of the microsomal (*mic*) fraction used as the starting material, and of the intracellular membrane (*icm*) fraction recovered from the lower phase (Fig. 5), are shown for comparison. The molecular weights of some prominent polypeptides clearly distingushing the plasma membrane from the bulk of intracellular membranes are indicated. The gel was stained with Coomassie brilliant blue. (From Kjellbom and Larsson 1984)

5 Protein and Lipid Composition

The plasma membranes of barley and spinach leaves contain about 30 and 40% protein, respectively, as revealed by elementary analyses (Kjellbom and Larsson 1984), assuming that all nitrogen is in protein and that protein contains 16% nitrogen. The corresponding figures for the microsomal fractions were 50 to 55%. Thus, the plasma membrane seems to contain relatively little protein. However, the polypeptide pattern of the spinach plasma membrane is still very complex (Fig. 10) and shows about 40 bands. Similar polypeptide patterns were obtained with plasma membranes from barley leaves (Kjellbom and Larsson 1984) and cauliflower inflorescences (P. Kjellbom and C. Larsson, unpublished) with a few high molecular weight bands around 100,000, several heavy bands around 50,000, and some additional prominent bands around 30,000. So far, only one of the polypeptides has been identified, namely the phosphorylated subunit of the K^+, Mg^{2+}-ATPase with a molecular weight of about 100,000 (Vara and Serrano 1983). Preliminary lipid analyses of oat root plasma membranes (Lundborg et al. 1983) indicate, similarly to the elementary analyses, that the plasma membrane is a lipid-rich membrane compared to the bulk of intracellular membranes. The plasma membrane was furthermore found to be rich in glycolipids.

6 Surface Properties of the Isolated Vesicles

Plasma membranes have an extreme affinity for the polyethylene glycol-rich upper phase compared to intracellular membranes (Figs. 3 and 4). Since the partition of a particle in the two-phase system is determined by the properties of the membrane surface exposed to the phase system (Albertsson 1971) the plasma membrane vesicles can be expected to have surface properties very different from those of the intracellular membranes.

The plasma membrane vesicles recovered from the upper phase are right-side-out and sealed as revealed by a high latency of the K^+, Mg^{2+}-ATPase on addition of detergent (Fig. 8).

This orientation is supported by the observations that protoplasts (Hallberg and Larsson 1981), and multiorganelle complexes surrounded by plasma membrane (Larsson et al. 1971; Larsson and Andersson 1979), also have a high affinity for the upper phase. In addition to being homogenous with respect to surface properties, the plasma membrane preparations are also homogenous with respect to density and give one band only in gradients of Percoll (Widell and Larsson 1983), sucrose (Yoshida et al. 1983), and urografin (Caubergs et al. 1983).

The isoelectric point of the outer surface of plasma membranes from spinach and barley is at pH 3.4 (Westrin et al. 1983) and 3.6 (Körner et al. 1985), respectively, as determined by cross-partition. This is 0.5 to 1 pH unit lower than for different surfaces of the spinach thylakoid membrane (Åkerlund et al. 1979) and for the chloroplast envelope (Westrin et al. 1983), and probably lower than for all intracellular membrane surfaces. The surface net charge density at neutral pH was in the range of -16 to -29 mC m^{-2} for the outer surface of plasma membranes from wheat and oat (Møller et al. 1984), and barley (Körner et al. 1985), whereas the bulk of the intracellular membranes, recovered from the lower phase, had a slightly less negative value (Bérczi et al. 1984).

The plasma membrane surface is known to have binding sites for 1-N-naphtylphtalamic acid, an inhibitor of auxin transport (Thomson et al. 1973). Plasma membranes from winter rye had a specific binding of naphtylphtalamic acid tenfold that of the intracellular membranes, with a pH optimum for the binding at pH 4 (Uemura and Yoshida 1983).

A drastic difference was found in the aggregation of plasma membranes and intracellular membranes by Zn^{2+}. Intracellular membranes were heavily aggregated by 10 mM $ZnCl_2$, whereas almost no aggregation was observed with plasma membranes (Uemura and Yoshida 1983).

Thus, the plasma membrane seems to posses surface properties which are relatively different from those of the intracellular membranes, and which may be related to the large differences in partition behavior (Figs. 3 and 4). These differences in surface properties may in turn be related to the low protein content of the plasma membrane compared to intracellular membranes (see Sect. 5).

References

Åkerlund H-E (1984) An apparatus for counter-current distribution in a centrifugal acceleration field. J Biochem Biophys Methods 9:133–141

Åkerlund H-E, Andersson B, Persson A, Albertsson P-Å (1979) Isoelectric points of spinach thylakoid surfaces as determined by cross partition. Biochim Biophys Acta 552:238–246

Albertsson P-Å (1971) Partition of cell particles and macromolecules, 2nd edn. Wiley, New York

Albertsson P-Å, Andersson B, Larsson C, Åkerlund H-E (1982) Phase partition – a method for purification and analysis of cell organelles and membrane vesicles. In: Glick D (ed) Methods of biochemical analysis, vol 28. Wiley, New York, pp 115–150

Andersson B, Åkerlund H-E (1978) Inside-out membrane vesicles isolated from spinach thylakoids. Biochim Biophys Acta 503:462–472

Bearden Jr JC (1978) Quantitation of submicrogram quantities of protein by an improved protein-dye binding assay. Biochim Biophys Acta 533:525–529

Bérczi A, Møller IM, Lundborg T, Kylin A (1984) The surface charge density of wheat root membranes. Physiol Plant 61:535–540

Brain RD, Freeberg JA, Weiss DV, Briggs WR (1977) Blue light-induced absorbance changes in membrane fractions from corn and *Neurospora*. Plant Physiol 59:948–952

Caubergs R, Widell S, Larsson C, De Greef JA (1983) Comparison of two methods for the preparation of a membrane fraction of cauliflower influorescences containing a blue light reducible b-type cytochrome. Physiol Plant 57:291–295

Hall JL (1983) Plasma membranes. In: Hall JL, Moore AL (eds) Isolation of membranes and organelles from plant cells. Academic Press, London, pp 55–81

Hallberg M, Larsson C (1981) Compartmentation and export of $^{14}CO_2$ fixation products in mesophyll protoplasts from the C_4-plant *Digitaria sanguinalis*. Arch Biochem Biophys 208:121–130

Hellergren J, Widell S, Lundborg T, Kylin A (1983) Frosthardiness development in *Pinus sylvestris*: the involvement of a K^+-stimulated Mg^{2+}-dependent ATPase from purified plasma membranes of pine. Physiol Plant 58:7–12

Hodges TK, Leonard RT (1974) Purification of a plasma membrane-bound adenosine triphosphatase from plant roots. Methods Enzymol 32:392–406

Johansson G (1970) Partition of salts and their effects on partition of proteins in a dextran-poly(ethylene glycol)-water two-phase system. Biochim Biophys Acta 221:387–390

Kjellbom P, Larsson C (1984) Preparation and polypeptide composition of chlorophyll-free plasma membranes from leaves of light-grown spinach and barley. Physiol Plant 62:501–509

Körner L, Kjellbom P, Larsson C, Møller IM (1985) Surface properties of right side-out plasma membrane vesicles isolated from barley roots and leaves. Plant Physiol (in press)

Larsson C, Andersson B (1979) Two-phase methods for chloroplasts, chloroplast elements and mitochondria. In: Reid E (ed) Plant organelles: methodological surveys B. Biochemistry, vol 9. Ellis Horwood, Chichester, pp 35–46

Larsson C, Collin C, Albertsson P-Å (1971) Characterization of three classes of chloroplasts obtained by counter-current distribution. Biochim Biophys Acta 245:425–438

Larsson C, Kjellbom P, Widell S, Lundborg T (1984) Sidedness of plant plasma membrane vesicles purified by partition in aqueous two-phase systems. FEBS Lett 171:271–276

Leong TY, Briggs WR (1981) Partial purification and characterization of a blue light-sensitive cytochrome flavin complex from corn membranes. Plant Physiol 67:1042–1046

Leong TY, Vierstra RD, Briggs WR (1981) A blue light-sensitive cytochrome-flavin complex from corn coleoptiles. Further characterization. Photochem Photobiol 34:696–703

Lundborg T, Widell S, Larsson C (1981) Distribution of ATPases in wheat root membranes separated by phase partition. Physiol Plant 52:89–95

Lundborg T, Sandelius AS, Widell S, Larsson C, Kylin A (1983) Characterization of root plasma membranes prepared by partition in an aqueous polymer two-phase system. In: Wintermans JFGM, Kuiper PJC (eds) Biochemistry and metabolism of plant lipids. Elsevier, Amsterdam, pp 183–186

Møller IM, Lundborg T, Bérczi A (1984) The negative surface charge density of plasmalemma vesicles from wheat and oat roots. FEBS Lett 167:181–185

O'Neill SD, Bennet AB, Spanswick RM (1983) Characterization of a NO_3^--sensitive H^+-ATPase from corn roots. Plant Physiol 72:837–846

Poff K, Butler WL (1975) Spectral characterization of the photoreducible b-type cytochrome of *Dictyostelium discoideum*. Plant Physiol 55:427–429

Pupillo P, Del Grosso E (1981) A possible plasma membrane particle containing malic enzyme activity. Planta 151:506–511

Quail PH (1979) Plant cell fractionation. Annu Rev Plant Physiol 30:425–484

Randall SK, Ruesink AW (1983) Orientation and integrity of plasma membrane vesicles obtained from carrot protoplasts. Plant Physiol 73:385–391

Ray PM (1979) Maize coleoptile cellular membranes bearing different types of glucan synthetase activity. In: Reid E (ed) Plant organelles: methodological surveys B. Biochemistry, vol 9. Ellis Horwood, Chichester, pp 135–146

Roland JC (1978) General preparation and staining of thin sections. In: Hall JL (ed) Electron microscopy and cytochemistry of plant cells. Elsevier, Amsterdam, pp 1–62

Senger H, Briggs WR (1983) The blue light receptor(s): primary reactions and subsequent metabolic changes. In: Smith K (ed) Photochemical and photobiological reviews, vol 6. Plenum, New York, pp 1–38

Sommarin M, Lundborg T, Kylin A (1985) Comparison of K,MgATPase in purified plasmalemma from wheat and oat – substrate specificity, effects of pH, temperature and inhibitors. Physiol Plant (in press)

Steck TL (1974) Preparation of impermeable inside-out and right-side-out vesicles from erythrocyte membranes. In: Korn EO (ed) Methods in membrane biology, vol 2. Plenum, New York, pp 245–281

Thomson K St, Hertel R, Müller S, Tavares JE (1973) 1-N-Naphtylphtalamic acid and 2,3,5-triiodobenzoic acid. In vitro binding to particulate cell fractions and action on auxin transport in corn coleoptiles. Planta 109:337–352

Uemura M, Yoshida S (1983) Isolation and identification of plasma membrane from light-grown winter rye seedlings (*Secale cereale* L. cv Puma). Plant Physiol 73:586–597

Vara F, Serrano R (1982) Partial purification and properties of the protontranslocating ATPase of plant plasma membranes. J Biol Chem 257:12826–12830

Vara F, Serrano R (1983) Phosphorylated intermediate of the ATPase of plant plasma membranes. J Biol Chem 258:5334–5336

Westrin H, Shanbag VP, Albertsson P-Å (1983) Isoelectric points of membrane surfaces of three spinach chloroplast classes determined by cross-partition. Biochim Biophys Acta 732:83–91

Widell S, Larsson C (1981) Separation of presumptive plasma membranes from mitochondria by partition in an aqueous polymer two-phase system. Physiol Plant 51:368–374

Widell S, Larsson C (1983) Distribution of cytochrome b photoreductions mediated by endogenous photosensitizer or methylene blue in fractions from corn and cauliflower. Physiol Plant 57:196–202

Widell S, Lundborg T, Larsson C (1982) Plasma membranes from oats prepared by partition in an aqueous polymer two-phase system. Plant Physiol 70:1429–1435

Widell S, Caubergs RJ, Larsson C (1983) Spectral characterization of light-reducible cytochrome in a plasma membrane-enriched fraction and in other membranes from cauliflower inflorescences. Photochem Photobiol 38:95–98

Yoshida S, Uemura M, Niki T, Sakai A, Gusta LV (1983) Partition of membrane particles in aqueous two-polymer phase system and its practical use for purification of plasma membranes from plants. Plant Physiol 72:105–114

Vacuoles

G. Wagner

1 Introduction

The mature-plant-cell vacuole is an organelle that is unique to plants, and one which plays critical roles in tissue support and movements, metabolite storage and sequestration, lytic processes, cyclosis, and organelle distribution. Efforts to elucidate the mechanisms involved in these processes have benefited greatly from the recent development of methods – an international effort – for isolating and purifying intact vacuoles and tonoplast from mature plant tissues. Several reviews which discuss these methods and their applications have appeared in the last few years (Leigh et al. 1979a; Marty et al. 1980; Wagner 1981b; Alibert and Boudet 1982; Boller 1982; Komor et al. 1982a; Matile 1982; D'Auzac et al. 1982; Marin et al. 1982; Wagner 1983; Ryan and Walker-Simmons 1983; Leigh 1983; Boudet et al. 1985). In some of these, particularly (Wagner 1981b; Alibert and Boudet 1982; Wagner 1983; Leigh 1983), the problems and promise of various isolation methods were discussed.

No one method for isolating vacuoles has been universally adopted. Therefore, it is appropriate to discuss all alternatives here so that the first-time investigator can select the method which seems most appropriate for his or her purpose.

Since methods developed prior to 1981 have been reviewed in detail elsewhere (Wagner 1983), these will only be summarized and highlighted here. Modifications and refinements of earlier protocols described since about 1981 will be considered in greater detail. Methods for isolating lutoids from latex of the rubber tree *Hevea brasiliensis*, and vacuoles from meristematic tissues of higher plants, yeast, and *Neurospora* will be briefly discussed. Also, techniques for isolating – from tissue homogenates – preparations containing proton pumping vesicles, which may in part be derived from tonoplast or vacuoles, will be described in an effort to consider the origin of these vesicles.

Finally, throughout this chapter, discussion will be made of certain aspects of the physiological functions of higher plant vacuoles in an effort to stimulate discussion of aspects which, in this author's view, are not considered in recent reviews on plant vacuoles.

2 Methods of Isolation

2.1 Isolation of Vacuoles from Meristematic Tissues

The first method for efficient isolation of plant vacuoles from fresh meristematic tissue was described by Matile in 1966 (see Matile 1975). This method was used and adapted by a number of investigators (Matile 1975, 1978) primarily during the years between 1966 and 1973. It is useful to briefly summarize and discuss this basic method here, since it was the first one described for isolating meristematic plant vacuoles and also to set the stage for later discussion of techniques used to prepare proton pumping vesicles, suggested by some to be tonoplast-vacuole derived.

According to Matile (Matile 1968), corn root tips (40 h post-germination) were homogenized in a mortar with sand in cold 50 mM Tris/HCl, pH 7.6, 1 mM EDTA containing 0.5 M sorbitol. The brei was centrifuged at 500 *g* for 10 min and the supernatant centrifuged at 20,000 *g* for 15 min to yield a "mitochondrial fraction". This pellet was washed and resuspended in buffered sorbitol (10 mM buffer) and layered on a discontinuous sucrose gradient having 15, 20, and 40% w/v sucrose steps. Centrifugation for 2 h at about 200,000 *g* yielded three acid hydrolase containing fractions having densities $\leqq 1.06$, $\leqq 1.09$ and $\leqq 1.18$ g cm^{-3}. The specific activities of hydrolases in the $\leqq 1.06$ g cm^{-3} fraction were about 12 times that observed in the cell free extract and higher than that observed in the more dense fractions. The last two fractions could be further separated into three populations of particles, each bearing hydrolases, including RNAse, acid phosphates, and acid protease. Matile has argued that these particles represent secondary (density, $\leqq 1.06$) and primary (more dense) lysosomes capable of autophagy and heterophagy (Matile 1975).

Heftmann (1971) used a medium containing 0.58 M sucrose, 1 mM EDTA, 100 mM Tris, pH 7.4, 1 g l^{-1} polyvinylpyrollidone (PVP) for homogenizing tomato fruit (differing degrees of maturity) and observed only one class of hydrolase containing particles (density, 1.1 g/cc) in 10 to 50% w/v linear sucrose gradients. Nakano and Asahi (1972) using a medium containing 0.5 M sucrose, but otherwise similar to that of Matile, prepared acid protease containing particles which had a density of about 1.1 g cm^{-3} from 1-day post-germination pea cotyledons. Acid phosphatase was low in these particles but antimycin A-insensitive NADH Cyt c reductase – a marker for endoplasmic reticulum – was abundant. In pea seedling roots, acid phosphatase and RNAse were primarily soluble, and particulate activities were primarily associated with particles having densities substantially greater than 1.1 g cm^{-3} (Hirai and Asahi 1973). Solubilization experiments suggested that phosphatase and RNAse were bound electrostatically to endoplasmic reticulum and ribosomes respectively (Hirai and Asahi 1973). Corbett and Price (1967) concluded that acid phosphate measured as para-nitrophenyl phosphatase (a widely used and reliable marker for animal lysosomes) lacked the properties expected of lysosome-like enzymes in a number of embryonic and mature plant tissues.

Pitt and Galpin (1973) isolated phosphatase, RNAse and esterase containing particles from dark-grown potato shoots which had a density of about 1.1, however, in this and other studies, total tissue activity of these enzymes which occurred in the particles was low relative to hydrolase enrichment observed in animal lysosomes. Also, unambiguous structure-linked latency (membrane enclosure of enzymes as demonstrated by efficient release of activity on treatment with detergents, freeze-thaw, etc.) which is an essential criterion for identifying animal lysosomes, has been more difficult to demonstrate in preparations of hydrolase containing particles isolated from meristematic tissues of higher plants (Pitt and Galpin 1973). Considerable variation in the soluble-particulate distribution, the degree of latency and the consistency of finding acid phosphatase, protease and nuclease activity in association with the same particles (particularly those having density ~ 1.1 g cm^{-3} has been observed from plant tissue to plant tissue. Perhaps the best higher plant system fulfilling these criteria is the original corn root system of Matile (1968). It is noted that in many studies using this procedure, particles having density 1.1 g cm^{-3} were recovered from a sample zone/gradient interface, leaving open the possibility that certain activities occurring in this region may represent enzyme aggregates and not membrane-bound particles.

Which acid hydrolase activity, in a membrane-bound particle having what density is most consistently observed in homogenates of meristematic and embryonic tissues prepared by the basic method of Matile? The answer appears to be acid protease activity localized in particles having a density of about 1.1 g cm^{-3}. Phosphohydrolases have been observed to be soluble as well as bound to various membranes in various plant tissues (Dauwalder et al. 1969; Wilson 1975; Wagner et al. 1981) and acid esterase is often soluble and also distributed among particles having a spectrum of densities from heavy to light (Matile et al. 1965; Heftmann 1971; Pitt and Galpin 1973). Phosphohydrolase and esterase have primarily been used to study the genesis of meristematic vacuoles and are seen in ultrastructural studies throughout the continuum of membranes referred to as GERL by Marty (Marty et al. 1980). This may explain their broad distribution in gradients prepared from a number of meristematic and embryonic tissues. Plant glycosidases are relatively little studied but generally do not have acid pH optima. Also, glycosidases are known to function in post-translational processing of glycoproteins in animal Golgi (Rothman 1981) and are expected, and possibly have been observed (Wagner 1981b; Wagner and Hrazdina 1984), in plant Golgi.

Tonoplast prepared from mature plant vacuoles and lutoids of *Hevea* have a density of about 1.1 g cm^{-3} and mature vacuoles from various sources are shown to contain the bulk of intracellular acid protease (Wagner 1981b, 1983; Boller 1982). Proton pumping vesicles recovered from root tips, cultured cells and storage roots also have a density of about 1.1 g cm^{-3} (see latter discussion). These vesicles apparently have not been examined for the possible presence of acid hydrolases. The correlation of density and hydrolase content of meristematic and mature vacuoles is consistent with morphological evidence which suggests that tonoplast is primarily formed by fusion of meristematic vacuoles during cell maturation (Matile 1975; Marty et al. 1980; Vian 1981). However, as pointed out by Leigh (1979), the presence of hydrolases in mature vacuoles does not prove that they function in pre-terminal senescence. A portion of abundant vacuolar hydrolases may serve a nonhydrolytic function. In summarizing our current understanding of plant nucleases, Wilson (1975) concluded that no direct correlation has been established between RNAse activity (one of the best-studied plant hydrolyases) and RNA metabolism in plants. Direct evidence for autophagy and heterophagy in mature plant vacuoles is somewhat lacking (Gahan 1968; Marty et al. 1980) but vacuoles and their hydrolases are clearly involved in terminal senescence (Matile and Winkenback 1971; Matile 1975, 1978) and turnover of storage constituents in seeds (Nishimura and Beevers 1979; Leigh 1979; Leigh et al. 1979b; Van der Wilden et al. 1980). The case for autophagy in meristematic vacuoles and protein bodies (the latter shown to contain acid lipase as well as other hydrolases) is perhaps stronger but it is largely based on ultrastructural evidence (Matile 1975, 1978; Van der Wilden et al. 1980; Pernollet 1982). The limitations of such ultrastructural evidence, especially where large vacuoles are concerned, has been discussed previously (Butcher et al. 1977). Also, as pointed out by Leigh (1979), turnover in cotyledons represents a reserve substance utilization process in a specialized tissue, a process which may not be analogous to autophagy.

Recently, Van der Wilden et al. (1980) reported the preparation of protein bodies from cotyledons of germinating mung beans. Protoplasts were prepared and purified using standard methods, and protein bodies were isolated from protoplasts using a modification of the method used by Lorz et al. (1976) to prepare higher plant vacuoles. Yields of protein bodies were reported to the 70 to 80% and they were shown to contain the bulk of the acid hydrolase present in protoplasts. The successful use of protoplasts as a source for isolating protein bodies suggests that this approach may also have potential for isolating vacuoles from meristematic tissue.

2.2 Isolation of Vacuoles from Mature Plant Tissue

2.2.1 Isolation of Mature Vacuoles from Protoplasts – Methods pre 1981

Two basic methods for efficient isolation of vacuoles from mature plant tissues were introduced in 1975–1976. One involves isolation from plant protoplasts (Wagner and Siegelman 1975; Wagner 1983, Lorz et al. 1976) and is high-yielding and can be applied to various plant tissues. In the other (Leigh and Branton 1976; Leigh 1983), fresh or plasmolyzed tissue is sliced to yield vacuoles directly. This method is direct but lower-yielding, and may be limited to root storage tissue. Since about 1976, several variations of basic methods have been described. Protocols developed prior to 1981 have been reviewed in some detail (Wagner 1983) and are summarized in Table 1 along with more recent protocols.

As summarized in Table 1, vacuoles can be prepared by osmotic destabilization of protoplasts, achieved by transferring them from mannitol or sorbitol to phosphate buffer having lower osmolality or by reducing the concentration of the protoplast suspension medium. Osmolality has been reduced by as much as 60% and as little as 10%. Alternatively, protoplasts are destabilized under isotonic conditions by mechanical perturbation or exposure to polybase. Assuming that

Table 1. Basic methods for efficient isolation of vacuoles from mature plants

Source	Basic method for destabilization or release	Reference[a]
Protoplasts	Osmotic shock	
	– Transfer from mannitol or sorbitol to phosphate buffer; final osmolality 90 to 40% of original	Group A
	– Reduced osmolality of protoplast suspension medium; final osmolality 71 to 27% of original	Group B
	Destabilization under isotonic conditions	
	– Sheer due to ultracentrifugation or mechanical perturbation	Group C
	– Polybase induced lysis	Group D
Tissue	– Mechanical slicing of fresh tissue	Group E
	– Mechanical slicing of plasmolyzed tissue	Group F

[a] Group A – Wagner and Siegelman (1975); see Wagner (1983); Boller and Kende (1979); Lin and Wittenbach (1981); Waters et al. (1982); Granstedt and Huffaker (1982); Wittenbach et al. (1982); Jonsson et al. (1983); Schmitt and Sandermann (1982).
Group B – Walker-Simmons and Ryan (1977); Saunders and Conn (1978); Loffelhardt et al. (1979); Mettler and Leonard (1979); Sasse et al. (1979); Saunders (1979); Fujiwake et al. (1980); Kringstad et al. (1980); Ohlrogge et al. (1980); Lancaster and Collin (1981); Keller and Wiemken (1982); Wagner et al. (1983).
Group C – Lorz et al. (1976); Nishimura and Beevers (1978); Guy et al. (1979); Heck et al. (1981); Martinoa et al. (1981); Thom et al. (1982a); Kaiser et al. (1982); Moskowitz and Harzdina (1981).
Group D – Buser and Matile (1977); Schmidt and Poole (1980); Boudet et al. (1981); Alibert et al. (1982); Boller and Alibert (1983).
Group E – Leigh and Branton (1976); Leigh et al. (1979a); Admon et al. (1981); Salyaev et al. (1981).
Group F – Doll et al. (1979); Willenbrink and Doll (1979); Grob and Matile (1979); Helmlinger et al. (1983).

each protocol listed in Table 1 was established as the one optimal for the particular tissue being studied, the wide variation in the degree of osmotic shock required may reflect species variation in characteristics of the plasma membrane, differential occurrence of residual cell wall on protoplasts (Wagner et al. 1981) or differential damage to plasma membranes during enzymatic digestion of tissues.

Reported vacuole yields for various protocols averaged 10 to 30% (Wagner 1983) but some procedures were reported to provide yields as high as 50% (see Wagner 1983). The suggestion has been made (Wagner 1983; Thom et al. 1982a) that vacuoles recovered in low yield may represent a specific cell population not generally by characteristic of the tissue from which protoplasts were derived. Early cytologists recognized that vacuoles and tonoplast characteristics varied with cell type within tissues and heterogeneity in sap constituents of vacuoles in adjacent and even within the same cell has long been recognized (Guilliermond 1941). Sap composition heterogeneity in populations of isolated vacuoles (especially leaf vacuoles) detected by differential neutral red staining (Guilliermond 1941) was recognized early (Wagner and Siegelman 1975), and has recently been quantitated (Kurkdjian et al. 1984).

Time required for isolation of protoplasts has ranged from 30 to 45 min (Kringstad et al. 1980) to about 20 h, depending on the tissue and isolation conditions. Time required for isolation and purification of vacuoles from protoplasts has ranged from 10 min to about 2 h (see Wagner 1983). Improved methods for isolating protoplasts can undoubtedly reduce the time required and perhaps the ease of vacuole release from protoplasts isolated from most or all the tissues studied thus far. A case in point is the use of Pectolyase Y23 from *Aspergillus japonicus* (Nagata and Ishii 1979) in conjunction with *Trichoderma* cellulase for rapid (1 h) isolation of protoplasts from tobacco leaves, a tissue which formerly required 3 to 4 h digestion at elevated temperature (Saunders 1979).

Purity of vacuoles prepared from protoplasts is difficult to assess and summarize since in many studies contamination was not extensively considered or was only estimated by visual examination. Where biochemical characterization was made, the most pure preparations contained less than 3% protoplasts (as percent of structures present) and about 1 to 3% of the total activity of microsome, mitochondria, and cytosol markers present in protoplast homogenates or in the sum of fractions recovered from protoplasts (see Wagner 1983).

Methods of isolation do not appear to greatly influence the permeability of the tonoplast since neutral, anionic, and cationic microsolutes have been shown to be largely retained in vacuoles isolated from protoplasts using various procedures (Wagner 1981b; Leigh 1983; Boudet et al. 1985). However, the proton content of isolated vacuoles is reduced, particularly in those prepared in high potassium medium (Lin et al. 1977). Another factor contributing to pH equilibration during isolation is the need for maintaining vacuole isolation media near pH 8. The average pH of isolation media used for preparation of vacuoles from protoplasts and tissue computed from 24 reports is pH 7.7 ± 0.05.

The biochemical competence of vacuoles within protoplasts – ability to undergo in vivo – like cytosol–vacuole interactions – is presumed to be dependent on the viability of protoplasts from which they are obtained and to preservation of the tonoplast-cytosol interface (Boudet et al. 1984). Illumination of tissue during digestion is required for obtaining optimal rates of photosynthesis in protoplasts and illumination is said to make tissue digestion more efficient, presumably by making enzyme solution penetration of tissue more efficient (Leegood and Walker 1983).

2.2.2 Isolation of Mature Vacuoles from Protoplasts – Methods post 1981

A number of modifications of earlier methods have been described since 1981. These are also included in Table 1. Granstedt and Huffaker (1982) prepared vacuoles from barley leaf-derived protoplasts (tissue digestion for 2 h at 26 °C) which had been equilibrated with 0.6 M mannitol by mixing protoplasts with 5 vol 60 mM KPO_4, pH 8 containing 12% (w/v) Ficoll and 1 mM DTT. After gentle mixing for 5 min, the solution was overlayered with buffered (pH 7) 0.6 M mannitol containing 1 M EDTA and the system was centrifuged at 1000 *g* for 15 min. The vacuole preparation recovered from the top of the overlayer lacked chlorophyll and was reported to lack cytoplasmic contamination, but marker enzyme analysis was not made. Vacuoles were estimated to contain 58% of the nitrate and perhaps 80% of the acid phosphatase and α-mannosidase of protoplasts. More nitrate than α-mannosidase was lost during a single washing of vacuoles. The yield from this procedure was 27%, and isolation of vacuoles was made in about 20 min.

The intracellular location of protein hydrolases in wheat leaf protoplasts was studied by Waters et al. (1982) by examining both isolated vacuoles and other organelles (obtained from ruptured protoplasts) separated on sucrose gradients. For vacuole isolation, protoplasts (tissue digestion for 16 h at 20 °C) equilibrated with 0.5 M sorbitol were suspended in 0.08 M sorbitol, 0.1 M KPO_4, pH 8, 1 mM DTT, 0.1% (w/v) soluble (PVP), 0.5% (w/v) BSA – the last two inclusions were said to reduce clumping of vacuoles – and gently shaken for 5 min at 28 °C. The preparation was cooled to 4 °C and protoplasts broken by passage through a 60 μM mesh. Sucrose was added to a concentration of 0.5 M and the mixture overlayered with three solutions containing buffer (pH 7) BSA, DTT, and PVP – increasing concentrations of sorbitol and decreasing concentrations of sucrose. After centrifugation for 20 min at 200 *g* the most highly purified vacuoles accumulated at the surface of the upper zone (no sucrose). The preparation contained 3% protoplasts and lacked the cytosol marker glucose-6-phosphate dehydrogenase. No further marker analysis was made. Carboxypeptidase and hemaglobin-degrading activities were established as primarily vacuolar, while phosphatase, α-mannosidase, and glucosaminidase were equally distributed between vacuole and extra vacuole space. Aminopeptidase was equally distributed between chloroplast and extra vacuolar space and dipeptidase was primarily cytosolic. Vacuoles were prepared from protoplasts in about 25 min and yield was not reported.

Keller and Wiemken (1982) isolated vacuoles from root protoplasts of *Gentiana lutea* to study the vacuole/extravacuole distribution of gentianose, gentinobiose, fructose, glucose, and sucrose in this storage tissue. Protoplasts were isolated using standard procedures (digestion overnight in the dark) except that 0.2 mM polyethylenegylcol (PEG) and BSA were included during tissue digestion and protoplasts were washed in 1 M betaine, 5 mM NaMES pH 5.5, 0.1 mM PEG 4000, and 1 mM $CaCl_2$. Addition of PEG and BSA were said to ensure viability and stability of protoplasts. The use of betaine as osmoticum prevented interference from osmoticum during sugar analysis. Protoplasts were lysed by suspension in buffered (pH 8.2) 0.8 M betaine (containing PEG and BSA as above, and 10 mM EDTA) and gentle swirling for 2 to 3 min at 40 °C. Vacuoles were purified by settling (30 min) through zones of 0.9 M and 1.0 M betaine, each containing PEG, BSA, and EDTA. Marker analysis indicated that vacuoles were contaminated with mitochondrial membrane (at least 13%) but not soluble mitochondrial or cytosol markers. Other membrane markers were not examined. The advisability of using even low concentrations of PEG is questioned, since PEG is an effective membrane adhesion agent. This factor may account for the unusually high contamination with mitochondrial membrane. Seventy to 80% of the acid hydrolases in the protoplast lysate were recovered in the vacuole sediment indicating a high yield (75%). Preparation of vacuoles from protoplasts required 40 min. The vacuole/extravacuole distribution of sugars (calculated by assuming that 100% of the acid phosphatase, α-mannosidose and β-N-acetylglucosaminidase are vacuolar in this tissue) were 100 to 0, 100 to 0, 84 to 16, 76 to 24, and 49 to 51 for gentianose, gentinobiose, fructose, glucose, and sucrose, respectively.

The role of vacuoles in fructan metabolism in barley leaves was studied by W. Wagner et al. (1983). In this system polyfructosylsucrose was made to accumulate in large amounts by subjecting plants to cold stress or excised leaves to continuous illumination. Protoplasts were isolated (digestion for 2 h at 25 °C) using standard methods (used Pectolyase Y23) were purified and were equilibrated with 0.7 M sorbitol. Resuspension in buffered (pH 8) 0.4 M sorbitol containing 5 mM EDTA, 1 mg ml^{-1} BSA, and gentle stirring resulted in lysis of protoplasts. Released vacuoles were purified by flotation (1000 *g* for 10 min) through Percoll – 0.6 M betaine steps. Marker analysis indicated that vacuole preparations contained about 1.5 to 0.1% of the mitochondrial inner membrane, soluble mitochondrial and cytosol markers present in protoplast lysates. About 80 to 90% of protoplast acid phosphatase, glucosaminidase, and invertase and 112% of protoplast α-mannosidase were found to be vacuolar. Yield was about 10% and the time of preparations from protoplasts about 15 min. As found in an earlier study with *Gentiana* (see above), essentially all the fructose, glucose, and fructan of barley protoplasts were found to be vacuolar, while sucrose was found to be equally divided between extravacuole and vacuole space. Interestingly, sucrose-fructosyltransferase which catalyzes fructan synthesis was vacuolar. The authors reported that fructan-fructan-fructosyltransferase is also vacuolar in *Helianthus tuberosus* (Frehner et al. 1984). These are the first reports of transferases occurring in mature vacuoles. Curiously, the pH optimum of the sucrose-fructosyltransferase is reported to be 5.7. It is difficult to rationalize the functioning of acid hydrolases and neutral transferase in the same compartment. It would be interesting to know if vacuolar sap from this system can digest the transferase in vitro. Perhaps transferases are located on the cytoplasmic face of the tonoplast or this tissue has a relatively neutral vacuolar sap. It is an interesting system with which to probe coordination of fructan metabolism, microsolute transport, vacuolar pH regulation, and lytic functions, since fructan accumulation can be easily manipulated.

Jonsson et al. (1983) prepared vacuoles from petals of *Petunia hybrida* to determine the vacuole/extravacuole location of anthocyanin methyltransferase, an activity near the end of the anthocyanin biosynthetic sequence. Protoplasts were prepared using a standard digestion mixture (incubation overnight at 4 °C then 3 h within shaking at 30 °C) and equilibrated with 0.6 M mannitol. To isolate vacuoles, protoplasts were diluted with 0.1 M K_2HPO_4/HCl, pH 8, 1 mM DTT and 8 mg ml^{-1} BSA. After gentle mixing, vacuoles were purified by flotation through Ficoll-mannitol steps. Contamination with protoplasts was 10–15% and aggregates of membraneous material adhered to vacuoles. Contamination with glucose-6-phosphate dehydrogenase, usually a reliable cytosol marker, varied from 9 to 53%. Despite the impurity and the limited and confusing results of marker analysis, the authors found little indication that methyltransferase was associated with the vacuole in this system.

It may be concluded that recent modifications of earlier methods for isolating vacuoles using osmotic lysis of protoplasts represent relatively minor advances in methodology. It is clear that there is still no single protocol for preparing vacuoles with good purity in high yields from various tissues. One notable, possible exception is the method of Keller and Wiemken (1982) which introduced the use of betaine as an osmoticum where sugar analysis is to be made and which used PEG to stabilize protoplasts and vacuoles. The 75% yield they observed is the highest yet reported, however, the contamination of their preparations with membraneous components may be excessive for many purposes. Yet the method deserves further study.

Several modifications of vacuole isolation procedures which utilize isotonic conditions to lyse protoplasts have been reported in recent years. Thom et al. (1982a) used a modification of the method of Lorz et al. (1976) to lyse protoplasts (isolated in 2 h at 30 °C) prepared from sugarcane suspension cells. Osmoticum in protoplast and vacuole isolation media was 0.4 M mannitol and vacuoles were isolated in 20 to 60 min. Marker enzyme assay indicated that contamination with cytosol, mitochondrial and microsomal markers were 11 to 22, 11 and 11%, respectively. Vacuoles recovered from ^{3}H-Con A-labeled protoplasts contained about 19% of the probe which was specifically bound to protoplasts. Fluorescamine-labeled Con A labeling suggested that 20% of the vacuoles were contaminated with the probe, which was associated with a small portion of the tonoplast surface. The authors concluded that plasma membrane did not envelope the vacuoles as suggested by Lorz et al. (1976), but was present as nonuniform surface contamination. Extensive surface contamination was evident from light micrographs and heterogeneity in vacuole sap composition, as evidenced by variation in neutral red staining. The authors suggest that they selected their isolation procedure – perhaps at the expense of purity – in order to preserve the integrity of the cytoplasmic face of the tonoplast so that meaningful transport studies could be made. Acid phosphatase, peroxidase, carboxypeptidase, protease, RNAse, acid invertase, and protein were said to be vacuolar to the extent of 96, 71, 49, 48, 34, 9, and 16%, respectively. Lysis of vacuoles and sedimentation of membrane suggested that carboxypeptidase and protease were associated with tonoplast to the extent of 87 and 52%, respectively. However, given the contamination in these preparations, the fact that tonoplast was not examined on gradients, and the lack of evidence for membrane-bound protease in other systems (Boller and Kende 1979; Wagner, unpublished), localization data are suspect. Isolated vacuoles were capable of sucrose uptake (Thom et al. 1982b), assayed by rapidly (1-min) sedimenting vacuoles from the uptake medium containing labeled sugars through a silicone oil layer. The uptake of 3-0-methyl glucose in vacuoles was twofold higher (V_{max}) than that in protoplasts, further suggesting the lack of the "vacuoplast" structure suggested by Lorz et al. (1976). In a study of the electrical properties of isolated surgarcane vacuoles (Komor et al. 1982b), it was concluded that the tonoplast energization mechanism or sucrose transport mechanism was disrupted, possibly due to loss of a critical cytosol-tonoplast interface.

Kaiser et al. (1982) prepared protoplasts and vacuoles from barley leaves and studied vacuolar accumulation of photosynthetic products using a rapid vacuole isolation procedure. Protoplasts were prepared using standard methods (incubation 2 h, 28 °C in the dark) with BSA and 0.05% PVP 10,000 present during tissue digestion, and were purified on a sorbitol-Percoll step gradient. Protoplasts in buffered (pH 7.8) 0.5 M sorbitol, 10% Percoll were incubated with NaH $^{14}CO_3$ and neutral red for 10 min in the dark to allow neutral red uptake. They were then stirred in the light at 20 °C to monitor photosynthesis. Aliquots were withdrawn at intervals and forced (15 s) through a syringe needle under a silicone oil/sorbitol layer. This rapidly lysed about 40% of the protoplasts and most vacuoles were said to remain intact. Samples were centrifuged for 15 s at 9700 *g*, resulting in flotation of intact vacuoles. The yield was 10 to 15% and contamination by soluble cytosol, chloroplast, and mitochondrial markers was $\leqq$5% and α-mannosidase was determined to be 98% vacuolar. However, markers for membrane contamination were not assessed. The authors concluded that soluble photosynthetic products were rapidly sequestered in the vacuole.

The adivisability of monitoring metabolic processes in protoplasts equilibrated with acidophylic substances such as neutral red is questioned (see deDuve 1984). So-called viability marker dyes are known to be less toxic than others, but are nevertheless known to be toxic (Guilliermond 1941). The method of rapid vacuole separation is nevertheless interesting, even though it may select for a specific population of vacuoles (Kaiser et al. 1982).

In a similar study using different methods of protoplast and vacuole isolation and the leaves of *Melilotus alba*, Boller and Alibert (1983) found that the bulk of photosynthetic products occur in the cytosol and are only slowly transferred to the vacuole. Protoplasts were prepared in the light according to the method of Boudet et al. (1981), were purified and incubated with NaH $^{14}CO_3$. Neutral red was not used. After incubation, protoplasts were immediately lysed and purified using the polybase-induced lysis procedure of Boudet et al. (1981). Yield was 40% and while vacuoles were said to be floated in <1 min, purification gradients were centrifuged for 30 min to completely remove chloroplasts. However, no analysis for contamination was made. Final preparations contained 4 to 9% protoplasts.

Alibert et al. (1982) utilized essentially the procedure of Boudet et al. (1981) to prepare protoplasts (incubation for 4 h at 36 °C) and vacuoles from *Acer pseudoplantanus* culture cells. Contamination with soluble mitochondrial and microsomal markers was <5% and that by RNA and DNA was <6%. Membrane marker assays were not utilized. A protoplast to vacuole distribution ratio for neutral red, α-mannosidase, ^{14}C benzylamine and ^{14}C nicotine led to the conclusion that protoplasts of young cells and old cells contained two vacuoles and one vacuole each respectively. Determinations were made by counting structures in a counting chamber, and light micrographs of a vacuole preparation indicated substantial structure-size variation. Therefore, the meaning of these data is questioned. The vacuole/extravacuole distribution of amino acids, malate, and protein were said to vary with cell age.

2.2.3 Isolation of Mature Vacuoles Directly from Tissue – Methods pre 1981

The basic method of Leigh and Branton (1976) for efficient isolation of vacuoles by slicing fresh beetroot tissue continues to be the best method for direct vacuole isolation. While yields are low (about 0.2%) vacuoles are obtained in about 1 h without plasmolysis of tissue; vacuoles are, however, plasmolyzed after release from tissue. One perhaps noteworthy modification of this basic procedure has been made recently (Salyaev et al. 1981), but it has yet to be shown to be useful for large-scale isolation of vacuoles from diverse tissues.

2.2.4 Isolation of Mature Vacuoles Directly from Tissue – Methods post 1981

Salyaev et al. (1981) suggest that their modification of the method of Leigh and Branton (1976) allows isolation of vacuoles from various tissues of various plants. In this method fresh tissue is sliced (using an apparatus similar to that of Leigh and Branton 1976) into KCl solutions of high concentration. Yields of vacuoles from beetroot tissue using this method were about 1%. Application of the method to tissues other than beetroot was not described in detail. Therefore, assessment of the uniqueness and utility of this method must await further study.

Bennet et al. (1984) isolated beetroot vacuoles using a minor modification of the procedure of Leigh and Branton (1976). In this case fresh tissue was sliced by hand into cold 1 M sorbitol, 5 mM EDTA, 4 mM DTT, 0.5%, PVP 40, 50 mM Tris/Mes, pH 8. The brei was filtered and a 13,000 *g* supernatant fractionated by flotation on sodium diatrizoate gradients containing 1.2 M sorbitol, 1 mM EDTA, 2 mM DTT, and 25 mM Tris/Mes, pH 7. Vacuoles were recovered from a 0 to 10% interface, diluted tenfold with 5 mM buffer, pH 6.5, 2 mM DTT and membrane vesicles recovered by sedimentation at 80,000 *g* for 30 min. Marker analysis showed the lack of mitochondrial and plasma mebrane contamination. Mg ATPase and proton transport in recovered vesicles (density 1.09 g cm^{-3}) was vanadate-insensitive and nitrate-sensitive.

Kinetics of vacuolar accumulation of indol-3-methyl glucosinolates was studied after isolating vacuoles from horseradish root tissue (Helmlinger et al. 1983). The method for isolating vacuoles directly from plasmolyzed tissue was essentially that of Grob and Matile (1979). Vacuoles were purified on Percoll-NaCl gradients. The most purified vacuole preparations were said to lack cytochrome c reductase and catalase, but were contaminated with soluble cytosol and mitochondrial markers. Other membrane markers were not examined. Acid phosphatase was used as a vacuolar marker and the correlation method (Leigh et al. 1979b) was used to show that indolglucosinolates and L-tryptophane were vacuolar in this tissue.

2.2.5 Preparation of Lutoids from Hevea Latex

Small vesicles, called lutoids, were first isolated from latex of the rubber tree *Hevea brasiliensis* by Pujarniscle (1968). Latex harvested by tapping contains lutoids (15% of latex volume), 35% rubber (cis-poly isoprene) and a small number of nuclei, mitochondria and other organelles which remain in the adult latex tubes (d'Auzac et al. 1982). Large quantities of lutoids are pelleted from freshly collected, diluted latex at 39,000 *g*, the pellet is resuspended, washed, lyophylized, and stored at −30 °C (Marin and Trouslot 1975). After suspension and gentle centrifugation, vesicles are sedimented and used. Suspending medium and latex diluant consists of 0.3 M mannitol, 10 mM KCl, 10 mM KCl, 10 mM $MgCl_2$, 100 mM Tris/HCl, pH 7.3.

Purity of the material prepared in this manner has apparently not been established. The first fractions of latex obtained after tapping are discarded because they are enriched in bacteria (Marin and Trouslout 1975). In certain lutoid preparations bacterial RNA was found to contribute 7.0% of the total RNA present (Marin and Trouslot 1975). Therefore, simple resuspension, particularly after lyophylization, may not remove residual bacteria, ribosomes, and other organelles which may be present in latex. More than one population of lutoids (at least in terms of sap composition) may be present in preparations made as described. After uptake of neutral red, two distinct populations of lutoids were separated on sucrose gradients (Marin and Trouslot 1975).

Despite questions about purity, extensive work has led to a good understanding of the composition, structure and the role of lutoids in ion accumulation (d'Auzac et al. 1982), much of it through the efforts of B. Marin, J. d'Auzac and colleagues. Lutoids contain an array of acid hydrolases which show structural latency. It would be interesting to know if lutoids are capable of immobilizing and digesting latex bacteria. The possible role of lutoids in autophagy and heterophagy has not been unequivocally established.

Lutoid membrane bears a vanadate-insensitive, nitrate-sensitive MgATPase and proton pump (Marin 1983). Its lipid composition (DuPont et al. 1976) is substantially different from that of beetroot vacuoles (Marty and Branton 1980) in that it has high phosphatidic acid and saturated fatty acid content. The extent to which *Hevea* lutoids can serve as a model for transport and other activities of the central vacuole of higher plants remains to be established. However, their avidity for citrate accumulation (Marin et al. 1982) makes them an ideal system for studying the biochemical mechanism of secondary transport of organic acids and other microsolutes.

2.2.6 Comments on Methods for Isolating Vacuoles from Higher Plants

The discussion in Section 2 indicates that while recent modifications of basic isolation procedures have not yielded more highly purified vacuoles, they have in certain cases reduced the time required for both protoplast and vacuole isolation. Using the most rapid methods for protoplast and vacuole isolation it is now possible to prepare vacuoles in high yield from tissue via protoplasts in about the same time required for low yield preparation from fresh tissue. Rapid isolation of vacuoles from protoplasts during metabolic studies enhances the value of such efforts and allows kinetic studies (Kringstad et al. 1980; Kaiser et al. 1982; Boller and Alibert 1983; see Wagner 1983).

The use of α-mannosidase as a vacuole sap marker has become widespread and while the general way in which it is applied is prone to perhaps a 15 to 20% error (Table 2) it is, within this margin of error, generally valid once verified for the particular tissue being examined. Verification is generally made by counting structures (the probable source of the error) and by comparing protoplast and vacuole α-mannosidase content on a per-structure basis. Structure size variation

Table 2. Experimentally determined vacuole/extravacuole distribution of the most-studied plant acid hydrolases

Source of vacuoles	Relative activity in vacuole (percent)[a]			Reference
	α-Mannosidase	Acid phosphatase	Acid protease	
Tomato leaf proto.	–	30	–	Walker-Simmons and Ryan (1977)
Castorbean endosperm proto.	–	29	77	Nishimura and Beevers (1978)
Tobacco suspension cell proto.	120–71	86	105	Boller and Kende (1979); Boller (1982)
Pineapple leaf proto.[b]	130	97	98–120	Boller and Kende (1979); Boller (1982)
Tulip petal proto.[b]	85	94	–	Boller and Kende (1979); Boller (1982)
Beet root tissue[c]	–	100	–	Leigh et al. (1979b)
Tobacco suspension proto.	–	100	–	Mettler and Leonard (1979)
Carrot suspension proto.[c]	–	55	–	Sasse et al. (1979)
Tobacco leaf proto.	–	93	–	Saunders (1979)
Horseradish root tissue	–	90–106	–	Grob and Matile (1979)
Bean cotyledon proto.	90	–	–	Van der Wilden and Chrispeels (1983)
Barley leaf proto.	109	59– 72	82– 87	Heck et al. (1981)
Wheat and corn leaf proto.	–	–	106–107	Lin and Wittenbach (1981)
Wheat leaf proto.	–	–	112–103	Wittenbach et al. (1982)
Barley leaf proto.[b]	100	–	85	Martinoa et al. (1981)
Wheat leaf proto[d]	–	–	87	Wagner et al. (1981)
Hippeastrum petal proto.[d]	–	–	15– 29	Wagnet et al. (1981)
Melilotus leaf proto.	116	50	–	Boudet et al. (1981)
Sugar cane suspension proto.	–	96	48– 49	Thom et al. (1982a)
Barley leaf proto.	98	–	–	Kaiser et al. (1982)
Gentiana root proto.	97	123	–	Keller and Wiemken (1982)
Barley leaf proto.	84–89	82	–	Grandstedt and Huffaker (1982)
Wheat leaf proto.	47	52	100–102	Waters et al. (1982)
Barley leaf proto.	112	77	–	Wagner et al. (1983)
Barley leaf proto.	85	–	95	Thayer and Huffaker (1984)

[a] Unless otherwise noted, quantitation was made by using the error prone method of counting numbers of protoplasts and vacuoles (estimated error ±15 to 20%)
[b] Assumed α-mannosidase was 100% vacuolar
[c] Quantitation by correlation method using vacuolar pigment
[d] Quantitation by total sap volume determination

of protoplasts and vacuoles is usually large (Wagner 1981 b; Wagner 1983) and is still generally ignored as a possible source of error.

It is unfortunate that only soluble marker enzymes have been assessed in many recent studies. This seems foolhardy, since membrane contaminants are generally encountered. Various basic methods such as polybase-induced, versus shear-induced, versus osmotic lysis have not been compared directly to assess the advantages and disadvantages of each in studies of transport phenomena. Studies of microsolute transport in vacuoles, where purity is perhaps less important, would benefit from such a comparison. It has been suggested that polybase-induced lysis may be more damaging to the cytosol–tonoplast interface than other methods (Thom et al. 1982b), however, there is no direct evidence to support this suggestion. Whether exposure of vacuoles to substances such as DEAE dextran (see Gimmler and Lotter 1982) high ionic strength (KPO_4, KCl, NaCl) chelators, PVP, BSA, PEG, KI, etc. damages or preserves the cytosol–tonoplast interface is not known. If actin filaments are shown to be associated with tonoplast (as expected), these may serve as a marker for the integrity of this presumably fragile interface.

Regarding the stability of isolated vacuoles, it would appear that a medium which is about 525 to 730 milliosmolal (the equivalent of 0.5 to 0.7 M mannitol), pH 7.5 to 8 and 4 °C is optimal for stabilizing vacuoles isolated from most tissues. No single vacuole-stabilizing agent has yet been identified, but low concentrations of PEG (Keller and Wiemken 1982) may prove useful for this purpose. However, it would not be surprising to find that PEG adversely affects tonoplast transport functions.

2.2.7 Isolation of Proton-Pumping Vesicles

In 1980 Sze introduced a method for isolating, from plant tissues, vesicles capable of MgATP-stimulated proton transport (Sze 1980). This basic method, with minor modifications, has been used to recover sealed vesicles from callus tissue (Sze 1982), corn and oat root tips (Dupont et al. 1982; O'Neill et al. 1983; Stout and Cleland 1982; Churchill et al. 1983) beet root storage tissue (Briskin and Poole 1983; Poole et al. 1984a; Bennet et al. 1984) and corn coleoptiles (Hager et al. 1980, Mandala et al. 1982). All of these may represent meristematic or storage tissues. Basically, fresh tissue is ground in a chilled mortor in low concentration (0.25 M) sucrose, mannitol or sorbitol medium (usually 1 to 3 ml g^{-1} tissue) containing 25–70 mM buffer, pH 7.2 to 8, 2 to 3 mM EDTA or EGTA and 1 to 4 mM DTT. The homogenate is centrifuged to recover a 10,000 *g* soluble, 60,000 or 80,000 *g* pellet which is subsequently purified by sedimentation of vesicles into dextran or sucrose step gradients containing 0.25 M sucrose. Two fractions of vesicles have been prepared in this manner and after additional separation of vesicles in linear dextran or sucrose gradients (Dupont et al. 1982; Churchill et al. 1983; Poole et al. 1984; Bennet et al. 1984). One fraction (density about 1.1 g cm^{-3}) is enriched in vanadate insensitive, nitrate sensitive Mg ATPase and proton pump activity and the other (density about 1.17 g cm^{-3}) in vanadate-sensitive nitrate-insensitive Mg ATPase and proton pump activity. It was concluded that these fractions were derived from tonoplast and plasma membrane vesicles, respectively. In certain cases 0.2 to 1 mM PMSF was added during homogenization (Stout and Cleland 1982; Churchill et al. 1983), and for beetroots, 0.5% PVP-40 was included (Briskin and Poole 1983) and 25 mM potassium metabisulfite was substituted for mercaptoethanol (Poole et al. 1984).

In studies with corn roots, Dupont et al. (1982) observed that vesicles obtained from the meristematic zone of the root had maximal ATPase, proton pump and NADH Cyt c reductase activity and specific activity. The specific activities of proton transport and NADH Cyt c reductase were maximal in the 10,000 *g* soluble, 80,000 *g* pellet fraction of the homogenate, and in vesicles recovered from the dextran step gradient. Nucleoside diphosphatase, a marker for Golgi, was not purified in these fractions. As previously discussed (Section 2.1) meristematic vacuoles having density about 1.1 g cm^{-3} contained NADH Cyt c reductase and vacuoles were more efficiently isolated from the tip region of corn roots (Matile 1968). Mature vacuoles also bear this activity (Marty et al. 1980; Wagner et al. 1983). This property may reflect the common origin of these membranes.

Could proton pumping vesicles isolated from corn root tips (and other embryonic tissues) be the vesicles isolated by Matile in 1966? To date, examination of proton pumping vesicle for structurally linked acid hydrolase content has not been reported. The method of Matile is very similar to those used to prepare proton pumping vesicles, differing only in the fact that 0.5 M sucrose was used in the former and 0.25 M sucrose in the latter. However, Pitt and Galpin (1973) used Ficoll containing about 0.25 M sucrose to isolate hydrolase containing particles which had a density of about 1.1 g cm^{-3}. It is therefore essential to examine proton pumping vesicles for the basic properties (structually latent hydrolase and potassium stimulated pyrophosphatase) of isolated meristematic and mature vacuoles to clarify the origin of the vesicles.

The density of light proton pumping vesicles (~1.08 to 1.15 g cm^{-3}), the insensitivity of proton pump and ATPase to vanadate and their sensitivity to nitrate suggest that they are tonoplast derived. Careful study of these characteristics and study of markers during density gradient centrifugation indicate that the ATPase and proton pump activities are not due to Golgi, plasma membrane, or mitochondrial contamination. However, their identification is only based on inhibitor characteristics, the physiological functions of which are not understood. Also, vanadate-sensitive, nitrate-insensitive ATPase was found in association with membrane recovered from isolated vacuoles of tulip petals (Wagner and Mulready 1983; Wagner unpublished), vanadate sensitivity being variable with the condition of the tissue (unpublished). These preparations lack glucan synthase II, a plasma membrane marker. The calcium status of membranes and ATPase are known to influence vanadate binding (Medda and Hasselbach 1983), and nitrate sensitivity may be associated with the nitrogen a status of the tissue. A tissue such as petals which is relatively deficient in primary metabolism and low in nitrate (unpublished), may lack nitrate sensitivity, if this has a regulatory function. Therefore caution should be exercised in interpretations based solely on inhibitor effects since sensitivities may vary from tissue to tissue and with the concentration of inhibitors and ions.

Bennet et al. (1984) compared beetroot vesicle preparations and membrane vesicles obtained from isolated beetroot vacuoles and found that their ATPase and proton pump properties were similar with regard to vanadate and nitrate. Earlier, Leigh and Walker (see Leigh 1983) showed that vanadate-insensitive, nitrate-inhibited ATPase is associated with vacuoles isolated from this tissue. As discussed earlier (Wagner 1983) when vacuoles are obtained in very low yield ($<1\%$) it is possible that a select population are obtained, a population which is not representative of the tissue as a whole. Whole tissue characteristics may not necessarily be represented by these vacuoles or vesicles. Recent studies of Poole et al. (1984 b) indicate that less than 10% of sealed vesicles recovered from beetroot homogenates respond to ATP.

Another property of tonoplast recovered from isolated vacuoles of tulip petals (Wagner 1983) and beetroots (Walker and Leigh 1981) is the presence of potassium-stimulated alkaline pyrophosphatase (PPiase). While investigators who have studied homogenate-derived proton-pumping vesicles have examined nucleotide specificity in detail (primarily in concern for the presence of nonspecific phosphatase), they have generally not examined PPiase. This activity should also be monitored if the origin(s) of homogenate-derived proton pumping vesicles is to be established.

In summary, the basic method of Sze has been extensively used in the past few years to isolate and study sealed vesicles capable of proton transport. While the presence in these preparations of chloroplast and mitochondrial outer membranes, secretory vesicle membranes and smooth ER membranes has not been excluded, the evidence is mounting that these vesicles may be, at least in part, tonoplast-derived. These preparations should now be examined for the presence of structurally latent acid hydrolases, particularly acid protease and, in the red beetroot system, acid invertase (Leigh et al. 1979 b). The possible presence of potassium stimulated, alkaline pyrophosphatase and NADH Cyt c reductase should also be examined. It would be interesting to know if meristematic vacuoles isolated from corn root tips exactly as described by Matile (1968) are capable of vanadate-insensitive, nitrate-sensitive, MgATP-dependent, proton transport.

2.2.8 Preparation of Vacuoles from Yeast and Neurospora

Vacuoles have been isolated from wall-less yeast cells (prepared in 20 to 120 min) by osmotic shock, mechanical rupture under isotonic conditions, and polybase-induced lysis (Matile 1975; Schwencke 1977; Matile 1978). Wall-less cells are produced using microbial and snail gut enzymes and pretreatment of cells with DTT, EDTA, and Tris is said to improve wall digestion. Lysis conditions at or near isotonic (0.6 to 1 M sorbitol) appear to produce the best vacuoles in terms of retention of microsolutes and their ability to transport metabolites (Schwencke 1977). The polybase-induced lysis procedure developed for yeast (Durr et al. 1975) was adapted for isolating higher plant vacuoles from protoplasts (see Sect. 2.2.2 and 2.2.3). Yeast vacuoles are generally purified by flotation in Ficoll-sorbitol-sucrose gradients (Schwencke 1977). Preparations are devoid of cytosol and mitochondrial soluble markers and cytochrome oxidase but are enriched in NADH Cyt c reductase. In a recent modification of the method of Matile and Wiemken (1967), Ohsumi and Anraku (1981) lysed yeast spheroplasts which were equilibrated in

1 M sorbitol by suspension in 0.14 M sorbitol, 10 mM Mes/Tris, pH 6.9, 0.1 mM $MgCl_2$, 12% Ficoll. After gentle homogenization in a loose-fitting Dounce homogenizer, vacuoles were recovered and purified by flotation in a Ficoll gradient. Yeast vacuoles prepared in this manner lacked soluble cytosol and mitochondrial markers and chitin synthetase (a plasma membrane marker). They were enriched in Mg ATPase, α-mannosidase, and acid phosphatase, the last two showing structural latency (Kakinuma et al. 1981). Membrane vesicles derived from vacuoles contained vanadate-insensitive ATPase and were capable of proton-pumping and energy-dependent basic amino acid accumulation.

Cramer et al. (1980), Vaughn and Davis (1981) and Bowman and Bowman (1982) isolated vacuoles from wall-less cells of *Neurospora crassa* – cells produced by digestion for 30 min at 30 °C with β-glucuronidase. Cells equilibrated with 1 M sorbitol were lysed in 1 M sorbitol, 10 mM TES/NaOH, pH 7.5, 1 mM EDTA using a Teflon glass homogenizer. The homogenate was filtered, and a 600 *g* soluble, 15,000 *g* insoluble fraction was recovered. Vacuoles were purified by sedimentation through a 30 to 60% sucrose gradient and the pellet was fractionated in a sorbitol-metrizamide gradient. Vacuoles equilibrated at 1.3 g cm^{-3} and were enriched in arginine, polyphosphate, α-mannosidase and acid phosphatase. The last two appear to be membrane-bound. α-mannosidase is membrane-bound in yeast vacuoles but not in higher plant vacuoles (Wagner 1981 b, 1983; Boller 1982). Membrane also had vanadate-insensitive, nitrate-sensitive ATPase which differed from ATPase of mitochondria and plasma membrane of *Neurospora*. Interestingly, this ATPase was solubilized by repeated washing with dilute medium containing EDTA and its DCCD-binding protein has characteristics similar (but not identical) to that of mitochondrial ATPase. ATPase of tulip vacuoles (Wagner and Mulready 1983) and *Hevea* lutoids (d'Auzac 1977) require detergent for solubilization. Comparison of the molecular weights of DCCD-binding proteins (Bowman 1984) of ATPases derived from intact vacuoles, proton-pumping vesicles, meristematic vacuoles and *Hevea* lutoids should be very useful for establishing similarities and dissimilarities between these enzymes.

3 Isolation of Tonoplast and Tonoplast Markers

Membrane recovered from isolated vacuoles of various sources has a density of about 1.1 g cm^{-3} (see Wagner 1983). In addition, a lighter density component (≦1.05 g cm^{-3}) has been observed in vacuole preparations from *Hippeastrum*, tobacco cultured cells and tulip (see Wagner 1983). Generally, preparation of tonoplast from vacuoles involves osmotic lysis, in some cases followed by freezing and thawing, homogenization or sonication. Sonication of *Hippeastrum* vacuoles results in a more homogeneous population of vesicles but one which is increased in acid phosphatase (Wagner 1981 a). Tonoplast from beetroot (Marty et al. 1980), tulip petals (Wagner et al. 1983) and *Hevea* lutoids (Moreau et al. 1975) retain antimycin A-insensitive NADH Cyt c reductase. While it is not established that this activity is not due to smooth ER contamination, its presence is not surprising considering tonoplast ontogeny. Antimycin A-insensitive NADH Cyt c reductase is enriched in ER membranes and cannot be considered a tonoplast marker. Enzymes formerly considered to be ER markers are shown to also occur in cis-Golgi as a result of ER-Golgi membrane flow (Howell et al. 1978). A nucleotide specific Mg ATPase having a pH optimum of about pH 7.5 to 8 has been observed in association with membrane recovered from vacuoles of beetroot (see Leigh 1983), *Hippeastrum* and tulip (Wagner 1981 a; Wagner and Mulready 1983) and *Hevea* lutoids (see d'Auzac et al. 1982). Membrane recovered from tulip vacuoles lacks nonspecific phosphatase, while that from beetroot and lutoids bears activity which can be selectively inhibited with 100 μM ammonium molybdate (see Leigh 1983); d'Auzac et al. 1982). Briskin and Poole (1983) and Bennet et al. (1984) used 25 mM KI to partially remove phosphatase from membranes of beetroot.

Another membrane bound activity common to beetroot and tulip petal vacuoles is potassium-stimulated, alkaline pyrophosphatase (PPiase). The function of this activity is unknown. Moyle et al. (1972) suggested that *Rhodospirillum rubrum* PPiase may be a proton pump. We found no evidence for PPi energization of proton transport in intact tulip vacuoles (Wagner and Lin 1982; Wagner unpublished) and Bennet et al. (1984) found PPiase in vesicles recovered from red beet roots but PPi supported only low levels of proton transport. It is possible that PPiase is associated with a pyrophosphorylase reaction which functions near the vacuole. Pyrophosphatase is unlikely to have specific membrane distribution and is therefore unlikely to be useful as a specific membrane marker (Wagner and Mulready 1983).

Buser-Suter et al. (1982) studied malate transport in isolated *Bryophyllum* leaf vacuoles and proposed the existence of a permease for catalyzing the exchange diffusion of malate across the vacuole membrane. Such a mechanism can be considered to be a uniport (Rosen and Kashket 1978), dependent on an internal proton gradient for sequestering internalized acid in the protonated-trapped form (Luttge and Ball 1979). Export of malate may occur by a specific proton symport. In yeast vacuoles, arginine accumulation may be catalyzed by both a uniport associated with polyphosphate sequestration (Matile 1978) and/or an arginine-proton antiport coupled to primary proton transport catalyzed by a vanadate insensitive, Mg ATP-dependent proton pump (Ohsumi and Anraku 1981). Recent results of Cramer et al. (1980) show that under certain growth conditions polyphosphate levels of *Neurospora* may be reduced by 90% without effect on arginine accumulation. This suggests that large amounts of polyphosphate are not obligatory to arginine accumulation in this organism which, like yeast, accumulates arginine and polyphosphate. Until porter proteins are identified, transport per se cannot serve as a marker for tonoplast or vacuoles.

Recently, Thom et al. (1982a) reported that carboxypeptidase and acid protease were bound to membrane sedimented from lysed sugarcane suspension cell vacuoles. However, preparations were impure, membrane was not purified by isopycnic centrifugation, and only 50% of these activities were judged to be vacuolar, so it is difficult to assess the significance of these findings. It is interesting to note that in yeast, carbohydrate-free carboxypeptidase Y is as efficiently transferred from Golgi to the vacuole as the normally glycosylated form (Schwaiger et al. 1982). This process is unlike that for transfer of hydrolases from Golgi to animal lysosomes, which occurs only ofter glycosylation in the Golgi. Perhaps the difference is due to a membrane-bound nature of carboxypeptidase in yeast and higher plants as compared with the soluble nature of lysosomal enzymes. Boller (1982) concluded that higher plant vacuole hydrolases are not membrane-bound, but carboxypeptidase has not been extensively studied in this regard.

The enzyme α-mannosidase has been utilized as a general vacuole sap marker, however, considerable variation has been observed in the vacuole/extravacuole distribution of this activity from plant to plant and even in the barley leaf system where it was studied in several different laboratories (Table 2). While it is clear that this enzyme is primarily vacuolar in most systems, the researcher is obliged to demonstrate its distribution using a reliable counting or correlative method (Wagner 1981 b; Leigh 1983) if it is to be used as a standard in vacuole/extravacuole localization studies. α-Mannosidase has been localized in plant Golgi and cell wall (Van der Wilden and Chrispeels 1983) and in animals a cytosolic enzyme has been demonstrated (Opheim and Touster 1978). The evidence for a specific vanadate-insensitive, nitrate-sensitive Mg ATPase and proton pump in tonoplast has been discussed in Section 2.2.7. This author feels that evidence for its specific tonoplast location is inadequate and consideration of this activity as a tonoplast marker must await further study.

4 Comments on Physiological Functions

As discussed in Section 2.1 and elsewhere, meristematic vacuoles and especially mature plant vacuoles isolated from a number of plant tissues contain various acid hydrolases. Yet clear evidence for autophagy and heterophagy in mature plant vacuoles is not easily obtained. A number of recent studies have focused on the question of the mechanism of turnover of the abundant chloroplast protein ribulose bisphophate carboxylase. This protein constitutes 50% or more of the soluble protein in leaves of many plants and is subject to early turnover under stress and early senescence conditions (Thomas and Stoddart 1980). Up to 80% of the chlorophyll and protein in chloroplasts may be degraded, yet the chloroplasts can be isolated with outer membranes intact (Wittenbach et al. 1982). Ultrastructural evidence indicates that dissolution of inner membranes can occur while outer membranes are intact and chloroplasts are extravacuolar. Yet acid endo-proteases are largely vacuolar and only neutral amino peptidase and dipeptidases are known to occur extravacuolarly (Lin and Wittenbach 1981; Waters et al. 1982). However, as noted in Section 3 and Table 2, variation has been observed in acid protease distribution and methods used to determine these distributions were generally the error prone counting methods. Even a few percent of the total cellular protease (or other hydrolase) in a small sap-volume compartment such as the chloroplast could represent a potentially relevant hydrolytic potential.

Recent results may provide an answer to the dilemma of intrachloroplast turnover. Dalling et al. (1983) isolated a unique chloroplastic endoproteas(es) which is capable of digesting carboxylase. It is possible that the endosymbiont chloroplast has retained its own mechanism for autolysis and is independent of extrachloroplastic mechanisms. Chloroplasts are the first organelles to undergo senescence followed by the endomembrane system and finally mitochondria and nuclei (Thomas and Stoddard 1980).

The suggested development of peroxisomes from glyoxysomes in fatty cotyledons which mature into cotyledonary leaves may represent extravacuolar, independent autolysis in an organelle which is a component of the endomembrane system (see Tolbert 1980). Here glyoxysomal enzymes are turned over (catalase turnover is said to be more rapid than glyoxysome disappearance) while peroxisomal enzymes are newly synthesized. An alternative explanation for this phenomenon is that peroxisomes and glyoxysomes are distinct but inseparable organelles, the latter being formed de novo during greening (see Tolbert 1980). Another possible

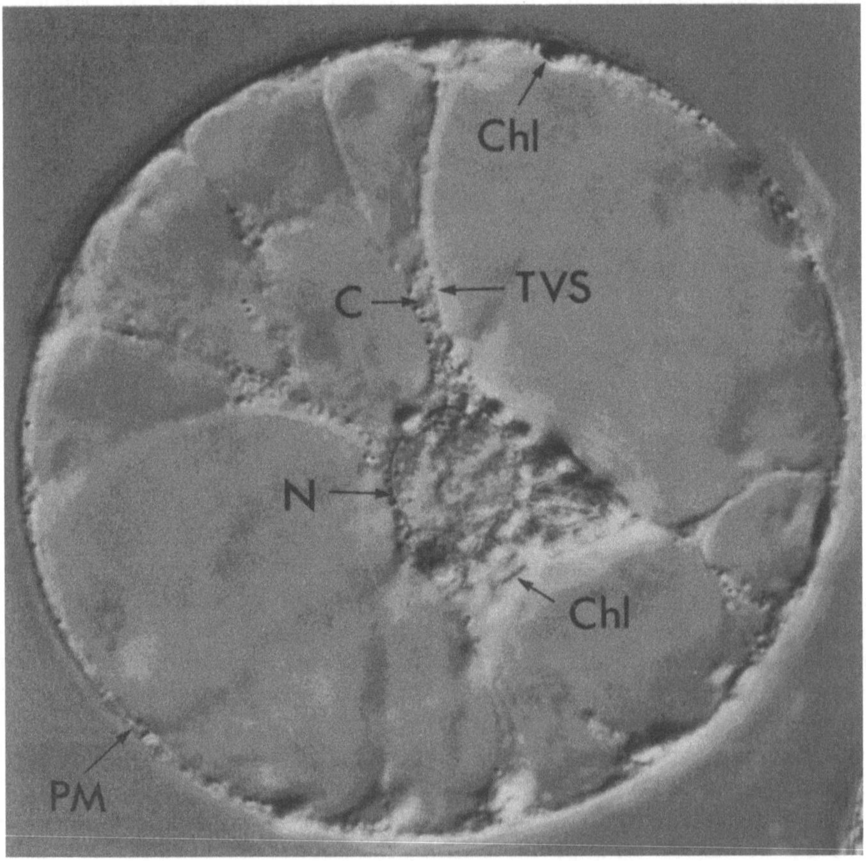

Fig. 1 A. Light micrograph of a live, nonpigmented *Hippeastrum* petal protoplast ×600. *TVS* transvacuolar strand; *C* cytosol; *PM* plasma membrane; *N* nucleus; *Chl* chloroplast. (By permission of Bioscience)

example of extravacuolar turnover of endomembrane system components may be suggested from the work of McKersie and Thompson (1975), who concluded that functionally distinct types of microsomal membranes, or regions of the same membrane, may undergo focal autolysis. It would appear that the evidence for both vacuole-mediated and vacuole-independent autolysis is weak and further study is needed to resolve the mechanism(s) of pre-terminal-senescence turnover.

Evacuolated protoplasts (Griesbach and Sink 1983) may be useful in determining the role of the mature vacuole in stress-induced, chloroplast turnover, in autolysis, in senescence in general, and in vacuole function in general.

A characteristic of mature plant vacuoles which is often ignored and is usually not observed in thin-section electron micrographs of fixed, stained and embedded tissue is the transvacuolar strand network (Fig. 1 A). Fragile, dynamic transvacuolar strands can be seen under the light microscope to break and reform in a healthy living cell. They may be partially preserved in frozen specimens observed by scanning electron microscopy (Fig. 1 B). Transvacuolar strands consist of tunnels of tonoplast which are filled with cytosol and traverse the cell. Vigor-

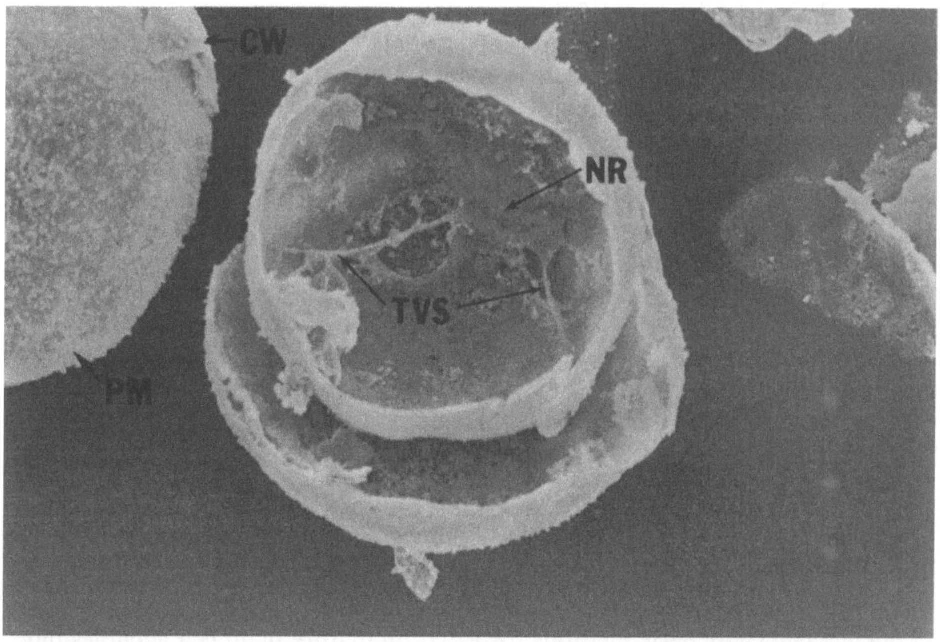

Fig. 1 B. Scanning electron micrograph of *Hippeastrum* petal protoplasts showing a unique view from inside the vacuole looking out. One fractured protoplast sits within another (×1500). Transvacuolar strands (*TVS*) are seen radiating from the region of the nucleus (*NR*) outward to the periphery of the protoplast. The outer surface of another protoplast-bearing residual cell wall (*CW*) is seen in the *upper left*. Protoplasts were fixed with cold alcohol, frozen in liquid nitrogen, fractured, critical point dried, lightly coated with gold and examined. (Specimen prepared and examined by M. C. Ledbetter and K. Garten)

ous cytoplasmic streaming through these strands (usually directed in and around the nucleus) is easily observed in healthy cells which lack a peripheral layer of chloroplasts, that can obscure viewing of the interior of a protoplast. Nonpigmented petal protoplasts are particularly useful for such viewing. Streaming of particles and even small chloroplasts (tulip petals) through transvacuolar strands is ordered and usually bidirectional within a single strand, indicating the presence of a directing influence on the tonoplast. It is not known if a layer of actin filaments, like that found in the subcortical layer in giant algal cells (Allen 1980), is responsible for directing cytoplasmic streaming in transvacuolar strands. Condeelis (1974) observed actin filaments in protoplasts of *Amaryllis*. We know nothing about the role of tonoplast in cytoplasmic streaming, yet this process may be one of the most important functions of mature plant cell vacuoles. The extensibility of tonoplast (Wagner 1981 a), as evidenced by the dynamic nature of the transvacuolar strand system, is probably responsible for the survival of vacuoles during protoplast lysis and therefore for the successful isolation of intact vacuoles from plant cells.

Defining the importance of mature vacuoles in internal membrane flow in plant cells is an exciting and challenging area of future research. Much effort in recent years has elucidated the complexities of membrane traffic and recycling now known to occur in animal cells (Farquhar 1983). It is known that ER-Golgi, post-translational processing of glycoproteins proceeds in plants as it does in animals (Chrispeels 1983). Recent results of Tanchak et al. (1983) suggest the occurrence of pathways for membrane flow from outside the cell (isolated protoplast, at least) to the vacuole and Golgi. Clathrin-coated vesicles and coated pits have been identified in plants and are thought to be involved in these endocytotic routes (Mersey et al. 1982). Elucidation of the routes of intracellular membrane flow can help to clarify the roles of the vacuole. For example, identification in mature plant cells of a mannose-6-phosphate-like receptor system – like that which is obligatory for directing Golgi-processed hydrolases to animal lysosomes (Shepherd et al. 1983) – would aid in elucidating the roles of the mature vacuole in lytic functions.

The rapidity and efficiency with which membrane vesicle recycling is known to occur (Brown et al. 1983) makes it possible to reconsider the suggestion that vacuolar protons may contribute to extracellular acid in cell-wall elongation.

5 Concluding Remarks

In the last few years basic methods for isolating mature plant cell vacuoles have been improved, to yield vacuoles more quickly and allow more precise kinetic studies of vacuole function. Purity of preparations has not improved in recent years and purity is often a minor consideration in studies relating to characterization of sap composition and transport activities. Quantitation of vacuoles in preparations is often made by visually counting structures, without concern for errors inherent in this method.

α-Mannosidase is generally regarded as general marker for higher plant vacuole sap, yet its vacuole/extravacuole distribution varies from report to report. Therefore it continues to be essential for each investigator to demonstrate the validity of this or other sap markers using a reliable method for quantitating total preparation sap volume. Acid protease may be an equally reliable general vacuole sap marker as α mannosidase. Phosphatase and esterase are less reliable.

No unequivocal tonoplast marker has been demonstrated. Although vanadate-insensitive, nitrate-sensitive Mg ATPase is associated with vacuoles from both higher and lower plants, this has not been established for many tissues. Comparing the chemical character of DCCD-binding proteins obtained from tonoplast of meristematic and mature higher and lower plant vacuoles and preparations of proton pumping vesicles should be very useful in comparing ATPases derived from these sources. ATPases which undoubtedly occur in ER, Golgi and other endomembranes of plants have yet to be characterized and reliance on inhibitor sensitivity of an ATPase – particularly where physiological relevance of the sensitivity is not understood – is inadequate. Known vacuole-like characteristics (structurally latent hydrolase content, secondary product content, pyrophosphatase) must be examined if homogenate-derived vesicles are to be equated with vacuoles. It must not be forgotten that homogenate-derived vesicles having density about 1.1 g cm^{-3} may be contaminated with Golgi, smooth ER, outer etioplast, chloroplast, and mitochondrial and secretory vesicle membranes. Proton transport functions are also expected in as yet uncharacterized endomembrane components (Golgi, endosomes, see deDuve 1983). Also, the level of tonoplast ATPase and proton pump activity may vary with the cell type, tissue and species in proportion to the tissue vacuolar pH, which can vary from perhaps pH 1.0 to 7.1 from tissue to tissue (Drawert 1955).

A challenge for the future is to develop isolation methods which can preserve or reconstitute the cytosol-tonoplast interface which is undoubtedly critical in vacuolar function (Boudet et al. 1984). The importance of this interface in transport is perhaps expressed in the difference between the pH optima for vacuole stability and sugar uptake in isolated sugar cane vacuoles (Thom et al. 1982 b).

Mature vacuoles were first isolated in small numbers from digested tomato fruit (Gregory and Cocking 1966) and Heftmann (1971) recovered acid hydrolase-bearing vesicles from mature as well as immature tomato fruit. Perhaps this system is an interesting one for examining the possibility that both types of particles may occur in the same cell.

Finally, the apparent dilemma presented by the fact that the endomembrane system is a membrane continuum (Marty et al. 1980) with parts that are functionally distinct (Palade 1983) is especially apparent in the case of plant vacuoles and animal lysosomes. Biosynthetic (ER, Golgi, etc.) compartments and those containing functional hydrolases must be kept separate. Plant vacuoles, but not animal lysosomes, also serve storage osmotic and other functions. We can expect to find different mechanisms for regulating hydrolytic activities of mature plant vacuoles. Coordination of storage, lytic and other functions of vacuoles is undoubtedly complex, as this is a multifunctional compartment.

Acknowledgements. The author thanks colleagues who provided preprints and reprints of their work so that the most recent information could be presented and incorporated.

References

Admon A, Jacoby B, Goldschmidt EE (1981) Some characteristics of the Mg-ATPase of isolated red beet vacuoles. Plant Sci Lett 22:89–96

Alibert G, Boudet A (1982) Progress, problems and perspectives in obtaining and utilizing isolated vacuoles. Physiol Veg 20:289–302

Alibert G, Carrasco A, Boudet AM (1982) Changes in biochemical composition of vacuoles isolated from *Acer pseudoplatanus* during cell culture. Biochim Biophys Acta 721:22–29

Allen NS (1980) Cytoplasmic streaming and transport in the characean alga Nitella. Can J Bot 58:786–796

Bennet AB, O'Neill SO, Spanswick RM (1984) H^+-ATPase activity from storage tissue of *Beta vulgaris*. Plant Physiol 74:538–544

Boller T (1982) Enzymatic equipment of plant vacuoles. Physiol Veg 20(2):247–257

Boller T, Alibert G (1983) Photosynthesis in protoplasts from *Melilotus alba*: distribution of products between vacuole and cytosol. Z Pflanzenphysiol 110:231–238

Boller T, Kende H (1979) Hydrolytic enzymes in the central vacuole of plant cells. Plant Physiol 63:1123–1132

Boudet AM, Canut H, Alibert G (1981) Isolation and characterization of vacuoles from *Melilotus alba* mesophyll. Plant Physiol 68:1354–1358

Boudet AM, Alibert G, Marigo G (1985) Vacuoles and tonoplast in regulation of cellular metabolism. Physiol Veg (in press)

Bowman EJ (1984, in press) Comparison of the vacuolar membrane ATPase of *Neurospora crassa* with the mitochondrial and plasma membrane ATPases. J Biol Chem

Bowman EJ, Bowman BJ (1982) Identification and properties of an ATPase in vacuolar membranes of *Neurospora crassa*. J Bacteriol 151:1326–1337

Briskin DP, Poole RJ (1983) Characterization of a K-stimulated adenosine triphosphatase associated with the plasma membrane of red beet. Plant Physiol 71:350–355

Brown MS, Anderson RGW, Goldstein JL (1983) Recycling receptors: the round trip itinerary of migrant membrane proteins. Cell 32:663–667

Buser C, Matile P (1977) Malic acid in vacuoles isolated for *Bryophyllum* leaf cells. Z Pflanzenphysiol 82:462–466

Buser-Suter C, Wiemken A, Matile B (1982) A malic acid permease in isolated vacuoles of a crassulacean acid metabolism plant. Plant Physiol 69:456–459

Butcher HC, Wagner GJ, Siegelman HW (1977) Localization of acid hydrolases in protoplasts. Plant Physiol 59:1098–1103

Corbett JR, Price CA (1967) Intracellular distribution of p-nitrophenyl-phosphatase in plants. Plant Physiol 42:827–830

Chrispeels MJ (1983) The golgi apparatus mediates the transport of phytohemagglutinin to the protein bodies in bean cotyledons. Planta 158:140–151

Churchill KA, Holaway B, Sze H (1983) Separation of two types of electrogenic H^+-pumping ATPases from oat roots. Plant Physiol 73:921–928

Condeelis JS (1974) Identification of factin in the pollen tube and protoplast of *Amaryllis belladonna*. Exp Cell Res 88:435–439

Cramer CL, Vaughn LE, Davis RH (1980) Basic amino acid and inorganic pyrophosphates in *Neurospora crassa*: independent regulation of vacuolar pools. J Bacteriol 142:945–952

Dalling MG, Tang AB, Huffaker RC (1983) Evidence for the existence of peptide hydrolase activity associated with chloroplasts isolated from barley mesophyll protoplasts. Z Pflanzenphysiol 111:311–318

Dauwalder M, Whaley WG, Kephart JE (1969) Phosphatases and differentiation of the golgi apparatus. J Cell Sci 4:455–497
d'Auzac J (1977) Membrane ATPase of lysosomal vacuoles: the lutoids of *Hevea brasiliensis* latex. Phytochemistry 16:1881–1885
d'Auzac J, Cretin H, Marin B, Lioret C (1982) A plant vacuolar system: the lutoids from *Hevea brasiliensis* latex. Physiol Veg 20:311–331
deDuve C (1984) Lysosomes revisited. Eur J Biochem 137:391–397
Doll S, Rodier R, Willenbrink J (1979) Accumulation of sucrose in vacuoles isolated from red beet tissue. Planta 144:407–411
Drawert H (1955) Der pH-Wert des Zellsaftes. In: Ruhland W (ed) Encyclopedia of plant physiology, vol 1. Springer, Berlin Heidelberg New York, pp 627–639
DuPont J, Moreau F, Lance C, Jacob JL (1976) Phospholipid composition of the membrane of lutoids from *Hevea brasiliensis*. Phytochemistry 15:1215–1217
DuPont FM, Bennet AB, Spanswick RM (1982) Localization of a protontranslocating ATPase on sucrose gradients. Plant Physiol 70:1115–1119
Durr M, Boller T, Wiemken A (1975) Polybase induced lysis of yeast spheroplasts. A new gentle method for preparation of vacuoles. Arch Microbiol 105:319–327
Farquhar MG (1983) Multiple pathways of exocytosis, endocytosis, and membrane recycling: validation of a golgi route. Fed Proc 42:2407–2413
Frehner M, Keller F, Wiemken A, Matile P (1985) Fructan metabolism in *Helianthus tuberosus*: compartmentation in protoplasts and vacuoles isolated from tubers. Z Pflanzenphysiol (in press)
Fujiwake H, Suzuki T, Iwai K (1980) Intracellular localization of capsicin and its analogues in *Capsicum* fruit. Plant Cell Physiol 21:1023–1030
Gahan PB (1968) Lysosomes. In: Pridham JB (ed) Plant cell organelles. Academic Press, London, pp 228–238
Gimmler H, Lotter G (1982) DEAE-dextran induced increase of membrane permeability and inhibition of photosynthesis in *Dunaliella parva*. Z Naturforsch 37C:609–619
Granstedt RC, Huffaker RC (1982) Identification of the leaf vacuole as a major nitrate storage pool. Plant Physiol 70:410–413
Gregory DW, Cocking EC (1966) Studies on isolated protoplasts and vacuoles. J Exp Bot 17:57–67
Griesbach RJ, Sink KC (1983) Evacuolation of mesophyll protoplasts. Plant Sci Lett 30:297–301
Grob K, Matile P (1979) Vacuolar location of glucosinolates in horseradish root cells. Plant Sci Lett 14:327–335
Guilliermond A (1941) The cytoplasm of the plant cell. Chronica Botanica, Waltham, Massachusetts, pp 1–223
Guy M, Reinhold L, Michaeli D (1979) Direct evidence for a sugar transport mechanism in isolated vacuoles. Plant Physiol 64:61–64
Hager A, Frenzel R, Laible D (1980) ATP-dependent proton transport into vesicles of microsomal membranes of *Zea mays* coleoptiles. Z Naturforsch 35C:783–793
Heck U, Martinoia E, Matile P (1981) Subcellular localization of acid proteinase in barley mesophyll protoplasts. Planta 151:198–200
Heftmann E (1971) Lysosomes in tomatoes. Cytobios 3:129–136
Helmlinger J, Rausch T, Hilgenberg W (1983) Localization of newly synthesized indole-3-methylglucosinolate in vacuoles from horseradish. Physiol Plant 58:302–310
Hirai M, Asahi T (1973) Membranes carrying acid hydrolases in pea seedling roots. Plant Cell Physiol 14:1019–1029
Howell KE, Ito A, Palade GE (1978) Endoplasmic reticulum marker enzymes in golgi fractions – what does it mean? J Cell Biol 79:581–589
Jonsson LMV, Donker-Koppman WE, Uitslager P, Schram AW (1983) Subcellular localization of anthocyanin methyltransferase in flowers of *Petunia hybrida*. Plant Physiol 72:287–290
Kaiser G, Martinoa E, Wiemken A (1982) Rapid appearance of photosynthetic products in the vacuoles isolated from barley mesophyll protoplasts by a new fast method. Z Pflanzenphysiol 107:103–113

Kakinuma Y, Ohsumi Y, Anraku Y (1981) Properties of H^+-translocating adenosine triphosphatase in vacuolar membranes of *Saccharomyces cerevisiae*. J Biol Chem 256:10859–10863
Keller F, Wiemken A (1982) Differential compartmentation of sucrose and gentianose in cytosol and vacuoles of storage root protoplasts from *Gentiana lutea*. Plant Cell Reports 1:274–277
Komor E, Thom M, Maretzki A (1982a) Sugar transport by sugarcane vacuoles. Physiol Veg 20(2):277–287
Komor E, Thom M, Maretzki A (1982b) Vacuoles from sugarcane cells III. Protonmotive potential difference. Plant Physiol 69:1326–1330
Kringstad R, Kenyon WH, Black CC (1980) The rapid isolation of vacuoles from leaves of crassulacean acid metabolism plants. Plant Physiol 66:379–382
Kurkdjian A, Barbier-Brygoo H, Manigault J, Manigault P (1984) Distribution of vacuolar pH values within populations of cells, protoplasts and vacuoles isolated from suspension culture and plant tissues. Physiol Veg 22:193–198
Lancaster JE, Collin HA (1981) Presence of allinase in isolated vacuoles and of alkyl cysteine sulphoxides in the cytoplasm of bulbs of onion. Plant Sci Lett 22:169–176
Leegood RC, Walker DA (1983) Chloroplasts. In: Hall JL, Moore AL (eds) Isolation of membranes and organelles from plant cells. Academic Press, London, pp 185–210
Leigh RA (1979) Do plant vacuoles degrade cytoplasmic components? Trends Biochem Sci 4:N37–N38
Leigh RA (1983) Methods, progress and potential for the use of isolated vacuoles in studies of solute transport in higher plant cells. Physiol Plant 57:390–396
Leigh RA, Branton D (1976) Isolation of vacuoles from root storage tissue of *Beta vulgaris*. Plant Physiol 58:656–662
Leigh RA, Branton D, Marty F (1979a) Methods for isolation of intact vacuoles and fragments of tonoplast. In: Reid E (ed) Plant organelles: methodological surveys B. Biochemistry, vol 9. Ellis Horwood, Chichester, pp 69–80
Leigh RA, Rees T, Fuller WA, Banfield J (1979b) The location of acid invertase activity and sucrose in vacuoles of storage roots of beetroot. Biochem J 178:539–547
Lin W, Wittenbach VA (1981) Subcellular localization of proteases in wheat and corn leaf mesophyll protoplasts. Plant Physiol 67:969–972
Lin W, Wagner GL, Siegelman HW, Hind G (1977) Membrane-bound ATPase of intact vacuoles and tonoplasts isolated from mature plant tissue. Biochim Biophys Acta 465:110–117
Loffelhardt W, Kopp B Kubelka W (1979) Intracellular distribution of cardiac glycosides in leaves of *Convallaria majalis*. Phytochemistry 18:1289–1291
Lorz H, Harms CT, Potrykus I (1976) Isolation of vacuoplasts from protoplasts of higher plants. Biochem Physiol Pflanz 169:617–620
Luttge U, Ball E (1979) Electrochemical investigation of active malic acid transport at the tonoplast into the vacuoles of the CAM plant *Kalanchoe*. J Membr Biol 47:401–422
Mandala S, Mettler IJ, Taiz L (1982) Localization of the proton pump of corn microsomal membranes by density gradient centrifugation. Plant Physiol 70:1743–1747
Marin B (1983) Sensitivity of tonoplast-bound adenosine-triphosphatase from *Hevea* to inhibitors. Plant Physiol 73:973–977
Marin B, Trouslot P (1975) The occurrence of ribonucleic acid in the lutoid fraction from *Hevea brasiliensis* latex. Planta 124:31–41
Marin B, Cretin H, D'Auzac J (1982) Energization of solute transport and accumulation at the tonoplast in *Hevea* latex. Physiol Veg 20:333–346
Martinoia E, Heck U, Wiemken A (1981) Vacuoles as storage compartments for nitrate in barley leaves. Nature 289:292–294
Marty F, Branton D (1980) Analytical characterization of beet root vacuole membrane. J Cell Biol 87:72–83
Marty F, Branton D, Leigh RA (1980) Plant vacuoles. In: Stumpf P, Conn E (eds) The biochemistry of plants, vol 1. Academic Press, New York, pp 625–658
Matile P (1968) Lysosomes of root tip cells in corn seedlings. Planta 79:181–196

Matile P (1975) The lytic compartment of plant cells. Springer, Berlin Heidelberg New York, pp 1–175 (Cell biology monographs, vol 1)
Matile P (1978) Biochemistry and function of vacuoles. Annu Rev Plant Physiol 29:193–213
Matile P (1982) Vacuoles come of age. Physiol Veg 20(2):303–310
Matile P, Wiemken A (1967) The vacuole as the lysosome of the yeast cell. Arch Microbiol 56:148–155
Matile P, Winkenbach F (1971) Function of lysosomes and lysosomal enzymes in the senescing corolla of the morning glory. J Exp Bot 22:759–771
Matile P, Balz JP, Semadeni E, Jost M (1965) Isolation of spherosomes with lysosome characteristics from seedlings. Z Naturforschung 20B:693–698
McKersie BD, Thompson JE (1975) Cytoplasmic membrane senescence in bean cotyledons. Phytochemistry 14:1485–1491
Medda P, Hasselbach W (1983) The vanadate complex of the calcium-transport ATPase of the sarcoplasmic reticulum, its formation and dissociation. Eur J Biochem 137:7–14
Mettler IJ, Leonard RT (1979) Isolation and partial characterization of vacuoles from tobacco protoplasts. Plant Physiol 64:1114–1120
Mersey BG, Fowke LC, Constabel F, Newcomb EH (1982) Preparation of a coated vesicle-enriched fraction from plant cells. Exp Cell Res 141:459–463
Moreau F, Jacob JL, DuPont J, Lance C (1975) Electron transport in the membrane of lutoids from the latex of *Hevea brasiliensis*. Biochim Biophys Acta 396:116–124
Moskowitz AH, Hrazdina G (1981) Vacuolar contents of fruit subepidermal cells from *Vitis* species. Plant Physiol 68:686–692
Moyle J, Mitchell R, Mitchell P (1972) Proton translocating pyrophosphatase of *Rhodospirillum rubrum*. FEBS Lett 23:233–236
Nagata T, Ishii S (1979) A rapid method for isolation of mesophyll protoplasts. Can J Bot 57:1820–1823
Nakano M, Asahi T (1972) Subcellular distribution of hydrolase in germinating pea cotyledons. Plant Cell Physiol 13:101–110
Nishimura M, Beevers H (1978) Hydrolases in vacuoles from castor bean endosperm. Plant Physiol 62:44–48
Nishimura M, Beevers H (1979) Hydrolysis of protein in vacuoles isolated from higher plant tissue. Nature 277:412–413
Ohlrogge JB, Garcia-Martinez JL, Adams D, Rappaport L (1980) Uptake and cellular compartmentation of gibberellin A_1 applied to leaves of barley and cowpea. Plant Physiol 66:422–427
Ohsumi Y, Anraku Y (1981) Active transport of basic amino acids driven by a proton motive force in vacuole membrane vesicles of *Saccharomyces cerevisiae*. J Biol Chem 256:2079–2082
O'Neill SD, Bennet AB, Spanswick RM (1983) Characterization of a NO_3-sensitive H^+-ATPase from corn roots. Plant Physiol 72:837–846
Opheim DJ, Touster O (1978) Lysosomal α mannosidase of rat liver. Purification and comparison with the golgi and cytosolic α mannosidases. J Biol Chem 253:1017–1023
Palade GE (1983) Membrane biogenesis: an overview. Methods Enzymol 96:xxix–iv
Pernollet JC (1982) Seed protein bodies, transient stage of specialized vacuoles. Physiol Veg 20(2):259–276
Pitt D, Galpin M (1973) Isolation and properties of lysosomes from darkgrown potato shoots. Planta 109:233–258
Poole FJ, Briskin DP, Kratky Z, Johnstone RM (1984a) Density gradient localization of plasma membrane and tonoplast from storage tissue of growing and dormant red beet. Plant Physiol 74:549–556
Poole RJ, Mehlhorn RJ, Packer L (1984b) A study of transport in tonoplast vesicles using spin-labelled probes. In: Marin BP (ed) Biochemistry and function of vacuolar adenosine-triphosphatase in fungi and plants. Springer, Berlin Heidelberg New York, pp 114–118

Pujarniscle S (1968) Characterization of lysosomes of lutoids of *Hevea brasiliensis*. Physiol Veg 6:27–46
Rosen BP, Kashket ER (1978) Energetics of active transport. In: Rosen BP (ed) Bacterial transport. Dekker, New York, pp 559–620
Rothman JE (1981) The golgi apparatus: two organelles in tandem. Science 213:1212–1219
Ryan CA, Walker-Simmons M (1983) Plant vacuoles. Methods Enzymol 96:580–589
Salyaev RK, Kuzevanov VY, Khaptagaev SB, Kopytchuk VN (1981) Isolation and purification of vacuoles and vacuolar membranes from plant cells. Physiol Plant (Russian) 28:1295–1305
Sasse F, Backs-Huesemann D, Barz W (1979) Isolation and characterization of vacuoles from cell suspension cultures of *Daucus carota*. Z Naturforsch 34C:848–853
Saunders JA (1979) Investigations of vacuoles isolated from tobacco. Plant Physiol 64:74–78
Saunders JA, Conn EE (1978) Presence of the cyanogenic glucoside dhurrin in isolated vacuoles from sorghum. Plant Physiol 61:154–157
Schmidt R, Poole RJ (1980) Isolation of protoplasts and vacuoles from storage tissue of red beet. Plant Physiol 66:25–28
Schmitt R, Saundermann H (1982) Specific localization of β-O-glucoside conjugate of 2,4-dichlorophenoxyacetic acid in soybean vacuoles. Z Naturforsch 37c:772–777
Schwaiger H, Hasilik A, von Figura K, Wiemken A, Tanner W (1982) Carbohydrate-free carboxypeptidase Y is transferred into the lysosomelike yeast vacuole. Biochem Biophys Res Comm 104:950–956
Schwencke J (1977) Characteristics and integration of the yeast vacuole with cellular functions. Physiol Veg 15:491–517
Shepherd V, Schlesinger P, Stahl P (1983) Receptors for lysosomal enzymes and glycoproteins. Curr Top Membr Transp 18:317–338
Stout R, Cleland R (1982) Evidence for a Cl^--stimulated Mg ATPase proton pump in oat root membranes. Plant Physiol 69:798–803
Sze H (1980) Nigericin-stimulated ATPase activity in microsomal vesicles of tobacco callus. Proc Natl Acad Sci 77:5904–5908
Sze H (1982) Characterization of nigericin-stimulated ATPase from sealed microsomal vesicles of tobacco callus. Plant Physiol 70:498–505
Tanchak M, Griffing LR, Mersey BG, Fowke LC (1983) Functions of coated vesicles in plant protoplasts: endocytosis of cationized ferritin and transport of peroxidase. J Cell Biol 97:177a
Thayer SS, Huffaker RC (1984) Vacuolar endoproteinase EP1 and EP2 in barley mesophyll cells. Plant Physiol 75:70–73
Thom M, Maretzki A, Komor E (1982a) Vacuoles from sugarcane suspension cultures. I. Isolation and partial characterization. Plant Physiol 69:1315–1319
Thom M, Komor E, Maretzki A (1982b) Vacuoles from sugar cane suspension cultures. II. Characterization of sugar uptake. Plant Physiol 69:1320–1325
Thomas H, Stoddart JL (1980) Leaf senescence. Annu Rev Plant Physiol 31:83–111
Tolbert NE (1980) Microbodies – peroxisomes and glyoxysomes. In: Tolbert NE (ed) The biochemistry of plants, vol 1. Academic Press, New York, pp 359–388
Van der Wilden W, Chrispeels MJ (1983) Characterization of the isozymes of α-mannosidase located in the cell wall, protein bodies, and endoplasmic reticulum of *Phaseolus vulgaris* cotyledons. Plant Physiol 71:82–87
Van der Wilden W, Herman EM, Chrispeels MJ (1980) Protein bodies of mung bean cotyledons as autophagic organelles. Proc Natl Acad Sci USA 77:428–432
Vaughn LE, Davis RH (1981) Purification of vacuoles from *Neurospora crassa*. Mol Cell Biol 1:797–806
Vian B (1981) Tonoplast differentiation of immature vacuoles in elongating plant cells visualized by selective cytochemical reactions. Biol Cell 41:239–242
Wagner GJ (1981a) Enzymic and protein character of tonoplast from *Hippeastrum* vacuoles. Plant Physiol 68:499–503

Wagner G (1981 b) Compartmentation in plant cells: the role of the vacuole. In Creasy LL, Hrazdina G (eds) Cellular and subcellular localization in plant metabolism. Plenum, New York, pp 1–45 Recent advances in phytochemistry, vol 16

Wagner G (1983) Higher plant vacuoles and tonoplasts. In: Hall JL, Moore AL (eds) Isolation of membranes and organelles from plant cells. Academic Press, London, pp 83–118

Wagner G, Hrazdina G (1984) Endoplasmic reticulum as a site of phenylpropanoid and flavonoid metabolism in *Hippeastrum*. Plant Physiol (in press)

Wagner GJ, Lin W (1982) An active proton pump of intact vacuoles isolated from *Tulipa* petals. Biochim Biophys Acta 689:261–266

Wagner GJ, Mulready P (1983) Characterization and solubilization of nucleotide-specific Mg^{2+}-ATPase and Mg^{2+}-pyrophosphatase of tonoplast. Biochim Biophys Acta 728:267–280

Wagner GJ, Siegelman HW (1975) Large-scale isolation of intact vacuoles and isolation of chloroplasts from protoplasts of mature plant tissue. Science 190:1298–1299

Wagner GJ, Mulready P, Cutt J (1981) Vacuole/extravacuole distribution of soluble protease in *Hippeastrum* petal and *Triticum* leaf protoplasts. Plant Physiol 68:1081–1089

Wagner W, Keller F, Wiemken A (1983) Fructan metabolism in cereals: induction in leaves and compartmentation in protoplasts and vacuoles. Z Pflanzenphysiol 112:359–372

Walker RR, Leigh RA (1981) Mg-dependent, cation-stimulated inorganic pyrophosphatase associated with vacuoles isolated from storage roots of red beet. Planta 153:150–155

Walker-Simmons M, Ryan CA (1977) Immunological identification of proteinase inhibitors I and II in isolated tomato leaf vacuoles. Plant Physiol 60:61–63

Waters SP, Noble ER, Dalling MJ (1982) Intracellular localization of peptide hydrolases in wheat leaves. Plant Physiol 69:575–579

Willenbrink J, Doll S (1979) Characteristics of the sucrose uptake system of vacuoles isolated from red beet tissue. Planta 147:159–162

Wittenbach VA, Lin W, Hebert RR (1982) Vacuolar localization of proteases and degradation of chloroplasts in mesophyll protoplasts from senescing primary whent leaves. Plant Physiol 69:98–102

Wilson CM (1975) Plant nucleases. Annu Rev Plant Physiol 26:187–208

Protein Bodies

A. H. C. HUANG

1 Introduction

Protein bodies are subcellular organelles occurring in the storage tissues of seeds. Depending on the plant species, they range in sizes from 1 to 20 μm in diameter, and they vary in structural inclusions. The spherical organelles contain an amorphous protein matrix bounded by a single membrane. Structural inclusions embedded in the matrix may be one to several protein crystalloids and phytin globoids (Fig. 1). The protein crystalloids are composed of globulin storage proteins, whereas other storage proteins and nonstorage proteins, such as hydrolytic enzymes, protease inhibitors, and lectins, are present in the amorphous matrix. Phytin globoids are the sites of storage cations and inositol phosphate. In a few spe-

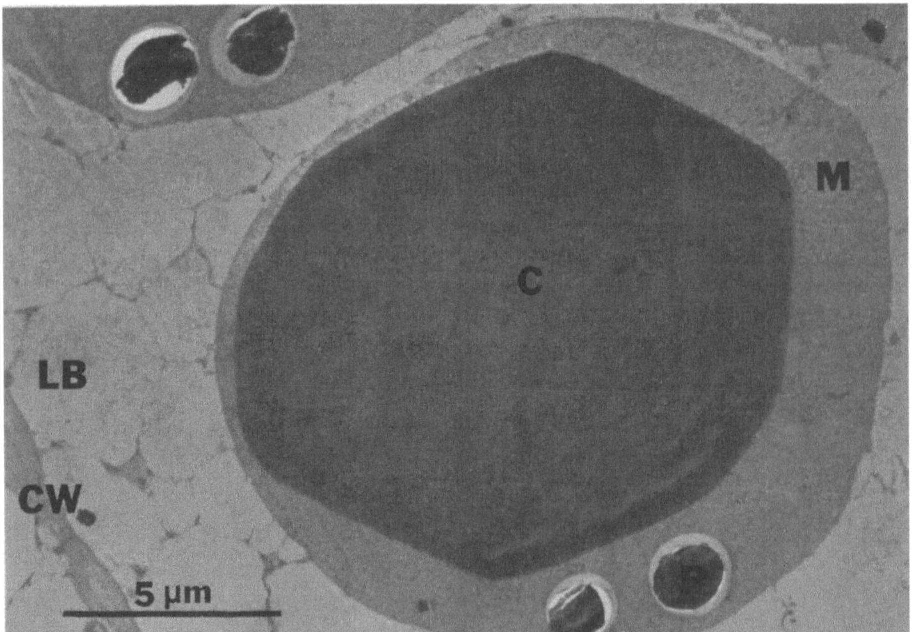

Fig. 1. Electron micrograph of a section of an endosperm cell of dry castor bean (Youle 1977), showing a protein body with a large crystalloid (*C*), matrix (*M*), and two phytin globoids (*P*). A densely packed background of lipid bodies (*LB*) and part of a cell wall (*CW*) are also visible

cies, druse crystals thought to be calcium oxalate are present; presumably they are the site for the deposition of cations. Several review articles covering protein bodies directly or indirectly have been published recently (Brown et al. 1982; Dieckert and Dieckert 1982; Larkin 1981; Lott 1980; Pernollet 1978).

Protein bodies and the cell vacuoles bear close ontogenic relationship (Bollini et al. 1983; Brown et al. 1982; Dieckert et al. 1962; Pernollet 1978). During seed maturation, the proteins in the protein bodies are synthesized by the ribosomes on the endoplasmic reticulum. The newly synthesized protein is co-translationally inserted into the lumen of the endoplasmic reticulum where processing occurs (proteolysis and glycosylation). Membrane vesicles containing the newly synthesized proteins are detached from the endoplasmic reticulum either to become new protein bodies directly or to serve as carrier vehicles. In the latter case, the vesicles fuse with either enlarging protein bodies or the central vacuoles, which in turn split into smaller protein bodies. The transfer of the vesicles to the vacuoles or enlarging protein bodies may proceed directly or indirectly via the Golgi apparatus, depending on the plant species. When the seed becomes mature, water is removed, and the numerous protein bodies in each cell are densely packed with storage and other components. In seed germination, as water is imbibed, the protein bodies enlarge and become less dense. The enlarging protein bodies acquire hydrolytic enzyme activities and become autophagic. Eventually, the enlarging protein bodies within one expanding cell fuse together to form a large central vacuole. Thus, the protein bodies in the mature seed are unequivocally authentic protein bodies, whereas those in the maturing and germinating seeds can also be looked upon as vacuoles. The extent of applicability of the latter terminology depends on the ontogenic stage of the organelle.

2 Special Consideration in Isolation of Protein Bodies

The choice of methods to isolate protein bodies from a certain seed tissue depends on the ontogenic stage of the organelles within the tissue. Protein bodies from mature seeds have been isolated successfully only by nonaqueous methods. Presumably, the extremely low water potential in the mature storage tissues causes osmolysis of the organelles when isolation is carried out in an aqueous medium. Although aqueous methods have been used (e.g., Donhowe and Peterson 1983; Pernollet et al. 1982), the purity and intactness of the isolated protein bodies are uncertain. Often, the seeds are allowed to imbibe for a short period of time to increase the cellular osmotic potential before aqueous isolation is performed. Understandably, the protein bodies thus isolated have undergone ontogenic changes and are no longer identical with those in the mature seeds. Protein bodies from maturing and germinating seeds can be isolated by aqueous procedures, since the osmotic difference between the tissue and the isolation medium can be eliminated. The aqueous procedure employing gradient centrifugation is similar to those for isolation of vacuoles and peroxisomes (microbodies). The protein bodies/vacuoles usually have equilibrium densities in sucrose density gradient centrifugation higher than those of other subcellular organelles, except nuclei and starch grains. Undoubtedly, their equilibrium densities vary, depending on the ontogenic stage of the organelles.

Three major methods for isolation of protein bodies have been used successfully and will be described. Two of these methods involve nonaqueous preparations, one utilizing glycerol as the medium in differential centrifugations, and another employing carbon tetrachloride and a low density organic solvent in density gradient centrifugation. The two nonaqueous methods are recommended for isolation from seeds at maturity (i.e., dormant or dry), or shortly before maturity and after germination. The third method is an aqueous procedure utilizing sucrose density gradient centrifugation. It should be used for seeds before maturity and during active germination. For tissues at the early stage of maturity and at the late stage of germination when the protein bodies resemble closely the cell vacuoles, the readers should refer to Wagner Chapter 6, this volume.

3 Nonaqueous Preparation in Glycerol

The procedure for isolation of protein bodies by differential centrifugation in glycerol was originated from Yatsu and Jack (1968), and has been applied successfully to many seeds (e.g., Tully and Beevers 1976; Youle and Huang 1979). The separation is achieved by taking advantage of the high density of protein bodies. The procedure is performed in cold (4 °C). Dehulled mature cottonseeds are ground in glycerol without buffer (10 g seed in 30 ml glycerol). Homogenization is carried out with a mortar and pestle for several minutes or a VirTis homogenizer at low speed for about 30 s. The homogenate is filtered through four layers of cheesecloth, and the filtrate is centrifuged at 1000 *g* for 5 min to remove unbroken cells and cell debris, etc. The supernatant is recentrifuged at 41,000 *g* for 20 min. The pellet is resuspended in 10 ml glycerol and recentrifuged. The final pellet contains the protein bodies and is resuspended in glycerol. When working with seed tissues at the late stage of maturity and the early stage of germination, the water in the tissues is removed before homogenization. The tissue is cut into small pieces (0.5 cm or less) with a razor blade in glycerol on a Petri dish, and the materials are gently stirred for about half an hour in order to extract the water from the tissue. A larger volume of glycerol to tissue (e.g., 10 ml to 1 g) is used so that the glycerol will not be diluted substantially by the extracted water. After water extraction, the tissues is transferred to fresh glycerol, and homogenization is initiated as described. The defects of the glycerol procedure include the handling of the highly viscous glycerol and the inherent problem of contamination in procedures involving differential centrifugation. The contaminants are generally cell walls and starch grains; the latter may be substantial in preparations from starchy seeds.

4 Nonaqueous Preparation in Hexane and Carbon Tetrachloride

Protein bodies can be isolated in a relatively pure preparation by nonaqueous density gradient centrifugation. The original work was carried out using cottonseed oil (density 0.92 g cm^{-3}) and carbon tetrachloride (density 1.59 g cm^{-3}) gradients (Dieckert et al. 1962). In my laboratory (Youle and Huang 1976), we use hexane (density 0.66 g cm^{-3}) instead of cottonseed oil in order to avoid the uncertainties of components and impurities in commercial cottonseed oil. The procedure is performed in cold (4 °C). Dehulled mature castor bean are ground in pure hexane (1 g per 3 ml hexane) with a mortar and pestle for several minutes or a VirTis homogenizer at low speed for about 30 s. The homogenate is filtered through a Nitex cloth of 35 μm × 35 μm pore size (or eight layers of cheesecloth). Two ml of the filtrate are layered onto a 32-ml linear gradient of hexane and carbon tetrachloride from density of 1.20 g cm^{-3} to 1.50 g cm^{-3} in a 37-ml Beckman cellulose nitrate or polyallomer centrifuge tube. The refractive indexes of hexane, carbon tetrachloride, and mixtures of the two solvents at varying proportions are determined with a refractometer, and a standard curve of refractive index versus density is constructed. The gradient is prepared immediately before use, and is covered with an aluminum foil until use to prevent evaporation. Cellulose nitrate tubes can withstand the organic solvents well; however, they are currently un-

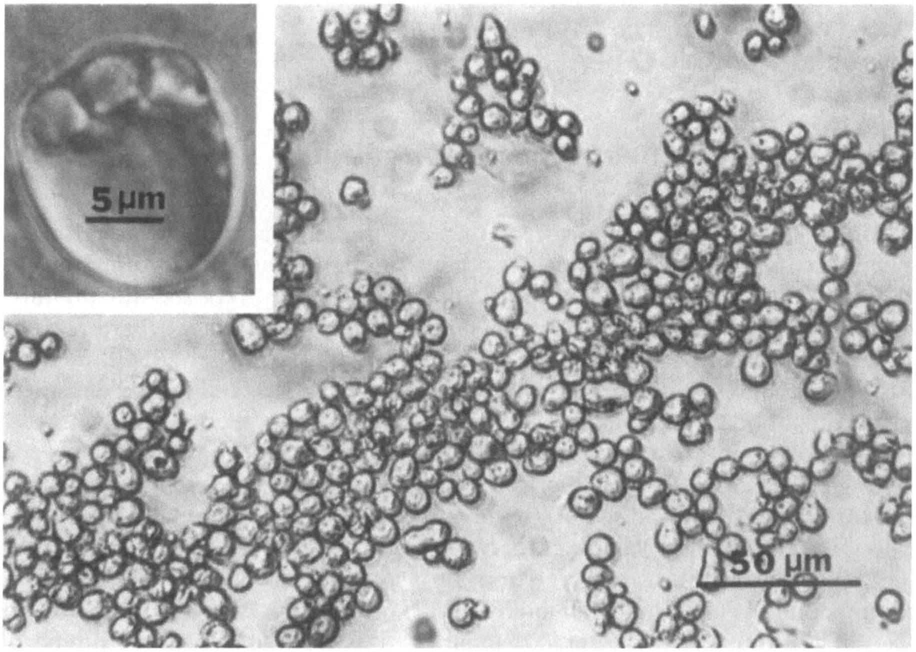

Fig. 2. Light micrograph of protein bodies isolated by successive nonaqueous gradient centrifugation. *Inset* shows an enlarged protein body with its inclusions. (Youle and Huang 1976)

available commercially. Polyallomer tubes can withstand the organic solvents for a few hours, especially in cold. Gradients prepared in polyallomer tubes should be loaded with the filtered homogenate, centrifuged, and fractionated as soon as possible. The gradient former and the connecting tubes should be made of glass and not plexiglas. The gradient is centrifuged at 21,000 rpm for 2 h in a Beckmann L2-65B ultracentrifuge using a SW-27 rotor. After centrifugation, a cloudy band of protein bodies can be observed in the middle (density 1.36 g cm^{-3}) of the gradient. The band is collected by puncturing a hole in the bottom of the centrifuge tube. The protein body fractions can be examined under the light microscope (Fig. 2). It is slightly contaminated by broken cell walls, the majority of which sediment to a density of 1.41 g cm^{-3}.

Approximately 80% of the total seed protein can be recovered at the fraction of protein bodies, and the rest is present at the top of the gradient. The protein components at the top of the gradient resemble those of the isolated protein bodies in sodium dodecyl sulfate gel electrophoresis (Youle and Huang 1976) and contain no enrichment of any specific protein components. Thus, the proteins at the top of the gradient are probably derived from broken protein bodies. For further purification the 3–5 ml fraction of protein bodies (1.36 g cm^{-3}) is made to 1.50 g cm^{-3} with carbon tetrachloride, resulting in a final volume of about 10 ml. The preparation is then put in a centrifuge tube and a similar gradient of 25 ml hexane and carbon tetrachloride is layered upon it. Centrifugation is performed as before, and the protein bodies again float to their equilibrium density of 1.36 g cm^{-3}. The band is collected by puncturing a hole at the bottom of the centrifuge tube.

5 Aqueous Preparation in Sucrose Gradients

The aqueous preparation using sucrose density gradient centrifugation has been applied successfully to obtain protein bodies from the storage tissues of maturing and germinating seeds of a great variety of species. Recent reports utilizing this procedure to isolate protein bodies include using seeds of wheat, barley, pea, corn, and cucumber (Chrispeels et al. 1982; Kara and Kindl 1982; Miflin et al. 1981; Vitale et al. 1982). Depending on the plant species, the equilibrium densities of the protein bodies range from 1.23 to 1.29 g cm^{-3}, but are consistently higher than the mitochondria and peroxisomes in the same tissue. The procedure for germinating pea (Huang 1973) is described here. Pea seeds are soaked in running tap water for 24 h and allowed to germinate in moist vermiculite for 3 days. The cotyledons are chopped with a new razor blade in grinding medium (7.5 g per 15 ml) on a Petri dish on top of crush ice until a fine mince is obtained. The grinding medium contains 0.15 M Tricine-KOH, pH 7.5, 0.4 M sucrose, and 1 mM EDTA. The homogenate is filtered through a Nitex cloth of 35 µm × 35 µm pore size (or eight layers of cheesecloth). Five ml of the filtrate are layered onto a linear sucrose density gradient consisting of 28 ml 20% (w/w) to 65% sucrose in a 37-ml Beckman centrifuge tube for Rotor SW-27. The gradient contains 1 mM EDTA-

NaOH, pH 7.5 throughout. The gradient is centrifuged at 21,000 rpm for 4 h at 4 °C in a Beckman L2-65B ultracentrifuge using a SW-27 rotor. After centrifugation, a conspicuous and dominant band of protein bodies can be observed about $^1/_5$ distance from the bottom (density 1.26–1.28 g cm^{-3}). Starch grains are present in the pellet at the bottom of the gradient, whereas the intact peroxisomes (density 1.22 g cm^{-3}, not visible) and mitochondria (a visible band at density 1.19 g cm^{-3}) are present at lower densities. The protein body band is collected by puncturing a hole from the bottom of the centrifuge tube.

6 Subfractionation of Isolated Protein Bodies

The four major suborganelle components (Fig. 1) of the isolated protein bodies (Fig. 2) can be separated from one another in one sucrose density gradient after centrifugation. These four components are the membrane, the amorphous protein matrix, the protein crystalloids, and the phytin globoids. The procedure to be described has been used successfully to subfractionate protein bodies isolated from mature castor bean (Youle and Huang 1976).

Twenty ml of protein bodies isolated from four hexane-carbon tetrachloride density gradients as described earlier (corresponding to organelles obtained from about 3 g of dehulled seeds) are mixed with 2 ml of water, and all of the organic solvents are allowed to evaporate under nitrogen. The protein bodies are broken by this procedure. The preparation is applied to a sucrose gradient consisting of, from the bottom, 1 ml 68% (w/w) sucrose, a 9-ml gradient from 68% to 30% sucrose, and 2 ml 15% sucrose. The gradient is constructed immediately to a few hours before use in a 17-ml Beckman centrifuge tube for Rotor SW-27.1. The gradient is centrifuged in a SW-27.1 rotor at 24,000 rpm for 4 h. After centrifu-

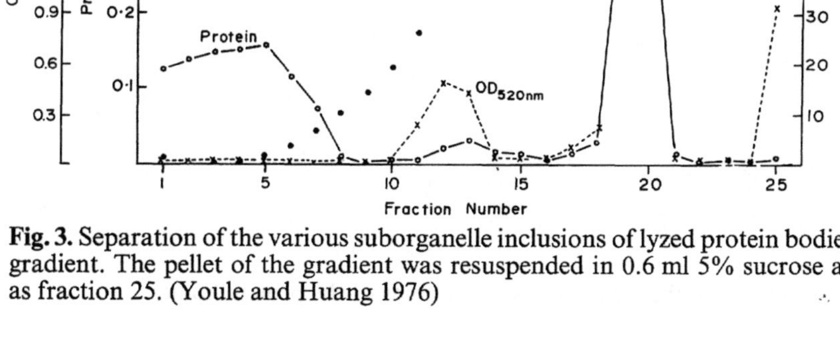

Fig. 3. Separation of the various suborganelle inclusions of lyzed protein bodies in a sucrose gradient. The pellet of the gradient was resuspended in 0.6 ml 5% sucrose and presented as fraction 25. (Youle and Huang 1976)

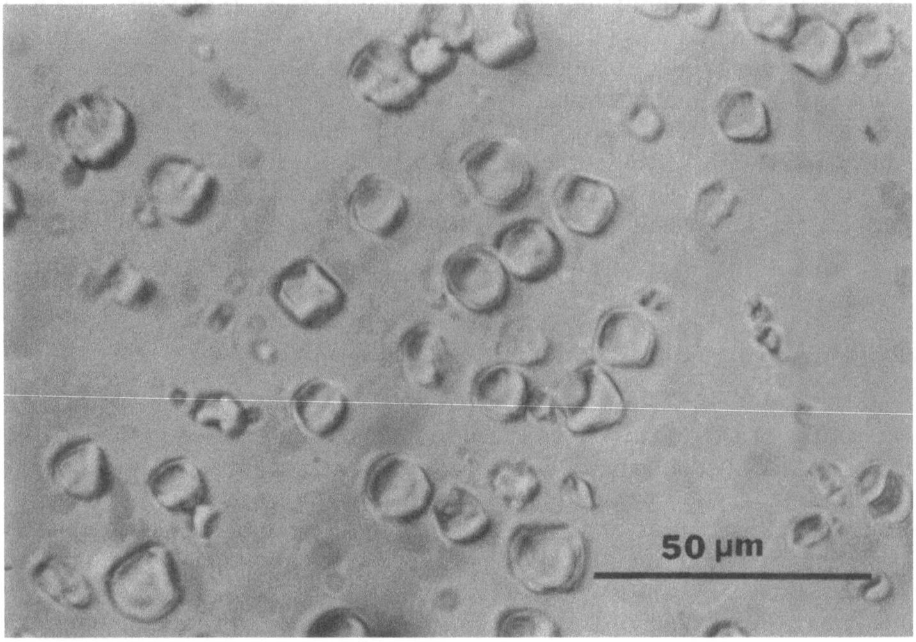

Fig. 4. Light micrograph of the isolated protein crystalloids of the protein bodies

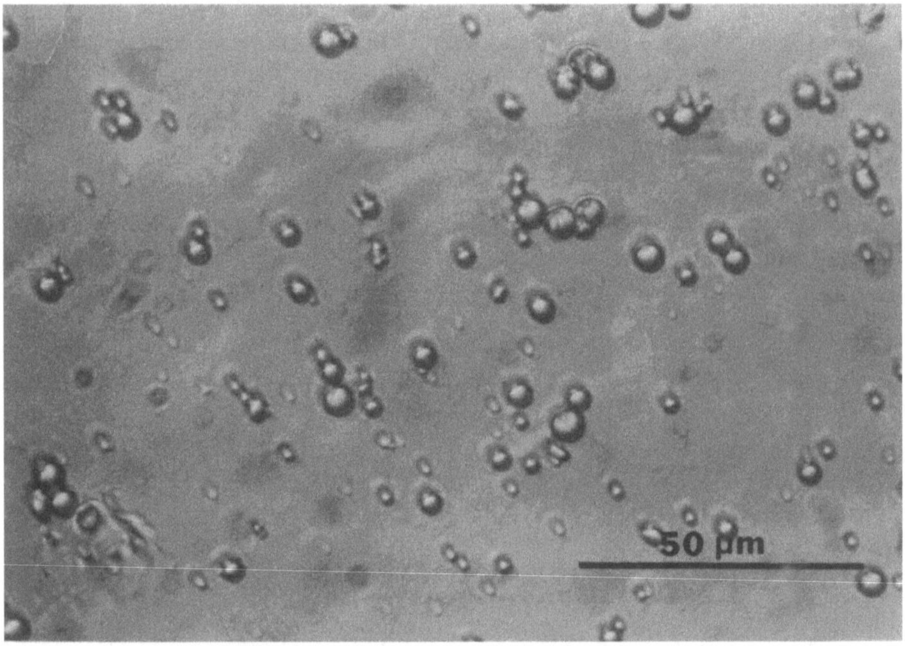

Fig. 5. Light micrograph of the isolated phytin globoids of the protein bodies

gation, two bands are visible in the gradient. One dense and dominant band occurs at about $^1/_5$ the distance from the bottom (density 1.30 g cm^{-3}) and a faint band is present at the junction between the 15% and 30% sucrose solutions (density 1.15 g cm^{-3}). In addition, a small pellet is present at the bottom of the gradient. The gradient is fractionated from the bottom into 0.6-ml fractions. The fractionation is performed with the centrifuge tube slightly tilted, so that after the drainage of the whole gradient, the pellet can be resuspended into a 0.6-ml fraction using 5% sucrose solution.

A profile of the sucrose gradient showing protein content (by the Lowry or Bradford method) and diffraction of light at 520 nm is presented in Fig. 3. Protein content and light diffraction at 520 nm are determined from an aliquot of each fraction. The four fractions – gradient supernatant, the fraction at density 1.15 g cm^{-3}, the fraction at density 1.30 g cm^{-3}, and the pellet – represent the four different components of the protein bodies.

The matrix proteins are soluble in water (i.e., albumins, by definition) and remain at the top of the sucrose gradient. This fraction represents about half of the total protein of the protein bodies. The protein components of this fraction are different from those of other fractions, as revealed in sodium dodecyl sulfate gel electrophoresis (Youle and Huang 1976). They include the 2S albumin storage proteins (the allergens), the lectins, and hydrolytic enzymes (Youle and Huang 1978).

The fraction at density 1.15 g cm^{-3} has a high light diffraction at wavelength 520 nm, but a low protein content. This fraction contains the membrane of the protein bodies, as revealed by its content of most of the phospholipids in the gradient (Youle and Huang 1976). A substantial amount of the protein in this fraction is due to the presence of contaminating protein crystalloids, which presumably are trapped in the membrane vesicles during organelle breakage. These protein crystalloids can be removed by sonication after gradient centrifugation (Mettler and Beevers 1979).

The fraction at density 1.30 g cm^{-3} contains protein crystalloids of the protein bodies, and accounts for about half of the total protein of the protein bodies. The size and polyhedral shape of the isolated crystalloids are the same as those observed in situ (Fig. 4). The crystalloids, as well as their protein components, are insoluble in water but soluble in 0.5 M NaCl; the proteins are globulins by definition.

The pellet at the bottom of the sucrose gradient represents phytin globoids that have a density greater than 1.46 g cm^{-3}. The spherical shape and size of the globoids are the same as those in situ (Fig. 5). They are not soluble in water or dilute salt solution (0.5 M NaCl), but are soluble in dilute acid (2.5% trichloroacetic acid).

7 Analyses

There is no well-established marker for protein bodies, especially those isolated from mature seeds. Generally, the organelle fraction is identified by its high protein content. In the three isolation methods described in this article, the recovery of intact protein bodies is always 60–90%, and the protein body fraction contains a very high proportion of the total protein in the seed homogenate. Because protein bodies contain mostly protein and very little membrane structure, they have an equilibrium density in sucrose gradients higher than those of other organelles composed of mostly lipid and protein (e.g., endoplasmic reticulum, mitochondria, plastids without starch grains, and peroxisomes). Furthermore, the organelles are large enough (1–20 μm) to be seen under the light microscope.

Activities of acid hydrolases may be present in isolated protein bodies. These activities generally are present in the protein bodies of germinating rather than maturing seeds, and thus are likely to be involved in the mobilization of storage protein, glycoprotein, and phytin. The activities are present in low amounts in the mature seeds, and increase during germination. The assays of the three major hydrolases are described. These hydrolases are protease, which has been shown to occur in the protein bodies of mung bean (Van der Wilden et al. 1980), and phytase and glycosidase, which are known to be present in the protein bodies of castor bean (Youle 1977).

β-N-acetylglucosaminidase activity is assayed using p-nitrophenylglucosamine and measuring the absorbance of the released p-nitrophenol at 400 nm (Li and Li 1972). The 1-ml reaction mixture contains 50 mM sodium acetate, pH 4.5, 2 mM p-nitrophenol-β-N-acetylglucosamine (from Sigma Corp., St. Louis, USA), and enzyme preparation. The reaction is allowed to proceed at 37 °C. At time intervals (15 min to 2 h), a 0.2-ml aliquot is added to 0.6 ml 0.2 M sodium borate buffer, pH 9.8. The absorbance of the released p-nitrophenol at 400 nm is read. A standard curve of p-nitrophenol in 0.2 M sodium borate buffer at pH 9.8 is constructed. Alternatively, the reported extinction coefficient of p-nitrophenol, $1.77 \times 10^4\ M^{-1}\ cm^{-1}$, is used. The assay is also suitable to measure the activity of other glycosidases using different p-nitrophenol glycosides and phosphatase using p-nitrophenolphosphate. These p-nitrophenol derivatives can be obtained from Sigma Corp.

Phytase activity is assayed by its catalytic release of inorganic phosphate from sodium phytate (Peers 1953). The 1-ml reaction mixture contains 0.1 M sodium acetate buffer, pH 5.15, 4 mM Mg_2SO_4, 1.6 mM sodium phytate, and enzyme preparation. The reaction is allowed to proceed at 37 °C. At time intervals (15 min to 2 h), an aliquot of 0.2 ml reaction mixture is added to 0.2 ml 10% (w/v) trichloroacetic acid. After centrifugation to remove the precipitate, the supernatant is assayed for inorganic phosphate (Dittmer and Wells 1969). A 0.1-ml aliquot of the supernatant is added to 0.2 ml 70% perchloric acid for hydrolysis in an oil bath until the sample is clear. If the supernatant contains a large amount of sucrose or other organic materials, 0.1 ml of 2% ammonium molybdate is added to accelerate the hydrolysis. Add successively 1.2 ml of an ammonium molybdate reagent (0.44 g ammonium molybdate, 25 ml water, 1.4 ml concentrated H_2SO_4, and make to 100 ml with water) and 1.2 ml of a reducing agent (grind with a mortar and pestle 6 g sodium bisulfite, 1.2 g sodium sulfite, and 0.1 g 1, 2, 4,-aminonaphthol sulfonic acid. Dissolve the powder in water to make a volume of 50 ml. Let the solution stand in dark for 3 h and filter into an amber bottle. The solution is stable for 6–8 weeks under refrigeration. Dilute 1 to 12 with water immediately before use). Mix and heat in a boiling water bath for 10 min. Cool and read absorbance at 735 nm. A standard curve of KH_2PO_4 (0 to 0.5 mM) solution substituting the 0.1 ml supernatant is constructed.

Protease activity is assayed with the chromogenic substrate Azocoll (from Sigma Corp., St. Louis, MO, USA). Azocoll is an insoluble powdered collagen bound with azo dye. The 1.5-ml reaction mixture contains 0.1 M sodium acetate, pH 5.0, 0.1% Triton X-100, 20 mM Azocoll, and enzyme preparation. The reaction mixture is shaken vigorously at 37 °C for 30 min to 2 h. The tube is cooled to 4 °C and centrifuged at 5000 *g* for 10 min and the absorbance of the supernatant at 520 nm is determined. The activity is expressed on a relative basis of change in absorbance at 520 nm per unit time.

References

Bollini R, Vitale A, Chrispeels MJ (1983) In vivo and in vitro processing of seed reserve protein in the endoplasmic reticulum. Evidence for 2 glycosylation steps. J Cell Biol 96:999–1007

Brown JWS, Ersland DR, Hall TC (1982) Molecular aspects of storage protein synthesis during seed development. In: Khan AA (ed) The physiology and biochemistry of seed development dormancy and germination. Elsevier Biomedical Press, Amsterdam, pp 3–42

Chrispeels MJ, Higgins TJV, Craig S, Spencer D (1982) Role of the endoplasmic reticulum in the synthesis of reserve proteins and the kinetics of their transport to protein bodies in developing pea cotyledons. J Cell Biol 93:5–14

Dieckert JW, Dieckert MC (1972) The deposition of vacuolar proteins in oilseeds. In: Inglett GE (ed) Symp: seed proteins, pp 53–85

Dieckert JW, Snowden JE, Moore AT, Heinzelman DC, Altschul AM (1962) Composition of some subcellular fractions from seeds of *Arachis hypogaea*. J Food Sci 27:321–325

Dittmer JC, Wells MA (1969) Quantitative and qualitative analysis of lipid and lipid components. Methods Enzymol 14:482–530

Donhowe ET, Peterson DM (1983) Isolation and characterization of oat aleurone and starchy endosperm protein bodies. Plant Physiol 71:519–523

Huang AHC (1973) Studies on plant microbodies. PhD Thesis, Univ California at Santa Cruz

Kara UAK, Kindl H (1982) Membranes of protein bodies I. Isolation from cotyledons of germinating cucumber seeds. Eur J Biochem 121:533–538

Larkin B (1981) Seed storage proteins. Characterization and biosynthesis. In: Stumpf PK, Conn EE (ed) The biochemistry of plants, vol 6. Academic Press, New York, pp 449–489

Li YT, Li SC (1972) α-Mannosidase β-acetylhexosaminidase β-galactase from jack bean meal. Methods Enzymol 28:702–713

Lott JNA (1980) Protein bodies. In: Stumpf PK, Conn EE (ed) The biochemistry of plants, vol 1. Academic Press, New York, pp 589–623

Mettler IJ, Beevers H (1979) Isolation and characterization of the protein body membranes of castor beans. Plant Physiol 64:506–511

Miflin BJ, Burgess SR, Shewty PR (1981) The development of protein bodies in the storage tissues of seeds: subcellular separations of homogenates of barley maize and wheat endosperms and of pea cotyledons. J Exp Bot 32:199–219

Peers FG (1953) The phytase of wheat. Biochem J 53:102–110

Pernollet JC (1978) Protein bodies of seeds: ultrastructure biochemistry biosynthesis and degradation. Phytochemistry 17:1473–1480

Pernollet JC, Kim SI, Mosse J (1982) Characterization of storage proteins extracted from *Avena sativa* seed protein bodies. J Agric Food Chem 30:32–36
Tully RE, Beevers H (1976) Protein bodies of castor bean endosperm. Isolation fractionization and the characterization of protein components. Plant Physiol 58:710–716
Van der Wilden W, Gilkes NR, Chrispeels MJ (1980) The endoplasmic reticulum of mung bean cotyledons role in the accumulation of hydrolases in protein bodies during seedling growth. Plant Physiol 66:390–394
Vitale A, Smaniotto E, Longhi R, Galante E (1982) Reduced soluble proteins associated with maize endosperm protein bodies. J Exp Bot 33:439–448
Yatsu LY, Jack TJ (1968) Association of lysosomal activity with aleurone grains in plant tissue. Arch Biochem Biophys 124:466–471
Youle RJ (1977) The proteins of *Ricinus* and other oilseeds: protein bodies storage proteins lectins and allergens. PhD Thesis, Univ South Carolina
Youle RJ, Huang AHC (1976) Protein bodies from the endosperm of castor bean. Subfractionation protein components lectins and changes during germination. Plant Physiol 58:703–709
Youle RJ, Huang AHC (1978) Evidence that the castor bean allergens are the albumin storage proteins in the protein bodies of castor bean. Plant Physiol 61:1040–1042
Youle RJ, Huang AHC (1979) Albumin storage proteins and allergens in cottonseed. J Agric Food Chem 27:500–503

Lipid Bodies

A. H. C. HUANG

1 Introduction

Most seeds contain storage lipids in the form of triacylglycerols, which usually comprise 20–50% of the total seed dry weight (Appelquist 1975; Gurr 1980; Roughan and Slack 1982). This lipid reserve is rapidly mobilized to provide energy and carbon skeleton for the growth of the embryo during germination. The triacylglycerols are densely packed in subcellular organelles called lipid bodies (oleosomes, spherosomes, oil bodies). The spherical lipid body is about 1 μm in diameter, and is surrounded by a half-unit membrane of about 3 nm thickness (Fig. 1; Yatsu and Jacks 1972). The fatty acid moieties of the membrane phospholipids are believed to orient themselves toward the matrix so that they can form hydrophobic interaction with the internal triacylglycerols.

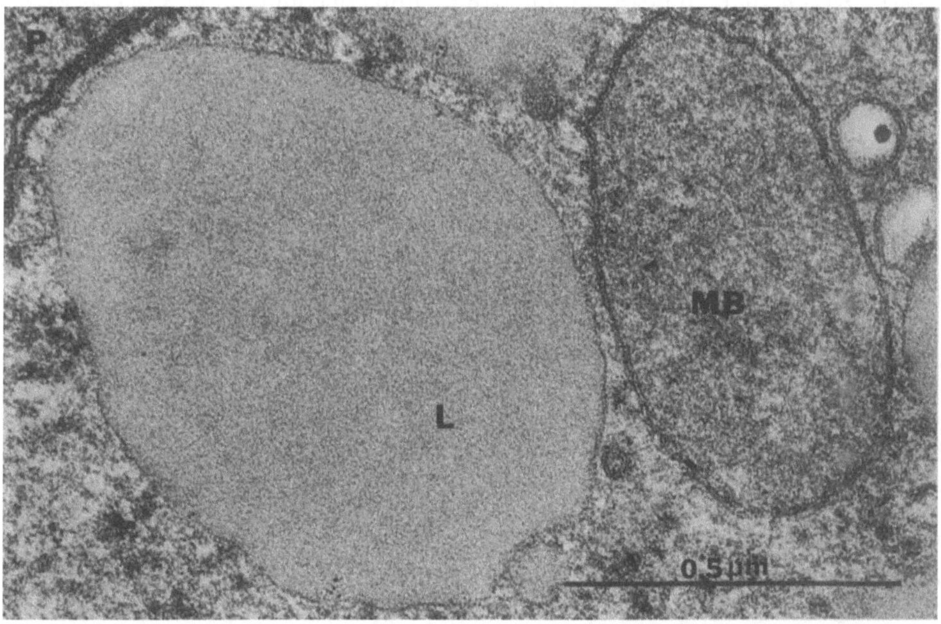

Fig. 1. Lipid body (*L*), microbody (*MB*), and small portion of a plastid (*P*), representing organelles surrounded by "half-unit" membrane, single membrane (one "unit" membrane, tripartite), and double membrane, respectively, in the subapical zone of a 1-day-old shoot apex of corn seedling. (Courtesy of R. N. Trelease 1969)

2 Ontogeny

The ontogeny of the lipid bodies is still unclear. In maturing seed, the lipid bodies do not contain any enzyme for triacylglycerol biosynthesis. Instead, the fatty acids are synthesized in the plastids (Appelquist 1975; Gurr 1980; Roughan and Slack 1982). The subsequent formation of mono-, di-, and tri-acylglycerols from activated fatty acids occurs in the endoplasmic reticulum. The mechanism of transport of fatty acids from the plastids to the endoplasmic reticulum, and of triacylglycerol from the endoplasmic reticulum to the lipid bodies is unknown. The origin of the lipid body membrane is also unclear. It has been suggested that the newly synthesized triacylglycerols in the endoplasmic reticulum are sequestered between the two phospholipid layers of the membrane at a particular region so that the hydrophobic triacylglycerols are stabilized by hydrophobic interactions (Schwarzenbach 1971, Wanner and Theimer 1978). Continuous deposition of the newly synthesized triacylglycerols at the same region eventually generates a budding vesicle of triacylglycerol surrounded by a half-unit membrane. The vesicle is then detached to become a lipid body. An alternative postulation states that the lipid bodies arise directly in the cytoplasm by condensation of triacylglycerol following by formation of the surrounding membrane (Bergfeld et al. 1978).

In seed germination, the triacylglycerols are hydrolyzed to fatty acids and glycerol in the initial step of mobilization (Galliard 1980; Huang 1984). In many seeds, lipase activity appears during germination concomitant with the decrease in the triacylglycerols. The lipase is associated with the membrane of the lipid bodies. The fate of the membrane components after the depletion of triacylglycerols is still unknown.

3 Isolation

Lipid bodies have an equilibrium density of less than 1.09 $g \cdot cm^{-3}$, and therefore can be isolated by flotation centrifugation of the tissue homogenate. When isolation is performed with mature seeds in which the lipid bodies contain no detectable lipase activity (see Sect. 4), the procedure can be performed at room temperature. When germinating or mature seeds are used, the procedure should be performed on ice or in a cold room (4 °C). The endosperm of castor bean, or the storage tissues of various other oil seeds in mature or germinating stage can be used.

The grinding medium contains 0.6 M sucrose, 1 mM EDTA, 10 mM KCl, 1 mM $MgCl_2$, 2 mM DTT, and 0.15 M Tricine buffer adjusted to pH 7.5 with KOH (Moreau et al. 1980). Dehulled castor bean is chopped with a new razor blade in grinding medium (15 g in 30 ml) on a Petri dish, and ground gently with a mortar and pestle. The homogenate is filtered through a piece of Nitex cloth (Petko, Elmsford, N.Y., USA) of pore size 20 μm × 20 μm or eight layers of cheesecloth. Each 15 ml of the filtrate is placed in one 40-ml centrifuge tube, and 15 ml flotation medium (grinding medium containing 0.5 M instead of 0.6 M sucrose) is layered on top. After centrifugation at 10,000 *g* for 10 min, the lipid bodies float to the top and are removed with a spatula. The lipid pad is carefully resuspended in 15 ml grinding medium with a rubber policeman by gentle stirring and pressing the unresuspended lumps against the tube wall. The 15-ml resuspended lipid bodies is again placed in a centrifuge tube, and a similar flotation centrifugation is performed. The procedure is repeated one more time (a total of three times flotation centrifugation). The final lipid pad is resuspended in 3-ml grinding medium.

4 Markers of Lipid Bodies

Naturally, triacylglycerols are the most important marker of the lipid bodies. There are no unusual phospholipid components in the membrane of the lipid bodies. The major phospholipids are phosphatidylcholine, phosphatidylethanolamine, and phosphatidylinositol (Moreau et al. 1980). On the contrary, the lipid body membrane contains unique protein components as revealed in SDS polyacrylamide gel electrophoresis (Bergfeld et al. 1978; Moreau and Huang 1979; Roughan and Slack 1982). If necessary, these protein components can be used as markers of the membrane, although the assay is tedious and nonquantitative.

Lipase activities may be present in the membrane of lipid bodies isolated from germinating seeds. The activity has been found in castor bean, corn, mustard, rape, and cotton (Huang 1984; Lin and Huang 1983; Moreau et al. 1980; Ory 1969; Ory et al. 1960). In other seeds such as soybean, sunflower, peanut, and cucumber (Huang 1984), the lipase activity is not detectable, and its absence may be due to the occurrence of inhibitors or components interfering with the enzyme assay. In those seeds with lipid bodies having detectable lipase activities, the enzyme may be tightly (castor bean, corn) or loosely (rape, mustard, and cotton) associated with the membrane. In the latter case, most of the enzyme in association with a small fraction of the membrane can be readily washed away from the isolated lipid bodies by simple grinding media. The removable lipase-containing membrane fraction has been postulated to represent either an extension of the lipid body membrane on which nascent lipase is synthesized (Wanner and Theimer 1978), or remanets of those activated lipid bodies in which the triacylglycerols have been hydrolyzed (Bergfeld et al. 1978).

Two exceptions are known in which the lipid bodies isolated from mature (desiccated) seeds contain lipolytic enzyme activities. In castor bean, the lipase is synthesized together with the lipid bodies in seed maturation, and persists throughout desiccation and germination until the triacylglycerols have been depleted (Moreau et al. 1980; Muto and Beevers 1974; Ory 1969; Ory et al. 1960). In soybean, an active acyl hydrolase which acts on monoacylglycerols but not di- and tri-acylglycerols is present in the mature seed (Lin and Huang 1982). During germination, its activity rapidly disappears before the depletion of the triacylglycerols. The function of the soybean hydrolase is unknown.

Several other enzymes are present in the lipid bodies of seeds. However, they are not unique to the lipid bodies or they have very restricted occurrence. NADH-

Cytochrome c oxidase is present in trace amounts in the lipid body membranes (Moreau et al. 1980). This enzyme does not seem to serve any metabolic role; its presence probably reflects the origin of the lipid body membrane directly or indirectly from the endoplasmic reticulum. Jojoba seed is unique among oil seeds because it is the only oil seed known to contain intracellular wax esters instead of triacylglycerols as food reserve. In the membrane of the lipid bodies (wax bodies) from jojoba seedlings, a wax esterase, a fatty alcohol oxidase, and a NADH-fatty aldehyde dehydrogenase are present (Moreau and Huang 1979). These enzymes catalyze the hydrolysis and conversion of storage wax esters to fatty acids.

The lipid body membrane can be obtained from the isolated lipid bodies (Jack et al. 1967; Moreau et al. 1980; Ory et al. 1960). Diethyl ether apparently can penetrate the membrane and solubilize the matrix triacylglycerol. To 6 ml of the resuspended lipid bodies (from 30 g castor bean) in a tube equipped with a Teflon screw cap, 6 ml diethyl ether is added. After shaking, the ether is removed by aspiration. The extraction is repeated two more times. After the third extraction, a stream of nitrogen is passed into the preparation in order to evaporate the remaining ether. After the extraction, about 50% of the membrane remains intact and the rest is solubilized (Moreau et al. 1980). The whole preparation can be used for further study. Alternatively, the preparation can be centrifuged at 100,000 *g* for 2 h to separate the membrane in the pellet and the solubilized components in the supernatant. If desired, the membrane fraction can be viewed under the electronmicroscope (Fig. 2; Jack et al. 1967; Trelease 1969).

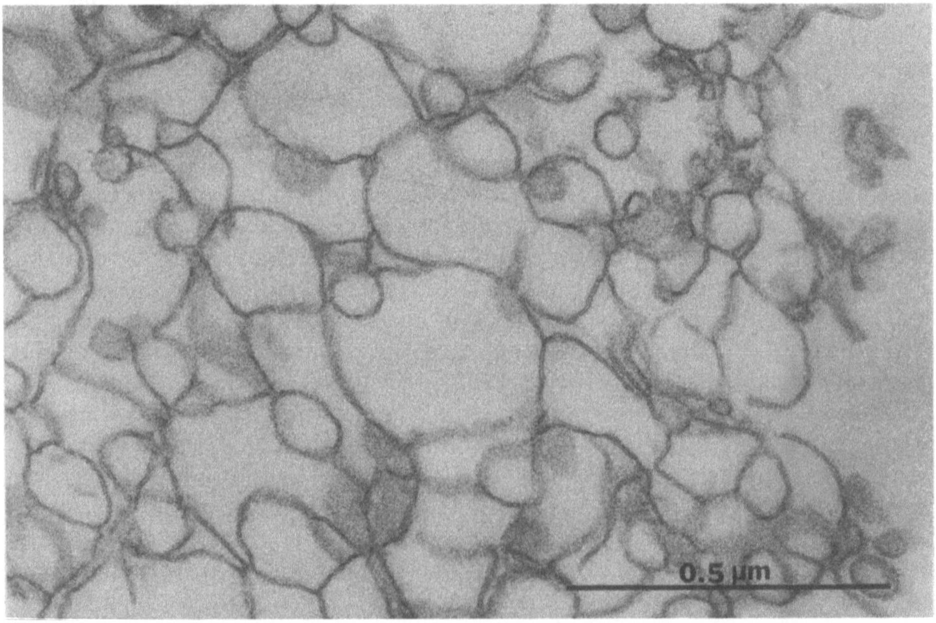

Fig. 2. Isolated lipid body membrane of corn scutella. The membranes appear as tripartite structures which are interpreted to represent closely appressed half-unit membranes of lipid bodies after the matrix triacylglycerols had been extracted. (Courtesy of R. N. Trelease 1969)

5 Assays

Two methods of assaying the marker enzyme lipase are described. The first method is a fluorometric assay of acyl hydrolase activity using an artificial substrate (Lin and Huang 1983; Muto and Beevers 1974), and the other method is a colorimetric assay of true lipase activity (Lin and Huang 1983; Nixon and Chen 1979). A true lipase (EC 3.1.1.3) is a glycerol ester hydrolase that releases fatty acids from insoluble triacylglycerols. The fluorometric method is a very convenient assay for the activities of lipase, as well as nonlipase acyl hydrolase, which is present in other compartments of the tissue. Thus, the assay is not specific, and should only be used on lipid bodies after they have been isolated (e.g., in subfractionation of the protein components). The colorimetric method is relatively time-consuming but measures only the true lipase activity.

5.1 Fluorometric Assay

Acyl hydrolase activity is measured at room temperature in a reaction mixture of 4 ml, containing 0.1 M Tris-HCl buffer, pH 7.5 (depending on specific enzyme), 2 mM dithiothreitol, and enzyme preparation. The reaction is initiated by the addition of 0.1 ml of 3.3 mM (final concentration, 0.83 mM) N-methylindoxylmyristate (from I.C.N. Pharmaceuticals, Cleveland, Ohio, USA) dissolved in ethylene glycol monomethyl ether. Fluorescence measurements are made with a Turner Model 111 fluorometer with excitation filter No. 405 (405 nm maxima) and emission filter No. 2A–12 (>510 nm) attached to a recorder. The reaction is usually linear for the first 10 min. Although unnecessary, a more sophisticated fluorometer can be used. The activity is expressed on a relative basis of ΔF unit per unit time.

In the past few years, the commercial supply of N-methylindoxylmyristate was unreliable. If it cannot be obtained, fatty esters of 4-methylumbelliferone are used (Hasson and Laties 1976). Many fatty derivatives of 4-methylumbelliferone are available commercially (from Sigma Corps., St. Louis, Mo, USA, and several other biochemical companies). Lipase from a particular seed species may exhibit some degree of specificity towards the fatty derivatives. Nevertheless, 4-methylumbelliferyl laurate is probably an active substrate of lipases from most if not all sources. In the assay, 1 mM of 4-methylumbelliferone laurate is used instead of 0.83 mM N-methylindoxylmyristate, and excitation filter No. 7–60 (365 nm Maxima) is used instead of No. 405. If desirable, the actual activity in nmol min^{-1} can be calculated from a standard curve of fluorescence units versus 4-methylumbelliferone (stable and commercially available) concentrations in the assay system minus enzyme at the same pH.

5.2 Colorimetric Assay

The true lipase from an individual seed species exhibits high activities on the native substrates which are present in the same seed species (Huang 1984). Most seeds contain triacylglycerols of oleic or linoleic acids, and the lipase from these seeds are generally active on triolein or trilinolein. Therefore, either of the two substrates is used for the lipase assay. The substrate should be free of contaminating monoacylglycerols which are also substrates for non-lipase acyl hydrolases in the same tissues. The substrate preparation obtained commercially should be checked by thin layer chromatography for purity irrespective of the claims of purity made by the manufacturers. The substrate is dissolved in chloroform (5–10 μg/10 μl for each spot) and spotted onto a TLC plate coated with 250 μm of Silica Gel G (Brinkman Instruments, Inc., Westbury, N.Y., USA). The plate is developed in 50:50:1 (v/v/v) hexane:diethyl ether:acetic acid, and allowed to react with iodine vapor (put some crystals of iodine in a TLC tank). Standards of tri-, di-, and mono-olein/linolein and the free acid (5–10 μg/10 μl for each spot) are run on the same TLC plate. The mobilities of the components in descending order are triacylglycerol, fatty acid, 1,3-diacylglycerol, 1,2-diacylglycerol, and monoacylglycerols. A small percent of free fatty acid in the triacylglycerol preparation will not interfere with the assay. If monoacylglycerol is present, the triacylglycerol preparation is purified on a similar TLC plate. The triacylglycerol preparation is applied as a line across the whole plate (200 μg per one 20 cm × 20 cm plate). A similar but smaller plate (20 cm × 5 cm) is run in the same tank as a marker. After development, the small plate is allowed to react with iodine vapor. By making a comparison between the two plates, the position of triacylglycerol in the large plate can be identified. The silica gel containing the triacylglycerol line is scrapped and extracted with chloroform. After centrifugation to remove the silica gel, the chloroform supernatant is obtained, and the chloroform is evaporated.

The activity of true lipase is measured by a colorimetric method. Although the released fatty acids can be measured more rapidly using an automatic titrator, the colorimetric method has the advantage of requiring no specific equipment or setup. Furthermore, if many enzyme samples are to be assayed, the colorimetric method probably consumes about the same amount of time per assay, and all the assays can be performed simultaneously for a more uniform quantitation. In the colorimetric assay, the fatty acids produced are converted to copper soaps and measured using 2,2′-diphenylcarbazide. The reaction is performed at room temperature in a 5-ml tube. The 1-ml reaction mixture contains 0.1 M Tris-HCl, pH 7.5 (depending on specific enzyme), 5 mM dithiothreitol, 5 mM substrate, and enzyme preparation. Triolein or trilinolein (50 mM) is first emulsified in 2 ml of 5% gum acacia for 1 min at low speed with a Bronwill Biosonic IV ultrasonic generator fitted with a microprobe. The reaction is stopped at time intervals (ranging from 5 min, to 2 h, depending on the amount of enzyme used) to ensure that proper kinetics are observed. Each 0.1 ml aliquot of the reaction mixture is put in a 7-ml tube and boiled in a boiling water bath for 5 min. After cooling to room temperature, 4 ml of chloroform heptane methanol (4:3:2, v/v/v) is added. The tube is closed with a Teflon screw cap, and shaken horizontally for 15 min. Two ml of 0.1 M sodium phosphate, pH 2.5 is added, and the tube is shaken horizontally for 3 min. After centrifugation in a table-top centrigufe, the upper layer of methanol water is pipetted out and discarded. One ml of 0.01 M HCl is added, and the tube is shaken horizontally for 3 min. After contrifugation, the upper layer of HCl solution is pipetted out and discarded. One and half ml of copper reagent (0.1 M $Cu(NO_3)_2$, 0.2 M triethanolamine, 0.06 N NaOH, and 6 M NaCl) is added. The tube is closed with a Teflon screw cap, and shaken horizontally for 30 min. After centrifugation in a table-top centrifuge, 2 ml of the chloroform layer is transfered to a tube, and 0.1 ml of color reagent (10 ml freshly prepared 0.4%, 2,2′-diphenylcarbazide solution in 100% ethanol, plus 0.1 ml 1 M triethanolamine added immediately before use) is added. After 5 min or more, the absorbance is read at 550 nm. Oleic acid or linoleic acid is used to produce a standard curve that is linear up to a concentration of 0.05 μmol per 2 ml chloroform.

References

Appelqvist LA (1975) Biochemical and structural aspects of storage and membrane lipids in developing oilseeds. In: Galliard T, Mercer EI (eds) Recent advances in the chemistry and biochemistry of plant lipids. Academic Press, London, pp 247–283

Bergfeld R, Hong YN, Kühnl T, Schopfer P (1978) Formation of oleosomes (storage lipid bodies) during embryogenesis and the breakdown during seedling development in cotyledons of *Sinapis alba* L. Planta 143:297–307

Galliard T (1980) Degradation of acyl lipids: hydrolytic and oxidative enzymes. In: Stumpf PK, Conn EE (ed) The biochemistry of plants, vol 4. Academic Press, New York, pp 85–116

Gurr MI (1980) The biosynthesis of triacylglycerols. In: Stumpf PK, Conn EE (ed) The biochemistry of plants, vol 4. Academic Press, New York, pp 205–248

Hasson EP, Laties GG (1976) Separation and characterization of potato lipid acylhydrolases. Plant Physiol 57:142–147

Huang AHC (1984) Plant lipases. In: Brockman HL, Borgstrom B (eds) Lipolytic enzymes. Elsevier, Amsterdam, pp 419–442

Jack TJ, Yatsu LY, Altschul AM (1967) Isolation and characterization of peanut spherosomes. Plant Physiol 42:585–597

Lin YH, Huang AHC (1982) Involvement of glyoxysomal lipase in the hydrolysis of storage triacylglycerols in the cotyledons of soybean seedlings. Plant Physiol 70:108–112

Lin YH, Huang AHC (1983) Lipase in lipid bodies of cotyledons of rape and mustard seedlings. Arch Biochem Biophys 225:360–369

Moreau RA, Huang AHC (1979) Oxidation of fatty alcohol in the cotyledons of Jojoba seedlings. Arch Biochem Biophys 194:422–430

Moreau RA, Liu KDF, Huang AHC (1980) Spherosomes in castor bean endosperm. Membrane components, formation, and degradation. Plant Physiol 65:1176–1180

Muto S, Beevers H (1974) Lipase activities in castor bean endosperm during germination. Plant Physiol 154:23–28

Nixon M, Chen SHP (1979) A simple and sensitive colorimetric method for the determination of long-chain free fatty acids in subcellular organelles. Anal Biochem 97:403–409

Ory RL (1969) Acid lipase of the castor bean. Lipids 4:177–185

Ory RL, St. Angelo AJ, Altschul AM (1960) Castor bean lipase: action on its endogenous substrate. J Lipid Res 1:208–213

Roughan PG, Slack CR (1982) Cellular organization of glycerolipid metabolism. Annu Rev Plant Physiol 33:97–132

Schwarzenbach AM (1971) Observations on spherosomal membranes. Cytobiologie 4:145–147

Trelease RN (1969) Changes and characteristics of lipid bodies during development. PhD Thesis, University of Texas, Austin

Wanner G, Theimer RR (1978) Membranous appendices of spherosomes (oleosomes). Possible role in fat utilization in germinating oilseeds. Planta 140:163–169

Yatsu LY, Jacks TJ (1972) Spherosome and membranes. Half unit membranes. Plant Physiol 49:937–943

Chloroplasts as a Whole

G. A. BERKOWITZ and M. GIBBS

1 Introduction

Since the publication of the last edition of this series, the scope and depth of our understanding of the nature and regulation of photosynthesis has taken a rapid leap forward. Advancements in three areas are most noteable. The compartmentalization of photosynthetic processes has been elucidated with regards to different plant types (C_3, C_4, CAM, and intermediates of these classes). The regulatory interaction of the photochemical activities of the thylakoid and the functioning of the enzymes involved in carbon metabolism in the stroma; both in terms of photochemical mediated pH, Mg^{2+}, and light regulation of photosynthetic enzymes and the effects of carbon metabolism in the stroma on the redox poising of the photochemical apparatus have been characterized. Also significant is the advancement in our understanding of the regulation of plastic localized carbon metabolism by extra-chloroplastic milieu parameters such as pH, and the level of cation, Pi, energy charge and reducing equivalents; most likely representing an interdependence of chloroplast, cytoplasmic, and mitochondrial activities in situ.

These advancements have been primarily due to elegant work with isolated structurally and *functionally* intact chloroplasts prepared from a variety of higher plant tissue and algae. Not coincidentally, then, in the last several years many new techniques involved in chloroplast isolation, stabilization, and use in photosynthetic studies have been developed.

It is the goal of this article to present a working knowledge of some of the more recently developed and widely employed isolation procedures, and to acquaint the reader with the wide variety of techniques available to isolate and purify plastids from a range of photosynthetic organisms and tissue types. Therefore, in addition to the "standard" C_3 chloroplast from higher plants, sections will be devoted to isolation procedures for chloroplasts from C_4, and CAM plants, and from marine and freshwater algae. The wide range of plant types which can now be used to isolate chloroplasts, and the purity and integrity of these preparations is due in great part to the application in the last decade of two techniques to the field of chloroplast isolation. These techniques are the formation and subsequent rupture of protoplasts from parent plant tissue, and the purification of chloroplasts from plant cell extracts using iso-osmotic density gradients of silica sols. Therefore, these two subjects will be briefly discussed. Also included is a section on innovative and recently developed concepts concerning the optimization of isolation procedures.

2 Considerations of Integrity and Purity

In terms of intactness, most researchers have adopted the nomenclature of Hall (1972) in describing plastids that are structurally intact and fully functional as Type A chloroplasts, as compared with plastids whose limiting membranes have broken and resealed during isolation (Type B), and broken chloroplasts (Type C). Structurally intact chloroplasts are surrounded by an outer membrane which is freely permeable to (ionic and uncharged) molecules of a molecular weight up to 10,000 (Heldt and Sauer 1971), and an inner membrane – the area between these membranes is often referred to as the sorbitol permeable space. The inner envelope of an intact chloroplast, enclosing the stroma, is impermeable to the free diffusion of most cations, anions, and small neutral molecules such as sugars (Walker 1976), has functioning specific anion (Heldt 1980) and cation (Maury et al. 1981) translocators, regulates the transport of externally synthesized chloroplast proteins (Chua and Schmidt 1979), and has factors which catalyze the formation of galactolipids (Douce and Joyard 1979) and carotenoids (Costes et al. 1979). One standard measurement of structural integrity is the ferricyanide assay (Heber and Santarius 1970).

The rate of ferricyanide reduction can be assayed spectrophotometrically (A_{420}) or by concomitant O_2 evolution in chloroplasts before and after osmotic lysis in a reaction mixture with 1–3 mM ferricyanide and an uncoupler (5–10 mM NH_4Cl). Chloroplasts are most effectively lysed in $\leqq$25 mM buffer and 10 mM $MgCl_2$, although some algal chloroplasts have been reported to be resistant to rupture in hypotonic medium (Giles and Sarafis 1974; Trench et al. 1973). The ferricyanide assay does not distinguish between Types A and B plastids – it is only a measure of structural intactness and can overestimate functional integrity. Also, it has the problem of underestimating structural intactness when monitored in an oxygen electrode, as background O_2 evolution supported by ambient CO_2 cannot be accounted for. This latter problem can be alleviated by employing a "modified" ferricyanide assay which includes 10 mM D,L-glyceraldehyde to inhibit completely CO_2-supported O_2 evolution (Stokes and Walker 1972).

Most isolation procedures used today routinely yield chloroplast preparations that are at least 80% intact according to the ferricyanide assay. Appearance under phase contrast microscopy is another measure of intactness, although it is more laborious and a much smaller population of plastids must be examined. Intact plastids (both Type A and B) appear bright and highly refractive under phase contrast illumination (Leech 1964), although the presence of starch grains may confound the assay, as they also appear phase-bright (Ellis and Hartley 1982).

Functional integrity is often monitored by quantitating the extent of retention of exclusively plastid-localized enzymes during the isolation procedure; this technique distinguishes Type A chloroplasts. Chloroplast marker enzymes that can be assayed in various fractions during isolation and in the chloroplast resuspension before and after osmotic shock are: RuBP carboxylase using a radiochemical assay (e.g., Winter et al. 1982), $NADP^+$-dependent GAP dehydrogenase (Wolosuik and Buchanan 1979), nitrite reductase (Bourne and Miflin 1973), and alkaline FBPase (Latzko and Gibbs 1974). Care should be taken in determining enzymatic activity in various fractions such that the marker enzymes are "activated" fully in all assays. Although RuBP carboxylase is the most widely accepted chloroplast marker enzyme, it might be wise to examine the activity of several chloroplast marker enzymes. This is recommended because in some cases, a portion of RuBP carboxylase activity has been found to be associated with the pyrenoid of the algal chloroplast (Holdsworth 1971; Wright and Grant 1978) and in stromal crystalline bodies of higher plants (Willison and Davey 1976), resulting in retention of the enzyme by osmotically lysed chloroplasts (Wright and Grant 1978). Additionally, high reaction medium Mg^{2+} may cause artifactual association of soluble stromal proteins with lysed plastids to varying degrees (Kow and Gibbs 1982).

Chloroplasts with a high degree of functional integrity should be able to carry out CO_2 fixation (or concomitant CO_2-dependent O_2 evolution) at rates of 50–200 μmol mg^{-1} Chl h^{-1}, and should additionally display ribonuclease and protease-resistant RNA and protein synthesis (Ellis 1977) when supplied with proper substrates. They should be unable to perform a variety of activities, in addition to ferricyanide reduction as described above, such as; (1) reduction of exogenous NADP+, (2) exogenously supplied ferredoxin-dependent O_2 uptake, (3) dark CO_2 fixation when supplied with RuBP or R5P and ATP, and (4) hydrolysis of exogenous PPi. The inability of chloroplasts to perform these activities can be used to quantitate intactness (Walker 1980a).

Ideally, chloroplast preparations should be free of contaminating cell debris. Chloroplast preparations are standardly examined for cytoplasmic, mitochondrial, and microbody (i.e., peroxisome) contamination. Cytoplasmic contamination of an isolated chloroplast preparation can be quantified by assaying PEP carboxylase as described by Robinson and Walker (1979) or acid phosphatase using p-nitrophenyl phosphate as substrate as described by Lindhardt and Walter (1963). Mitochondrial contamination can be ascertained by assaying fumarase (Hampp 1980), NAD-dependent glutamate dehydrogenase (Miflin 1974), or Cyt c oxidase (Hampp 1979). Microbody contamination can be quantified by assaying catalase according to the procedures of Van Ginkel and Brown (1978). Endoplasmic reticulum contamination can be estimated by assaying NADPH-Cyt c reductase according to the protocol of Takabe et al. (1979). It should be noted that preparations of chloroplasts isolated from some *Panicum* species (Brown et al. 1983) and from siphonous algae such as *Caulerpa*, *Codium*, and *Acetabularia* (Grant and Wright 1980) actually envelope mitochondria and/or cytoplasm, making purification of chloroplasts from these contaminants impossible using standard isolation procedures.

3 Chloroplasts from Protoplasts

As opposed to initial procedures which involved mechanical maceration of leaf tissue (e.g., Edwards and Black 1971), virtually all protoplast isolation procedures used today for the eventual purification of chloroplasts rely on cell-wall digesting enzymes. Various commercial preparations of cellulase and pectinase are standardly used together in most procedures. However, Hampp and Ziegler (1980) have reported that a specific cellulase preparation (cellulysin, from Calbiochem) worked better alone with oat leaves (particularly when leaves were etiolated or only partially greened) than combined preparations of cellulase and pectinase from other sources. Horvath et al. (1978) used a third enzyme preparation (Helicase, from Industrie Biologique Française) which is particularly rich in callose-degrading (β-1,3-glucanase) activity (Evans and Cocking 1977), along with cellulase and pectinase to prepare protoplasts from maize bundle-sheath strands. As callose formation in cell walls is known to occur in response to biological stresses [such as wound injury; (Aist 1976)], it is interesting to speculate that the prior stress history of the parent plant material used for protoplast formation be considered when developing digestion media.

The isolation of chloroplasts from protoplasts has many advantages over direct isolation from mechanically macerated plant tissue, which can be summarized as follows: (1) an obvious reduction in the mechanical stresses imposed on chloroplast membranes which should increase the ratio of Type A to Type B plastids obtained in the intact plastid fraction, (2) a greater proportion of the total chloroplasts found in the parent tissue recovered as intact plastids after purification, (3) chloroplasts can be isolated from parent tissue which has a higher starch content [however, see Walbot (1977) and Walbot and Hoisington (1982) for effects of starch content on chloroplast isolation from protoplasts], (4) a reduction in the degree of exposure of chloroplasts to potentially harmful substances found in the cell walls and vascular strands of parent tissue, and (5) protoplasts remain functionally viable even after long periods of storage (Nishimura and Akazawa 1975; Rathnam and Edwards 1976).

Investigators should consider the prior preparation of protoplasts when developing chloroplast isolation procedures for plant material not previously dealt with by other investigators. The work of Edwards and co-workers (Edwards et al. 1978a, 1979; Huber and Edwards 1975; Rathnam and Edwards 1976) should be consulted with regards to the optimization of isolation parameters for the particular plant material used. A brief summary of experimental parameters which affect both the yield and activity of chloroplasts isolated from higher plant protoplasts is as follows: (1) The parent plant material; the most active chloroplasts are generally obtained from healthy leaf tissue. (2) Leaf and plant age is important. Generally, young tissue is more susceptible to enzymatic digestion [although Leegood and Walker (1979) have successfully applied the technique to flag leaves of mature wheat plants]. (3) A method of disrupting the epidermis is sometimes employed. Leaves can be stripped of one epidermis (removing the cuticle) by peeling [this can be laborious, and works best when large dicot leaves such as tobacco, spinach, and pea are used; (Leegood and Walker 1983)] or lightly scraped with a nylon brush, as is practiced with sunflower leaves (Edwards et al. 1978a). An alternative to disruption of the epidermis is vacuum infiltration of the digestion medium into the intracellular leaf space (Edwards et al. 1979). Leaves are then cut into thin strips of between 0.5 and 1 mm in width to increase the surface area exposed to the digestive enzymes. The width of the slices and whether the leaves are cut longitudinally or transversally affects the degree of protoplast formation

(Nagata and Ishii 1979). (4) Duration of the digestive treatment affects yield and quality (Huber and Edwards 1975; Nagata and Ishii 1979) and should be as short as possible (Maury et al. 1981). (5) Milieu parameters of the leaf slice incubation (i.e., digestion) process that have been documented as affecting the final yield and quality of chloroplasts are; temperature (Jensen 1979), illumination (Leegood and Walker 1983), pH (Chapman et al. 1980; Huber and Edwards 1975), osmoticum concentration (Otsuki and Takebe 1969), and the presence of reducing agents (Rathnam and Edwards 1976), cations (Rathnam and Edwards 1976), PPi (Edwards et al. 1978 b), and chelating agents such as EDTA (Edwards et al. 1978).

Chloroplasts are released from pre-formed protoplasts by passage through nylon mesh nets or syringe needles with varying orifice diameters. PVP is sometimes included in the breaking medium to sequester phenolics (Spalding et al. 1979).

Although the advantages of isolating chloroplasts from protoplasts are many, the disadvantages of the technique are not insignificant. The cost of the digestive enzymes is high. Also, most procedures involve a lengthy incubation period of several hours [although Nagata and Ishii (1979) have reported that the use of a pectinase purified from *Aspergillus japonicus* which contains endo-polygalacturonase and endo-pectin lyase activity greatly reduced incubation time]. The protoplast technique is most often used exclusively when the parent plant material is specialized tissue such as bundle-sheath cells of C_4 plants, CAM tissue, or guard cells of the epidermis. Most investigators use spinach or pea as parent material, primarily due to the wealth of information in the literature regarding chloroplasts isolated from these plants. It is recommended here that when spinach or pea is used as parent plant material, a chloroplast isolation technique involving mechanical tissue maceration and purification of intact plastids using iso-osmotic silica sols be employed. This is because the technically simple method of mechanical maceration works well in releasing active chloroplasts from spinach and pea laminae, and purification using silica sol gradients can quantitatively separate these active chloroplasts from contaminating organelles and broken plastids.

The previous discussion has focused on the use of protoplast formation to obtain chloroplasts from higher plants. It should be noted that newly developed procedures to isolate intact chloroplasts from algae of the genera *Euglena*, *Chlamydomonas*, and *Dunaliella* all involve the use of "protoplasts" in the sense that the starting material for plastid isolation in these cases is cell-wall-less.

4 The Use of Silica Sols in Density Gradient Purification of Chloroplasts

The information presented here is essentially condensed from the incisive and thorough discussions presented by Price (1982, 1983) and Schmitt and Herrman (1977). These references are recommended to the reader for a more technical and extensive review of the use of silica sols and density gradients for chloroplast purification.

Most early chloroplast isolation procedures were based on the techniques developed by Jensen and Bassham (1966) and Walker (1964). These procedures involved the use of differential centrifugation to purify chloroplasts from homogenized leaf brei. Differential centrifugation separates particles on the basis of their sedimentation coefficients, and generally cannot resolve particles whose sedimentation coefficients differ by less than a factor of 10. In practical terms, differential centrifugation can substantially resolve chloroplasts from other cell organelles, but only partially resolves intact plastids from chloroplasts which have been stripped of their limiting membranes and have lost their stroma. Repeated centrifugation enhances the resolving power of differential centrifugation, but the plastid membranes are ruptured during the repeated pelleting. Therefore, differential centrifugation alone is not recommended for purification of intact chloroplasts.

A development of great import to the isolation of intact chloroplasts was the application of density gradient centrifugation techniques. Density gradient centrifugation allowed for much higher particle resolution than was obtained with differential centrifugation of cell extracts. Density gradient centrifugation involves centrifugation of subcellular components through a solution with a density greater than that of the contaminating particles in an intact chloroplast isolation procedure. Density gradient centrifugation provides two kinds of separation, rate-zonal and isopycnic. The two kinds of separation are really differentiated, in practical terms, by the fact that isopycnic centrifugation occurs when particles are centrifuged in a density gradient which encompassed all densities of the particles for a long enough time for the particles to reach their equilibrium position in the density gradient. If all the particles do not reach their equilibrium positions, then the technique is considered rate-zonal. Density gradient centrifugation can involve the use of continuous linear density gradients, step gradients, or the employment of a single "step" or cushion, which is of a sufficient density to allow intact plastids to pellet through, but is denser than all other subcellular organelles and broken plastids. Obviously, the single cushion is most easily prepared, and is recommended as adequate in many isolation procedures. A modification of the single-cushion technique is sometimes employed where leaf brei is centrifuged through two steps. The bottom step is of sufficient density that intact plastids cannot penetrate the cushion and therefore form a band between the two steps. This technique allows intact chloroplasts to be purified without deleterious exposure to pelleting.

Initial application of density gradient centrifugation to the isolation of intact chloroplasts involved the use of sucrose density gradients (Nishimura et al. 1976; Rathnam and Das 1974). However, the concentrations of sucrose employed in these procedures exposed plastids to severe hypertonic dehydration which irreversibly inhibits subsequent photosynthetic capacity (Kaiser et al. 1981; Plaut 1971). Ficoll and other organic polymers were also found to be unacceptable as gradient materials because of their high viscosity, which required long centrifugation times and high speeds and therefore exposed plastids to damaging hydrostatic pressures (cf. Price and Reardon 1982).

Silica sols such as Percoll and Ludox have a low viscosity, low osmotic strength due to their high particle weight, and are commercially available at sufficient densities to be used in gradients to purify intact chloroplasts. Therefore, silica sols have proved to be excellent components of density gradients, and have been and are today widely employed in the isolation of highly intact chloroplasts which are substantially free from contaminants and fix CO_2 at high rates. Morgenthaler et al. (1974, 1975) first detailed the use of Ludox AM density gradients to isolate chloroplasts which had a high degree of intactness and functional integrity. Their protocol involved the prior purification of silica sol with cation exchange resin and activated charcoal (Morgenthaler et al. 1974), and the addition of protective agents such as PEG, BSA, and Ficoll (Morgenthaler et al. 1975). Other protective agents which have been added to Ludox sols are antioxidants (thiol reagents and ascorbate), polyvinyl sufate, mercaptoethanol, and dextran (Hampp and Ziegler 1980, cf. Schmitt and Hermann 1977). Although purified Ludox is now available commercially, isolation procedures still involve the use of protective agents (Price and Reardon 1982).

More recently, Takabe et al. (1979) have demonstrated that density gradients of Percoll (with Ficoll and BSA) yield highly intact, active chloroplast preparations. Since then, dialyzed Percoll (50 ml against four changes of one l distilled H_2O over 24 h) alone has been demonstrated to function well as a density gradient material in the purification of highly active, intact plastids (Stitt and Heldt 1981). The simplicity of preparation of Percoll, then, makes its use in density centrifugation preferable over Ludox. Another advantage that Percoll has is its ability to form self-generated density gradients in 10–30 min.

With regards to the previous discussion, the choice of chloroplast isolation protocols subsequently presented is based on a combination of simplicity and speed of isolation procedures balanced against the functional quality and purity of the chloroplast preparations which result from these procedures.

5 General Notes on Isolation Procedures

The following recommendations pertain to work with higher plants.

a) Leaf Material. Most plants yield "good" chloroplasts only when harvested fresh. The exception is field or greenhouse grown spinach, which can be kept at 0 °C–5 °C for a day or two to reduce starch content. Chloroplasts isolated from field or greenhouse grown spinach are often more active than preparations from plants grown in controlled environment cabinets (this is most likely due to a reduction in light quality and intensity). Sometimes leaf tissue is surface sterilized in weak (<5% v/v) solutions of Clorox with a wetting agent (e.g., 0.1% v/v Tween 80) and always rinsed with distilled water prior to use. For further details regarding growth of spinach, the reviews by Walker (1980a) and Leegood and Walker (1983) are recommended.

b) Preparation of Crude Protoplast Extract or Leaf Brei. Time and temperature are crucial factors. The grind medium for mechanical maceration of leaf tissue should ideally be chilled to a semi-frozen slush. This may be accomplished by stirring the medium in a bath of ethyl alcohol kept in a deep freeze until the medium has the correct consistency (Leegood and Walker 1983). Aside from the incubation of laminar material in digestion medium during protoplast formation, most chloroplast purification procedures stress that isolation media be chilled to 4 °C. This cold requirement is generally accepted as protecting against degradation of the chloroplast envelope by decompartmentalized (vacuole) enzymes released during cell rupture. Additionally, chloroplasts should be centrifuged away from crude leaf brei or protoplast extract during isolation procedures as quickly as possible. This is done to minimize exposure of plastids to quinones during cell disruption.

Standard instruments used to macerate leaf tissue are the Polytron, Virtis grinder, and Waring Blendor. Although most expensive, the Polytron is often recommended as producing the highest yield of intact plastids from a given quantity of leaves. Homogenization times should be kept under 5 s, as longer periods only increase the proportion of broken plastids (Leegood and Walker 1983).

c) Isolation Media. Isolation media should be kept essentially microbe-free. When making up stock solutions, vacuum microfiltration through a 0.45-µm filter allows refrigerated solutions to be usable for several months. After dialysis, Percoll should be sterilized. Price and Reardon (1982) recommend heat sterilization (although pressure-release times during autoclaving should be carefully monitored to avoid evaporation and subsequent change in density and crystallization of the silica sol). Forcing the silica sol through a 0.45-µm filter attached to a large-volume syringe is an option. Percoll residues carried over from the grind medium with the recovered intact plastid fraction may interfere with subsequent silicone oil centrifugation procedures (Berkowitz and Gibbs 1983), possibly due to increased chloroplast clumping. Therefore, the intact plastid fraction should be washed once after isolation using Percoll (e.g., Mourioux and Douce 1981). Tris should not be a first choice as a buffer in isolation media, as it has some biocidic effects (Perrin and Dempsey 1974). When BSA is used in isolation media, it should be defatted by washing 100 g in one l of ethanol and one l of acetone. The BSA is recovered by filtering on a Buchner funnel. All solutions containing BSA should be microfiltered and not autoclaved.

d) Protoplast Formation. The most important point to be considered here is that the commercially available preparations of "pectinase" and "cellulase" are crude extracts from fungal cultures. By the very definition of their ability to digest cell walls, they incorporate many different enzymatic activities. Therefore, as the fungal sources for the commercial preparations of these enzymes varies, and the composition of leaf cell wall varies according to plant type (e.g., monot versus dicot), developmental stage, and prior exposure to various stresses, the choice of the optimum digestion medium for a particular leaf material is by its very nature empirical. If digestion media suggested in this chapter are not found adequate for a given application, preparations from other commercial sources, or titration of enzymes and particular leaf tissue can be tried. Most protocols do not recommend either desalting or purifying the digestion enzymes.

e) Phenolics. Most work with chloroplasts has been done with plastids isolated from spinach and pea; two plants which are relatively free from phenolics. High phenolic content has posed great problems during the isolation of functional chloroplasts from other plants. Problems arise when vacuoles are broken during tissue maceration, releasing a wide range of previously compartmentalized phenolic compounds into the tissue brei, where they are exposed to the action of diphenol oxidase [which is reported to be mitochondrial or chloroplast membrane-bound, or cytoplasmic; (Walker 1980 b)]. Diphenol oxidase (DPO) catalyzes the oxidation of o-dihydroxy-phenols to their corresponding o-quinones. These quinones polymerize with themselves or co-polymerize with proteins. This quinone condensation is the organelle-deactivating process (Walker 1980 b). In addition to the action of DPO, high pH's favor phenol auto-oxidation, which may be why chloroplast grind media are characteristically lower in pH than resuspension or reaction media. Protection of chloroplasts against phenolics and o-quinones during tissue maceration can be sought by several strategies. Tissue can be homogenized under inert atmospheres (although this is technically difficult). Use can be made of polymers such as BSA, PEG, PVP, and polycaprolactam, which afford protection by their high degree of binding to quinones and phenolics. Other control measures include DPO inhibitors such as copper chelators, DIECA, or substrate analogs such as cinnamic acid (Walker and McCallion 1980). Walker (1980 b) reports that the most widely used DPO inhibitors are MBT, metabisulfite, cysteine, and ascorbic acid. Thiols which are thought to inhibit DPO include thioglycolate, DTT, and mercaptoethanol. For a more extensive review of the phenolic problem in chloroplast isolation, the review of Loomis (1974) is recommended.

f) Storage of Intact Chloroplasts. After isolation, intact chloroplasts should be stored as a concentrated suspension at 0 °–4 °C in the dark. After final pelleting during isolation procedures, plastids can be resuspended in a small volume of appropriate medium by gentle swirling in a circular motion, or with either a soft paintbrush, frayed nylon netting attached to a glass rod, or a Teflon homogenizer.

g) Chlorophyll Analysis. Plastid concentration is usually expressed on a Chl basis. Chl is routinely determined by the method of Arnon (1949). An aliquot (e.g., 50 μl) of chloroplast stock suspension is added to 80% acetone (e.g., 10 ml final volume), incubated for 5 min (in the dark) and then centrifuged at 500 *g* for 5 min. Chl is calculated (after subtracting out background absorbance at 730 nm) by the equation: $\mu g\ ml^{-1}$ $(A_{645} \times 20.2) + (A_{663} \times 8.02)$.

6 Specific Isolation Protocols

6.1 Higher Plants

6.1.1 C_3 Plants

1. Pea [method of Mills and Joy (1980)]: Leaves and upper shoots of 9–12-day-old seedlings are diced into grind medium (0.33 M sorbitol, 50 mM Tricine-KOH pH 7.9, 2 mM EDTA, 1 mM $MgCl_2$ and 0.1% BSA) with a tissue: volume ratio of 1:4 and homogenized for 5 s with a Polytron PT20. Brei is filtered through two layers of muslin and two layers of Miracloth into 50-ml centrifuge tubes (30 ml/tube) and underlayered with approximately 14 ml (3.5 cm) of grind medium which has 40% (v/v) Percoll and no EDTA or $MgCl_2$. Tubes are centrifuged in a swinging bucket rotor (HB-4, Sorvall) at 2500 *g* for 1 min. Intact chloroplasts are recovered as a pellet after aspiration of the medium in the tubes.

Notes: Mills and Joy (1980) have indicated that this procedure can be completed in 5 min and yields preparations that are 93% intact. The intact plastids (10% of original Chl recovered) have photosynthetic rates over 100 μmol (either CO_2 fixed or O_2 evolved) mg^{-1} Chl h^{-1} and are contaminated by 0.2% of original peroxisomes and 0.2% of the original mitochondria. They found that BSA was essential to avoid plastid clumping during isolation. Leegood and Walker (1983) indicate that small amounts of Pi in isolation medium also prevent pea chloroplast clumping. Due to the action of a PPi/adenosine nucleotide antiporter (Robinson and Wiskich 1977), PPi should always be excluded from isolation medium. Other isolation procedures for pea chloroplasts include the technique of Mills and Joy (1980), where chloroplasts are prepared from isolated protoplasts, the techniques of Cline et al. (1981), which uses a preformed Percoll gradient after tissue homogenization, and the relatively simple and inexpensive technique of Mills et al. (1980) which involves mechanical maceration of leaves and purification of plastids by differential centrifugation, and yields preparations which are 86% intact, have photosynthetic rates of 85 μmol mg^{-1} Chl h^{-1} and show a 1% and 0.4% contamination by mitochondria and peroxisomes, respectively.

2. Spinach [method of Mourioux and Douce (1980), with minor changes]: Deribbed leaves (10–15 g) from 3–8 week-old spinach plants are sliced into 50 ml grind medium (0.33 M mannitol or sorbitol, 30 mM Mops-NaOH (pH 7.8, 2 mM EDTA and 0.15% BSA) amd homogenized in a Polytron for 3–5 s. Brei is filtered through four layers of muslin and two layers of Miracloth and centrifuged either in an MSE bench top centrifuge for 30 s at 2200 *g* (total time, from start to hand-braking is 90 s) or in a SS-34 (Sorvall) rotor in a refrigerated centrifuge for 50 s at 750 g. Pellets are resuspended in grind medium and layered (2 ml/tube) on two pre-formed Percoll gradients (34 ml of grind medium with 50% Percoll is centrifuged at 3 °C in an SS-90 vertical rotor for 100 min at 10,000 g). After centrifugation at 5000 *g* for 10 min, a green band containing intact chloroplasts is recovered near the bottom of the gradients, diluted with grind medium (10:1) and centrifuged at 3500 *g* for 90 s in an SS-34 rotor. After aspiration of the supernatant, the pellet is gently resuspended with grind medium.

Notes. This procedure can be completed in 25 min, and yields a plastid preparation which is 98% intact, has no detectable contamination from other cell constituents, and demonstrates photosynthetic rates over 100 μmol mg^{-1} Chl h^{-1} (Mourioux and Douce 1981). Alternatives to this procedure are; the method of Berkowitz and Gibbs (1982a, b), which involves a mechanical grind and short (60 s) centrifugation of crude plastid preparations through a 40% Percoll cushion, and yields preparations which are 90% intact, demonstrate rates over 100 μmol mg^{-1} Chl h^{-1} and have no detectable peroxisome or mitochondrial contamination, and the method of Stitt and Heldt (1981), which involves protoplast formation and subsequent purification of both protoplasts and subsequently formed chloroplasts on Percoll step gradients. Suggested protocols for special applications are the isolation techniques of Demmig and Gimmler (1983) and of Price and Reardon (1982). The procedure of Demmig and Gimmler, which involves mechanical leaf maceration and purification of plastids by differential centrifugation, yields plastid preparations which show relatively high degrees of intactness (85–95%) and rates of photosynthesis (over 150 μmol mg^{-1} Chl h^{-1}) for a procedure which does not involve density gradient purification. Due to the low cost [expensive silica sols are not involved, and the cost of materials can be reduced further by replacing the Mes buffer and NaCl by 10 mM $Na_4P_2O_7$ brought to pH 6.5 with HCl as suggested by Leegood and Walker (1983)] and simplicity, this procedure is recommended for instructional classes. The isolation technique of Price and Reardon (1982), which involves processing kilogram quantities of leaves and purification using isopycnic separation on Percoll/PEG/BSA/Ficoll density gradients using continuous flow centrifugation through a zonal rotor, is recommended when large quantities of intact plastids are needed.

3. Wheat [method of Edwards et al. (1978), as modified by Leegood and Walker (1979)]: Six to 7 g of mature leaves (e.g., flag leaves removed from plants following anthesis) are sliced into 0.5 to 1.0 mm wide strips with razors and incubated in a 19 cm diameter, 10-cm-high crystallizing dish for 3 h in the light at 28 °C in 40 ml of extraction medium (0.5 M sorbitol, 1 mM $CaCl_2$, 0.5% (w/v) BSA, 2.5% (w/v) cellulase (Onozuka 3S), 0.3% (w/v) pectinase (Macerozyme) and 5 mM Mes, pH 5.5). The solution, which should contain no released protoplasts, is discarded and the protoplasts are released from the partially digested strips by gently chopping them with the blade of a spatula. The tissue is washed three times by shaking in and collecting 20 ml aliquots of 0.5 M sorbitol, 1 mM $CaCl_2$, 5 mM Mes (pH 6.0). The pooled washed medium is filtered through two nylon nets (a tea strainer with 1 mm apertures and a nylon cloth with 195 μm apertures) and centrifuged in six 12 × 125 mm tubes at 100 *g* for 5 min. The pellets are resuspended in about 0.1 ml/pellet of 0.5 M sucrose, 1 mM $CaCl_2$, and then 5 ml of the same solution is mixed into each tube. Two ml of 0.4 M sucrose, 0.1 M sorbitol, and then 1 ml of 0.5 M sorbitol (both solutions also contain 1 mM $CaCl_2$ and 5 mM Mes-KOH, pH 6.0) are carefully layered into each tube. After centrifugation at 200 *g* for 5 min in a swinging bucket rotor, the band at the interface of the upper two layers is collected from each tube. These pooled protoplast fractions are then washed by adding 0.4 to 1 ml of protoplast fraction to 5 ml of 0.5 M sorbitol, 1 mM $CaCl_2$ and centrifuging at 250 *g* for 2 min. The protoplast pellets are resuspended in 0.4 to 1.0 ml of breaking medium (0.33 M sorbitol, 25 mM tricine, pH 8.4, 10 mM $NaHCO_3$, and 10 mM EDTA), and sucked up 3 times into a 1-ml disposable syringe which had the tip cut to leave a 2 mm opening over which a 20 μm aperture nylon net is attached. The resulting protoplast extract is centrifuged at 250 *g* for 40 s and resuspended in breaking medium.

Notes. This technique yields chloroplast preparations (5% yield from parent tissue) which are 90% intact and have rates of photosynthesis exceeding 100 μmol mg^{-1} Chl h^{-1}. Leegood and Walker (1979) did not specify the level of contaminating organelles in the plastid preparation. As the plastids are purified from protoplast extracts by differential centrifugation, there may be a relatively high level of contamination. Therefore, investigators wishing to obtain purer preparations may wish to alter the protocol detailed here (and other isolation procedures involving protoplasts) by centrifuging protoplast extracts through a Percoll cushion [e.g., 40% (v/v)] to reduce organelle contamination. Other published protocols for isolating chloroplasts from agronomic cereal plants [oat (Hampp 1980; Nagata and Ishii 1979; Pilwat et al. 1980); barley (Rathnam and Edwards 1976); rice (Nagata and Ishii 1979); and wheat (Edwards et al. 1978a, b)] involve the use of seedlings (i.e., less than 3 weeks of growth) and so therefore may not be as useful for some applications. Nagata and Ishii (1979) have published procedures for chloroplast isolation involving protoplast formation from a number of C_3 species which involve very short incubation times (30 min). This protocol should be considered if time is a factor.

4. Other Species. Good protocols (Edwards et al. 1978a) have been published for chloroplasts prepared from such agronomic plants as sunflower (chloroplast preparations are 75 to 95% intact, with photosynthetic rates of 140 μmol mg^{-1} Chl h^{-1}) and tobacco (Nagata and Ishii 1979; Rathnam and Edwards 1976) where protoplast formation is involved. The protoplast method has been used to prepare chloroplasts from *Agropyron repens* and *Panicum bisulcatum* (Rathnam and Edwards 1976). Protocols have been published for isolating chloroplasts from agronomic species such as cucumber (Walden and Leaver 1981) and lettuce (Bamberger and Avron 1975), although high photosynthetic capacity, intactness, and degree of purity have not been established for these preparations. Chloroplasts have also been isolated from *Brassica* (i.e., wild turnip) weeds (Burke et al. 1982), *Senecio vulgaris* (i.e., common groundsel) weeds (Radosevich and Devilliers 1976), fescue grass (Krueger and Miles 1981), and *Lupinus polyphyllus* [i.e., lupins (Wink and Hartmann 1982)]. In addition to the above-mentioned leaf material, pea roots and castor been endosperm (Miflin and Beevers 1974) have been used as parent material for plastid purification. Schnabl and Hampp (1980) have assayed photochemical activity in crude extracts of guard cell protoplasts from *Vicia fava*. It seems likely that the simple purification of this protoplast extract on Percoll gradients (e.g., Mourioux and Douce 1981) will allow researchers in the future to work with preparations of guard cell chloroplasts which are quantitatively separated from mesophyll cell constituents and other guard cell organelles for the first time.

6.1.2 C_4 Plants [bundle-sheath and mesophyll cell chloroplasts isolated from fully expanded leaves of corn according to a modification of the procedures of Walbot and Hoisington (1982)]: Plants are suitable for plastid isolation at the six-leaf stage and beyond. Leaves can be "destarched" prior to use by a 24–28-h dark treatment if required. The level of starch in the leaves is crucial to the isolation procedure; mesophyll plastids should be devoid of starch, and the bundle-sheath plastids should contain a large number of small starch grains. As variety and growth conditions affect the size of starch grains accumulated in corn bundle-sheath plastids, these factors should be altered if needed (Walbot and Hoisington 1982). The size and number of bundle-sheath starch grains can be determined by viewing iodine-stained plastids under high power magnification. Leaf starch content is checked by making a few transverse sections of a sample leaf, and staining the sections with $KI\text{-}I_2$ in 70% ethanol for several minutes. Viewing the sections end-on under 30 × magnification should indicate that the bundle sheath is darkly stained and the mesophyll is unstained. Deribbed leaves (20–100 g) are cut into 1-cm segments with scissors into a sevenfold excess of grind medium (0.33 M sorbitol, 50 mM HEPES, pH 7.6, 5 mM $MgCl_2$, 0.1% (w/v) BSA, and 0.1 mM β-mercaptoethanol) and homogenized at full speed five times for 2–3 s in a Waring Blendor. The leaf brei is filtered through 64-μm and 20-μm aperture nylon netting. The material retained on the nylon is put in the same volume of grind medium as used initially, and homogenized three times for 10 s with a Polytron (PT20 probe type) at full speed and then filtered through the same size nylon nets. The first filtrate contains mostly mesophyll plastids, and the second contains some bundle-sheath plastids. The two fractions are centrifuged separately at 800 g for 15 min, resuspended in small volumes of grind medium, and quickly loaded on Ludox step gradients. The step gradients are prepared from Ludox HS-40 which is purified according to the method of Morgenthaler et al. (1974). The density of the purified Ludox is determined empirically by refractometry. This is done by making up 3%, 10%, and 30% (w/v) purified Ludox solutions in grind medium. A 10-ml aliquot of each solution is weighed in a volumetric flask and the refractive index of each solution determined to make a standard curve for that preparation of Ludox. The silica sol content of the stock, purified Ludox is then determined from the standard curve. PEG (Carbowax 6000) is then added (10% w/v) to the purified Ludox stock. Gradient steps of 3, 5, 7, 14, and 20% (w/v) final Ludox concentration are prepared from the stock Ludox, distilled water, and a 5 × gradient buffer stock solution of 25 mM $MgCl_2$, 250 mM Tris-HCl ph 7.8, 5% BSA, and 50 mM β-mercaptoethanol. Gradients are made in 15-ml glass centrifuge tubes using 3 ml of each step solution (the step with the highest % Ludox is put in the tube first). After quickly layering the crude plastid preparations on the step gradients, the tubes are centrifuged in a swinging bucket rotor (Sorvall HB-4) at 2600 *g* for 10 min. After the rotor comes to a stop (without braking), the tubes are inspected to see if the plastids have reached the 14/20% interface; if not, the tubes are then centrifuged for an additional 2–10 min at 2600 *g*. Mesophyll plastids are recovered from the 7/14% interface, and bundle-sheath plastids are recovered from the 14/20% interface. The partitioned bundle-sheath and mesophyll plastids are washed by diluting with two volumes of grind medium, and pelleting through a 5-ml cushion of 0.5 M sucrose, 50 mM HEPES (pH 7.6), 5 mM $MgCl_2$, and 0.1% (w/v) BSA.

Notes. This isolation procedure is the only protocol published to date for isolation of both bundle-sheath and mesophyll chloroplasts from fully expanded leaves of mature (i.e., past the six-leaf stage) plants of an agronomically important C_4 species. Walbot and Hoisington's (1982) original protocol had a yield of 10–20% extracted plastids, and the cross-contamination of plastid types was relatively high: 5–10%. Both the bundle-sheath and mesophyll plastids isolated by the original protocol were photosynthetically competent (V. Walbot, personal communication). Several important changes have been made in the protocol reported hare. The Tris buffer was replaced with HEPES, and the grind medium pH lowered slightly. Also, in the original protocol, the plastids are resuspended in grind medium with no sorbitol after the first centrifugation. Here, standard grind medium is used to prevent the breaking and resealing of plastids during isolation due to exposure to hypotonic conditions. Therefore, the crude chloroplast preparation layered on the step gradients is denser than in the original protocol, and might mix into the first step of the gradient. This is unimportant as long as the third step (containing 7% Ludox) is intact. The above-mentioned changes in the isolation protocol are suggested for the preparation of plastids to be used in photosynthetic studies. If the purified plastids are to be used for subsequent DNA purification, the original protocol (Walbot and Hoisington 1982) should be followed. An additional change in the protocol which is suggested here is the subsitution of Percoll for the Ludox in the step gradients. As indicated by Price (1982), the densities of Ludox HS-40 and Percoll are 1.295 and 1.13, respectively. Based on these densities, the 3, 5, 7, 14, and 20% Ludox steps can be replaced by 6.81, 11.35, 15.88, 31.77, and 45.38% Percoll, respectively. However, the exact Percoll concentration of each step (particularly the one replacing the 14% Ludox, which is crucial for the separation of contaminating plastid types in the gradient) should be determined empirically.

Other procedures have been developed for isolation of bundle-sheath and mesophyll chloroplasts from corn and C_4 plants, although these protocols require leaves from 3-week-old (or younger) seedlings. This is because the isolation procedures involve protoplast formation. With older tissue, bundle-sheath strands become resistant to enzymatic digestion. However, protocols which involve enzymatic digestion of seedling laminae for protoplast formation usually result in bundle-sheath and mesophyll chloroplast preparations with less cross-contamination than the procedure outlined above. Therefore, if the developmental status of the parent plant material is not critical to a particular investigation, then it is suggested that one of the following protocols be followed for isolation of plastids from C_4 plants. In addition to the protocol described above, Walbot and Hoisington (1982) detail a procedure for isolation of both mesophyll and bundle-sheath plastids with only 2% cross contamination from 3-week-old corn seedlings. Day et al. (1981) isolated mesophyll chloroplasts from corn which were 80% intact and were contaminated with only 1% bundle-sheath plastids. Horvath et al. (1978) have developed the only procedure to date which isolates functional bundle-sheath chloroplasts (along with mesophyll chloroplasts) from corn by an entirely enzymatic process; the action of Helicase aided cellulase and pectinase in digestion of the bundle-sheath strands. The methods of O'Neal et al. (1972) can be used to prepare plastids (from 4–6-day-old seedlings) which have not fully differentiated into typical mesophyll and bundle-sheath plastids. In addition to corn, excellent protocols have been developed for the isolation of chloroplasts from a range of C_4 plants, including plants of the NADP-malic, NAD-malic, and PEP-carboxylase groups (Edwards et al. 1979; Halberg and Larsson 1983; Rathnam and Edwards 1977; Walbot 1977).

There are reports in the literature (e.g., Morgan and Brown 1979) that several species of the Laxa group of the *Panicum* genus such as *P. miliodes*, *P. schenckii*, and *P. decipiens* display anatomical and physiological characteristics of an intermediate nature between C_3 and C_4 species of *Panicum*, and are collectively called the C_3–C_4 group. To date, there are no reports in the literature detailing procedures for the isolation of functional chloroplasts from these C_3–C_4 plants. Several groups of investigators have developed chloroplast isolation procedures for C_4-type *Panicum* species such as *P. milaceum* and *P. maximum* (Edwards et al. 1979; Rathnam and Edwards 1977; Walbot 1977). It seems likely that these techniques can be applied to C_3–C_4 species of the Laxa group of *Panicum*, providing future investigators with a new research tool; the isolated, intact chloroplast from a plant intermediate to C_3 and C_4 types.

6.1.3 CAM Plants [*Sedum praealtum*, isolated by the procedure of Piazza et al. (1982)]: Isolation is begun near the end of the solar day. Young, but fully expanded leaves from plants which had been vegetatively established for less than 8 months were thinly sliced (40 g) into cold 0.3 M sorbitol. Slices were washed once, gently vacuum-infiltrated for a few seconds, washed again with 0.3 M sorbitol, and transferred to 65 ml cooled digestion medium (0.3 M sorbitol, 0.9% (w/v) cellulase, 0.45% pectinase (Macerozyme), and 4% BSA at pH 5.6) in a large beaker on a bed of ice. After an overnight incubation, the medium was swirled, passed through a 210-µm nylon net, and centrifuged for 10 to 15 min at 20 *g* in a swinging bucket rotor. The pellets were resuspended in 40 ml 30% (w/v) dextran, 0.3 M sucrose, 1 mM $CaCl_2$, and 50 mM HEPES-NaOH (pH 7.8), swirled for 15 s in a small beaker, and placed in a centrifuge tube. Resuspension medium (except with 20% instead of 30% dextran) is then layered 1.5 cm above this solution in the centrifuge tube. A third layer, 0.5 cm deep, which contains resuspension medium with no dextran, is then added to the tube. The gradient is centrifuged at 165 *g* for 5 to 10 min in a swinging bucket rotor. Intact protoplasts are recovered at the top of the 20% dextran step, and resuspended in 40 ml breaking medium (0.2 M sorbitol, 5 mM EDTA, 5 mM $NaHCO_3$, 1 mM $MgCl_2$, 1% (w/v) BSA, and 0.2 M Tricine-NaOH, pH 8.2). Protoplasts are ruptured by drawing the medium into and expelling it quickly from a 5-ml disposable syringe fitted with a 22-gauge needle. This process is repeated, and the medium expelled into a centrifuge tube containing a 1.5 cm cushion of 8% (w/v) Ficoll, 0.3 M sucrose, 5 mM EDTA, 1 mM $MgCl_2$, 1 mM $MnCl_2$, and 50 mM HEPES-NaOH, pH 7.6. After centrifugation for 3 to 5 min at 100 g, intact plastids are recovered as a pellet and suspended in 2 ml of resuspension medium (0.33 M sorbitol, 5 mM EDTA, 1 mM $MnCl_2$, 1% (w/v) BSA, and 50 mM HEPES-NaOH, pH 7.6). This solution is layered on a preformed Percoll gradient and centrifuged at 165 *g* for 15 min. The Percoll gradient is prepared by centrifuging (31,000 *g* for 25 min) 10 ml of 80% Percoll in resuspension medium with no BSA in 100 × 16 mm tubes in a Sorvall SS-34 rotor. Intact, purified chloroplasts band near the bottom of the gradient.

Notes. This technique is modified from that of Spalding and Edwards (1980) to achieve higher yields of chloroplasts with a high degree of purity. Also, the overnight incubation at relatively low temperatures was done for convenience. However, plastids isolated by the method of Spalding and Edwards (1980) demonstrate a somewhat higher rate of CO_2 fixation. Nishida and Sanada (1977) isolated plastids from several CAM species, although only from leaves with below normal starch content.

Procedures for the isolation of functional chloroplasts from the inducible CAM plants, *Mesembryanthemum crystallinum* have recently been developed (Demmig and Winter 1983; Monson et al. 1983; Winter et al. 1982). If *M. crystallinum* plants are grown in low NaCl (20 mM) nutrient solution, the plants and chloroplasts isolated from the plants demonstrate typical C_3 photosynthetic characteristics. If 400 mM NaCl is introduced into the growing medium, then the plants exhibit CAM metabolism, and chloroplasts isolated from their leaves have metabolic characteristics typical of CAM rather than C_3 plastids. The CAM chloroplasts isolated from this plant according to the protocols developed by these investigators have significant advantages over the chloroplasts isolated from *Sedum praealtum* (Piazza et al. 1982; Spalding and Edwards 1980) and might prove useful to future investigators of CAM metabolism. As reported by Demmig and Winter (1983) and Winter et al. (1982), these plastids have far greater rates of photosynthesis than preparations of other CAM plastids (up to 150 μmol mg^{-1} Chl h^{-1}), have a very high degree of intactness and purity, and the yield is extremely high.

6.2 Algae

6.2.1 Volvocales

1. Chlamydomonas reinhardii [according to the procedure of Klein et al. (1983 b)]: Optimal results are obtained by using the strain 11-32/b which is grown in synchronized cultures, and harvested after 6 h in the dark period (when cell division has just been completed, and the algae are still in the sporangia form). The chloroplasts are small and compact, with a low starch content. About 100 ml of culture (30 μg Chl ml^{-1}) are harvested (600 *g*, 2 min) and resuspended in breaking medium (1 mM K-phosphate, pH 6.0, 0.5 mg ml^{-1} autolysine, 0.5 mg ml^{-1} BSA, 10 μg ml^{-1} cycloheximide) at a concentration of 50 μg Chl ml^{-1}. The sporangial cell walls are lysed during a 30 min incubation at 34 °C. Autolysine is prepared from *C. reinhardii* according to the procedure of Schlosser et al. (1976). Protoplasts are harvested by centrifugation at 600 *g* for 2 min and resuspended in ice cold lysing medium (35–40 μg Chl ml^{-1} in 5 mM K-phosphate, pH 6.5, 5% PEG 6000 (w/v), 0.004% digitonin (w/v), and 4 mg ml^{-1} BSA). The suspension is rapidly warmed in 30–45 s with gentle shaking to 30 ° to 32 °C and kept at this temperature for 15 s. The solution is then cooled as rapidly as possible by pouring into pre-cooled large beakers, centrifuged at 800 *g* for 2 min, resuspended in suspension media [Tricine-NaOH (pH 7.7), 0.15 M mannitol, 1 mM $MgCl_2$, and 2 mM EDTA at a concentration of 40 μg Chl ml^{-1}], and then centrifuged at 370 *g* for 1 min.

The pellets are resuspended (with gentle shaking) in a few ml of the suspension medium, resulting in a plastid concentration of about 250 μg Chl ml^{-1}. The plastids spontaneously form a dark green clump due to the absence of free Mg^{2+} ions. The dark green clumps are resuspended by forcing them several times through a plastic micropipette tip (o.d. 1 mm) until the solution appears homogenous. The crude plastid preparation is then centrifuged at 370 *g* for 1 min, taken up in a few ml of suspension medium, layered on top of a 1.5 cm high 50% Percoll cushion (in suspension medium) and centrifuged at 365 *g* for 1 min in a swinging bucket rotor. Intact plastids will band on top of the cushion (contaminating, unlysed protoplasts will pellet). The chloroplast band, together with most of the buffer, is resuspended and layered on top of a 3 to 3.5 cm high 40% Percoll cushion and centrifuged in a swinging bucket rotor for 1 min at 1250 *g*. Intact chloroplasts are recovered in the pellet.

Notes. The crucial aspect of this isolation procedure is the use of the digitonin under carefully controlled incubation times and temperatures. The digitonin is a nonspecific detergent, and will lyse the plastid membrane in addition to the plasmalemma under improper incubation conditions. The final preparation may be contaminated by 2–5% protoplasts (which do not break under hypotonic conditions while the plastids do – this is a convenient assay of protoplast contamination). The plastids may be contaminated by up to 21% peroxisomes and 12% mitochondria. They display rates of photosynthesis of 25–50 μmol mg^{-1} Chl h^{-1} and are stabilized during storage by DTT (Klein et al. 1983a).

Another procedure for the isolation of intact *C. reinhardtii* chloroplasts has recently been developed by Belknap (1983). The advantages of the Belknap protocol, which involves protoplast lysis by pressure shock with a Yeda press instead of with digitonin, is that the yield is higher than with the protocol of Klein et al. (1983b), and that batch cultures of algae can be used instead of synchronous cultures. However, the contamination of the plastid preparation by peroxisomes and mitochondria is greater than with the procedure of Klein et al.

2. Dunaliella marina [according to the procedure of Kombrink and Wober (1980)]: Cells are harvested after 3–8 days of growth under continuous light by centrifugation at 500 *g* for 10 min, and washed twice with 0.5 M sorbitol and 5 mM HEPES-NaOH (pH 7.5). Pellets are then resuspended in wash medium to a final concentration of 50–200 μg Chl ml^{-1}. DEAE-dextran (MW = 500,000) is added to the resuspension solution with stirring (final concentration of 1 mg mg^{-1} Chl) and allowed to adsorb to the cells for 1 min at 0 °C. The solution is then incubated at 30 °C for 5–6 min with occasional shaking. This treatment facilitated lysis of the plasmalemma in 99% of the cells in the preparation. After incubation, the liberated chloroplasts are centrifuged (1 min at 500 *g*), and the pellet carefully resuspended with a brush into a small volume of resuspension medium [30% (w/w) sucrose, 50 mM HEPES-NaOH, pH 7.5, 1 mM $MgCl_2$, 1 mM $MnCl_2$, and 2 mM EDTA]. This resuspension is then layered (4 mg/tube) on a linear gradient of 30–60% sucrose in resuspension medium. The gradient is then centrifuged (Beckman ultra-centrifuge, SW-27 rotor) with the lowest acceleration rates at 5,000, 10,000, 15,000 and 20,000 rpm for 15 min each, followed by 24,000 rpm for 4 h. The dark green band at the bottom of the gradient (d = 1.24 g cm^{-3}) contains intact plastids.

Notes. This procedure yields up to 40% of the chloroplasts in the cells as isolated, intact plastids. The lysis of the plasmalemma by exposure to polycations (poly-D,L-lysine, in addition to DEAE-dextran) is a novel approach to the difficult problem of the extraction of chloroplasts from algae, and could provide future researchers with another "tool" in the arsenal of isolation methods. The action of the polycation was optimized with regards to pH, concentration, and temperature. However, the purification of chloroplasts from the cell extract involved lengthy centrifugation in sucrose density gradients; the subsequently isolated plastids were not photosynthetically competent. Therefore, it is strongly suggested here that the use of the polycation lysis protocol be integrated with plastid purification using Percoll density gradient centrifugation by future researchers who seek *Dunaliella* plastids which display optimal photosynthetic activity.

6.2.2 (Ceramiaceae, Rhodophyta) – *Griffithsia monilis* [according to the procedure of Lilley and Larkum (1981)]: Young, pinkish-red cell strands (stored in aerated seawater no longer than 24 h) are pre-illuminated for 30 min in seawater at 15 °C. About 20 g (net weight) are then washed twice in a Petri dish with 20 ml of isolation medium [0.6 M glucose, 20 mM HEPES (pH 8.0), 1 mM $MgCl_2$, 1 mM $MnCl_2$, 2 mM EDTA, 100 mM KCl, 10 mM NaCl, 2 mM Na-isoascorbate, and 0.5% (w/v) BSA]. Strands are then chopped with a hand-held cutter (razors mounted in a block with 0.5 mm spaces between them) and the vacuolar sap discarded. Isolation medium (20 ml) is added to the Petri dish and the cells are crushed with a flat pad faced with coarse (No. 80) waterproof emery paper, releasing the plastids. The suspension is then filtered through a 100 µm nylon net and a layer of Miracloth. This process of crushing in isolation medium and filtering is repeated three more times and the pooled filtrates layered on density gradients in four 50-ml tubes. The density step gradients are prepared by sequentially underlaying at the bottom of each tube with a hypodermic syringe; 20 ml crude plastid preparation; then 3 ml isolation medium, then 1 ml isolation medium containing 9.5% Dextran T_{40} (from a stock made by adding 1.0 g Dextran to 10 ml isolation medium); and finally 3 ml of 13.8% Dextran T_{40} (1.5 g added to 10 ml medium) in isolation medium in which sucrose replaces glucose, and the isoascorbate and BSA are deleted. The tubes are centrifuged in a swinging bucket rotor (HB-4) using acceleration and deceleration control (60 s to accelerate to 1000 rpm) at 400 *g* for 2 min. Intact red plastids (rhodoplasts) are recovered from the third layer.

Notes. This isolation techniques yields rhodoplast preparations which are 78–85% intact. The maximum rates of photosynthesis achieved with these preparations was 180–225 μmol mg^{-1} Chl a h^{-1}; this rate was achieved from September to November (there is seasonal variability in plastid activity).

6.2.3 Siphonales

1. Caulerpa simpliciuscula [according to the procedure of Grant and Wright (1980)]: Algal fronds (stored less than 1 month) are pre-illuminated in seawater for 30 min. The material is then blotted dry, chopped into 1–2 cm pieces, and crushed into extracting medium [0.6 M NaCl, 50 mM Tricine, pH 7.6, 5 mM EDTA, 5 mM isoascorbate, 5 mM DTT, and 0.1% (w/v) BSA] with a tissue-: medium ratio of 1 : 3 (w/v). The extract is filtered through three layers of muslin and two of Miracloth, and centrifuged at 750 g for 1 min. The pellet is surface washed with suspension medium [1.2 M sorbitol, 10 mM Tricine, pH 7.6, 1 mM isoascorbate, 5 mM EDTA, 1 mM DTT, and 0.25% (w/v) BSA], and then gently resuspended using a glass rod and cotton wool pad in 1 ml of the suspension medium. The crude chloroplast suspension is then layered on a Percoll step gradient which has an upper step of 5% Percoll in suspension medium, and a lower cushion of 100% Percoll. The gradients are centrifuged at 750 *g* for 5 min, and intact chloroplasts are recovered at the interface of the two steps.

Notes. This procedure yields plastid preparations which are 70–90% intact and are photosynthetically competent. It is suggested here that the 100% Percoll cushion used in the discontinuous density gradient be replaced by a cushion with much lower Percoll content (e.g., 50%). It should be noted that electron micrographs and enzyme localization studies of *Caulerpa* and other siphonous algae (Grant and Wright 1980; Shepard and Bidwell 1973) indicate that chloroplasts isolated by standard crushing procedures appear as "cytoplasts" after purification, with a nonchloroplast [possibly derived from the tonoplast; (Cobb 1977)] membrane surrounding the plastid, along with small amounts of cytoplasm and mitochondria. Due to this encapsulation and the spatial relationship between the contaminants and the plastid, the small level of contaminants causes a great change in the range of photosynthetic products of "isolated" plastid preparations (Grant and Wright 1980). For example, sucrose is known to be a product of siphonous algal chloroplasts, while higher plant plastids do not form sucrose. These "cytoplasts" are also more resistant to osmotic lysis than true chloroplasts (Borowitzka 1976; Grant and Wright 1980; Grant et al. 1976), confounding ferricyanide assays of intactness. Although Grant and Wright indicate that this altered metabolite profile is unavoidable, Cobb (1977) employed gel filtration purification of chloroplasts which effectively removed contaminants from a preparation of siphonous algal chloroplasts. Therefore, it is suggested that this technique be employed to purify chloroplasts from siphonous algae (see next section).

Other published protocols for purification of plastids from *Caulerpa* are those of Grant et al. (1976) and Wright and Grant (1978), which employ differential centrifugation for plastid purification, and an additional protocol described by Grant and Wright (1980) which uses a continuous density Percoll gradient yielding a plastid preparation which is 93% intact. Chloroplasts prepared from siphonous algae such as *Caulerpa*, *Codium*, and *Acetabularia* are remarkably stable, as compared to plastids isolated from higher plants – retaining photosynthetic activity for up to 5 days after isolation (cf. Grant et al. 1976).

2. Codium fragile [according to the procedure of Cobb (1977)]: About 5 g of frond tips are homogenized for 10 s at full speed on a Virtis grinder in 50 ml grind medium [9% (w/v) sucrose, 0.2% (w/v) BSA, 50 mM Tris, pH 7.8], and the brei filtered through two layers of cheesecloth and one layer of nylon (Nytal). The filtrate is centrifuged at 1000 *g* for 7 min, and after pellet resuspension in grind medium, centrifuged again. After the second centrifugation, the pellet is resuspended in 2 ml grind medium and layered on a water-jacketed column (15 × 2 cm) of loosely packed Sephadex G50 coarse (4 g resin preswollen in 60 ml grind medium at 4 °C poured into the column just prior to use). A peristaltic pump (operated at 2–4 ml min^{-1}) can be used to facilitate flow through. The first 2 ml of the green band (intact plastid fraction) is collected.

Notes. Only the leading edge of the plastid fraction should be collected, as the latter portion of the fraction is contaminated with mitochondria. After gel filtration, the plastid preparation is contaminated by 6% mitochondria, and "cytoplasts" are disrupted to yield true chloroplasts. The plastids also are photosynthetically competent; they display rates of CO_2 fixation, (40–60 μmol mg^{-1} Chl h^{-1}) up to four times that of "pre-column" plastids, and are 80% intact. In addition to the isolation protocols detailed here for *Caulerepa simpliciuscula* and *Codium fragile*, procedures have been developed for the isolation of plastids from the siphonous alga *Acetabularia mediterranea* (Shepard and Bidwell 1973).

3. Bryopsis maxima [according to the procedure of Yamagishi et al. (1981)]: The long, multinucleate algal cells are sliced while submerged in extraction medium [1.0 M sorbitol, 50 mM HEPES (pH 6.7), 1 mM $MgCl_2$, 1 mM $MnCl_2$, 2 mM EDTA, and 2 mM $NaNO_3$. The extract is centrifuged at 1000 *g* for 90 s, and the pellet is resuspended in 1.1 M sorbitol which has been adjusted to pH 7.5 with Tris base. This suspension is then centrifuged at 100 *g* for 20 s, and the supernatant then centrifuged at 1000 *g* for 90 s. The pellet containing intact chloroplasts is washed once with 1.1 M sorbitol. Before the pellet containing intact plastids is resuspended, the upper, loose surface of the pellet should be discarded with the supernatant.

Notes. This isolation procedure, although involving differential centrifugation to purify plastids, yields preparations which are 87 to 95% intact, and displays photosynthetic rates of 60 μmol mg^{-1} Chl h^{-1}.

Table 1. Partial list of plant material and isolation methods for chloroplasts used in photosynthetic studies

Organism	Method [a]	Contamination [b] (%)	Intactness (%)	Rate [c]	Yield [d] (%)	Reference
C_3						
Pea	Mechanical/step	p-0,2,m-0.2	94	109	10.7	Mills and Joy (1980)
Pea	Mechanical/d.c.	p-0.4,m-1	86	85	7.8	Mills et al. (1980)
Spinach	Mechanical/p.g.	c,p,m-n.d.	98	108	–	Mourioux and Douce (1981)
Spinach	Mechanical/step	p,m-n.d.	92	106	0.01*	Berkowitz and Gibbs (1982b)
Wheat	Protoplast/d.c.	–	> 90	86	5	Leegod and Walker (1979)
Oat	Protoplast/d.c.	–	> 95	125	–	Hampp and Ziegler (1980)
Barley	Protoplast/d.c.	–	100	122	20–50	Rathnam and Edwards (1976)
Quackgrass	Protoplast/d.c.	–	100	123	20–50	Rathnam and Edwards (1976)
Sunflower	Protoplast/d.c.	–	75–85	140	0.03*	Edwards et al. (1978b)
Tobacco	Protoplast/d.c.	–	–	86	20–50	Rathnam and Edwards (1976)
Lettuce	Mechanical/d.c.	–	–	36	–	Bamberger and Avron (1975)
C_4						
Corn [e]	Mechanical/step	5–10	–	–	10–20	Walbot and Hoisington (1982)
Corn [e]	Mechanical/d.c.	0	–	–	–	Horrath et al. (1978)
Panicum milaceum [e]	Protoplast/d.c.	5	89–92	150	5	Edwards et al. (1979)
CAM						
Sedum praealtum	Protoplast/step	–	> 90	12	–	Piazza et al. (1982)
Sedum praealtum	Protoplast/step	–	80–99	50	–	Spalding and Edwards (1980)
Mesembryanthemum crystallinum	Protoplast/d.c.	–	90–98	150	–	Demmig and Winter (1983)
Algae						
Chlamydomonas reinhardii	Protoplast/step	p≦21,m≦12	≈ 85	50	–	Klein et al. (1983b)
Dunaliella marina	Protoplast/s.g.	m-5.5,p-7.6,c-n.d.	–	–	44	Kombrink and Wöber (1980)
Griffithsia monilis	Mechanical/step	–	78–85	225	–	Lilley and Larkum (1981)
Caulerpa simpliciuscula	Mechanical/step	m-9,c-6,p-2.5	70–90	29	35–44	Grant and Wright (1980)
Codium fragile	Mechanical/g.f.	m-14,p-n.d.,c-4	80	60	8.5	Cobb (1977)
Acetabularia mediterranea	Mechanical/step	–	–	50	–	Shepard and Bidwell (1973)
Bryopsis maxima	Mechanical/d.c.	–	87–95	60	–	Yamagishi et al. (1981)
Bumilleriopsis filiformis	Protoplast/d.c.	–	50–75	20	–	Spiller and Böger (1980)
Euglena gracilis	Protoplast/p.g.	–	–	–	0.2–7	Price and Reardony (1982)

6.2.4 (Xanthophyceae) *Bumilleriopsis filiformis* [according to the procedure of Spiller and Böger (1980)]: Two g of harvested cells are washed once, and then resuspended in 8 ml of 0.1 M K-phosphate (pH 5.8) and 0.2 M sorbitol. Hemicellulase (300 mg) and pectinase (100 mg) are added and the cell suspension incubated at 35 °C for 90 min. After digestion, the suspension is diluted to 20 ml and centrifuged at 1000 *g* for 3 min. Cells are resuspended in extraction medium [0.4 M sorbitol, 50 mM Mes-NaOH (pH 6.2), 5 mM $Na_4P_2O_7$, 2 mM isoascorbate, 10 mM NaCl, 0.5% BSA, and 1.5% Ficoll] and washed once. The pellet is resuspended in 17 ml of extraction medium, brought to 50 ml with 1 mm glass beads, and cooled to −5 °C. After disruption on a Merkenschlager homogenizer (Braun, Melsungen, FRG) for 5 s at 2000 rpm, the brei is quickly separated from the beads by dilution in 40 ml of 25 mM Tricine-NaOH (pH 7.0), 0.4 M sorbitol, 1 mM $Na_4P_2O_7$, 2 mM $MgCl_2$, 0.5% (w/v) BSA, 1.5% (w/v) Ficoll, and 0.5 mM K-phosphate. Intact cells can be partitioned away from the released plastids by centrifugation at 500 *g* for 90 s. Intact chloroplasts are pelleted at 1000 *g* for 3 min and resuspended in 2 ml of the Tricine medium.

Notes. Chloroplasts isolated according to this procedure are 50–75% intact, and display photosynthetic rates of 12–20 µmol mg^{-1} Chl h^{-1}. A greater percent intactness in the plastid preparations could probably be achieved by including a density gradient centrifugation step in the isolation protocol after the crude plastid preparation is separated from the beads.

6.2.5 (Euglenophyceae) *Euglena gracilis* [according to the procedure of Ortiz et al. (1980) as described by Price and Reardon (1982)]: Plastids are prepared from *Euglena* cultures grown in vitamin B_{12} (cyanocobalamin) deficient cultures. Six g of harvested cells are resuspended in deionized water, centrifuged at 250 *g* for 3 min, and washed in KP medium (50 mM K-phosphate, pH 7.0, and 30 mM sorbitol). The large cells (medium volume/cell should be approximately 3500 μm^3)

[a] The isolation methods described are for chloroplast release/purification. The abbreviations for release techniques are either "mechanical" when parent plant tissue is mechanically macerated, or "protoplast" when the procedure involves proplast formation for subsequent plastid isolation. All "protoplast" release procedures except *C. reinhardii* and *C. marina* involved mechanical rupture of protoplasts. The abbreviations for purification techniques are as follows: step, density centrifugation through step gradient; d.c., differential centrifugation; p.g., density centrifugation through continuous Percoll gradients; s.g., density centrifugation through continuous sucrose gradients; g.f., gel filtration

[b] % of total mitochondria (m), peroxisomes (p), or cytoplasm (c) present in purified plastid fraction; n.d. refers to "none detectable"

[c] µmol CO_2 fixed or O_2 evolved in the presence of CO_2 mg Chl^{-1} h^{-1} (except *G. monilis*, where rate is expressed on a Chl a basis)

[d] % plastids recovered in purified fraction, except when followed by a *, indicating yield expressed on a Chl recovered/fresh weight of laminae basis. Yield from isolation protocols involving protoplast formation is dependent on degree of digestion of leaf material, as recovery of intact plastids from released protoplasts is very high

[e] With C_4 plants, contamination refers to cross-contamination by chloroplast types (bundle-sheath or mesophyll), and rate is shown for bundle-sheath plastids

are then incubated at 0 °C in 10 ml of KP medium with 0.5% (w/v) trypsin (which should not be purified from chymotrypsin contamination) for 1 h with shaking. The pellicle digestion medium is then diluted twofold with KP medium, centrifuged at 227 *g* for 1 min, and resuspended in 25 ml of breaking medium [0.25 M sorbitol, 20 mM HEPES (pH 7.4), and 0.4 mM EDTA]. The suspension is then either vigorously stirred (with a magnetic stirrer), homogenized in a Waring Blendor for 3 s, or passed through a French press at 90 kg cm^{-3}. The brei is then diluted with 1.5 (v/v) of breaking medium and centrifuged (in 100 ml tubes) at 227 g for 3 min. The supernatant is then recentrifuged at 2400 g to pellet the chloroplasts. The pellet is resuspended in a minimum volume of gradient medium [0.3 M sorbitol, 1% w/v Ficoll, 5 mM HEPES (pH 6.8), 2 µg ml^{-1} PVS, 15 mM NaCl, and 5 mM *β*-mercaptoethanol added just prior to use], and layered on 10–80% linear Percoll gradients.

The Percoll gradients are made up with 0.33 sorbitol, 1 mM $MgCl_2$, 2 mM EDTA, 50 mM Tricine (pH 8.4), 5 mM glutathione, 1% PEG (Carbowax 6000, Union Carbide) and 0.1% BSA throughout in a gradient maker. The gradients are 30 ml in volume, and have a small 80% Percoll cushion in the bottom of the tubes. The gradients are centrifuged at 8632 *g* for 20 min in a swinging bucket rotor, and intact plastids recovered in the lower green band of the gradient. The plastids are then diluted with two volumes of gradient medium and centrifuged at 5905 *g* for 3 min (SS-34 rotor, Sorvall). The pellet is then resuspended in 20 ml of 0.33 M sorbitol, 1 mM $MgCl_2$, 2 mM EDTA, 50 mM Tricine (pH 8.4), and the pellet (containing intact plastids) is stored in 0.33 M sorbitol, 50 mM Tricine (pH 8.4).

Notes. The yield of this procedure ranges from 0.2 to 7% of plastids in cells recovered as photosynthetically competent intact chloroplasts. Price and Reardon (1982) indicate that if the pellicle-digesting trypsin is purchased as a purified enzyme (e.g., Sigma Type III or IV), 1 mg ml^{-1} chymotrypsin should be added along with the trypsin. The cells used for plastid isolation have to be grown in B_{12} deficient medium, as this treatment inhibits cytokinesis, causing the cells to be enlarged and the pellicle more susceptible to trypsin digestion (Carell 1969). The pellicle digestion step should be monitored by light microscopy; cells should be rounded (forming spheroplasts) when digested fully. Soybean trypsin inhibitor can be added to the cells after the trypsin step. Preparations isolated using the magnetic stirrer to lyse spheroplasts generally have a higher percent intactness than when the French press or Waring Blendor is used.

A compilation of some of the isolation procedures mentioned in this section is presented in Table 1.

7 Additional Comments on Chloroplast Isolation

In addition to the many points brought up in the preceding discussion regarding the optimization of isolation techniques, there are some newly developed concepts which should be mentioned. Although not widely employed, the technique of particle partition according to surface properties in aqueous polymer two-phase systems has been developed by Larsson and co-workers for the isolation of highly purified and intact chloroplast preparations from C_3 and C_4 plants (e.g., Halberg and Larsson 1983; Larsson et al. 1977). The fundamentals of this technique have been recently reviewed (Larsson 1983), and will not be discussed here. Briefly stated, the technique can employ differential affinity of organelles for the two phases [usually dextran (T500) and PEG (4000)] and separation by either counter-current distribution or batch procedures. Although these techniques have, in the past, been thought of as too lengthy to be used as a standard isolation protocol, Larsson (1983) points out that the procedure can take as little as 30 min. The advantage of the technique is that plastids can be partitioned away from "organelle clumps" which include bits of cytoplasm. This results, then, in altered metabolic profiles of chloroplast preparations. For example, Buchholz et al. (1979) indicated that phase partition isolation of plastids results in less than 0.02% of fixed carbon going into amino acids, as opposed to 11% when phase partition is not used.

The question of physiological changes of plastids during isolation should be addressed. Edwards et al. (1978b) have indicated that PPi in the isolation and storage medium decreases the lag during subsequently assayed CO_2-supported O_2 evolution in wheat chloroplasts by preventing loss of triose phosphates. Robinson and Wiskich (1977) have indicated that the presence of PPi in isolation and storage media inhibits photosynthesis of chloroplasts isolated from young pea and wheat plants by exchanging for stromal ATP. Piazza and Gibbs (1983) have demonstrated that Mg^{2+} and Mg-adenosine nucleotide complexes can enter pea and Sedum chloroplasts and affect photosynthetic capacity. Rathnam and Edwards (1976) indicated that Mg^{2+} in the isolation medium may have a beneficial effect on plastids by stabilizing RuBP carboxylase. The presence of HCO_3^- in the isolation and storage media also might be facilitating this stabilization, and results in a decreased lag during subsequently assayed photosynthesis (Edwards et al. 1978b). Chelators such as EDTA have been found to have beneficial effects on sunflower chloroplasts (Edwards et al. 1978a), possibly because membrane surface charges are stabilized during isolation. Mourioux and Douce (1981) have indicated that in addition to Pi flux across the chloroplast membrane via the action of the phosphate translocator, there is a constant passive diffusion of Pi out of the chloroplast during isolation and storage. They found that this Pi "leaking" can inhibit subsequent photosynthesis, and suggest that the presence of 10 mM Pi counters this loss of a stromal Pi and can greatly increase the time which plastids can be stored. Finally, Demmig and Gimmler (1983) have recently demonstrated that isolated chloroplasts lose monovalent cations during isolation and storage. They indicated that the isolation and reaction media used by most researchers do not account for this, and therefore, subsequent photosynthetic potential is not optimized (this is why the "low cation" isolation protocol of Nakatani and Barber (1977) is not suggested here for isolation of spinach chloroplasts). Demmig and Gimmler indicate that due to this loss of cations, reaction media for photosynthetic studies should include up to 100 mM K^+. However, in recommending this addition to reaction media, Demmig and Gimmler did not consider the indirect cation additions to reaction media in most protocols. We have determined (unpublished data) that whith typical reaction media such as 50 mM HEPES (brought to pH 7.6 with NaOH) and 50 mM Tricine (brought to pH 8.1 with NaOH), there can be 28 mM and 25 mM Na^+, respectively, in the media due to pH adjustment. Also, due to $NaHCO_3$ (up to 10 mM Na^+) and Na_2EDTA (up to 10 mM Na^+) additions, reaction media can have approximately 50 mM cations without specific additions of salts for the purpose of reversing cation leakage from the stroma.

8 Abbreviations

A	absorbance
ATP	adenosine 5-triphosphate
BSA	bovine serum albumin
CAM	Crassulacean acid metabolism
Chl	chlorophyll
Cyt	cytochrome
d	density
DEAE	diethylaminoethyl
DIECA	diethyldithiocarbamate
DTT	dithiothreitol
EDTA	ethylenediaminetetraacetate
FBP	fructose 1,6-bisphosphate
GAP	glyceraldehyde 3-phosphate
HEPES	N-2-hydroxyethylpiperazine-N′-2-ethanesulfonic acid
MBT	mercaptobenzothiazole
Mes	2-(N-morpholino) ethanesulfonic acid
NAD	nicotinamide adenine dinucleotide
NADP(H)	nicotinamide adenine dinucleotide phosphate and its reduced form
PEG	polyethylene glycol
PEP	phosphoenolpyruvate
Pi	orthophosphate
PPi	pyrophosphate
PVP	polyvinylpyrrolidone
RNA	ribonucleic acid
R5P	ribose 5-phosphate
RuBP	ribulose 1,5-bisphosphate
Tricine	N-tris(hydroxymethyl) methylglycine
Tris	tris (hydroxymethyl) aminomethane

Acknowledgement. This paper was supported in part by Grant PCM 76-82157 from the National Science Foundation.

References

Aist JR (1976) Papillae and related wound plugs of plant cells. Annu Rev Phytopathol 14:145–163

Arnon DI (1949) Copper enzymes in chloroplasts. Polyphenoloxidases in *Beta vulgaris*. Plant Physiol 24:1–14

Bamberger E, Avron M (1975) Site of action of inhibitors of carbon dioxide assimilation by whole lettuce chloroplasts. Plant Physiol 56:481–485

Belknap WR (1983) Partial purifications of intact chloroplasts from *Chlamydomonas reinhardtii*. Plant Physiol 72:1130–1132

Berkowitz GA, Gibbs M (1982a) Effect of osmotic stress on photosynthesis studied with the isolated spinach chloroplast. Generation and use of reducing power. Plant Physiol 70:1143–1148

Berkowitz GA, Gibbs M (1982b) Effect of osmotic stress on photosynthesis studied with the isolated spinach chloroplast. Site-specific inhibition of the photosynthetic carbon reduction cycle. Plant Physiol 70:1535–1540

Berkowitz GA, Gibbs M (1983) Reduced osmotic potential effects on photosynthesis. Identification of stromal acidification as a mediating factor. Plant Physiol 71:905–911

Borowitzka MA (1976) Some unusual features of the ultrastructure of the chloroplasts of the green algal order *Caulerpales* and their development. Protoplasma 89:129–147
Bourne WF, Miflin BJ (1973) Studies on nitrite reductase in barley. Planta 111:47–56
Brown RH, Rigsby LL, Akin DE (1983) Enclosure of mitochondria by chloroplasts. Plant Physiol 71:437–439
Buchholz B, Reupke B, Bickel H, Schultz G (1979) Reconstitution of amino acid synthesis by combining spinach chloroplasts with other leaf organelles. Phytochemistry 18:1109–1111
Burke JJ, Wilson RF, Swafford JR (1982) Characterization of chloroplasts isolated from triazine-susceptible and triazine resistant biotypes of *Brassica campestris* L. Plant Physiol 70:24–29
Carell EF (1969) Studies on chloroplast development and replication in *Euglena* I. Vitamin B_{12} and chloroplast replication. J Cell Biol 41:431–440
Chapman KSR, Berry JA, Hatch MD (1980) Photosynthetic metabolism in bundle sheath cells of the C_4 species *Zea Mays*: sources of ATP and NADPH and the contribution of Photosystem II. Arch Biochem Biophys 202:330–341
Chua NH, Schmidt GW (1979) Transport of proteins into mitochondria and chloroplasts. J Cell Biol 81:461–483
Cline K, Andrews J, Mersey B, Newcomb EH, Keegstra K (1981) Separation and characterization of inner and outer envelope memebranes of pea chloroplasts. Proc Natl Acad Sci USA 78:3595–3599
Cobb AH (1977) The relationship of purity to photosynthetic activity in preparations of *Codium fragile* chloroplasts. Protoplasma 92:137–146
Costes C, Burghoffer C, Joyard J, Block M, Douce R (1979) Occurrence and biosynthesis of violaxanthin in isolated spinach chloroplast envelope. FEBS Lett 103:17–21
Day DA, Jenkins CLD, Hatch MD (1981) Isolation and properties of functional mesophyll protoplasts and chloroplasts from *Zea Mays*. Aust J Plant Physiol 8:21–29
Demmig B, Gimmler H (1983) Properties of the isolated intact chloroplast at cytoplasmic K^+ concentrations. Plant Physiol 73:169–174
Demmig B, Winter K (1983) Photosynthetic characteristics of chloroplasts isolated from *Mesembryanthemum crystallinum* L., a halophilic plant capable of Crassulacean acid metabolism. Planta 159:66–76
Douce R, Joyard J (1979) Structure and function of the plastid envelope. Adv Bot Res 7:1–116
Edwards GE, Black CC (1971) Isolation of mesophyll cells and bundle sheath cells from *Digitaria sanguinalis* (L.) Scop. leaves and a scanning microscopy study of the internal leaf cell morphology. Plant Physiol 47:149–156
Edwards GE, Robinson SP, Tyler NJC, Walker DA (1978a) A requirement for chelation in obtaining functional chloroplasts of sunflower and wheat. Arch Biochem Biophys 190:421–433
Edwards GE, Robinson SP, Tyler NJC, Walker DA (1978b) Photosynthesis by isolated protoplasts, protoplast extracts, and chloroplasts of wheat. Influence of orthophosphate, pyrophosphate, and adenylates. Plant Physiol 62:313–319
Edwards GE, Lilley McCR, Craig S, Hatch MD (1979) Isolation of intact and functional chloroplasts from mesophyll and bundle sheath protoplasts of the C_4 plant *Panicum* miliaceum. Plant Physiol 63:821–827
Ellis RJ (1977) Protein synthesis by isolated chloroplasts. Biochim Biophys Acta 463:185–215
Ellis RJ, Hartley MR (1982) Preparation of higher plant chloroplasts active in protein and RNA synthesis. In: Edelman M, Hallick RB, Chua N-H (eds) Methods in chloroplast molecular biology. Elsevier Biomedical, Amsterdam, pp 169–188
Evans PK, Cocking EC (1977) Isolated plant protoplasts. In: Street HE (ed) Plant tissue and cell culture. Univ California Press, Berkeley
Giles KL, Sarafis V (1974) Implictions of rigescent integuments as a new feature of some algal chloroplasts. Nature 248:512–513
Grant BR, Wright SW (1980) Purity of chloroplasts prepared from the siphonous green alga, *Caulerpa simpliciuscula*, as determined by their ultrastructure and their enzymic content. Plant Physiol 66:130–138

Grant BR, Howard RJ, Gayler KR (1976) Isolation and properties of chloroplasts from the siphonous green alga *Caulerpa simpliciuscula*. Aust J Plant Physiol 3:639–651

Halberg M, Larsson C (1983) Highly purified intact chloroplasts from mesophyll protoplasts of the C_4 plant *Digitaria sanguinalis*. Inhibition of phosphoglycerate reduction by orthophosphate and by phosphoenolpyruvate. Physiol Plant 57:330–338

Hall DO (1972) Nomenclature for isolated chloroplasts. Nature 235:125–126

Hampp R (1979) Kinetics of mitochondrial phosphate transport and rates of respiration and phosphorylation during greening of etiolated *Avena* leaves. Plants 144:325–332

Hampp R (1980) Rapid separation of the plastid, mitochondrial, and cytoplasmic fractions from intact leaf protoplasts of *Avena*. Determination of in vivo ATP pool sizes during greening. Planta 150:291–298

Hampp R, Ziegler H (1980) On the use of *Avena* protoplasts to study chloroplast development. Planta 147:485–494

Heber U, Santarius KA (1970) Direct and indirect transfer of ATP and ADP across the chloroplast envelope. Z Naturforsch 25b:718–728

Heldt HW (1980) Measurement of metabolite movement across the envelope and of the pH in the stroma and the thylakoid space in intact chloroplasts. Methods Enzymol 69:604–613

Heldt HW, Sauer F (1971) The inner membrane of the chloroplast envelope as the site of specific metabolite transport. Biochim Biophys Acta 234:83–91

Holdsworth RH (1971) The isolation and partial characterization of the pyrenoid protein of *Eremosphaera viridis*. J Cell Biol 51:499–513

Horvath G, Droppa M, Mustardy LA, Faludi-Daniel A (1978) Functional characteristics of intact chloroplasts isolated from mesophyll protoplasts and bundle sheath cells of maize. Planta 141:239–244

Huber SC, Edwards GE (1975) An evaluation of some parameters required for the enzymatic isolation of cells and protoplasts with CO_2 fixation capacity from C_3 and C_4 grasses. Physiol Plant 35:203–209

Jensen RG (1979) The isolation of intact leaf cells, protoplasts, and chloroplasts. In: Gibbs M, Latzko E (eds) Photosynthesis II: photosynthetic carbon metabolism and related processes. Encyclopedia of plant physiology, new ser. Springer, Berlin Heidelberg New York

Jensen RG, Bassham JA (1966) Photosynthesis by isolated chloroplasts. Proc Natl Acad Sci USA 4:1095–1101

Kaiser WM, Kaiser G, Schoner S, Neimanis S (1981) Photosynthesis under osmotic stress. Differential recovery of photosynthetic activities of stroma enzymes, intact chloroplasts, protoplasts, and leaf slices after exposure to high solute concentrations. Plants 153:430–435

Klein U, Chen C, Gibbs M (1983 a) Photosynthetic properties of chloroplasts from *Chlamydomonas reinhardii*. Plant Physiol 72:488–491

Klein U, Chen C, Gibbs M, Platt-Aloia KA (1983 b) Cellular fractionation of *Chlamydomonas reinhardii* with emphasis on the isolation of the chloroplast. Plant Physiol 72:481–487

Kombrink E, Wöber G (1980) Preparation of intact chloroplasts by chemically induced lysis of the green alga *Dunaliella marina*. Planta 149:123–129

Kow YW, Gibbs M (1982) Characterization of a photosynthesizing reconstituted spinach chloroplast preparation. Regulation by primer, adenylates, ferredoxin, and pyridine nucleotides. Plant Physiol 69:179–186

Krueger RW, Miles D (1981) Photosynthesis in fescue I. High rate of electron transport and phosphorylation in chloroplasts of hexaploid plants. Plant Physiol 67:763–767

Larsson C (1983) Partition in aqueous polymer two-phase systems. In: Hall JL, Moore AL (eds) Isolation of membranes and organelles from plant cells. Academic Press, New York, pp 277–309

Larsson C, Andersson B, Roos G (1977) Scanning electron microscopy of different populations of chloroplasts isolated by phase partition. Plant Sci Lett 8:291–298

Latzko E, Gibbs M (1974) "Alkaline" C_1-fructose-1,6-diphosphatase. In: Bergmeyer HU (ed) Methods of enzymatic analysis. Chemie, Weinheim

Leech RM (1964) The isolation of structurally intact chloroplasts. Biochim Biophys Acta 79:637–639
Leegood RC, Walker DA (1979) Isolation of protoplasts and chloroplasts from flag leaves of *Triticum aestivum* L. Plant Physiol 63:1212–1214
Leegood RC, Walker DA (1983) Chloroplasts. In: Hall JL, Moore AL (eds) Isolation of membranes and organelles from plant cells. Academic Press, New York
Lilley RMC, Larkum AWD (1981) Isolation of functionally intact rhodoplasts from *Griffithsia monilis* (Ceramiaceae, Rhodophyta). Plant Physiol 67:5–8
Lindhardt K, Walter K (1963) Phosphatases. In: Bergmeyer HU (ed) Methods of enzymatic analysis. Academic Press, New York, pp 779–785
Loomis WD (1974) Overcoming problems of phenolics and quinones in the isolation of plant enzymes and organelles. Methods Enzymol 31A:524–544
Maury WJ, Huber SC, Moreland DE (1981) Effects of magnesium on intact chloroplasts II. Cation specificity and involvements of the envelope ATPase in (sodium) potassium/proton exchange across the envelope. Plant Physiol 68:1257–1263
Miflin BJ (1974) The location of nitrite reductase and other enzymes related to amino acid biosynthesis in the plastids of roots and leaves. Plant Physiol 54:550–555
Miflin BJ, Beevers H (1974) Isolation of intact plastids from a range of plant tissues. Plant Physiol 53:870–874
Mills WR, Joy KW (1980) A rapid method for isolation of purified, physiologically active chloroplasts, used to study the intracellular distribution of amino acids in pea leaves. Plants 148:75–83
Mills WR, Lea PJ, Miflin BJ (1980) Photosynthetic formation of the aspartate family of amino acids in isolated chloroplasts. Plant Physiol 65:1166–1172
Monson RK, Rumpho ME, Edwards GE (1983) The influence of inorganic phosphate on photosynthesis in intact chloroplasts from *Mesembryanthemum crystallinum* L. plants exhibiting C_3 photosynthesis or Crassulacean acid metabolism. Plants 159:97–104
Morgan JA, Brown RH (1979) Photosynthesis in grass species differing in carbon dioxide fixation pathways II. A search for species with intermediate gas exchange and anatomical characteristics. Plant Physiol 64:257–262
Morgenthaler J-J, Price CA, Robinson JM, Gibbs M (1974) Photosynthetic activity of spinach chloroplasts after isopycnic centrifugation in gradients of silica. Plant Physiol 54:532–534
Morgenthaler J-J, Marsden MPF, Price CA (1975) Factors affecting the separation of photosynthetically competent chloroplasts in gradients of silica sols. Arch Biochem Biophys 168:289–301
Mourioux G, Douce R (1981) Slow passive diffusion of orthophosphate between intact isolated chloroplasts and suspending medium. Plant Physiol 67:470–473
Nagata T, Ishii S (1979) A rapid method for isolation of mesophyll protoplasts. Can J Bot 57:1820–1823
Nakatani HY, Barber J (1977) An improved method for isolating chloroplasts retaining their outer membranes. Biochim Biophys Acta 461:510–512
Nishida K, Sanada Y (1977) Carbon dioxide fixation in chloroplasts isolated from CAM plants. In: Mujachi S, Katoh S, Jujita Y, Shibuta K (eds) Photosynthetic organelles, structure and function. Special issue of Plant Cell Physiol 3:341–346
Nishimura M, Akazawa T (1975) Photosynthetic activities of spinach leaf protoplasts. Plant Physiol 55:712–716
Nishimura M, Graham D, Akazawa T (1976) Isolation of intact chloroplasts and other cell organelles from spinach leaf protoplasts. Plant Physiol 58:309–314
O'Neal D, Hew CS, Latzko E, Gibbs M (1972) Photosynthetic carbon metabolism of isolated corn chloroplasts. Plant Physiol 49:607–614
Ortiz W, Reardon EM, Price CA (1980) Preparation of chloroplasts from *Euglena* highly active in protein synthesis. Plant Physiol 66:291–294
Otsuki Y, Takebe I (1969) Isolation of intact mesophyll cells and their protoplasts from higher plants. Plant Cell Physiol 10:917–921
Perrin DD, Dempsey B (1974) Buffers for pH and metal ion control. Chapman and Hall, London

Piazza G, Gibbs M (1983) Influence of adenosine phosphates and magnesium on photosynthesis in chloroplasts from peas, *Sedum*, and spinach. Plant Physiol 71:680–687
Piazza G, Smith MG, Gibbs M (1982) Characterization of the formation and distribution of photosynthetic products by *Sedum praealtum* chloroplasts. Plant Physiol 70:1748–1758
Pilwat G, Hampp R, Zimmermann U (1980) Electrical field effects induced in membranes of developing chloroplasts. Planta 147:396–404
Plaut Z (1971) Inhibition of photosynthetic carbon dioxide fixation in isolated spinach chloroplasts exposed to reduced osmotic potentials. Plant Physiol 48:591–595
Price CA (1982) Centrifugation in density gradients. Academic, New York
Price CA (1983) General principles of cell fractionation. In: Hall JL, Moore AL (eds) Isolation of membranes and organelles from plant cells. Academic Press, New York, pp 1–24
Price CA, Reardon EM (1982) Isolation of chloroplasts for protein synthesis from spinach and *Euglena gracilis* by centrifugation in silica sols. In: Edelman M, Hallick RB, Chua N-H (eds) Methods in chloroplast molecular biology. Elsevier Biomedical, Amsterdam, pp 189–210
Radosevich SR, Devilliers OT (1976) Studies on the mechanism of S-triazine resistance in common groundsel. Weed Sci 4:229–232
Rathnam CKM, Das VSR (1974) Nitrate metabolism in relation to the aspartate-type C-4 pathway of photosynthesis in *Eleusine coracana*. Can J Bot 52:2599–2605
Rathnam CKM, Edwards GE (1976) Protoplasts as a tool for isolating functional chloroplasts from leaves. Plant Cell Physiol 173:177–186
Rathnam CKM, Edwards GE (1977) C_4 acid decarboxylation and CO_2 donation to photosynthesis in bundle sheath strands and chloroplasts from species representing three groups of C_4 plants. Arch Biochem Biophys 182:1–13
Robinson GP, Walker DA (1979) Rapid separation of the chloroplast and cytoplasmic fractions from intact leaf protoplasts. Arch Biochem Biophys 196:319–323
Robinson SP, Wiskich JT (1977) Pyrophosphate inhibition of carbon dioxide fixation in isolated pea chloroplasts by uptake in exchange for endogenous adenine nucleotides. Plant Physiol 59:422–427
Schlosser UG, Sachs H, Robinson DG (1976) Isolation of protoplasts by means of a "species-specific" autolysine in *Chlamydomonas*. Protoplasma 88:51–64
Schmitt JM, Hermann RG (1977) Fractionation of cell organelles in silica sol radients. In: Prescott DM (ed) Methods in cell biology, vol 15. Academic Press, New York, pp 177–200
Schnabl H, Hampp R (1980) Vicia guard cell protoplasts lack photosystem II activity. Naturwissenschaften 67:465–466
Shepard DC, Bidwell RGS (1973) Photosynthesis and carbon metabolism in a chloroplast preparation from *Acetabularia*. Protoplasma 76:289–307
Spalding MH, Edwards GE (1980) Photosynthesis in isolated chloroplasts of the Crassulacean acid metabolism plant *Sedum praealtum*. Plant Physiol 65:1044–1048
Spalding MH, Schmitt MR, Ku SB, Edwards GE (1979) Intracellular localization of some key enzymes of Crassulacean acid metabolism is *Sedum praealtum*. Plant Physiol 63:738–743
Spiller H, Böger P (1980) Photosynthetically active algal preparations. Methods Enzymol 69:105–121
Stitt M, Heldt HW (1981) Physiological rates of starch breakdown in isolated intact spinach chloroplasts. Plant Physiol 68:755–761
Stokes DM, Walker DA (1972) Photosynthesis by isolated chloroplasts. Inhibition by DL-glyceraldehyde of carbon dioxide assimilation. Biochem J 128:1147–1157
Takabe T, Nishimura M, Akazawa T (1979) Isolation of intact chloroplasts from spinach leaf by centrifugation in gradients of the modified silica "Percoll". Agric Biol Chem 43:2137–2142
Trench RK, Boyle JE, Smith DC (1973) Association between chloroplasts of *Codium fragile* and the mollusc *Elysia viridis* I. Characteristics of isolated *Codium* chloroplasts. Proc R Soc Lond Ser B 184:51–61

Van Ginkel G, Brown JS (1978) Endogenous catalase and superoxide dismutase activities in photosynthetic membranes. FEBS Lett 94:284–286

Walbot V (1977) Use of silica sol step gradients to prepare bundle sheath and mesophyll chloroplasts from *Panicum maximum*. Plant Physiol 60:102–108

Walbot V, Hoisington DA (1982) Isolation of mesophyll and bundle sheath chloroplasts from maize. In: Edelman M, Hallick RB, Chua N-H (eds) Methods in chloroplast molecular biology. Elsevier Biomedical, Amsterdam, pp 211–220

Walden R, Leaver CJ (1981) Synthesis of chloroplast proteins during germination and early development of cucumber. Plant Physiol 67:1090–1096

Walker DA (1964) Improved rates of carbon dioxide fixation by illuminated chloroplasts. Biochem J 92:22–23

Walker DA (1976) CO_2 fixation by intact chloroplasts: photosynthetic induction and its relation to transport phenomena and control mechanisms. In: Barber J (ed) The intact chloroplast. Elsevier, Amsterdam, pp 235–278

Walker DA (1980a) Preparation of higher plant chloroplasts. Methods Enzymol 69:94–104

Walker JRL (1980b) Enzyme isolation from plants and the phenolic problem. What's New in Plant Physiol 11:33–36

Walker JRL, McCallion RF (1980) The selective inhibition of *ortho*- and *para*-diphenol oxidases. Phytochemistry 19:373–377

Willison JHM, Davey MR (1976) Fraction I protein crystals in chloroplasts of isolated tobacco leaf protoplasts: a thin section and freeze etch morphological study. J Ultrastruct Res 55:303–311

Wink M, Hartmann T (1982) Localization of the enzymes of quinolizidine alkaloid biosynthesis in leaf chloroplasts of *Lupinis polyphyllus*. Plant Physiol 70:74–77

Winter K, Foster JG, Edwards GE, Holtum JAM (1982) Intracellular localization of enzymes of carbon metabolism in *Mesembryanthemum crystallinum* exhibiting C_3 photosynthetic characteristics of performing Crassulacean acid metabolism. Plant Physiol 69:300–307

Wolosuik RA, Buchanan BB (1979) Studies on the regulation of chloroplast NADP-linked glyceraldehyde-3-phosphate dehydrogenase. J Biol Chem 251:6456–6461

Wright SW, Grant BR (1978) Properties of chloroplasts isolated from siphonous algae. Effects of osmotic shock and detergent treatment on intactness. Plant Physiol 61:768–771

Yamagishi A, Satoh K, Katoh S (1981) The concentrations and thermodynamic activities of cations in intact *Bryopsis* chloroplasts. Biochim Biophys Acta 637:252–263

Purification of Inner and Outer Chloroplast Envelope Membranes

K. CLINE

1 Introduction

Chloroplasts of higher plants are enclosed by a pair of closely spaced bilayer membranes called the envelope. The outer membrane of the pair is in contact with the cell cytoplasm, whereas the inner membrane bounds the stroma. In addition to its structural role, the envelope mediates a number of important chloroplast functions. It regulates the flow of ions and metabolites between the stroma and the cell cytoplasm (Heber and Heldt 1981). It is also involved in various aspects of chloroplast biogenesis (Cline and Keegstra 1983; Chua and Schmidt 1979; Douce and Joyard 1979; Ellis 1983; Roughan and Slack 1982).

Thorough study of the functions and properties of the envelope requires the ability to isolate it in a purified state. In 1970, Mackender and Leech (1970) reported the first method for purifying the envelope. Their protocol involved preparing intact chloroplasts from *Vicia faba* tissue, rupturing the chloroplasts by osmotic shock, and purifying the envelopes by differential centrifugation. The resulting preparation was an enriched envelope fraction, contaminated by a small but significant amount of thylakoids, the internal chloroplast membranes. Since this initial report, more than 15 methods of envelope purification have appeared in the literature. Modifications in the method of isolating intact chloroplasts, the ionic conditions used during chloroplast lysis, and the centrifugation schemes used to fractionate the ruptured chloroplasts have led to improvements in both yield and purity of the resulting membranes. Recently published methods now enable one to readily prepare reasonable quantities of envelope membrane, i.e., up to 5 mg of protein per kg of leaves, free of contamination by thylakoids or other cellular membranes (Douce and Joyard 1979, 1982).

In certain respects, the envelope is a relatively easy membrane system to isolate. First, it is abundant in plant cells. Forde and Steer (1976) estimated by quantitative electron microscopy that the chloroplast envelope accounts for about 15% of the total membrane area in expanding cucumber leaf cells. Second, because recent developments in chloroplast isolation procedures allow for rapid purification of highly purified intact chloroplasts (Mills and Joy 1980; Morgenthaler et al. 1975; Tahabe et al. 1979), the envelope can be obtained free of contamination by extrachloroplastidic membranes.

On the other hand, envelope purification has presented some difficulties. For example, structural characteristics of the envelope have hindered attempts to purify separately the inner and outer envelope membranes. Until recently all methods of envelope purification yielded a mixture of the two (Douce and Joyard 1979, 1982), making it impossible to determine the properties of the individual membranes. In 1981 we reported a purification procedure that does succeed in resolving the inner from the outer envelope membrane of pea chloroplasts (Cline et al. 1981). The following article describes this procedure in more detail, gives some of the properties of the isolated membranes, and discusses possible modification toward an improved procedure.

2 General Considerations

Purification of envelope membranes involves growth and preparation of the plant tissue, purification of intact chloroplasts, and subfractionation of the chloroplasts. We recommend starting with healthy plant tissue and purifying the chloroplasts to a high degree of intactness. In our experience tissue obtained from the local market or grown under suboptimal conditions frequently yields envelope membranes of poor quality. A protocol for culturing pea plants under controlled

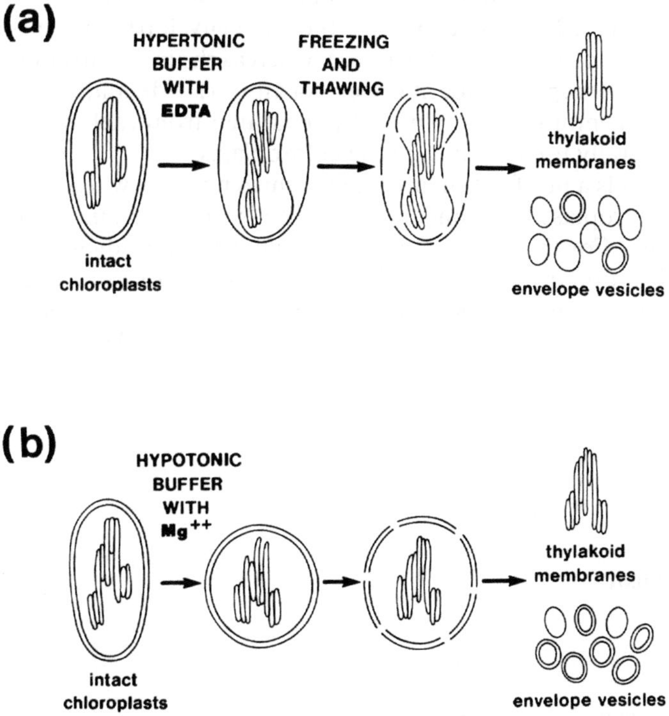

Fig. 1. Schematic representation of two methods of rupturing chloroplasts. **a** Rupture under hyperosmotic conditions (Fig. 4). Chloroplasts, preincubated in hyperosmotic sucrose to separate the outer and inner envelope membranes, are ruptured by a freeze–thaw cycle. **b** Rupture by hypoosmotic shock. Chloroplasts, suspended in a low osmolarity medium, take up water, swell, and burst when inner and outer envelope membranes are tightly appressed

conditions is described below. Poor results are also obtained if the chloroplast preparations contain a high proportion of broken chloroplasts. To purify chloroplasts, a combination of differential centrifugation and density gradient centrifugation on silica sol (Percoll) gradients is employed. This procedure allows rapid isolation of chloroplasts with a high degree of intactness.

There are two important requirements of a successful scheme for subfractionating chloroplasts. The first is a method of chloroplast rupture which avoids the irreversible mingling of inner and outer envelope membranes. Previous methods of envelope purification used osmotic shock to lyse chloroplasts (Douce and Joyard 1979, 1982). Figure 1 b is a schematic representation of the events of such a lysis. Chloroplasts in hypoosmotic medium take up water, swell, and burst. Under these conditions, the two envelope membranes are pressed tightly against each other at the time of lysis. This results in the production of numerous double membranes vesicles in which one envelope membrane is enclosed within the other (Cline et al. 1981) and probably also results in fusion between inner and outer membrane fragments. To minimize the formation of fused hybrids and double membrane vesicles, we utilize a rupture procedure in which chloroplasts are first preincubated in hyperosmotic sucrose solution (shown diagramatically in Fig. 1 a). The purpose of the preincubation is to maximize the space between inner and outer envelope membranes. This occurs because the outer envelope is nonspecifically permeable to small molecules such as sucrose, whereas the inner envelope membrane is impermeable except via specific translocators (Heldt and Sauer 1971). Once the envelope membranes are physically separated from each other, a freeze–thaw cycle is generally used to rupture the chloroplasts.

The second requirement is a purification procedure which not only will isolate envelopes from other chloroplast components, but will also resolve inner and outer envelope membranes. For this, we use two sequential density gradient centrifugations. The first, flotation centrifugation, isolates the envelopes (a mixture of inner and outer) from thylakoids and stroma. The second centrifugation, sedimentation through a linear density sucrose gradient, resolves inner and outer envelope membranes.

3 The Procedure

3.1 Reagents and Equipment

3.1.1 Solutions

1. 2 × HS – per liter
   0.1 M HEPES 23.8 g
   0.66 M Sorbitol 120.3 g
   pH to 7.3 at 20 °C with 5 M NaOH
2. 1 × HS
   mix 1 volume 2 × HS with 1 volume distilled water.
3. 2 × HSB – per liter
   dissolve 2 g of bovine serum albumin (Sigma fraction V) in 1 liter of 2 × HS.

4. 1 × HSB
   mix one volume of 2 × HSB with one volume distilled water.
5. 10 × TE – per 500 ml

0.1 M Tricine	8.96 g
0.02 M Na_2EDTA	3.36 g

   pH to 7.2 at 20 °C with 5 M NaOH
6. 1 × TE
   mix one volume 10 × TE with 9 volumes of distilled water.
7. Sucrose solutions (250 ml)

	Sucrose	10 × TE
(a) 0.3 M	25.7 g	25 ml
(b) 0.6 M	51.3 g	25 ml
(c) 1.2 M	102.7 g	25 ml
(d) 2.6 M	222.5 g	25 ml

   Adjust to a final volume of 250 ml with distilled water using volumetric flasks
8. Percoll gradients
   Mix 60 ml Percoll (Sigma or Pharmacia) with 60 ml 2 × HSB. Distribute 30 ml aliquots of this mixture into 4, 50 ml Sorvall tubes. Centrifuge in an SS-34 rotor at 20,000 rpm for 30 min at 4 °C.

3.1.2 Materials

1. Cheesecloth and sterile cotton
2. Homogenizer (Polytron fitted with a PT35K probe)
3. Superspeed centrifuge (Sorvall or equivalent) with the following rotors:
   SS-34 – fixed angle, 8 × 50 ml
   HS-4 – swinging bucket, 4 × 250 ml
   HB-4 – swinging bucket, 4 × 50 ml
4. Preparative ultracentrifuge (Beckman L-8 or equivalent) with the following rotors:
   SW-27 – swinging bucket, 4 × 39 ml
   Type 30 – fixed angle, 6 × 25 ml
5. Refractometer
6. Gradient fractionator

3.2 Growth of Peas and Purification of Intact Chloroplasts

Dwarf pea seeds, e.g., Laxton's Progress No. 9, are imbibed in running water overnight and then planted in vermiculite. Approximately 125 g (dry wt.) of seeds are planted 1.5 to 3 cm deep per 30 × 50 × 6 cm flat. Seedlings are grown at 18 °C–21 °C under 11 h of fluorescent light (400 ft-C ≅ 170 µE m^{-2} s^{-1}) and 13 h darkness. The pea seedlings appear healthier and yield better envelope membranes when the relative humidity is kept low. Seedlings are watered with tap water and are harvested after 2 weeks by cutting off the majority of the above-soil portion and then rinsing the cut shoots several times with distilled water. Approximately 100 g fresh weight of tissue is obtained per 30 × 50 cm^2 flat.

The protocol for purifying intact chloroplasts from pea seedlings is shown in Fig. 2. Homogenization is carried out on 200 g batches of tissue with a Polytron.

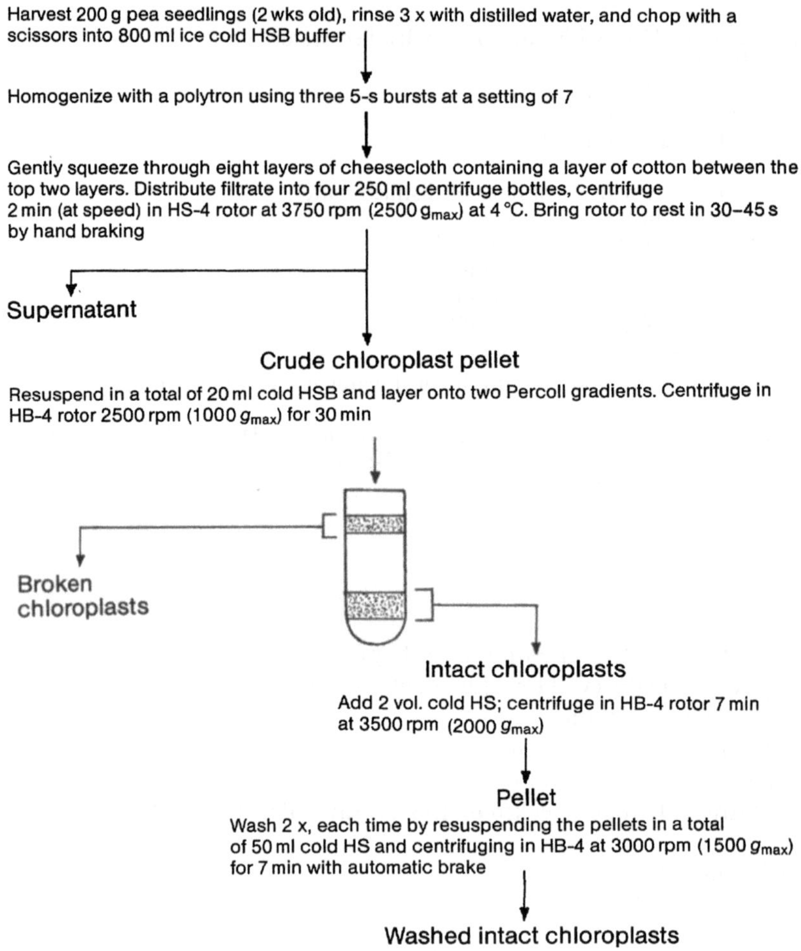

Fig. 2. Purification of intact chloroplasts

Other types of homogenizer, e.g., Waring blendor, can be used, but have been reported to be less efficient (Grossman et al. 1980). An important consideration here is to limit the duration of homogenization, as extensive homogenization leads to breakage of many or most of the chloroplasts.

Following homogenization, filtration and differential centrifugation are carried out rapidly to minimize the time that the chloroplasts are exposed to the disrupted cellular contents. The resulting crude chloroplast pellets are then resuspended in HSB buffer by repipeting with a disposable pasteur pipet. The pellets are kept covered with medium during this operation and foaming is avoided.

Intact chloroplasts are separated from broken chloroplasts on Percoll gradients. Figure 3 shows such a gradient containing chloroplasts obtained from approximately 25 g of tissue. The upper band (broken chloroplasts) and lower band (intact chloroplasts) are easily distinguishable. In practice, we load each

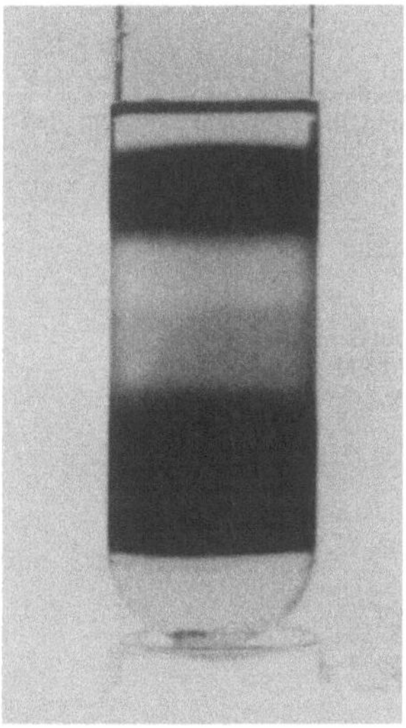

Fig. 3. Purification of intact chloroplasts by Percoll gradient centrifugation as described in Fig. 2. The gradient pictured above was loaded with a crude chloroplast preparation obtained from approximately 25 g of pea tissue. The *upper band* contains broken chloroplasts and the *lower band* contains intact chloroplasts

gradient with chloroplasts obtained from 100 g tissue. In this case, it is more difficult to distinguish the two bands and it is helpful to illuminate the tube from the back when fractionating.

Using the procedure outlined in Fig. 2, intact chloroplasts containing approximately 7 mg of chlorophyll (Chl) as measured by Arnon (1949) are obtained per 100 g of tissue. The chloroplasts are greater than 90% intact (Cline et al. 1981) and have been purified from the filtered homogenate 100-fold relative to mitochondria as estimated by Chl/cytochrome c oxidase ratios, 100-fold relative to peroxisomes as estimated by Chl/catalase ratios, and at least fourfold relative to endoplasmic reticulum as estimated by Chl/NADPH cytochrome reductase ratios (Cline et al. 1981).

3.3 Purification of Inner and Outer Envelope Membranes

Figure 4 is a protocol for purifying the envelope membranes from intact pea chloroplasts containing 20 to 30 mg Chl, the amount which can be obtained from 300 to 400 g of pea seedlings. With appropriate scaling of the procedure, we have prepared envelopes from as little as 100 g to as much as 2000 g of seedlings.

As described above, chloroplasts are ruptured by incubation in hyperosmotic sucrose followed by a cycle of freeze–thaw. This method breaks about 70% of the chloroplasts as judget by phase contrast light microscopy. Envelope membranes are then purified by a combination of flotation centrifugation and linear density gradient centrifugation. During the flotation centrifugation, the envelopes (unfractionated) rise to the 0.3 M/1.2 M interface and appear as a fluffy yellow material. Thylakoids, stroma, and any unbroken chloroplasts remain in the lower (1.3 M) layer. The concentration of sucrose in the lower layer was chosen to match the buoyant density of thylakoids. A refractometer is used to adjust this

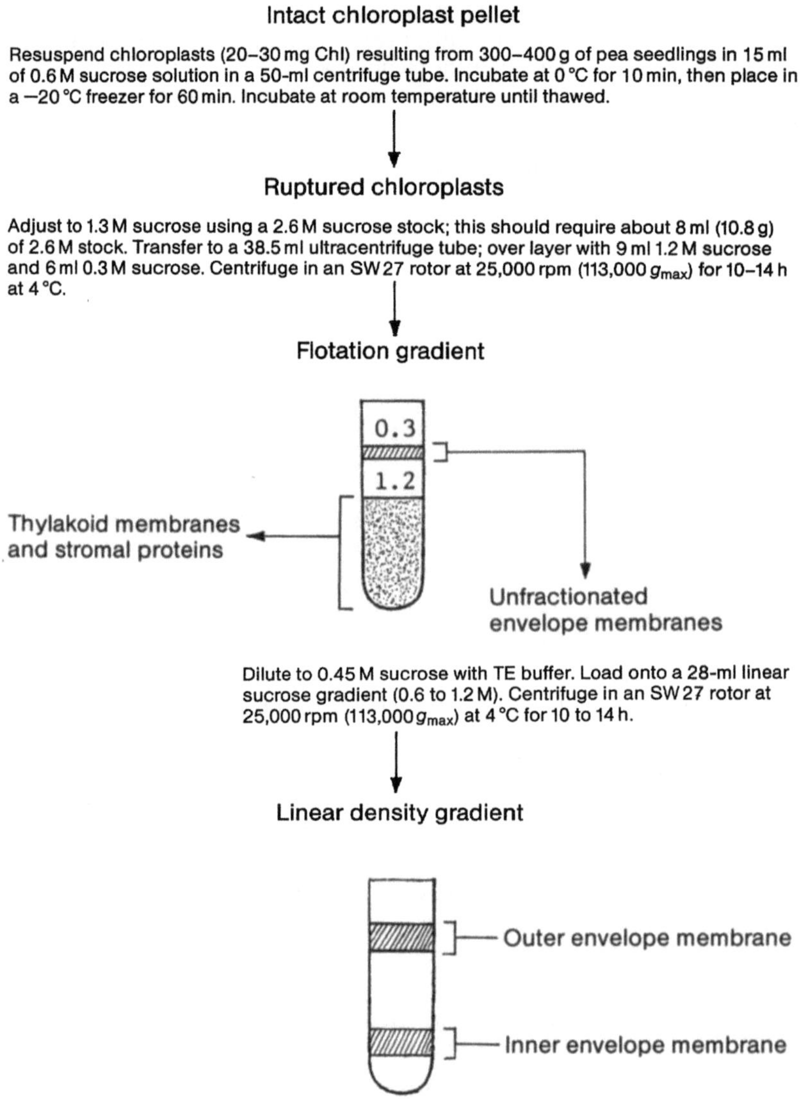

Fig. 4. Purification of inner and outer envelope membranes

layer to 1.3 M sucrose because if the density is too low, the thylakoids pellet during centrifugation; if it is too high they rise to the 1.3 M/1.2 M interface and form a thick mat. In our experience, either situation results in significantly lower envelope yields. The envelopes obtained by flotation centrifugation can then be prepared for the linear sucrose gradient by (a) diluting to about 0.45 M sucrose (Fig. 4) or (b) diluting fourfold with TE, pelleting 20,000 rpm in the SS-34 Sorvall rotor for 1 h, and resuspending in 0.45 M sucrose.

The optical density profile of a typical linear density gradient used to resolve inner and outer envelope membranes is shown in Fig. 5. Two peaks are obtained;

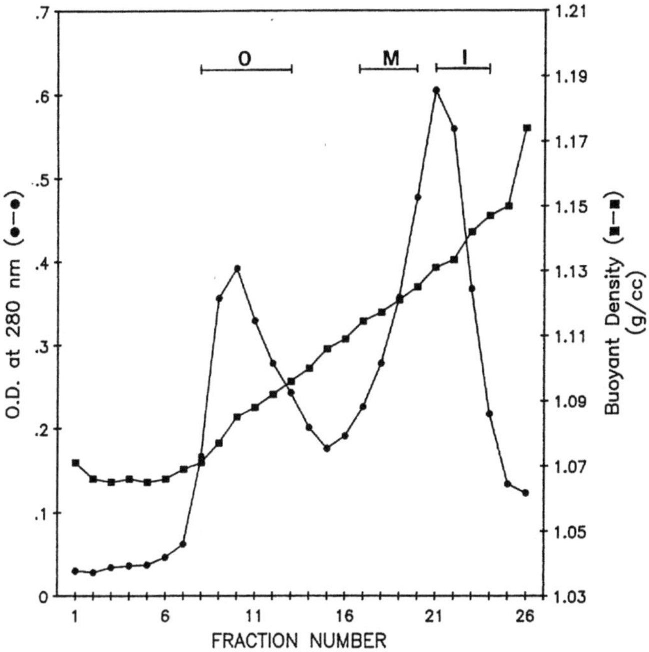

Fig. 5. Purification of inner and outer envelope membranes by linear density centrifugation as described in Fig. 4. Unfractionated envelope membranes, isolated by flotation centrifugation from 24 mg Chl of frozen and thawed chloroplasts, were sedimented through a 28 ml linear sucrose gradient. The gradient was fractionated with an ISCO fractionator and 1.5 ml fractions collected. Pools *O*, *M*, and *I* were made as shown

one centering at 1.08 g cm^{-3} has been identified as the outer envelope membrane (Cline et al. 1981, 1984, 1985); the other at 1.13 g cm^{-3} is enriched inner envelope membrane. A compositional analysis of the fractionated gradient (see below) has led us to routinely make three separate pools: pool O is the outer envelope membrane, pool M contains a mixture of outer and inner membrane, and pool I is predominantly inner membrane. The membranes in these pools are recovered by diluting threefold with TE and centrifuging at 90,000 g_{max} for 60 min. It is best to remove the supernatant carefully by aspiration because the envelope pellets can be very soft and are easily lost if the supernatant is decanted.

The yield of envelopes obtainable by this procedure is shown in Table 1. Because both the total yield and the ratio of inner to outer membrane subfractionations varies somewhat with the preparation, Table 1 gives the range and average amounts of the various fractions for five different preparations. At present, the percentage recovery is uncertain because the total amount of envelope per chloroplast is unknown. However, it is known that during chloroplast fractionation a certain proportion of the envelopes copurify with the thylakoid membranes. Estimates for the proportion of envelopes in thylakoids come from assays of thylakoid fractions for the presence of envelope enzymes. Assays for UDP-Gal : diacylgylcerol galactosyltransferase (Cline and Keegstra 1983), an outer envelope en-

Table 1. Yield and purity of envelope membrane fractions obtained by Fig. 4

Membrane fraction	Yield[a] (mg protein/100 mg Chl)		Outer membrane protein[b] (%)
	Range	Average	
Unfractionated[c]	3 –7	4.3	28
Outer membrane[d] (Pool O)	0.4 –1	0.7	100
Mixed membranes[d] (Pool M)	0.6 –1.3	0.9	31
Inner membrane[d] (Pool I)	0.65–1.5	1.1	9

[a] Values are the range and average from five different preparations
[b] The percent outer membrane protein in various envelope fractions was estimated by quantitative SDS-polyacrylamide gel electrophoresis as described (Cline and Keegstra 1983) on three of the five preparations
[c] Unfractionated envelope membranes described in Fig. 4
[d] Pools O, M, and I are shown in Fig. 5

zyme, indicates that the average recovery of unfractionated envelopes represents about 45% of the total, or chloroplasts contain 90–100 μg envelope protein per mg chlorophyll. However, this may represent a minimum value. Andrews and Keegstra (1983) have presented evidence that a greater percentage of inner membrane is found in the thylakoid fraction. In addition, the stereological analyses conducted by Forde and Steer (1976) suggest a much greater amount of envelope membrane per chloroplast than is indicated by the enzyme assays.

4 Properties of the Isolated Membranes

4.1 Purity

Because the envelope membranes are obtained from highly purified chloroplasts, there is little contamination by extrachloroplastidic membranes. This conclusion is supported by the fact that envelope subfractions possess only trace amounts of phosphatidylethanolamine (Table 2), a major polar lipid constituent in other cell membranes (Mazliak 1977); and by the fact that envelope membrane preparations are devoid of glycoproteins (Keegstra and Cline 1982), a distinguishing feature of the endomembrane system (Hughes 1976). On the other hand, there is a significant possibility of contamination by components of the chloroplast itself.

4.1.1 Cross-Contamination by Envelope Membranes

Much of the information concerning cross-contamination of envelopes comes from analysis of sodium dodecyl sulfate (SDS)-polyacrylamide gels of the membrane subfractions. A representative gel is shown in Fig. 6. Examination of the polypeptide profiles reveals that inner and outer envelope membranes possess very different polypeptide patterns (lanes I and O, Fig. 6), i.e., in general, the inner

Table 2. Properties of inner and outer pea chloroplast envelope membrane fractions

	Outer membrane[a] (Pool O)	Inner membrane[a] (Pool I)
Buoyant density (g cm^{-3})	1.08	1.13
Pigments[b]	Carotenoids	Carotenoids
Intramembranous particles, EF+PF[c]	~1400 per μm^2	~6500 per μm^2
Enzymes:		
UDP-Gal: diacylglycerol galactosyltransferase[d] (nmol min^{-1} mg $protein)^{-1}$	40.4	6.7
Acyl-CoA synthetase[d] (μmol h^{-1} mg $protein)^{-1}$	13.8	1.5
Acyl-CoA thioesterase[d] (μmol h^{-1} mg $protein)^{-1}$	0.005	0.357
Acyl-ACP: lysophosphatidic acid acyltransferase[d] (nmol min^{-1} mg $protein)^{-1}$	0.049	10.39
Mg^{2+}-dependent ATPase[d] (nmol min^{-1} mg $protein)^{-1}$	4.5	71.3
Polar lipids[e]: (% of total)		
MGDG	6.4	44.5
DGDG	32.7	31.3
TGDG	1	1.3
SL	3.3	2.2
PE	1.5	0.9
PG	5.5	6.7
PC	43.9	10.1
PI	5.1	2.2

[a] Pools O and I are as shown in Fig. 5
[b] Unpublished results of the author
[c] Obtained from Cline et al. 1985
[d] Data on UDP-Gal: diacylglycerol galactosyltransferase from Cline and Keegstra 1983, Acyl-CoA synthetase and Acyl-CoA thioesterase from Andrews and Keegstra 1983, Acyl-ACP: lysophosphatidic acid acyltransferase (personal communication Dr. Jaen Andrews, 1984). Mg^{+2}-dependent ATPase specific activities were taken from peak fractions of the sucrose gradient (personal communication D. McCarty, 1984)
[e] Obtained from Cline et al. 1981. MGDG, monogalactosyldiacylglycerol; DGDG, digalactosyldiacylglycerol; TGDG, trigalactosyltriacylglycerol; SL, sulfoquinovosyldiacylglycerol; PE, phosphatidylethanolamine; PG, phosphatidylglycerol; PC, phosphatidylcholine; PI, phosphatidylinositol

membrane subfraction has different major polypeptides than the outer. There are several polypeptide bands which are present in substantial quantities in both membranes, e.g., those at mW 86,000, 55,000, and 16,000. However, the 86,000 band actually represents two polypeptides; one an inner membrane polypeptide, and the other an outer membrane polypeptide (Cline et al. 1984; Werner-Washburne et al. 1983). The 55,000 and 16,000 polypeptides are stromal contaminants (see below).

A closer examination of the profiles shows that the inner membrane subfractions (pools I and M) contain minor amounts of outer membrane polypeptides,

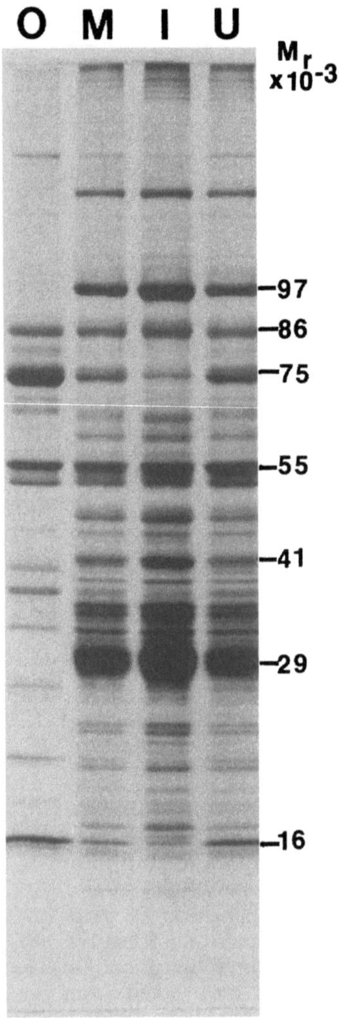

Fig. 6. An SDS-electrophoretogram of envelope membrane fractions. Electrophoresis is carried out on 7.5% to 15% acrylamide gels using the Laemmli buffer system (Laemmli 1970). Envelopes are generally prepared by pelleting and then dissolving in sample buffer at room temperature. Methods of preparation involving organic extraction of samples or heating of the dissolved membranes should be avoided. Such treatments result in the disappearance of several prominant polypeptide bands, most notably the 29,000 mw band of the inner membrane (Keegstra and Cline 1982). Lanes *O, M, I* are the envelope pools *O, M,* and *I* shown in Fig. 5; *U* unfractionated envelope membranes obtained by flotation centrifugation

e.g., the 75,000 polypeptide (Fig. 6). On the other hand, the outer membrane subfraction is devoid of major inner membrane polypeptides, e.g., the 29,000 band. We interpret this to indicate that the inner membrane subfractions are contaminated by outer membrane, whereas the outer membrane fraction is free of inner membrane. This type of analysis of polypeptide profiles is routinely used to assess the success of each envelope purification.

In order to obtain an inner membrane preparation with minimal outer membrane contamination, the inner membrane peak in the linear density sucrose gradient (Fig. 5) is subfractionated into two pools as described above. The pooling locations were chosen based on an SDS polyacrylamide gel analysis of fractions across the peak (Cline et al. 1985). This analysis showed that the contamination by outer membrane is greatest on the low bouyant density side of the peak, i.e., fractions 17 through 20 in Fig. 5, and decreases with increasing density until the peak fraction (number 21 of Fig. 5). This fraction, and those at a higher bouyant density, appear to contain a constant and relatively low amount of outer membrane. By pooling as shown in Fig. 5 it is possible to obtain in pool I the greatest yield of inner membrane with the least amount of outer membrane contamination. Densitometric analysis of the SDS gels provides estimates of the percentage of outer membrane in the various pools (Table 1). These estimates are in good agreement with values determined using specific activities of outer membrane enzymes, i.e., Acyl CoA synthetase (Andrews and Keegstra (1983) and UDP-Gal : diacylglycerol galactosyltransferase (Cline and Keegstra 1983), and inner membrane enzymes, Acyl CoA thioesterase (Andrews and Keegstra 1983) and Acyl ACP : phosphatidic acid acyl transferase (J. Andrews, personal communication, 1984), in the various subfractions.

The cross-contamination of envelopes may result from interactions between the outer and the inner membrane in situ. Electron microscopic (EM) studies of intact chloroplast have provided evidence that there are certain regions, called contact sites, in which the outer membrane adheres to the inner (Carde et al. 1982; Cline et al. 1985; Heldt and Sauer 1971). In a recent freeze-fracture EM study (Cline et al. 1985), we found that even under hyperosmotic conditions, contact sites are present. Additionally, the contact sites appear to survive the envelope purification procedures because they are observed in certain double membranes found in pool M. It seems likely that during chloroplast rupture and envelope purification, the contact sites stabilize the association of fragments of outer membrane with inner membrane, causing such fragments to co-sediment with inner membrane in the sucrose density gradients. Until a means of disrupting the contact sites is found, it is unlikely that inner membrane free of outer membrane will be obtainable. At present, the problem of outer membrane contamination is minimized by conducting analyses on pool I, which contains the least amount of outer membrane. Thus, the envelopes are purified on a linear sucrose gradient, not a step gradient, in order to achieve the resolution necessary for a careful subfractionation of the inner membrane peak.

4.1.2 Contamination by Thylakoids

Contamination of the envelopes by thylakoids can be conveniently detected by centrifuging the membranes and visually examining the pellets for traces of green. The contamination can be quantified by measuring the Chl level. In the procedure described above, the unfractionated envelopes obtained by flotation generally contain a small amount of thylakoids, whereas purified inner and outer membranes do not. When thylakoid contamination occurs, it is always in the inner membrane, never the outer.

Thylakoid contamination seems to result from the association of plastoglobuli with the thylakoids and/or from mechanical fragmentation of thylakoids. Both situations produce thylakoid-derived membranes which have buoyant densities similar to that of the inner envelope membrane. The best way to minimize such contamination is to use healthy, young tissue (plastoglobuli increase with the age of the tissue), and to avoid shearing forces during chloroplast rupture and envelope purification.

4.1.3 Contamination by Stroma

Despite the fact that stromal components are water-soluble, they are invariably present in envelope membrane subfractions, especially the outer membrane. The most predominant contaminants are the large (mw 55,000) and small subunits (mw 16,000) of ribulose bisphosphate carboxylase (Rubisco). These polypeptides have been identified in envelopes both by immunological procedures and by two-dimensional gel electrophoresis (Werner-Washburne et al. 1983). The 2-D gel analysis reveals that a number of other stromal polypeptides are also present in envelope preparations, indicating that the association of Rubisco with envelopes probably results from general stromal contamination. At present we have been unable to eliminate stromal contamination despite a variety of treatments designed to remove proteins absorbed to the surface of membranes (Werner-Washburne et al. 1983). Therefore, it is important to make certain that properties associated with envelope preparations are not the result of stromal contamination.

4.2 Other Properties

Properties of the two envelope membrane fractions are summarized in Table 2. Clearly, the two membranes are very different from each other. They possess different buoyant densities and their own distinctive set of polypeptides and enzymes. Even in polar lipid composition, the membranes are distinct. Although the two membranes have qualitatively similar lipid compositions, i.e., galactolipid/phospholipid, they differ in the relative amounts of the individual species. For example, the outer membrane has a small amount of monogalactosyldiacylglycerol and a large amount of phosphatidylcholine. The reverse is true of the inner membrane.

5 Modifications of the Procedure

Several modifications of the described protocol have been attempted in an effort to improve yields and purity and to reduce the time of isolation. In order to provide framework for future improvement of the method, successful and unsuccessful modifications are discussed below. However, we must caution that at present we have not fully characterized the membranes resulting from these modifications. They have been analyzed only for their quantity and their polypeptide profiles.

5.1 Alternate Methods of Chloroplast Rupture

Methods which are successful involve preincubation of chloroplasts in hyperosmotic sucrose or sorbitol solutions followed by mechanical lysis, e.g., 20 to 30 strokes with a Dounce homogenizer, several min in a bath sonicator (probe sonicators deliver too much energy), or forcing the chloroplast suspension through a nylon screen. The major disadvantage of these methods is more contamination by thylakoids. Rupture methods with do not work are either osmotic shock in hypoosmotic solutions or rapid freeze-thaw, i.e., freezing hyperosmotic chloroplasts in liquid nitrogen followed by thawing in a 25 °C water bath.

The ionic conditions under which chloroplasts are ruptured are also important. Chloroplasts are normally incubated in 10 mM Tricine/NaOH, 2 mM EDTA. Omitting the EDTA does not have serious consequences on the resulting separation. However, in our hands, inclusion of divalent metal ions such as Mg^{2+} dramatically impairs the ability to separate inner from outer membranes.

5.2 Purification Subsequent to Rupture

Several more rapid methods have been explored to shorten the current procedure which involves two 10 to 14 h centrifugations. We have been able to eliminate the 2nd centrifugation step and purify inner and outer envelopes by flotation centrifugation simply by pouring a linear sucrose gradient (0.5 M to 1.2 M) above the ruptured and adjusted chloroplast suspension. This method gives adequate separation of inner from outer membranes, but leads to more thylakoid contamination of the inner membrane.

A promising method which is presently being optimized in our laboratory is to use differential centrifugation to replace flotation centrifugation as a means of obtaining unfractionated envelopes. The ruptured chloroplast suspension is diluted to 0.2 M sucrose, thylakoids are removed by centrifugation at 4500 g_{max} for 15 min, and envelopes are recovered from the supernatant by centrifugation at 40,000 g_{max} for 30 min. After a wash step, inner and outer membranes are resolved by sedimentation through a 0.5 to 1.2 M sucrose gradient (A. Yousif and K. Keegstra, personal communication, 1984). This procedure reduces the purification time dramatically without increasing contamination by thylakoids or stroma.

5.3 Application to Other Tissues

Although the described procedure has been used in our laboratory with pea chloroplasts, there are two reports that it is applicable to spinach chloroplasts. Soll and Buchanan (1983) were able to separate the two envelope membranes of spinach chloroplasts and localize a protein kinase to the outer membrane. Van't Riet et al. (1983) have used this method to purify inner and outer spinach chloroplast envelope membranes. However, they report that significant amounts of inner membrane can only be obtained if the chloroplasts are frozen and thawed twice.

6 Other Procedures

In 1983, another method of purifying inner and outer chloroplast envelopes was reported by Block et al. (1983a, b), using spinach chloroplasts. Their method is similar to our procedure in that they rupture chloroplasts in a hyperosmotic mannitol solution. However, there are some important differences. Block et al. include Mg^{2+} ions in their lysis and gradient buffers, use a Yeda press to rupture the chloroplasts, and employ a step sucrose gradient to resolve fractions enriched in inner and outer membranes. The resulting inner and outer envelope membrane preparations have properties very similar to inner and outer pea chloroplast envelope membranes as shown in Table 2, with a single exception. Block et al. (1983b) report that the enzyme UDP-Gal: diacylglycerol galactosyltransferase is located in the inner spinach chloroplast envelope membrane, whereas our analysis with pea chloroplasts indicates that it is an outer envelope enzyme. Although this difference may reflect a species difference, definitive experiments to test this possibility have not been carried out at this writing.

An advantage of the method by Block et al. over the procedure reported here is that it is rapid, the isolation being performed in 1 day. However, the limitations of this method are that much lower yields of envelopes are obtained, with a significantly higher level of cross-contamination. Thus, at this time, choosing a method for preparing inner and outer envelope membranes means deciding between rapidity of isolation and yield and purity of the resulting preparations. It is our hope that in the near future, technical problems can be overcome to produce a method with minimal limitations. Such a method would greatly benefit studies of the interesting functional and structural properties of the chloroplast envelope.

Acknowledgements. The author thanks Jaen Andrews, Ken Keegstra, Craig Lending, Tom Lubben, Peter Quail, and Maggie Werner-Washburne for helpful discussions and critical review of the manuscript. A special thanks to Ken Keegstra in whose laboratory this work was conducted. Supported in part by a grant from the United States Department of Agriculture Competitive Research Grants Office.

References

Andrews J, Keegstra K (1983) Acyl-CoA synthetase is located in the outer membrane and Acyl-CoA thioesterase in the inner membrane of pea chloroplast envelopes. Plant Physiol 72:735–740

Arnon DI (1949) Copper enzymes in isolated chloroplast. Phenoloxidase in *Beta vulgaris*. Plant Physiol 24:1–15

Block MA, Dorne A-J, Joyard J, Douce R (1983a) Preparation and characterization of membrane fractions enriched in outer and inner envelope membranes from spinach chloroplasts. Electrophoretic and immunochemical analyses. J Biol Chem 258:13272–13280

Block MA, Dorne A-J, Joyard J, Douce R (1983b) Preparation and characterization of membrane fractions enriched in outer and inner envelope membranes from spinach chloroplasts. Biochemical characterization. J Biol Chem 258:13281–13286

Carde J-P, Joyard J, Douce R (1982) Electron microscopic studies of envelope membranes from spinach plastids. Biol Cell 44:315–324

Cline K, Keegstra K (1983) Galactosyltransferases involved in galactolipid biosynthesis are located in the outer membrane of pea chloroplast envelopes. Plant Physiol 71:366–372

Cline K, Andrews J, Mersey B, Newcomb EH, Keegstra K (1981) Separation and characterization of inner and outer envelope membranes of pea chloroplasts. Proc Natl Acad Sci USA 78:3595–3599

Cline K, Keegstra K, Staehelin LA (1985) Freeze-fracture electron microscopic analysis of envelope membranes on intact chloroplasts and after purification. Protoplasma 125:111–123

Cline K, Werner-Washburne M, Andrews J, Keegstra K (1984) Thermolysin is a suitable protease for probing the surface of intact pea chloroplasts. Plant Physiol 75:675–678

Chua NH, Schmidt GW (1979) Transport of proteins into mitochondria and chloroplasts. J Cell Biol 81:461–483

Douce R, Joyard J (1979) Structure and function of the plastid envelope. Adv Bot Res 7:1–116

Douce R, Joyard J (1982) Purification of the chloroplast envelope. In: Edelman M, Hallick RB, Chua N-H (eds) Methods in chloroplast molecular biology. Elsevier Biomedical Press, Amsterdam, pp 239–256

Ellis RJ (1983) Chloroplast protein synthesis: principles and problems. Subcell Biochem 9:237–261

Forde J, Steer MW (1976) The use of quantitative electron microscopy in the study of lipid composition of membranes. J Exp Bot 27:1137–1141

Grossman AR, Bartlett SG, Schmidt GW, Chua N-H (1980) Posttranslational uptake of cytoplasmically synthesized proteins by intact chloroplasts in vitro. Ann NY Acad Sci 343:266–274

Heber U, Heldt HW (1981) The chloroplast envelope: structure, function and role in leaf metabolism. Annu Rev Plant Physiol 32:139–168

Heldt HW, Sauer F (1971) The inner membrane of the chloroplast envelope as the site of specific metabolite transport. Biochim Biophys Acta 234:83–91

Hughes RC (1976) Membrane glycoproteins. A review of structure and function. Butterworths, Boston

Keegstra K, Cline K (1982) Evidence that envelope and thylakoid membranes from pea chloroplasts lack glycoproteins. Plant Physiol 70:232–237

Laemmli UK (1970) Cleavage of structural proteins during the assembly of the head of bacteriophage T4. Nature 227:680–685

Mackender RO, Leech RM (1970) Isolation of chloroplast envelope membranes. Nature 228:1347–1348

Mazliak P (1977) Glyco- and phospholipids of biomembranes in higher plants. In: Tevini M, Lichtenthaler HK (eds) Lipids and lipid polymers in higher plants. Springer, Berlin Heidelberg New York, pp 48–74

Mills WR, Joy KW (1980) A rapid method for isolation of purified, physiologically active chloroplasts, used to study the intracellular distribution of amino acids in pea leaves. Planta 148:75–83

Morgenthaler J-J, Marsden MPF, Price CA (1975) Factors affecting the separation of photosynthetically competent chloroplasts in gradients of silica sols. Arch Biochem Biophys 168:289–301

Roughan PG, Slack CR (1982) Cellular organization of glycerolipid metabolism. Annu Rev Plant Physiol 33:97–132

Soll J, Buchanan BB (1983) Phosphorylation of chloroplast ribulose bisphosphate carboxylase/oxygenase small subunit by an envelope-bound protein kinase in situ. J Biol Chem 258:6686–6689

Tahabe T, Nishimuru M, Akazawa T (1979) Isolation of intact chloroplasts from spinach leaf by centrifugation in gradients of modified silica "Percoll". Agric Biol Chem 43:2137–2142

Van'T Riet J, Kemp F, Abraham PR (1983) Isolation and characterization of outer and inner chloroplast envelope membranes of *Spinacia oleracea*. Advances in Photosynthesis Research. Sytesma C. (ed) pp 31–34. Sixth International Congress on Photosynthesis Brussels, 1983, Martinus Nijhoff/Dr. W. Junk, The Hague

Werner-Washburne M, Cline K, Keegstra K (1983) Analysis of pea chloroplast inner and outer envelope membrane proteins by two dimensional gel electrophoresis and their comparison with stromal proteins. Plant Physiol 73:569–575

The Major Protein of Chloroplast Stroma, Ribulosebisphosphate Carboxylase

C. PAECH

1 Introduction

The previous edition of *Modern Methods of Plant Analysis* contained a summary of characteristics of a "carboxylating enzyme... also known as carboxydismutase" (Losada and Arnon 1964). Today, this enzyme is identified as D-ribulose-1,5-bisphosphate carboxylase/oxygenase [3-phospho-D-glycerate carboxyl-lyase (dimerizing), EC 4.1.1.39]. It was not until 1970 that the synonymy of Fraction I protein with carboxydismutase and RuBP carboxylase[1] was documented (Akazawa 1970; Kawashima and Wildman 1970). The reviews illustrated how our perception of Fraction I protein evolved, from a protein presenting the singularly largest component among soluble plant proteins (Singer et al. 1952; Wildman and Bonner 1947), to an enzyme catalyzing the conversion of CO_2 and RuBP to D-glycerate 3-phosphate in photosynthesis (Jakoby et al. 1956; Racker 1957; Weissbach et al. 1956).

In 1971, Ogren and Bowes suggested yet another function for Fraction I protein, the catalysis of RuBP oxygenation (Ogren and Bowes 1971). Acceptance, and later proof, of the hypothesis that phosphoglycolate emerging in the RuBP oxygenase reaction is the source of photorespiratory CO_2 loss gave rise to a vision that has become a theme with variations centering around the questions: Will net carbon uptake (and hence food and fiber production by plants) increase if the uBP oxygenase reaction is suppressed? If not, what is the purpose of photorespiration?

When purified to homogeneity from a variety of plant and algal sources, RuBP carboxylase has a molecular weight of 530,000 and consists of eight large (L, 53,000) and eight small (S, 14,000) subunits. In higher plants, the biosynthesis of these two subunits is specified by two different genetic systems (Criddle et al. 1970; Kawashima 1970). The large subunit is encoded by the chloroplast genome and synthesized within the organelle. The small subunit is encoded by the nuclear DNA and synthesized in the cytoplasm as a precursor molecule. As this precursor enters the chloroplast, it is processed into the mature subunit and subsequently

1 Abbreviations: Bicine, N,N-bis(2-hydroxyethyl)glycine; BSA, bovine serum albumin; CABP, 2-carboxy-D-arabinitol 1,5-bisphosphate; cDNA, cloned DNA; cpDNA, chloroplast DNA; DTT, dithiothreitol; CRBP, 2-carboxy-D-ribitol 1,5-bisphosphate; Na_2-EDTA, ethylenediaminetetraacetic acid, disodium salt; kbp, kilo base pair; mw, molecular weight; PCMB, p-hydroxymercuribenzoate; PGA, D-glycerate 3-phosphate; PVP, polyvinylpolypyrrolidone; PVPP, polyvinylpolypyrrolidone (insoluble); rbcL, gene coding for the large subunit of RuBP carboxylase; RuBP, D-ribulose 1,5-bisphosphate; RuBP carboxylase, D-ribulose 1,5-bisphosphate carboxylase/oxygenase, EC 4.1.1.39; SDS-PAGE, polyacrylamide gel electrophoresis in the presence of sodium dodecyl sulfate

assembled with large subunits into the holoenzyme (L_8S_8). Obviously, the complexity in biosynthesis, structure, and function of RuBP carboxylase has stirred the imagination of many researchers ranging from geneticists to crystallographers. The topic RuBP carboxylase has developed into a fascinating challenge. RuBP carboxylase has become the most abundantly discussed protein in current literature.

In this chapter, information will be presented pertaining to the physical structure of this protein, the kinetic characteristics of this enzyme, and the biosynthesis and in vivo interaction of this chloroplast component with other organelle constituents (e.g., membranes, soluble proteins, protons, metal ions, and metabolites). This background is prerequisite to designing experiments to answer such questions as: What is the reason for the extraordinarily vast amount of this enzyme (protein) in the stromal space of chloroplasts? Does RuBP carboxylase impose a rate-limiting step on net photosynthesis? Is RuBP carboxylase a useful parameter for modeling photosynthesis? Can the ratio of the two enzyme activities of RuBP carboxylase be altered to favor the carboxylation without impairing the plant as a whole? Can genetic engineering increase the production of RuBP carboxylase and advance terrestrial plants to an economically feasable resource for a nutritionally valuable protein?

Since investigators for many years to come will be concerned with assaying and isolating this enzyme from a variety of plant material, laboratory instructions pertaining to these tasks will conclude this article. Reading of a number of recent reviews, from both a factual as well as an interpretative point of view, is recommended to do full justice to the many facets of RuBP carboxylase (Akazawa 1979; Akazawa et al. 1984; Broglie et al. 1984; Ellis 1979, 1981; Jensen and Bahr 1977; Lorimer 1981; McFadden 1980; Miziorko and Lorimer 1983; Siegelman and Hind 1978; Wildman 1979).

2 Characteristics of RuBP Carboxylase

2.1 Molecular Arrangement and Physical Structure of Subunits

All higher plants, green algae and prokaryotic photoautotrophs assemble an oligomer of 16 polypeptides to form catalytically functional RuBP carboxylase (L_8S_8). The holoenzyme of higher plants, under nondenaturing conditions, does not dissociate into subunits or smaller aggregates unless treated with PCMB followed by an adjustment of pH to 11 and a rise in temperature (Nishimura et al. 1973). In contrast, the enzyme from several prokaryotic sources appears to form a rather loose although strictly stoichiometric aggregate of the 16 polypeptides. The small subunits can be removed from an A_8 core of the enzyme from *Aphanothece* at low ionic strength and neutral pH (Asami et al. 1983; Takabe et al. 1984a, b), from *Synechococcus* at high ionic strength and mildly acidic pH (Andrews and Abel 1981; Andrews and Ballment 1983, 1984), and from *Chromatium* at pH 9 (Jordan and Chollet 1985; Takabe and Akazawa 1975). Thus, reports on

RuBP carboxylase of smaller aggregation, such as L_4, L_6, L_8, and L_nS_n ($n<8$), must be viewed with caution despite an appealing correlation between developmental level of the organism and degree of subunit aggregation of RuBP carboxylase (Akazawa et al. 1978; McFadden 1973, 1980). Failure to detect small subunits in RuBP carboxylase, particularly from prokaryotic sources, by SDS-PAGE may be attributed to variability in staining intensity as seen with the enzyme from *Synechococcus* (Andrews and Ballment 1983) and with higher plant enzymes (Paech, unpublished observation), and should not be construed as evidence of absence of small subunits.

A well-documented case of RuBP carboxylase with fewer than 16 subunits is the enzyme from *Rhodospirillum rubrum* which exists as dimer of only large subunits (Tabita and McFadden 1974). An L_6 form has been identified in *Rhodopseudomonas sphaeroides* (Gibson and Tabita 1977; Weaver and Tabita 1983). Using the gene from *R. rubrum* as a hybridization probe a DNA fragment was isolated and cloned into *E. coli*. The expressed protein showed the characteristics of the *Rhodopseudomonas* enzyme and thus substantiated the claim for the existence of a distinct L_6 form (Quivey and Tabita 1984).

The large subunit shows a high degree of similarity in amino acid composition among plants (Akazawa 1979). Moreover, sequence analysis of the polypeptide and of relevant DNA fragments revealed a large extent of sequence homology (>80%) in the polypeptide chain even for enzymes from rather distant eukaryotic and prokaryotic sources (Curtis and Haselkorn 1983; Dron et al. 1982; McIntosh et al. 1980; Poulsen 1979, 1981; Poulsen et al. 1979; Reichelt and Delaney 1983; Shinozaki and Sugiura 1982; Zurawski et al. 1981). Segments with highly conserved homology are found particularly in the vicinity of lysine-201 which is shown to participate in the activation reaction (Lorimer 1981 a), and in the vicinity of lysine-175 and -335 which were labeled with active-site-directed irreversible inhibitors (Hartman et al. 1978). In contrast, an extraordinary sequence divergence of amino acids is displayed by the subunit of the *R. rubrum* enzyme (only 31% homology with the large subunit of the spinach enzyme). Yet, some critical sections (around lysine-175, -201, and -335) are essentially unaltered (Hartman et al. 1984). This offers a unique opportunity for focusing attention on a small fraction of amino acid residues when investigating structure-function relationships or evaluating X-ray data.

The small subunit of RuBP carboxylase from various plant sources displays a significantly higher degree of amino acid differences than the large subunit, albeit the primary structure reflects some conserved areas (Berry-Lowe et al. 1982; Broglie et al. 1983 a, b; Coruzzi et al. 1983 a, b; Martin 1979; Müller et al. 1983). Seemingly there should be at least one structural element common to all small subunits of higher plants. This is the recognition site for interaction with the large subunit in the holoenzyme. Indeed, the wheat small subunit precursor is accepted and correctly processed to RuBP carboxylase by pea chloroplasts and vice versa (Coruzzi et al. 1983 b).

The solubility of isolated small subunits is quite high due to a large proportion of hydrophilic, particularly basic, amino acid residues. The isolated large subunit, however, retains solubility in aqueous solution only in the absence of divalent metal ions. Addition of Mg^{2+}, e.g., for assays, renders the polypeptide insoluble

(Brown and Chollet 1979). Whether this is an artifact due to the isolation procedure or whether it is physiologically significant remains speculative. It should be noted, however, that the cloned and expressed large subunit of maize is insoluble, too (Gatenby 1984; Somerville et al. 1983). Thus, small subunits could have a structural and regulatory function in the assembly of RuBP carboxylase. The dissociation of any complex requires the components to become solvated. Since large subunits are insoluble in aqueous solutions containing millimolar quantities of Mg^{2+} (which is the situation encountered in the stromal space of chloroplasts) removal of the large subunit from its carrier and, in effect, transfer to the enzyme to be assembled, requires an acceptor molecule, e.g., the small subunit, which has a higher affinity for the large subunit and provides sufficient solvation capacity. This transfer process apparently is aided by $ATP \cdot Mg^{2+}$ (Milos and Roy 1984).

2.2 Molecular Structure

Viewed by electron microscopy, RuBP carboxylase appears to be a particle measuring 10–12 nm in diameter with a center hole or depression (Kawashima and Wildman 1970). With crystalline enzyme from tobacco leaves (Kawashima and Wildman 1971) Eisenberg's group developed from X-ray diffraction patterns of a series of crystal forms a highly symmetrical bilayer model (Baker et al. 1977). Each layer of this structure is formed of four large subunits peripherally surrounded by four small subunits. Two of these units are stacked in such a way that they form a channel with a fourfold symmetry axis in the center. The positioning of the small subunits relative to the large subunits was not possible. However, indirect evidence from cross-linking experiments (Roy et al. 1978) favors a symmetric arrangement by pairs. Even though Eisenberg's model may be criticized on grounds that it is based on crystals grown below pH 6 and in the absence of Mg^{2+} and CO_2 (i.e., from catalytically inactive enzyme) the interpretation of small angle neutron scattering by spinach RuBP carboxylase in solution (Donnelly et al. 1984) is consistent with Eisenberg's hypothesis.

Quite a different structural model was described for RuBP carboxylase from the chemoautotrophic bacterium *Alcaligenes* (Bowien et al. 1976, 1980; Meisenberger et al. 1984). The large subunit is U-shaped and the small subunits are located on top and below a bilayer of large subunits. However, the fourfold symmetry axis coincident with the channel in the center of the molecule is maintained. Moreover, the sedimentation coefficient of this enzyme changes from 17.5S to 14.3S (Bowien and Gottschalk 1982) and the globular structure flattens to an ellipsoid upon activation with CO_2 and Mg^{2+} (Meissenberger et al. 1984). No such change was observed with the enzyme from spinach and *R. rubrum* when investigated by small angle neutron scattering and sedimentation centrifugation (Donnelly et al. 1984). This striking difference is highly surprising, considering the close amino acid homology between the enzyme from spinach and *Alcaligenes* but the large difference in both polypeptide composition and primary structure between the enzyme from spinach and *R. rubrum*.

A landmark in structural studies on RuBP carboxylase was set when Johal and Bourque (1979), successfully crystallized the enzyme from spinach. This was later repeated from several other sources (Johal et al. 1980), including the enzyme from *R. rubrum* (Schloss et al. 1979). Since then several laboratories have embarked on the X-ray crystallography of quaternary complexes of RuBP carboxylase with Mg^{2+}, CO_2 and the reaction intermediate analog CABP (Andersson and Brändén 1984; Andersson et al. 1983; Schneider et al. 1984) or the ternary complex of Mg^{2+} and CO_2 with the enzyme from spinach (Barcena et al. 1983) and from *R. rubrum* (Janson et al. 1984). The ternary complex probably closely resembles the configuration of the unperturbed, active enzyme. Further support for this assumption could be gained from experiments which demonstrate, e.g., by microspectrophotometry (Schirch et al. 1981), that the crystalline complex $Enz \cdot CO_2 \cdot Mg^{2+}$ has catalytic properties similar to those of the enzyme in solution. In any event, X-ray studies on these crystalline ternary and quaternary complexes will provide much useful insight into the architecture and the functioning of RuBP carboxylase.

2.3 Biosynthesis and Assembly of Subunits

2.3.1 Large Subunit

In higher plants, the DNA sequence coding for the large subunit is a single copy on the circular chloroplast DNA (Bedbrook et al. 1979), which frequently contains two inverted repeat sequences (Whitfeld and Bottomley 1983). On cpDNA with inverted repeat sequence rbcL maps in the large single-copy region separated by about 50 kbp from the gene P32 for the 32,000 mw polypeptide (regulating photosystem II electron transport and binding triazine herbicides) which in turn is located at the margin of one repeat sequence. In soybean cpDNA, however, rbcL and P32 are separated by only 5 kbp (Spielmann et al. 1983). This is reminiscent of cpDNA of pea and mung bean (Palmer et al. 1982), which has lost the inverted repeat sequence during evolution.

From an initially small number of cpDNA's investigated [corn (McIntosh et al. 1980), mustard (Link 1981), and spinach (Zurawski et al. 1981)] it appeared that a single mRNA of relatively constant size, about 1.6 kbp, was transcribed from rbcL. In a more extensive survey, however, an unexpected variation in size, number, and relative proportion of transcripts surfaced in the family *Vigna* and *Phaseolus* (Palmer et al. 1982). There two mRNA's with 2.4 to 2.6 kbp dominate the 1.6 kbp mRNA usually found. More recently in corn, two mRNA's of different size (1.6 and 1.8 kbp) were discovered (Crossland et al. 1984). The relative proportion of these two mRNA's varies during the light-induced development of plastids. The reason for transcript multiplicity is not immediately obvious. Also, it is not known whether all different mRNA's for the large subunit in a chloroplast are translated.

A translated large subunit binds to a carrier protein which is composed of 10 to 12 identical polypeptides with a molecular weight of 60,000 each (Barraclough and Ellis 1980). The entire complex has a sedimentation coefficient of 29 S (Roy et al. 1982). Newly translated large subunits seem to appear also in a pool of particles with a sedimentation coefficient of 7 S (Roy et al. 1982). These aggregates are either large subunit dimers or heterodimers with an unidentified 60,000 polypeptide. Incorporation of unassembled large subunits into RuBP carboxylase proceeds only in light and can be enhanced by ATP (Milos and Roy 1984). Release of a large subunit from its 29 S carrier protein requires ATP·Mg^{2+} (Milos and Roy 1984). Although the assembly mechanism is still obscure, these findings are of interest from two points of view. One is the fact that the Mg^{2+} concentration in chloroplasts is 5 to 10 mM and is thought to fluctuate 1 to 3 mM during light–dark transitions (Portis 1981; Portis and Heldt 1976). Since isolated large subunits are rendered insoluble in the presence of Mg^{2+}, it is conceivable that one of the purposes of the large carrier protein is to ensure solubility of large subunits before assembly into the holoenzyme. This would suggest that the 7 S particle is either a heterodimer where the 60,000 polypeptide fulfills a function similar to that of the large carrier protein, or that the large subunit, when dimerized, becomes soluble (as is the enzyme from *R. rubrum*). The second interesting aspect of this ATP·Mg^{2+} involvement in processing large subunits is the observation that ATP, but not ADP or any other nucleotide, stabilizes RuBP carboxylase against heat denaturation (Li and Wang 1980). This effect is enhanced by Mg^{2+} (Paech, unpublished observation). Hence ATP·Mg^{2+} might be required for maintaining the proper active site configuration during subunit assembly.

2.3.2 Small Subunit

The small subunit is encoded by six to ten different genes on the nuclear DNA (Berry-Lowe et al. 1982; Coruzzi et al. 1983 a). Regulation of their expression appears to be more sophisticated than that of the large subunit. For example, in heat-treated barley large subunits and RuBP carboxylase are absent, while small subunits accumulate (Feierabend and Wildner 1978). This indicates a dominant role of the nuclear gene system and possibly accounts for the observation that RuBP carboxylase is not detectable in root cells, in dark-grown seedlings, or in mesophyll cells of mature corn leaves. The underlying principles for the tissue-specific expression are still unknown, but some clues exist for the light-dependent formation of small subunits. There the regulatory control is on the level of mRNA synthesis (Smith and Ellis 1981; Viro and Kloppstech 1983) and appears to be phytochrome-mediated (Sasaki et al. 1983; Tobin 1981).

The information on mRNA for the small subunit from wheat (Broglie et al. 1983 b; Coruzzi et al. 1983 b), soybean (Berry-Lowe et al. 1982), and two different varieties of pea (Bedbrook et al. 1980; Broglie et al. 1981, 1984) has been transcribed into cDNA. Nucleotide sequence comparison showed two exons (wheat) or three exons (pea, soybean), of which the first exon always coded for the transit peptide plus the first two amino acids of the mature small subunit. Sections of highly conserved amino acid sequences (Broglie et al. 1983 a, b) confirm that the transit peptide is a structural element essential for the uptake and processing of precursor molecule by the chloroplast envelope (Chua and Schmidt 1978). This is substantiated by the finding that wheat and pea chloroplasts incorporate precursors homologously and heterologously (Coruzzi et al. 1983 b), and lends credit to the hypothesis that exons encompass polypeptide sections with distinct function (Gilbert 1979). The question arises whether the additional intron on the pea and soybean small subunit genome reflects only the genetic diversion of monocots and dicots or whether evolution of dicots was favored by structural and/or functional features of the small subunit and this caused formation of two exons.

Translation of small subunit mRNA in the cytosol produces a precursor molecule approximately 5,000 mw larger than the mature subunit (Dobberstein et al. 1977). Transport of this precursor through the chloroplast envelope is energy-dependent and is accompanied by removal of the transit peptide (Grossman et al. 1980). Normally, the pool of free small subunits in chloroplasts is very small (Roy et al. 1982). This suggests either a rapid assembly of RuBP carboxylase, or the presence of a selective proteolytic enzyme system which prevents the formation of a large pool of free small subunits (Schmidt and Mishkind 1983). However, when chloroplast protein synthesis in rye was eliminated by heat treatment, small subunits continued to be rapidly imported (Feierabend and Wildner 1978). In order to reconcile these two conflicting reports, heat inactivation of the proteolytic enzyme system was invoked (Schmidt and Mishkind 1983).

2.3.3 Subunit Heterogeneity

How uniform is the polypeptide composition of RuBP carboxylase? In view of the multiplicity of nuclear genes coding for the small subunit, charge heterogeneity in isoelectric focusing patterns is plausible. This has been borne out by amino acid sequence analysis of small subunits of spinach (Martin 1979) and tobacco (Müller et al. 1983). However, no substitution of amino acids was found in the pea small subunit (Bedbrook et al. 1980; Takruri et al. 1981). As sequence analyses on both protein and nucleic acid level become routine, ambiguities of isoelectric focusing patterns, particularly in the detection of electrically neutral substitutions will be resolved. Already, the partial primary structure of the small subunit has been entered as a guide in taxonomic and phylogenetic studies (Martin and Dowd 1984; Yeoh et al. 1982).

Charge variants of the large subunit must be viewed with caution. More often than not, they are artifacts introduced during sample preparation (Johal and Chollet 1983; O'Connell and Brady 1982). For example, isoelectric focusing resolved large subunits of the tobacco enzyme into three distinct bands. Upon fingerprinting and amino acid analyses these large subunits became indistinguishable (Gray et al. 1978). Lack of multiple forms of the large subunit points to single gene mutation in chloroplasts. This explanation is reinforced by the finding that in *Chlamydomonas* a single nucleotide mutation in the gene for the large subunit rendered RuBP carboxylase inactive but did not prevent its synthesis (Dron et al. 1983).

2.3.4 Coordinate Control of Subunit Synthesis

When comparing kinetic and analytical complexities, it appears that the chloroplast contains 10 to 100 copies of cpDNA (Bedbrook and Kolodner 1979). Since there may be as many as 50 chloroplasts per cell (Possingham 1980), the nuclear genome faces an overwhelming excess of plastid genetic material. Do the cytoplasmatic and chloroplastic protein synthesis systems communicate to maintain a balanced production of RuBP carboxylase subunits? During development, but also in aging leaves, there appears to be a tight reciprocal control in the synthesis of both subunits (Brady 1981; de Heij et al. 1984; Kobayashi and Akazawa 1982; Pineau 1982; Sasaki et al. 1983; Tobin 1981). Although mRNA levels increase with illumination, the enzyme and its mRNA's are also present in dark-grown plants (Dean and Leech 1982; de Heij et al. 1984; Sasaki et al. 1981; Tobin 1981). This suggests that light modulates, but does not exclusively determine, the synthesis of RuBP carboxylase. Rather, the factor limiting total amount of RuBP carboxylase appears to be located on nuclear chromosomes (Jellings et al. 1983).

2.4 Catalytic Mechanism

2.4.1 Activation and Role of Mg^{2+}

It is useful to separate the mechanism of action of RuBP carboxylase into two distinct steps: activation and catalysis. Activation constitutes all events involved in converting the protein into a catalyst. Catalysis concerns the carboxylation of RuBP yielding two molecules of D-glycerate 3-phosphate, and oxygenation of RuBP resulting in one molecule each of D-glycerate 3-phosphate and glycolate 2-phosphate. The obligatory cofactor in both reactions is Mg^{2+}. CO_2 and O_2 are mutually competitive for RuBP.

Anomaly in kinetics and the various $K_m(CO_2)$-forms of RuBP carboxylase (Jensen and Bahr 1977) became understandable once the homotropic effect of CO_2 (Murai and Akazawa 1972; Pon et al. 1963) was fully recognized as activation of RuBP carboxylase (Badger and Lorimer 1976; Lorimer et al. 1976). In an equilibrium reaction, CO_2 forms a carbamate with the ε-amino group of lysine-201 and the subsequent rapid addition of Mg^{2+} is thought to displace this equilibrium towards the ternary complex $Enz \cdot CO_2 \cdot Mg^{2+}$, which is the catalytically functional enzyme form (Lorimer et al. 1976). Other divalent cations can substitute for Mg^{2+} but are less effective (Christeller 1981; Weissbach et al. 1956). The half-life of the ternary complex with Mg^{2+} is in the order of one minute (Badger and Lorimer 1981; Lorimer et al. 1976).

Partial proof of this model was obtained when methylation of the quaternary complex $Enz \cdot CO_2 \cdot Mg^{2+} \cdot CABP$ trapped the putative carbamate and identified lysine-201 as the binding site for the activator CO_2 (Lorimer and Miziorko 1980). Coordinative binding of Mg^{2+} to the carbamate is supported by circumstantial

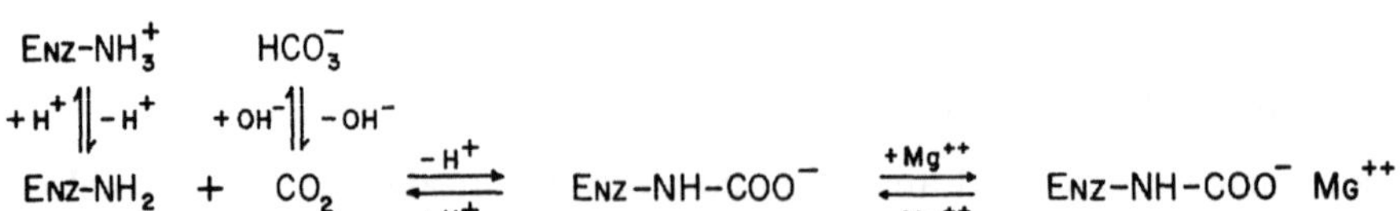

Fig. 1. Reaction sequence leading to activated RuPB carboxylase

evidence only: Mg^{2+} is an obligatory cofactor for RuBP carboxylase (Weissbach et al. 1956) and binds rapidly to an Enz·CO_2 complex (Lorimer et al. 1976).

Some organic phosphates (e.g., 6-phosphogluconate, ribose 5-phosphate, fructose 1,6-bisphosphate) will stabilize the activated enzyme (Enz·CO_2·Mg^{2+}) and form rather exchange-inert complexes (Badger and Lorimer 1981, Gutteridge et al. 1982; Jordan et al. 1983; McCurry et al. 1981; Vater et al. 1983). Concern was raised that the enzyme cannot be catalytically competent and activated simultaneously by one of the effectors if in both cases occupancy of the same site is involved (Lorimer 1981 b; Miziorko and Lorimer 1983). In regards to the in vivo situation of RuBP carboxylase this concern can be dispelled. Binding of either RuBP or effector to Enz·CO_2·Mg^{2+} depends only on their relative affinity and their relative concentration. In the chloroplast stroma phosphorylated metabolites with a dissociation constant (K_D) in the micromolar range (compared to $K_D < 0.1$ μM for RuBP) do not exceed RuBP in concentration. Therefore, severe interference with catalysis by these metabolites is not to be expected. The physiological significance of effector-mediated stabilization of activated RuBP carboxylase remains unclear. Neither the total concentration of phosphorylated compounds (Portis et al. 1977) nor the quantity of free Mg^{2+} (1–3 mM) in the stroma would suffice for an effective regulatory mechanism for RuBP carboxylase unless compartmentation of enzymes and metabolites is invoked.

Another unresolved problem is the observation that RuBP carboxylase in chloroplasts may attain full catalytic activity at levels of CO_2 (10 μM) and free Mg^{2+} (1–3 mM) completely insufficient for a similar degree of activity in vitro.

In search of a role of the obligatory divalent cation cofactor, the participation of Mg^{2+} not only in activation but also in catalysis has been invoked (Miziorko and Mildvan 1974; Rabin and Trown 1964). Probably the best support for this comes from the observation that Mn^{2+} (substituting for Mg^{2+}) alters the ratio of the two catalytic rates of RuBP carboxylase (Jordan and Ogren 1981; Wildner and Henkel 1979). However, only one tight-binding divalent metal ion per large subunit was detected in magnetic and spin resonance studies (Miziorko and Mildvan 1974). To resolve this discrepancy, Lorimer (1981 b) suggested that the metal ion must play a dual role and participate in both activation and catalysis. This interpretation was challenged by Miziorko and Sealy (1984) who, in extension of their ESR studies, could find no direct evidence for a cation-activator CO_2 interaction and tentatively placed the metal ion elsewhere on the protein surface of the active site. Thus, a precise assignment of ligands to Mg^{2+} and the issue of Mg^{2+} -mediated regulation of RuBP carboxylase remains for future research to determine.

2.4.2 Carboxylation of RuBP

A chemical model for the enzyme-catalyzed carboxylation of RuBP was predicted by Calvin (1954) before the enzyme and its properties were known.

The reaction is initiated by proton abstraction at C-3 of RuBP (I) followed by 2,3-enediol formation (II) (Fiedler et al. 1967). Nucleophilic attack on CO_2 – the actual substrate in the carboxylation reaction (Cooper et al. 1969) – occurs through either a concerted mechanism or via prior carbanion formation at C-2

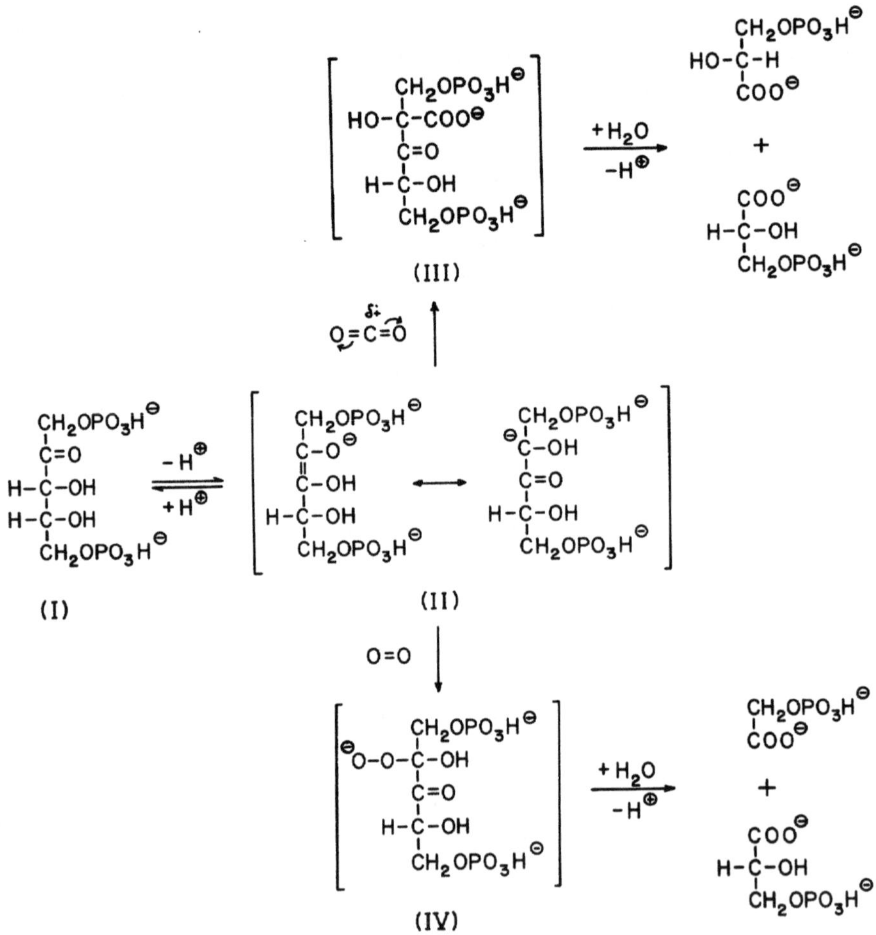

Fig. 2. Reactions catalyzed by ribulose-1,5-biphosphate carboxylase/oxygenase

(Bhagwat and McFadden 1983) and produces the 6-carbon reaction intermediate 2-carboxy-3-keto-D-arabinitol 1,5-bisphosphate (III) (Lorimer et al. 1984; Schloss and Lorimer 1982). Cleavage of (III) at C-2 and C-3 of the 5-carbon backbone (Müllhofer and Rose 1965) yields PGA from C-3 to C-5 and the aci-acid form of PGA from the remainder of the carbon skeleton. Since only D-phosphoglycerate is detectable in the reaction products, a stereospecific proton addition must complete the formation of two molecules of PGA from RuBP and CO_2.

Substantiation of the events during carboxylation is derived from several experiments. Release of tritium from 3-^{3}H-RuBP (Fiedler et al. 1967), an exchange of solvent protons with the hydrogen at C-3 of RuBP in the presence of activated enzyme (Saver and Knowles 1982; Sue and Knowles 1982a, b), the trapping of an intermediate with I_2 (Jaworowski et al. 1984), and the release of inorganic phosphate through β-elimination (Jaworowski et al. 1984; Mulligan and Tolbert

1983) are all properties expected of the 2,3-enediol (II in Fig. 2) of RuBP. The existence of a 6-carbon reaction intermediate was inferred from inhibition studies with reaction intermediate analogs (Pierce et al. 1980; Siegel and Lane 1973). The structure, 2-carboxy-3-keto-D-arabinitol 1,5-bisphosphate, was suggested because RuBP carboxylase binds the reaction intermediate analog 2-carboxy-D-arabinitol 1,5-bisphophate five orders of magnitude tighter than the epimeric 2-carboxy-D-ribitol 1,5-bisphosphate (Pierce et al. 1980), and it was assumed that the tightest binding inhibitor most closely resembles the structure of the reaction intermediate (Wolfenden 1972). The half-life of (III) when separated from the enzyme is several minutes. This longevity permitted reduction of (III) with $NaBH_4$. Structure and distribution of the reaction products are compatible with the existence of (III) (Schloss and Lorimer 1982). These data and earlier observations on the retention of oxygen at C-2 and C-3 during catalysis (Lorimer 1978; Sue and Knowles 1978) exclude the possibility that (III) intermediately forms a covalent bond with the enzyme (Rabin and Trown 1964). A further refinement of the reaction mechanism was introduced with the observation that (III) probably exists on the enzyme in the *gem*-diol form and not, as depicted, in the keto-form (Lorimer et al. 1984). This would imply that, in a concerted reaction, one molecule of water forms a bond to C-3 of the 2,3-enediol while CO_2 is being nucleophilically added to C-2.

Catalysis of a chemical reaction by an enzyme is the result of an interplay of amino acid residues, solvent, cofactors, and substrates. Thus, to fully appreciate the intricacies of biological catalysis, knowledge of the chemistry of the reaction must be complemented by experimental details on "functional" amino acid residues. To this end, a large body of information has accumulated, but "essentiality" of any of these amino acid residues has been demonstrated only for lysine-201 (Lorimer and Miziorko 1980). Arginyl groups, by virtue of their ubiquituous occurrence in the binding domain of enzymes metabolizing phosphorylated carbohydrates (Riordan et al. 1977), have been implied in the binding of RuBP, serving as counterion for the phosphate moiety (Chollet 1981; Lawlis and McFadden 1978; Schloss et al. 1978). A single histidine group per large subunit has been differentially labeled with diethylpyrocarbonate (Saluja and McFadden 1982). A molecular pK of 6.5 to 6.7 of this histidine group as determined from the pH-dependent rate of inactivation (Paech 1985) coincides with a pK value observed in pH-dependent V_{max}/K_m profiles (Schloss 1983). These data lend some credence to the claim that a histidyl group may be "essential" in the mechanism of RuBP carboxylase, e.g., for proton abstraction at C-3 of RuBP.

The most prominent and controversial among the essential amino acid residues is cysteine. The enzyme responds to SH-specific reagents with loss of catalytic activity, but RuBP affords protection (Rabin and Trown 1964), suggesting that cysteine residues are located at the active site. Active-site-directed irreversible inhibitors also modify cysteine residues (Hartman et al. 1978). But are they involved in catalysis? Hartman et al. (1984) have discounted this possibility on grounds of complete lack of commonality in the location of cysteinyl residues in the enzyme from *R. rubrum* and from higher plants. However, structural or regulatory functions for which the higher plant may have developed a need could be fulfilled by cysteinyl residues.

2.4.3 Oxygenation of RuBP

The mechanism for the RuBP oxygenase reaction as in Fig. 2 is appealing in its simplicity but not compatible with physical reality. The addition of molecular oxygen in its triplet state to the π-electron pair of the 2,3-enediol (or the carbanion form) (II) in the absence of any cofactor normally associated with oxygenases is, as pointed out earlier (Paech et al. 1978 a), spin-forbidden. Yet, RuBP oxygenase reaction can be measured and the reaction products can be analyzed. One oxygen atom of $^{18}O_2$ is incorporated into phosphoglycolate (Lorimer et al. 1973), RuBP is cleaved at C-2 and C-3 (Pierce et al. 1980), and the stoichiometry of reaction products, PGA and phosphoglycolate, is 1 : 1 (Jordan and Ogren 1981; Pierce et al. 1980).

The search for cofactors which promotes spin inversion and allows for subsequent formation of reaction products to occur, has produced only trace amounts of Cu^{2+} (Chollet et al. 1975; Johal et al. 1980; Lorimer et al. 1973; McCurry et al. 1978). Although Brändén et al. (1984 a, b) have detected by ESR measurements $Enz \cdot Cu^{2+} \cdot RuBP$ complexes and shown some phosphoglycolate formation, these results do not yet offer a satisfactory explanation for the oxygenase mechanism. Another approach to resolve the spin inversion problem invoked a "cage effect" (Lorimer 1981 b; Miziorko and Lorimer 1983). The active site and water molecules are thought to provide a rigid structure in which a radical pair, formed from RuBP and O_2, is held long enough for spin delocalization to occur and oxygenation to proceed.

2.4.4 Localization of Catalytic and Activator Site

Catalytic activity of RuBP carboxylase resides in the large subunit. First implications to this effect came from reconstitution experiments with the spinach enzyme (Nishimura et al. 1973). More recently, probing with active-site-directed irreversible inhibitors (Hartman et al. 1978) and with protein-modifying reagents (Paech and Tolbert 1978; Saluja and McFadden 1982) quite convincingly demonstrated the RuBP-binding site is located on the large subunit. In an elegant experiment, Miziorko (1979) exploited the properties of the reaction intermediate analog CRBP to form a stable quaternary of a 1 : 1 : 1 : 1 stoichiometry with CO_2, Mg^{2+}, and enzyme. Subsequently, the trapped CO_2 was identified as the activator CO_2 molecule which is covalently bound to the ε-amino group of lysine-201 of the large subunit (Lorimer 1981 a; Lorimer and Miziorko 1980). Similar quaternary complexes were found with A_8 cores of RuBP carboxylase from *Synechococcus* (Andrews and Ballment 1984), and from *Aphanothece* (Takabe et al. 1984 a, b). These data imply that the small subunit is not required for substrate binding, that the activator site is in close proximity to the catalytic site, and that both sites are located on the large subunit. Further support, independent of protein chemistry, kinetics, and binding studies, is derived from the following observations. A uniparental mutant of *Chlamydomonas* with a single nucleotide modification in the gene for the large subunit rendered RuBP carboxylase, although synthesized, catalytically inactive (Dron et al. 1983). The *R. rubrum* enzyme is completely devoid of small subunits, but shares almost all catalytic properties with the higher plant enzyme.

The small subunit in the higher plant enzyme posed a problem for many years. None of the known properties of RuBP carboxylase could be associated with the small subunit. Only recently, with isolated large subunit cores available, has a tentative picture emerged. It appears that the small subunit does not affect the binding of CO_2 to the activator site, but modulates V_{max} and $K_m(CO_2)$ (Andrews and Ballment 1984; Andrews and Lorimer 1985; Ogren 1984; Takabe 1984a, b).

3 Practical Aspects

3.1 Purification

3.1.1 Summary of Techniques

In reviewing the literature on purification of RuBP carboxylase essentially four different approaches became apparent. (a) Direct crystallization (Chan et al. 1972; Kung et al. 1980): the resulting crystals, typically from tobacco leave extracts, while electrophoretically homogeneous, suffer from low specific activity (around 0.5 U mg^{-1}) for which Bahr et al. (1981) offered some explanation. Higher plant enzymes crystallized *after* lengthy purification procedures show in general very poor specific activity (Johal and Bourque 1979; Johal et al. 1980) but can be obtained now with rather good activity from tobacco leaves (Servaites 1985). (b) Sucrose gradient centrifugation of crude extracts (Berhow and McFadden 1983; Berhow et al. 1982; Goldthwaite and Bogorad 1975): this was done initially in swinging bucket rotors. The total procedure required about 24 h (Goldthwaite and Bogorad 1975). Later, fixed angle rotors were used and very recently vertical rotors were introduced (Berhow and McFadden 1983; Berhow et al. 1982). Thus, the time span for the entire purification was reduced to 3–5 h, yielding an enzyme with exceptionally high specific activity [2.5 to 3 U mg^{-1} for the enzyme from spinach (Berhow et al. 1982; Paech, unpublished observation)]. (c) Purification by immunoabsorption (Gray and Wildman 1976): as expected an extremely pure enzyme was obtained, but subunits were separated. (d) Salt or PEG fractionation followed by column chromatography and occasionally interspersed with sucrose gradient centrifugation: this is a theme with variation threading through the entire literature on RuBP carboxylase (e.g., Hall and Tolbert 1978; McCurry et al. 1982; Paulsen and Lane 1966) and yields enzyme of variable quality. There are other techniques, too, but they are either in their infancy (HPLC and FPLC separation) or cannot be called truly purification procedures (e.g., the production of *R. rubrum* enzyme from clones).

3.1.2 Interfering Compounds

The preparation of pure RuBP carboxylase protein from almost any green plant is a relatively easy task due to the abundance of RuBP carboxylase. The isolation of pure protein with adequate enzymatic activity is more challenging. Since all purification strategies have some problems in common, they will be briefly mentioned.

Phenolic compounds and their oxidation products (due to polyphenoloxidases!) present a major source of contamination in isolated enzymes (Loomis 1974)[2]. PVPP (2–4%, w/v), occasionally soluble PVP (Khan and Malhotra 1982), ion exchange resin (Gray et al. 1978; Rejda et al. 1981), or borate buffer (Quayle et al. 1980) have been advantageously used to divert polyphenols from RuBP carboxylase. PVPP is most effective at low pH (6.4–7.2) and definitely requires proper hydration and degassing prior to use. PVP has the disadvantage that it may interfere with subsequent salt fractionation by co-precipitation. Ion exchange resins are highly efficient (Gray 1978), but their use is often prohibitive due to their expense. Polyphenoloxidases are strongly inhibited by diethyldithiocarbamate (1 mM) which chelates Cu^{2+} (Jones and Lyttleton 1972).

Low yields of RuBP carboxylase have been tolerated probably because of abundant plant material and enzyme. Little attention has been paid in the past to proteolytic degradation of RuBP carboxylase during purification. The action of proteases on RuBP carboxylase was seen in preparations from soybean leaves. As much as 90% of the initial activity in the crude extract was lost in a few hours during traditional salt fractionation followed by column chromatography. SDS-PAGE revealed that, as enzymic activity decreased, a polypeptide with a slightly lower molecular weight than that of the large subunit emerged while the quantity of the large subunit decreased (Paech, unpublished observation). This is corroborated by the observation that tryptic digestion of spinach RuBP carboxylase yields a small C-terminal peptide (approximately 3,000 mw) and is paralleled by loss of activity (Mulligan and Tolbert 1984).

For RuBP carboxylase of soybean, adsorption and batch elution from DEAE cellulose before salt fractionation (Jordan and Ogren 1981) partially alleviated the problem of proteolytic digestion. However, removal of degrading enzymes was not complete, as evidenced from the decay of enzymatic activity within 4 weeks. Protease inhibitors such as leupeptin (10 μM) and phenylmethyl sulfonylfluoride (1 mM) in the grinding buffer have ameliorated the quality of RuBP carboxylase, but did not solve entirely the problem of proteolytic degradation as judged by the stability of the enzyme after purification. Further stabilization of the enzyme during purification is afforded by the inclusion of casein (Rosichan and Huffaker 1984) or BSA into the extraction buffer. For RuBP carboxylase from spinach, soybean, and sunflower we find no significant difference between the effect of casein (1% w/v) or BSA (1% w/v), but prefer BSA because the crude extract, when subjected to centrifugation for removal of cell debris and membrane fragments, yields a hard pellet in a shorter period of time.

3.1.3 Choice of Extraction Buffer and Grinding Procedures

At times there was discussion about the choice of buffer compound to be used during the purification. Tris or Good buffers (Bicine, Hepes, etc.) have been substituted for phosphate since they do not interfere with Mg^{2+} in alkaline solution and do not compete in the enzymatic reaction. We have reintroduced phosphate

2 In fact the affinity of RuBP carboxylase for polyphenols has been exploited in an assay for tannins (Martin and Martin 1983)

for economic reasons and use the Penefsky technique (Penefsky 1977) to exchange Bicine for phosphate buffer in a sample in less than 2 min.

Once purified, RuBP carboxylase is a robust protein with broad pH stability (at least pH 5.5 to 9.3). This may explain why choice of pH, ionic strength, and buffer compound during purification have little effect on the overall yield and activity. Addition of mercapto compounds (10–50 mM 2-mercaptoethanol, 5 mM dithiothreitol or dithioerythritol) to buffers is indicated in view of 96 free sulfhydryl groups per molecule of enzyme. The inclusion of Mg^{2+} (5–10 mM) and $NaHCO_3$ (10 mM) in the buffer has been mentioned occasionally and may be justified on grounds that they maintain RuBP carboxylase in the activated state, and hence in a form possibly less susceptible to degradation.

The volume of grinding buffer in milliliter per gram of plant tissue to be extracted varies from 2:1 to 15:1 and depends on the texture and the water content of the leaf tissue. For example, the volume to weight ratio used was 2:1 for leaves from spinach (McCurry et al. 1982), 3:1 for soybean (Bowes and Ogren 1972), 4:1 for maize (Reger et al. 1983), 10:1 for jack pine needles (Khan and Malhotra 1982), and 15:1 for cotton (Quayle et al. 1980).

Tissue-grinding procedures depend mainly on the scale of operation. Small plant samples can be frozen conveniently in liquid nitrogen, pulverized, and then extracted with buffer. Homogenization of leaf material and buffer in a chilled mortar with a pestle is another quick method for extracting small plant samples. Large-scale preparations require the aid of a mechanical blender. Blending times are typically 45–60 s. The addition of n-octanol as antifoam reagent is not recommended because it will interfere with pelleting cellular and membraneous debris.

3.1.4 Protein Determination

RuBP carboxylase is not exempt from problems inherent to protein determinations. The biuret technique (Gornall et al. 1948), not very sensitive but convenient and fast, will give only approximate protein estimates in the course of purification owing to the presence of polyphenols. For pure and colorless RuBP carboxylase the biuret reaction is compatible with any other protein determination method.

The Bradford technique (Bradford 1976) relies on dye binding to proteins and is essentially independent on nonprotein impurities. However, the calibration of RuBP carboxylase with BSA standards overestimates RuBP carboxylase by as much a 1.7-fold. When using the Bradford technique routinely, it is worth establishing a correlation factor by an independent protein assay, e.g., Kjeldahl determination of total nitrogen.

The Lowry method, as modified by Bensadoun and Weinstein (1976), has been used with success in both crude extracts and with purified RuBP carboxylase. When calibrated with BSA, the data agree well with total nitrogen determinations.

Often protein quantitation relies on absorbance measurements at 280 nm, using a correlation factor of 0.61 mg ml^{-1} to calculate the protein concentration in the sample (McCurry et al. 1982; Paulsen and Lane 1966). This factor is subject to large variation due to contaminants which have a high extinction coefficient in the near UV but are inconspicuous to the naked eye. For a large number of preparations of RuBP carboxylase from spinach, the factor varied from 0.55 to 0.75 (Paech, unpublished observation). We feel the factor is not so much species-specific [0.61 for spinach (McCurry et al. 1982); 0.7 for tobacco (Jordan and Ogren 1981; McCurry et al. 1982)] as dependent on the purity of the enzyme. Hence it will give only approximate protein estimates.

3.1.5 Example: RuBP Carboxylase from Soybean Leaves

For RuBP carboxylase from soybean leaves, the following protocol has achieved the goal of obtaining an enzyme which is colorless, stable at room temperature, and has an acceptable specific activity. The procedure described here is a combination of previously published procedures (Berhow et al. 1982; Jakoby et al. 1956; Li and Wang 1980; Weissbach et al. 1956).

Deveined and washed soybean leaves (250 g) are ground in a Waring blendor for 45 s in 1 000 ml of 25 mM phosphate buffer, pH 6.9 containing 5 mM $MgCl_2$, 40 mM mercaptoethanol, 10 μM leupeptin, 1 mM diethyldithiocarbamate, 1 mM EDTA, 4% (w/v) PVPP. The resulting slurry is strained through cheesecloth. Ten ml of 0.5 M ATP is added, and the extract is brought quickly to 58 °C (e.g., with the aid of a microwave oven; when used in the temperature mode, the temperature probe should be immersed into a reference solution and not into the extract). The temperature of the extract is maintained for 4–10 min at 58–60 °C. As soon as coagulation is noticeable, the mixture is chilled and centrifuged at 40,000 *g* for 10 min. Proteins are precipitated from the combined yellowish supernatant by the addition of an equal volume of saturated ammonium-sulfate solution at neutral pH. The precipitate is collected by centrifugation at 10,000 *g* for 10 min. The pellet is resuspended in a total volume of 45 ml of 50 mM bicine buffer, pH 8.0, and desalted on Penefsky columns (Penefsky 1977). This procedure takes about 5 min and yields undiluted and essentially salt-free protein with >95% recovery. The resulting eluate is ready for gradient centrifugation on a 0.2–0.8 M linear sucrose gradient in 50 mM bicine, pH 8.0, for 2.5 h in a vertical rotor (VTi50) as described (Berhow et al. 1982). Total yield was 210 mg of protein with a specific activity 1.7 U mg^{-1} at pH 8.0, 30 °C. The enzyme was stored frozen as ammonium sulfate pellet (Hall et al. 1981).

3.2 Assay

3.2.1 Substrates

For most practical purposes the concentration of bicarbonate solutions is accurate enough when made from weighed amounts of sodium bicarbonate (reagent grade, stored desiccated). With $[^{14}C]NaHCO_3$, one has to rely on the specifications given by the manufacturer. However, before use, an aliquot of the batch should be checked for residual ^{14}C-activity when treated with a volatile acid and brought to dryness. Occasionally, it is necessary to have an exact account of CO_2 present, particularly in CO_2-free buffers. This can be accomplished with a spectrophotometric assay using phosphoenolpyruvate carboxylase (Hall et al. 1983).

Effective CO_2 concentration in a bicarbonate solution of known concentration may be calculated from the apparent pK′ of the CO_2/HCO_3^- equilibrium using the Henderson-Hasselbalch equation. When reporting data, it is important to indicate on which pK value the results are based. The apparent pK′ for any given temperature or ionic strength can be calculated from the thermodynamic value of pK = 6.317 at 38 °C (Shedlovsky and MacInnes 1935) and $dpK_a/dT = -0.005$ for increasing temperatures. A program written in Basic simplifies this task (Ellis and Morrison 1982).

RuBP may be purchased but can be synthesized in the laboratory at a fraction of the commercial cost by enzymatic synthesis from ribose-5-phosphate (Horecker et al. 1958). The yield depends largely on the quality of phosphoribulokinase used. If the phosphorylation step is too slow as judged by a drop in pH, synthesis of RuBP will be offset by nonenzymatic degradation of RuBP. The use of Ba^{2+} to precipitate RuBP should be avoided. Instead, RuBP should be purified by ion exchange chromatography (Pierce et al. 1980). Purity of RuBP may be checked chromatographically. Calibration of RuBP solutions should involve: determination of inorganic phosphate (Black and Jones 1983) and total phosphate (Leloir and Cardini 1957), determination of carbohydrate (Horecker et al. 1958), and enzymatic determination of RuBP using [^{14}C]$NaHCO_3$ of known specific activity.

Preparation of RuBP in large scale has been accomplished with an immobilized enzyme system (Wong et al. 1982).

A batch of RuBP should be subdivided and stored at −20 °C or below. Repetitive thawing and prolonged handling in solution may cause decomposition and will yield inhibitors for RuBP carboxylase (Paech et al. 1978 b). To avoid possible interference of these contaminants with the assay, RuBP may be synthesized in situ from ribose 5-phosphate, ATP, and auxiliary enzymes.

3.2.2 Activation of RuBP Carboxylase

RuBP carboxylase may be assessed with continuous and with discontinuous assay techniques. To observe maximum catalytic activity, the enzyme must be in its fully activated form, and this applies to both carboxylase and oxygenase assays (Lorimer et al. 1977, Pierce et al. 1982).

In the carboxylase assay, the enzyme may be presented in the assay solution with 10 mM $NaHCO_3$ and 10 mM $MgCl_2$ at pH 7.8–8.2. Activation is completed within 10 min at room temperature. Because the partial pressure of CO_2 in the assay solution is higher than pCO_2 of air, tightly sealed containers with a minimum of air space above the solution must be used to avoid escape of CO_2. Also a rigid time regime is recommended to ensure reproducibility of assays. Particularly for longer series of experiments, a more convenient control over the degree of activation is achieved by preparing activated enzyme in a "reactivial" as stock solution (typically 1–2 mg of protein per ml) with 10 mM $NaHCO_3$ and 10 mM $MgCl_2$ at pH 7.8–8.2. An appropriate aliquot is added to the assay solution thereby initiating the reaction. To avoid rate limitations imposed on RuBP carboxylase by the CO_2/bicarbonate equilibrium, the addition of carbonic anhydrase (100 U ml^{-1}) to the assay buffer is recommended.

3.2.3 Continuous Spectrophotometric Assay for RuBP Carboxylase Activity (Lilley and Walker 1974; Racker 1962)

The continuous method employs a coupled enzyme system which converts PGA into glyceraldehyde 3-phosphate or α-glycerophosphate. Thus, for each mol of RuBP carboxylated either 2 or 4 mol of NADH are oxidized (Racker 1962). This assay has a time lag of about 1 min, which can be overcome by the addition of an ATP-regenerating system (Lilley and Walker 1974). With NADH at 1 mM final concentration, cuvettes with 2 mm light path are required. The assay system is rather versatile but a point to bear in mind is that any change in assay parameters (pH, ionic strength, addition of regulators) may also effect the coupled enzyme system.

Assay solutions in a cuvette with 2 mm lightpath:

300 µl	166 mM Bicine/K^+, pH 8.0, containing 16.6 mM $MgCl_2$, 1.67 mM DTT, and 30 units of carbonic anhydrase
40 µl	62.5 mM ATP in water, approximately at neutral pH
40 µl	12.5 mM NADH in water
20 µl	creatine phosphokinase, 60 U ml^{-1} in water
20 µl	125 mM phosphocreatine in water
20 µl	glyceraldehyde 3-phosphate dehydrogenase (125 U ml^{-1}) and 3-phosphoglycerate kinase (200 U ml^{-1}) in water
20 µl	250 mM $NaHCO_3$ in water
20 µl	12.5 mM RuBP in water

This mixture is incubated at the desired temperature and monitored for absorbance changes at 340 nm (blank rate). After 2 min the reaction is initiated by the addition of 20 µl of activated enzyme (1 mg ml^{-1}). Absorbance change is recorded for 2–4 min. Initial rate can be calculated from the linear phase of the recorder trace. In this assay a change in optical density of 1 is equivalent to 190 nmol of RuBP carboxylated, assuming that the oxygenase reaction is negligible. The sensitivity of the assay may be doubled by the inclusion of triosephosphate isomerase and α-glycerophosphate dehydrogenase in the assay solution.

3.2.4 Discontinuous Assays for RuBP Carboxylase Activity

3.2.4.1 Radiochemical Assay with [^{14}C]$NaHCO_3$ (Jakoby et al. 1956; Lorimer 1982; Lorimer et al. 1977)

In this most widely used assay for the carboxylase activity, the rate of formation of [1-^{14}C]-PGA from $^{14}CO_2$ (supplied as [^{14}C]$NaHCO_3$) and RuBP is monitored in intervals over a period of time. At indicated times, reactions are terminated by the injection of a volatile acid (1–3 N HCl, 15–25% formic acid in water or methanol). The entire solution is evaporated to dryness under a hood (at room temperature in a stream of gas, or more rapidly, in a drying oven at 70 °–80 °C). The residue is resuspended in water and [1-^{14}C]-PGA is quantitated by liquid scintillation spectrometry.

The total assay volume may vary from 100 to 1 000 µl and is mainly dictated by the scarceness of material or the thrift of the experimentalist. In addition to uncontrolled escape of CO_2 from the assay solution two other variables have to be considered. (1) Isotope dilution of $^{14}CO_2$ by CO_2 from the air space above the solution. This can be remedied by flushing reaction vessels with N_2 prior to use. Alternatively, flushing with N_2 may be omitted when a tightly sealed vessel with a total volume not exceeding four times the assay volume is used. Then, the introduced experimental error becomes negligible. (2) Isotope dilution of $^{14}CO_2$ by CO_2 in buffers and particularly by CO_2 carried over with activated enzyme. The former is avoided by gassing the buffer solution with N_2 prior to pH adjustment with >10 N KOH in which potassium bicarbonate is insoluble. The latter must be taken into account arithmetically when cpm measured by liquid scintillation spectrometry is converted into quantity of PGA.

For serial assays all reagents except [^{14}C]$NaHCO_3$ and the enzyme may premixed and dispensed into vials with a repipet device. After temperature equilibration, an aliquot of a [^{14}C]$NaHCO_3$ solution followed by an aliquot of activated enzyme is added in timed intervals. When sodium bicarbonate and Mg^{2+} concentration and pH are similar in the activation and the assay mixture, the length of the assay is determined by the quantity of enzyme and the level of RuBP present, because degree of activation is maintained. If the assay calls for conditions different from that of activation the reaction time must be as short as possible (15 to 60 s) to insure that observed rates of product formation do not arise from an undefined superposition of rate of activation and rate of catalysis. In addition, it must be assured that observed rates are linear and that substrate depletion does not exceed 6–10%.

Assay mixture for radiometric carboxylase assay in 2-ml disposable autosampler cups equipped with caps containing a small hole.

150 µl 166 mM Bicine/K^+, pH 8.0, containing 16.6 mM $MgCl_2$, 1.67 mM DTT, and 15 units of carbonic anhydrase
70 µl water
10 µl 12.5 mM RuBP in water

When temperature equilibrium is reached add

10 µl 250 mM [^{14}C]$NaHCO_3$ (specific activity 0.16 to 2 Ci mol^{-1})
followed shortly, typically within 15–20 s, by
10 µl activated enzyme (1–2 mg of protein per ml)

Immediately after the addition of enzyme, a syringe equipped with a blunt needle and charged with 250 µl of 3 N HCl is inserted through the hole and is allowed to rest on the bottom of the vessel. Premature contact of acid with the assay solution is prevented by an air space in the needle. The needle should sit tightly in the cap. At a predetermined time the acid is injected into the reaction mixture. The quenched assay solution is transferred quantitatively into 7-ml scintillation vials and brought to dryness. The residue is suspended in 0.5 ml of water and 3.5 ml of a toluene-Triton X-100 cocktail (Patterson and Greene 1965) and is ready for liquid scintillation spectrometry.

3.2.4.2 *Radiochemical Assay Using [^{14}C]$NaHCO_3$ and [1-^3H] RuBP (Jordan and Ogren 1981)*

This method is a sensitive procedure for the simultaneous determination of the carboxylating and oxygenating activity of RuBP carboxylase over a wide range of conditions. In the carboxylase reaction [1-^{3}H] RuBP and $^{14}CO_2$ are converted to [1-^{3}H, 3-^{14}C]PGA and unlabeled PGA. ^{14}C-Label in PGA is determined directly by liquid scintillation spectrometry as described above. In the oxygenase reaction, the labeled product from [1-^{3}H]RuBP and O_2 is [2-^{3}H]phosphoglycolate. An aliquot of the quenched assay solution is neutralized and incubated with phosphoglycolate phosphatase. The resulting [2-^{3}H] glycolate is separated by ion exchange chromatography and quantitated by liquid scintillation spectrometry. This simultaneous assay removes ambiguities in the ratio of carboxylase and oxygenase activities which arise from separate assays, even when care was taken to maintain identical reaction conditions.

3.2.4.3 *Discontinuous Assay Using Nonlabeled Substrates (Di Marco and Tricoli 1983)*

For those without access to a scintillation counter and the resources to acquire radiolabeled compounds, elements of the continuous spectrophotometric assay and the radiochemical assay were combined into a discontinuous assay without the aid of radiotracers. The assay is set up as in Section 3.2.4.1 except that nonlabeled bicarbonate substitutes for [^{14}C]$NaHCO_3$. After the assay is quenched by addition of acid, the pH is raised again to 8.0 and the entire mixture is transferred to a cuvette (unless the assay is run directly in the cuvette). The mixture is incubated with the coupled enzyme system as described in Section 3.2.3. The change in absorbance at 340 nm is recorded after 10–15 min and then converted into quantities of PGA formed during the carboxylase reaction.

This assay has the advantage that the detection of product is independent on pH and ionic strength. But substrates and particularly effector may still interfere in the subsequent spectrophotometric measurement of PGA.

3.2.5 Assays for RuBP Oxygenase

RuBP oxygenase activity can be observed in several ways: by polarographic determination of oxygen consumption; by colorimetric determination of phosphoglycolate formation; by radiochemical determination of phosphoglycolate formation. The polarographic method is the most widely used technique (Lorimer 1982; Lorimer et al. 1977).

The enzyme requires to be activated by CO_2 and Mg^{2+} (Badger and Lorimer 1976; Lorimer et al. 1976). Since CO_2 and O_2 are mutually competitive for RuBP, bicarbonate/CO_2 must be diluted out before the oxygenase assay is begun. This, however, causes the $Enz \cdot CO_2 \cdot Mg^{2+}$ complex (half-life approximately 1 min) to decompose during the assay. Obviously, observed reaction rates result from a set of variable parameters, even if the recorder tracing appears to be linear for some time. Nonetheless, the polarographic assay has its merits when used consistently.

It will allow us to determine an approximate but not an exact oxygenase activity of RuBP carboxylase. Determination of oxygenase activity in correct relation to carboxylase activity is only possible with a simultaneous assay such as the radiochemical assay described by Jordan and Ogren (1981) (cf. Sect. 3.2.4.2).

Assay solution for RuBP oxygenase assay with an oxygen electrode:

600 μl 166 mM Bicine/K^+, pH 8.2, containing 16.6 mM $MgCl_2$, 1.67 mM DTT, saturated with air levels of oxygen
340 μl water
40 μl 12.5 mM RuBP in water

After temperature equilibration and recording of a base line (ca. 1–3 min)

20 μl activated enzyme (10 mg protein per ml) are added.

Except for a little spike at the time the syringe needle is introduced and the enzyme is injected into the assay solution, oxygen consumption will be recorded without lag time and proceeds linearly for at least 30 s.

The oxygen electrode is calibrated with air-saturated water (full scale pen deflection) and with oxygen-depleted water (with a grain of sodium dithionite; zero deflection). Thus full recorder span corresponds to 250 nmol of oxygen.

3.2.6 Kinetic Parameters of RuBP Carboxylase and RuBP Oxygenase

Michaelis-Menten parameters for many higher plants are (cf. Jordan and Ogren 1983, Yeoh et al. 1980, 1981):

$K_m(CO_2)$ (anaerobically)	15–35 μM	at pH 8.2, 25 °C
K_m(RuBP)	10–50 μM	at pH 8.2, 25 °C
Specific activity at V_{max}	0.4–2.4 $\mu mol \cdot min^{-1} \cdot mg^{-1}$	at pH 8.2, 25 °C
$K_i(O_2)$	300–400 μM	at pH 8.2, 25 °C
$K_m(O_2)$	500–800 μM	at pH 8.2, 25 °C
Specific activity at V_{max}	0.05–0.25 $\mu mol \cdot min^{-1} \cdot mg^{-1}$	at pH 8.2, 25 °C
$K_i(CO_2)$	50–80 μM	at pH 8.2, 25 °C

4 Conclusion

Spectacular progress has been made in elucidating the many facets of RuBP carboxylase. Some questions, however, have received only sporadic attention. For example, why this large number of catalytically active subunits if there is no cooperativity or allosteric interaction? How does RuBP carboxylase function in the chloroplast? The concentration of soluble proteins in chloroplasts is estimated at 40–50% (w/w) (Ellis 1979). By comparison with protein crystals, which contain 40–60% (w/w) water (Matthews 1968), the stroma is a closely packed matrix. The degree of freedom of macromolecules is reduced. Solvent water for metabolites and ions is limited because hydration water of macromolecules is osmotically inactive (Franks 1977). The formation of "enzyme clusters" is almost inevitable.

The advantage of aggregates consisting of enzymes belonging to one metabolic pathway is obvious: coordination and compartmentation. Effective substrate concentrations can be raised. Regulatory function can be introduced or modified (Arnold and Pette 1970; Masters et al. 1969). The transient time between different steady states is reduced (Welch 1977). Kinetic parameters of enzymes may be altered by homologous and heterologous protein–protein interactions (Frieden and Nichol 1981; Masters 1981; Minton 1981). Many years ago, Akazawa (1970) hypothesized on the existence of such a multi-enzyme complex for Calvin-cycle enzymes. However, no attempt to verify this compelling idea experimentally has met with success.

By necessity, proteins in the stroma interact with thylakoid membranes. It has been argued that reversible adsorption to membrane structures represents a possible mechanism for allosteric control of enzymes with more than one active site per molecule (Masters 1977). The structure of RuBP carboxylase is in agreement with this requirement and experimental evidence for an interaction of RuBP carboxylase with thylakoid membranes to which Akazawa alluded earlier (1979) is emerging (Mori et al. 1984). Mg^{2+} ions and the concentration of protein appear to be essential for the interplay between thylakoid membrane and RuBP carboxylase (McNeil and Walker 1981; Mori et al. 1984).

Light-mediated effects on RuBP carboxylase in vivo (Perchorowicz et al. 1981; Vu et al. 1984) could find their explanation in membrane interaction, and this could also involve sulfhydryl groups. Consequently, RuBP carboxylase in the stroma of chloroplasts would not be a homogeneous population of molecules with respect to catalytic competence. Only those molecules in proximity of thylakoids would be catalytically active. Change in pH and Mg^{2+} concentration, possibly redox reactions of sulfhydryl groups could effectively modulate RuBP carboxylase during light–dark transitions. RuBP carboxylase molecules in the center between thylakoid membranes may not be affected at all by pH and Mg^{2+}. To reconcile this model with observed rates of CO_2 fixation by intact leaves it must be assumed that RuBP carboxylase in vivo catalyzes the carboxylation of RuBP severalfold faster than in vitro. Preliminary data have shown that purified RuBP carboxylase, when assayed at 16 mg of protein per ml, has a lag phase of only 10 to 20 ms and proceeds at a rate two to three times faster than the steady-state rate (Paech, unpublished observation). Approximately after one turnover, the rate slows down and eventually becomes the steady-state rate as measured in Michaelis-Menten type assays. It is possible that RuBP forces the enzyme into a less favorable conformation, something that might not occur in vivo. It is also conceivable that at 16 mg protein per ml, a concentration which is at least one order of magnitude below the actual concentration in the stroma, the experiment is not a true in vivo simulation.

In summary, the stroma of chloroplasts is not only a bag of proteins but may be a highly organized matrix of macromolecules whose structural and functional elucidation represents a new dimension in research on RuBP carboxylase.

Acknowledgement. Unpublished results reported from the authors laboratory were derived from research supported by Grant 82-CRCR-1-1050 from the U.S. Department of Agriculture, Competitive Research Grants Office, and South Dakota Agricultural Experiment Station (Journal article No. 2030).

References

Akazawa T (1970) The structure and function of fraction I protein. Progr Phytochem 2:107–141

Akazawa T (1979) Ribulose-1,5-bisphosphate carboxylase/oxygenase. In: Gibbs M, Latzko E (eds) Photosynthesis II. Springer, Berlin Heidelberg New York, pp 208–229 (Encyclopedia Plant Physiol, New Series, Vol 6)

Akazawa T, Takabe T, Asami S, Kobayashi H (1978) Ribulosebisphosphate carboxylase from *Chromatium linosum* and *Rhodospirillum rubrum* and their role in photosynthetic carbon assimilation. In: Siegelman HW, Hind G (eds) Photosynthetic carbon assimilation. Plenum, New York, pp 209–226

Akazawa T, Takabe T, Kobayashi H (1984) Molecular evolution of ribulose-1,5-bisphosphate carboxylase/oxygenase. Trends Biochem Sci 9:380–383

Andersson I, Brändén CI (1984) Large single crystals of spinach ribulose-1,5-bisphosphate carboxylase/oxygenase for x-ray studies. J Mol Biol 172:363–366

Andersson I, Tjäder AC, Cedergren-Zeppezauer E, Brändén CI (1983) Crystallization and preliminary x-ray studies of spinach ribulose-1,5-bisphosphate carboxylase/oxygenase complexed with activator and a transition state analogue. J Biol Chem 258:14088–14090

Andrews TJ, Abel KM (1981) Kinetics and subunit interactions of ribulose bisphosphate carboxylase/oxygenase from the cyanobacterium *Synechococcus sp.* J Biol Chem 256:8445–8452

Andrews TJ, Ballment B (1983) The function of the small subunits of ribulosebisphosphate carboxylase/oxygenase. J Biol Chem 258:7514–7518

Andrews TJ, Ballment B (1984) Active-site carbamate formation and reaction-intermediate-analog binding by ribulosebisphosphate carboxylase/oxygenase in the absence of its small subunits. Proc Natl Acad Sci USA 81:3660–3664

Andrews TJ, Lorimer GH (1985) Catalytic properties of a hybrid between cyanobacterial large subunits and higher plant subunits of ribulose biphosphate carboxylase-oxygenase. J Biol Chem 260:4632–4636

Arnold H, Pette D (1970) Binding of aldolase and triosephosphate dehydrogenase to F-actin and modification of catalytic properties of aldolase. Eur J Biochem 15:360–366

Asami S, Takabe T, Akazawa T, Codd GA (1983) Ribulose-1,5-bisphosphate carboxylase from the halophilic cyanobacterium *Aphanothece halophytica*. Arch Biochem Biophys 225:713–721

Badger MR, Lorimer GH (1976) Activation of ribulose-1,5-bisphosphate oxygenase. Arch Biochem Biophys 175:723–729

Badger MR, Lorimer GH (1981) Interaction of sugar phosphates with the catalytic site of ribulose-1,5-bisphosphate carboxylase. Biochemistry 20:2219–2225

Bahr JT, Johal S, Capel M, Bourque DP (1981) High specific activity ribulose-1,5-bisphosphate carboxylase/oxygenase from *Nicotiana tabacum*. Photosynthesis Res 2:235–242

Baker TS, Suh SW, Eisenberg D (1977) Structure of ribulose-1,5-bisphosphate carboxylase/oxygenase: form III crystals. Proc Natl Acad Sci USA 74:1037–1041

Barcena JA, Pickersgill RW, Adams MA, Phillips DC, Whatley FR (1983) Crystallization and preliminary x-ray data of ribulose-1,5-bisphosphate carboxylase from spinach. EMBO J 2:2363–2367

Barraclough R, Ellis RJ (1980) Protein synthesis in chloroplasts. IX. Assembly of newly-synthesized large subunits into ribulosebisphosphate carboxylase in isolated intact pea chloroplasts. Biochim Biophys Acta 608:19–31

Bedbrook JR, Kolodner R (1979) The structure of chloroplast DNA. Annu Rev Plant Physiol 30:593–620

Bedbrook JR, Coen DM, Beaton AR, Bogorad L, Rich A (1979) Location of a single gene for the large subunit of ribulosebisphosphate carboxylase on the maize chloroplast genome. J Biol Chem 254:905–910

Bedbrook JR, Smith SM, Ellis RJ (1980) Molecular cloning and sequencing of cDNA encoding the precursor to the small subunit of chloroplast ribulose-1,5-bisphosphate carboxylase. Nature 287:692–697

Bensadoun A, Weinstein D (1976) Assay of protein in the presence of interfering materials. Anal Biochem 70:241–250

Berhow MA, McFadden BA (1983) A rapid and novel method for purification of ribulose-1,5-bisphosphate carboxylase from *Chromatium vinosum*. FEMS Microbiol Lett 17:269–272

Berhow MA, Saluja A, McFadden BA (1982) Rapid purification of ribulose-1,5-bisphosphate carboxylase by vertical sedimentation in a reoriented gradient. Plant Sci Lett 27:51–57

Berry-Lowe SL, McKnight TD, Shah DM, Meagher RB (1982) The nucleotide sequence, expression, and evolution of one member of a multigene family encoding the small subunit of ribulose-1,5-bisphosphate carboxylase in soybean. J Mol Appl Genet 1:483–498

Bhagwat AS, McFadden BA (1983) Evidence of a carbanion intermediate in catalysis by spinach ribulose-1,5-bisphosphate carboxylase/oxygenase. Arch Biochem Biophys 223:604–609

Black MJ, Jones ME (1983) Inorganic phosphate determination in the presence of a labile organic phosphate. Anal Biochem 135:233–238

Bowes G, Ogren WL (1972) Oxygen inhibition and other properties of soybean ribulose-1,5-diphosphate carboxylase. J Biol Chem 247:2171–2176

Bowien B, Gottschalk EM (1982) Influence of the activation state on the sedimentation properties of ribulose-1,5-bisphosphate carboxylase from *Alcaligenes eutrophus*. J Biol Chem 257:11845–11847

Bowien B, Mayer F, Codd GA, Schlegel HG (1976) Purification, some properties and quaternary structure of ribulose-1,5-bisphosphate carboxylase of *Alcaligenes eutrophus*. Arch Microbiol 110:157–166

Bowien B, Mayer F, Spiess E, Pähler A, Englisch U, Saenger W (1980) On the structure of crystalline ribulosebisphosphate carboxylase from *Alcaligenes eutrophus*. Eur J Biochem 106:405–410

Bradford MM (1976) A rapid and sensitive method for the quantitation of microgram quantities of protein utilizing the principle of protein-dye binding. Anal Biochem 72:248–254

Brady CJ (1981) A coordinated decline in the synthesis of subunits of ribulosebisphosphate carboxylase in aging wheat leaves. I. Analysis of isolated protein, subunits and ribosomes. Aust J Plant Physiol 8:591–602

Brändén R, Nilsson T, Styring S (1984) Ribulose-1,5-bisphosphate carboxylase/oxygenase incubated with Cu^{2+} and studied by electron paramagnetic resonance spectroscopy. Biochemistry 23:4373–4378

Brändén R, Nilsson T, Styring S (1984b) An intermediate formed by the Cu^{2+}-activated ribulose-1,5-bisphosphate carboxylase/oxygenase in the presence of ribulose-1,5-bisphosphate and O_2. Biochemistry 23:4378–4382

Broglie R, Bellemare G, Bartlett SG, Chua NH, Cashmore AR (1981) Cloned DNA sequences complementary to mRNAs encoding precursors to the small subunit of ribulose-1,5-bisphosphate carboxylase and a chlorophyll a/b binding polypeptide. Proc Natl Acad Sci USA 78:7304–7308

Broglie R, Coruzzi G, Lamppa G, Keith B, Chua NH (1983a) Structural analysis of nuclear genes coding for the precursor to the small subunit of wheat ribulose-1,5-bisphosphate carboxylase/oxygenase. Biotechnology 1:55–61

Broglie R, Coruzzi G, Lamppa G, Keith B, Chua NH (1983b) Monocot and dicot genes encoding the small subunit of ribulose-1,5-bisphosphate carboxylase: structural analysis and gene expression. Stadler Genet Symp 15:59–71

Broglie R, Coruzzi G, Fraley RT, Rogers SG, Horsch RB, Niedermeyer JG, Fink CL, Flick JS, Chua NH (1984) Light-regulated expression of a pea ribulose-1,5-bisphosphate carboxylase small subunit gene in transformed plant cells. Science 224:838–843

Brown HM, Chollet R (1979) Quantitative dissociation of crystalline tobacco RuBP carboxylase into subunits under non-denaturing conditions. Plant Physiol 63:S65

Calvin M (1954) Chemical and photochemical reactions of thioctic acid and related disulfides. Fed Proc, Fed Am Soc Exp Biol 13:697–711

Chan PH, Sakano K, Singh S, Wildman SG (1972) Crystalline fraction I protein: preparation in large yield. Science 176:1145–1146

Chollet R (1981) Inactivation of crystalline tobacco ribulosebisphosphate carboxylase by modification of arginine residues with 2,3-butanedione and phenylglyoxal. Biochim Biophys Acta 659:177–190

Chollet R, Anderson LL, Hovsepian LC (1975) The absence of tightly bound copper, iron, and flavin nucleotide in crystalline ribulose-1,5-bisphosphate carboxylase/oxygenase from tobacco. Biochem Biophys Res Commun 64:97–107

Christeller JT (1981) The effects of bivalent cations on ribulose-bisphosphate carboxylase/oxygenase. Biochem J 193:839–844

Chua NH, Schmidt GW (1978) Post-translational transport into intact chloroplasts of a precursor to the small subunit of ribulose-1,5-bisphosphate carboxylase. Proc Natl Acad Sci USA 75:6110–6114

Cooper TG, Filmer D, Wishnick M, Lane MD (1969) The active species of CO_2 utilized by ribulose diphosphate carboxylase. J Biol Chem 244:1081–1083

Coruzzi G, Broglie R, Cashmore A, Chua NH (1983a) Nucleotide sequence of two pea cDNA clones encoding the small subunit of ribulose-1,5-bisphosphate carboxylase and the major chlorophyll a/b-binding thylakoid polypeptide. J Biol Chem 258:1399–1402

Coruzzi G, Broglie R, Lamppa G, Chua NH (1983b) Expression of nuclear genes encoding the small subunit of ribulose-1,5-bisphosphate carboxylase. In: Ciferri O, Dure L (eds) Structure and function of plant genomes. Plenum, New York, pp 47–59

Criddle RS, Dan B, Kleinkopf GE, Huffaker RC (1970) Differential synthesis of ribulose-bisphosphate carboxylase subunits. Biochem Biophys Res Commun 41:621–627

Crossland LD, Rodermel SR, Bogorad L (1984) Single gene for the large subunit of ribulosebisphosphate carboxylase in maize yields two differentially regulated mRNAs. Proc Natl Acad Sci USA 81:4060–4064

Curtis SE, Haselkorn R (1983) Isolation and sequence of the gene for the large subunit of ribulose-1,5-bisphosphate carboxylase from the cyanobacterium *Anabaena 7120*. Proc Natl Acad Sci USA 80:1835–1839

Dean C, Leech RM (1982) Genome expression during normal leaf development. Plant Physiol 69:904–910

de Heij HT, Jochemsen AG, Willemsen PTJ, Groot GSP (1984) Protein synthesis during chloroplast development in *Spirodela oligorhiza*. Eur J Biochem 138:161–168

Di Marco G, Tricoli D (1983) RuBP carboxylase determination by enzymic estimation of D-3-PGA formed. Photosynthesis Res 4:145–149

Dobberstein B, Blobel G, Chua NH (1977) In vitro synthesis and procession of a putative precursor for the small subunit of ribulose 1,5-bisphosphate carboxylase of *Chlamydomonas reinhardii*. Proc Natl Acad Sci USA 74:1082–1085

Donnelly MI, Hartman FC, Ramakrishnan V (1984) The shape of ribulose bisphosphate carboxylase/oxygenase in solution as inferred from small angle neutron scattering. J Biol Chem 259:406–411

Dron M, Rahire M, Rochaix JD (1982) Sequence of the chloroplast DNA region of *Chlamydomonas reinhardii* containing the gene of the large subunit of ribulosebisphosphate carboxylase and parts of it flanking genes. J Mol Biol 162:775–793

Dron M, Rahire M, Rochaix JD, Mets L (1983) First DNA sequence of a chloroplast mutation: a missense alteration in the ribulosebisphosphate carboxylase large subunit gene. Plasmid 9:321–324

Ellis KJ, Morrison JF (1982) Buffers of constant ionic strength for studying pH-dependent processes. Methods Enzymol 87:405–426

Ellis RJ (1979) The most abundant protein in the world. Trends Biochem Sci 4:241–244

Ellis RJ (1981) Chloroplast proteins: synthesis, transport, and assembly. Annu Rev Plant Physiol 32:111–137

Feierabend J, Wildner G (1978) Formation of the small subunit in the absence of the large subunit of ribulose-1,5-bisphosphate carboxylase in 70S ribosome-deficient rye leaves. Arch Biochem Biophys 186:283–291

Fiedler F, Müllhofer G, Tebst A, Rose IA (1967) Mechanism of ribulose diphosphate carboxydismutase reaction. Eur J Biochem 1:395–399

Franks F (1977) Solvation interactions of proteins in solution. Phil Trans R Soc B 278:89–96
Frieden C, Nichol LW (eds) (1981) Protein-protein interactions. John Wiley, New York
Gatenby AA (1984) The properties of the large subunit of maize ribulosebisphosphate carboxylase/oxygenase synthesised in *Escherichia coli*. Eur J Biochem 144:361–366
Gibson JL, Tabita FR (1977) Different molecular forms of ribulose-1,5-bisphosphate carboxylase from *Rhodopseudomonas sphaeroides*. J Biol Chem 252:943–949
Gilbert W (1979) Introns and exons: playgrounds of evolution. In: Axel R, Maniatis T, Fox CF (eds) Eucaryotic gene regulation. Academic Press, New York, pp 1–14
Goldthwaite J, Bogorad L (1975) Ribulose-1,5-bisphosphate carboxylase from leaf. Methods Enzymol 42:481–484
Gornall AG, Bardawill CS, David MM (1948) Determination of serum proteins by means of the biuret reaction. J Biol Chem 177:751–766
Gray JC (1978) Absorption of polyphenols by polyvinylpyrrolidone and polystyrene resins. Phytochemistry 17:495–497
Gray JC, Wildman SG (1976) A specific immunoabsorbent for the isolation of fraction I protein. Plant Sci Lett 6:91–96
Gray JC, Kung SD, Wildman SG (1978) Polypeptide chains of the large and small subunits of fraction I protein from tobacco. Arch Biochem Biophys 185:272–281
Grossman A, Bartlett S, Chua NH (1980) Energy-dependent uptake of cytoplasmatically synthesized polypeptides by chloroplasts. Nature 285:625–628
Gutteridge S, Parry M, Schmidt CNG (1982) The reactions between active and inactive forms of wheat ribulosebisphosphate carboxylase and effectors. Eur J Biochem 126:597–602
Hall NP, Tolbert NE (1978) A rapid procedure for the isolation of ribulosebisphosphate carboxylase/oxygenase from spinach leaves. FEBS Lett 96:167–169
Hall NP, McCurry SD, Tolbert NE (1981) Storage and maintaining activity of ribulosebisphosphate carboxylase/oxygenase. Plant Physiol 67:1220–1223
Hall NP, Cornelius MJ, Keys AL (1983) The enzymatic determination of bicarbonate and CO_2 in reagents and buffer solutions. Anal Biochem 132:152–157
Hartman FC, Norton IL, Stringer CD, Schloss JV (1978) Attempts to apply affinity labeling techniques to ribulosebisphosphate carboxylase/oxygenase. In: Siegelman HW, Hind G (eds) Photosynthetic carbon assimilation. Plenum, New York, pp 245–269
Hartman FC, Stringer CD, Lee EH (1984) Complete primary structure of ribulosebisphosphate carboxylase/oxygenase from *Rhodospirillum rubrum*. Arch Biochem Biophys 232:280–295
Horecker BL, Hurwitz J, Weissbach A (1958) Ribulose diphosphate. Biochem Prep 6:83–90
Jakoby WB, Brummond DO, Ochoa S (1956) Formation of 3-phosphoglyceric acid by carbon dioxide fixation with spinach leaf enzymes. J Biol Chem 218:811–822
Janson CA, Smith WW, Eisenberg D, Hartman FC (1984) Preliminary structural studies of ribulose-1,5-bisphosphate carboxylase/oxygenase from *Rhodospirillum rubrum*. J Biol Chem 259:11594–11596
Jaworowski A, Hartman FC, Rose IA (1984) Intermediate in the ribulose-1,5-bisphosphate carboxylase reaction. J Biol Chem 259:6783–6789
Jellings AJ, Leese BM, Leech RM (1983) Location of chromosomal control of ribulosebisphosphate carboxylase amounts in wheat. Mol Gen Genet 192:272–274
Jensen RG, Bahr JT (1977) Ribulose-1,5-bisphosphate carboxylase/oxygenase. Annu Rev Plant Physiol 28:379–400
Johal S, Bourque DP (1979) Crystalline ribulose-1,5-bisphosphate carboxylase/oxygenase from spinach. Science 204:75–77
Johal S, Chollet R (1983) Analysis of catalytic subunit microheterogeneity in ribulosebisphosphate carboxylase/oxygenase from *Nicotiana tabacum*. Arch Biochem Biophys 223:40–50
Johal S, Bourque DP, Smith WW, Suh SW, Eisenberg D (1980) Crystallization and characterization of ribulose-1,5-bisphosphate carboxylase/oxygenase from eight plant species. J Biol Chem 255:8873–8880

Jones WT, Lyttleton JW (1972) The importance of inhibiting polyphenol oxidase in the extraction of fraction I leaf protein. Phytochemistry 11:1595–1596

Jordan DB, Chollet R (1985) Subunit dissociation and reconstitution of ribulose-1,5-bisphosphate carboxylase from *Chromatium vinosum*. Arch Biochem Biophys 236:487–496

Jordan DB, Ogren WL (1981) A sensitive procedure for simultaneous determination of ribulose-1,5-bisphosphate carboxylase and oxygenase activities. Plant Physiol 67:237–245

Jordan DB, Ogren WL (1983) Species variation in kinetic properties of ribulose-1,5-bisphosphate carboxylase/oxygenase. Arch Biochem Biophys 227:425–433

Jordan DB, Chollet R, Ogren WL (1983) Binding of phosphorylated effectors by active and inactive forms of ribulose-1,5-bisphosphate carboxylase. Biochemistry 22:3410–3418

Kawashima N (1970) Non-synchronous incorporation of $^{14}CO_2$ into amino acids of the two subunits of fraction I protein. Biochem Biophys Res Commun 38:119–124

Kawashima N, Wildman SG (1970) Fraction I protein. Annu Rev Plant Physiol 21:325–358

Kawashima N, Wildman SG (1971) Effect of crystallization of fraction I protein from tobacco leaves on ribulose diphosphate carboxylase activity. Biochim Biophys Acta 229:240–249

Khan AA, Malhotra SS (1982) Ribulosebisphosphate carboxylase and glycollate oxidase from jack pine: effects of sulphur dioxide fumigation. Phytochemistry 21:2607–2612

Kobayashi H, Akazawa T (1982) Biosynthetic mechanism of ribulose-1,5-bisphosphate carboxylase in the purple photosynthetic bacterium, *Chromatium vinosum*. Arch Biochem Biophys 214:540–549

Kung SD, Chollet R, Marsho TV (1980) Crystallization and assay procedure of tobacco ribulose-1,5-bisphosphate carboxylase/oxygenase. Methods Enzymol 69:326–336

Lawlis VB, McFadden BA (1978) Modification of ribulosebisphosphate carboxylase by 2,3-butadione. Biochem Biophys Res Commun 80:580–585

Leloir LF, Cardini CE (1957) Characterization of phosphorous compounds by acid lability. Methods Enzymol 3:840–850

Li LR, Wang WG (1980) The effect of adenosine triphosphate on heat stability of RuBP carboxylase from spinach. Acta Phytophysiol Sin 6:237–241

Lilley RM, Walker DA (1974) An improved spectrophotometric assay for ribulosebisphosphate carboxylase. Biochim Biophys Acta 358:226–229

Link G (1981) Cloning and mapping of the chloroplast DNA sequences for two mRNAs from mustard (*Sinapis alba* L.). Nucleic Acids Res 9:3681–3684

Loomis WD (1974) Overcoming problems of phenolics and quinones in the isolation of plant enzymes and organelles. Methods Enzymol 31:528–544

Lorimer GH (1978) Retention of the oxygen atoms at carbon-2 and carbon-3 during the carboxylation of ribulose-1,5-bisphosphate. Eur J Biochem 89:43–50

Lorimer GH (1981 a) Ribulosebisphosphate carboxylase: amino acid sequence of a peptide bearing the activator carbon dioxide. Biochemistry 20:1236–1240

Lorimer GH (1981 b) The carboxylation and oxygenation of ribulose-1,5-bisphosphate. Annu Rev Plant Physiol 32:349–383

Lorimer GH (1982) Activities of RuBP carboxylase/oxygenase. In: Edelman M, Hallick RB, Chua NH (eds) Methods in chloroplast molecular biology. Elsevier, Amsterdam New York Oxford, pp 803–808

Lorimer GH, Miziorko HM (1980) Carbamate formation on the ε-amino group of a lysyl residue as the basis for the activation of ribulosebisphosphate carboxylase by CO_2 and Mg^{2+}. Biochemistry 19:5321–5328

Lorimer GH, Andrews TJ, Tolbert NE (1973) Ribulose diphosphate oxygenase. II. Further proof of reaction products and mechanism of action. Biochemistry 12:18–23

Lorimer GH, Badger MR, Andrews TJ (1976) The activation of ribulose-1,5-bisphos-phate carboxylase by carbon dioxide and magnesium ions. Equilibria, kinetics, a suggested mechanism, and physiological implications. Biochemistry 15:529–536

Lorimer GH, Badger MR, Andrews TJ (1977) D-Ribulose-1,5-bisphosphate carboxylase/oxygenase: improved methods for the activation and assay of catalytic activities. Anal Biochem 78:66–75

Lorimer GH, Pierce J, Gutteridge S, Schloss JV (1984) Some mechanistic aspects of ribulosebisphosphate carboxylase. In: Sybesma C (ed) Advances in photosynthesis research, vol 3. Nijhoff/Dr W Junk, The Hague, pp 720–734

Losada M, Arnon DI (1964) Enzyme systems in photosynthesis. In: Linskens HF, Sanwal BD, Tracey MV (eds) Modern methods of plant analysis, vol 7. Springer, Berlin Heidelberg New York, pp 569–615

Martin PG (1979) Amino acid sequence of the small subunit of ribulose-1,5-bisphosphate carboxylase from spinach. Aust J Plant Physiol 6:401–408

Martin PG, Dowd JM (1984) The study of plant phylogeny using amino acid sequences of ribulose-1,5-bisphosphate carboxylase. Aust J Bot 32:301–309

Martin JS, Martin MM (1983) Tannin assays in ecological studies precipitation of ribulose-1,5-bisphosphate carboxylase/oxygenase by tannic acid, quebracho, and oak foliage extracts. J Chem Ecol 9:285–294

Masters CJ (1977) Metabolic control and the microenvironment. Curr Top Cell Regul 12:75–105

Masters CJ (1981) Interactions between soluble enzymes and subcellular structure. CRC Crit Rev Biochem 11:105–143

Masters CJ, Sheedy RJ, Winzor DJ, Nichol LW (1969) Reversible adsorption of enzymes as a possible allosteric control mechanism. Biochem J 112:806–808

Matthews BW (1968) Solvent content of protein crystals. J Mol Biol 33:491–497

McCurry SD, Gee R, Tolbert NE (1982) Ribulose-1,5-bisphosphate carboxylase/oxygenase from spinach, tomato, or tobacco leaves. Methods Enzymol 90:515–521

McCurry SD, Hall NP, Pierce J, Paech C, Tolbert NE (1978) Ribulose-1,5-bisphosphate carboxylase/oxygenase from parsley. Biochem Biophys Res Commun 84:895–900

McCurry SD, Pierce J, Tolbert NE, Orme-Johnson WH (1981) On the mechanism of effector-mediated activation of ribulosebisphosphate carboxylase. J Biol Chem 256:6623–6628

McFadden BA (1973) Autotrophic CO_2 assimilation and the evolution of ribulosebisphosphate carboxylase. Bacteriol Rev 37:289–319

McFadden BA (1980) A perspective of ribulosebisphosphate carboxylase/oxygenase, the key catalyst in photosynthesis and photorespiration. Acc Chem Res 13:394–399

McIntosh L, Poulsen C, Bogorad L (1980) Chloroplast gene sequence for the large subunit of ribulosebisphosphate carboxylase of maize. Nature 288:556–560

McNeil PH, Walker DA (1981) The effect of magnesium and other ions on the distribution of ribulose-1,5-bisphosphate carboxylase in chloroplast extracts. Arch Biochem Biophys 208:184–188

Meisenberger O, Pilz I, Bowien B, Pal GP, Saenger W (1984) Small angle x-ray study of the structure of active and inactive ribulosebisphosphate carboxylase from *Alcaligenes eutrophus*. J Biol Chem 259:4463–4465

Milos P, Roy H (1984) ATP-released large subunits participate in the assembly of RuBP carboxylase. J Cell Biochem 24:153–162

Minton AP (1981) Exchanged volume as a determination of macromolecular structure and reactivity. Biopolymers 20:2093–2120

Miziorko HM (1979) Ribulose-1,5-bisphosphate carboxylase. Evidence in support of the existence of distinct CO_2 activator and CO_2 substrate sites. J Biol Chem 254:270–272

Miziorko HM, Lorimer GH (1983) Ribulose-1,5-bisphosphate carboxylase/oxygenase. Annu Rev Biochem 52:507–535

Miziorko HM, Mildvan AS (1974) Electron paramagnetic resonance, ^{1}H, and ^{13}C nuclear magnetic resonance studies of the interaction of manganese and bicarbonate with ribulose-1,5-bisphosphate carboxylase. J Biol Chem 249:2743–2750

Miziorko HM, Sealy RC (1984) Electron spin resonance studies of ribulosebisphosphate carboxylase identification of activator cation ligands. Biochemistry 23:479–485

Mori H, Takabe T, Akazawa T (1984) Loose association of ribulose-1,5-bisphosphate carboxylase/oxygenase with chloroplast thylakoid membranes. Photosynthesis Res 5:17–28

Müller KD, Salnikow J, Vater J (1983) Amino acid sequence of the small subunit of ribulosebisphosphate carboxylase/oxygenase from *Nicotiana tabacum*. Biochim Biophys Acta 742:78–83

Müllhofer G, Rose IA (1965) The position of carbon-carbon bond cleavage in the ribulose diphosphate carboxydismutase reaction. J Biol Chem 240:1341–1346
Mulligan RM, Tolbert NE (1983) The lability of an intermediate of the ribulosebisphosphate carboxylase reaction. Arch Biochem Biophys 225:610–620
Mulligan RM, Tolbert NE (1984) Trypsin inactivation of ribulose-bisphosphate carboxylase/oxygenase. Plant Physiol 75:S159
Murai T, Akazawa T (1972) Homotropic effect of CO_2 in ribulose-1,5-bisphosphate carboxylase reaction. Biochem Biophys Res Commun 46:2121–2126
Nishimura M, Takabe T, Sugiyama T, Akazawa T (1973) Structure and function of chloroplast proteins. XIX. Dissociation of spinach leaf ribulose-1,5-bisphosphate carboxylase by p-mercuribenzoate. J Biochem 74:945–954
O'Connell PBH, Brady CJ (1982) Multiple forms of the large subunit of wheat ribulosebisphosphate carboxylase generated by excess iodoacetamide. Biochim Biophys Acta 670:355–361
Ogren WL (1984) Photorespiration: pathways, regulation and modification. Annu Rev Plant Physiol 35:415–442
Ogren WL, Bowes G (1971) Ribulosebisphosphate carboxylase regulates soybean photorespiration. Nature New Biol 230:159–160
Paech C (1985) Further characterization of an essential histidine residue of ribulose-1,5-biphosphate carboxylase/oxygenase. Biochemistry 24:3194–3199
Paech C, Tolbert NE (1978) Active site studies of ribulose-1,5-bisphosphate carboxylase with pyridoxal 5′-phosphate. J Biol Chem 253:7864–7873
Paech C, McCurry SD, Pierce J, Tolbert NE (1978a) Active site of ribulose-1,5-bisphosphate carboxylase/oxygenase. In: Siegelman HW, Hind G (eds) Photosynthetic carbon assimilation. Plenum, New York, pp 227–243
Paech C, Pierce J, McCurry SD, Tolbert NE (1978b) Inhibition of ribulose-1,5-bisphosphate carboxylase/oxygenase by ribulose-1,5-bisphosphate epimerization and degradation products. Biochem Biophys Res Commun 83:1084–1092
Palmer JD, Edwards H, Jorgensen RA, Thompson WF (1982) Novel evolutionary variation in transcription and location of two chloroplast genes. Nucleic Acids Res 10:6819–6832
Patterson MS, Greene RC (1965) Measurement of low energy beta-emitters in aqueous solution by liquid scintillation counting of emulsions. Anal Chem 37:854–857
Paulsen JM, Lane MD (1966) Spinach RuBP carboxylase. I. Purification and properties of the enzyme. Biochemistry 5:2350–2357
Penefsky H (1977) Reversible binding of P_i by beef heart mitochondrial adenosine triphosphatase. J Biol Chem 252:2891–2899
Perchorowicz JT, Raynes DA, Jensen RG (1981) Light limitation of photosynthesis and activation of ribulosebisphosphate carboxylase in wheat seedlings. Proc Natl Acad Sci USA 78:2985–2989
Pierce J, Tolbert NE, Barker R (1980) Interaction of ribulosebisphosphate carboxylase/oxygenase with transition-state analogues. Biochemistry 19:934–942
Pierce J, Tolbert NE, Barker R (1980) A mass spectrometic analysis of the reactions of ribulosebisphosphate carboxylase/oxygenase. J Biol Chem 255:509–511
Pierce JW, McCurry SD, Mulligan RM, Tolbert NE (1982) Activation and assay of ribulose-1,5-bisposphate carboxylase/oxygenase. Methods Enzymol 89:47–55
Pineau B (1982) Biosynthesis of ribulose-1,5-bisphosphate carboxylase in greening cells of *Euglena gracilis*. Planta 156:117–128
Pon NG, Rabin BR, Calvin M (1963) Mechanism of the carboxydismutase reaction. Biochem Z 338:7–19
Portis AR (1981) Evidence of a stromal Mg^{2+} concentration in intact chloroplasts in the dark. Plant Physiol 67:985–989
Portis AR, Heldt HW (1976) Light-dependent changes of the Mg^{2+} concentration in the stroma in relation to the Mg^{2+} dependency of CO_2 fixation in intact chloroplasts. Biochim Biophys Acta 449:434–446
Portis AR, Chon CA, Mosbach A, Heldt HW (1977) Fructose- and sedoheptulosebisphosphatase. The site of a possible control of CO_2 fixation by light-dependent changes of the stromal Mg^{2+} concentration. Biochim Biophys Acta 461:313–325

Possingham JV (1980) Plastid replication and development in the life cycle of higher plants. Annu Rev Plant Physiol 31:113–129
Poulsen C (1979) The cyanogen bromide fragments of the large subunit of ribulosebisphosphate carboxylase from barley. Carlsberg Res Commun 44:163–189
Poulsen C (1981) Comments on the structure and function of the large subunit of the enzyme ribulosebisphosphate carboxylase/oxygenase. Carlsberg Res Commun 46:259–278
Poulsen C, Martin B, Svendsen IB (1979) Partial amino acid sequence of the large subunit of ribulosebisphosphate carboxylase from barley. Carlsberg Res Commun 44:191–199
Quayle TJ, Katterman FR, Jensen RG (1980) Isolation of active ribulose-1,5-bisphosphate carboxylase from glanded cotton. Physiol Plant 50:233–236
Quivey RG, Tabita FR (1984) Cloning and expression in *E. coli* of the form II ribulose-1,5-bisphosphate carboxylase/oxygenase gene from *Rhodopseudomonas sphaeroides*. Gene 31:91–101
Rabin BR, Trown PW (1964) Mechanism of action of carboxydismutase. Nature 202:1290–1293
Racker E (1957) Reductive pentose phosphate cycle. I. Phosphoribulokinase and ribulose diphosphate carboxylase. Arch Biochem Biophys 69:300–310
Racker E (1962) Ribulose diphosphate carboxylase from spinach leave. Methods Enzymol 5:266–270
Reger BK, Ku MSB, Potter JW, Evans JJ (1983) Purification and characterization of maize ribulose-1,5-bisphosphate carboxylase. Phytochemistry 22:1127–1132
Reichelt BY, Delaney SF (1983) The nucleotide sequence for the large subunit of ribulose-1,5-bisphosphate carboxylase from a unicellular cyanobacterium, *Synechococcus* PCC6301. DNA 2:121–129
Rejda JM, Johal S, Chollet R (1981) Enzymic and physicochemical characterization of ribulose-1,5-bisphosphate carboxylase/oxygenase from diploid and tetraploid cultivars of perennial ryegrass. Arch Biochem Biophys 210:617–624
Riordan JF, McElvany KD, Borders CL (1977) Arginine residues: anion recognition sites in enzymes. Science 195:884–886
Rosichan JL, Huffaker RC (1984) Source of endoproteolytic activity associated with purified ribulosebisphosphate carboxylase. Plant Physiol 75:74–77
Roy H, Valeri A, Pope DH, Rueckert L, Costa KA (1978) Small subunit contacts of ribulose-1,5-bisphosphate carboxylase. Biochemistry 16:665–668
Roy H, Bloom M, Milos P, Monroe M (1982) Studies on the assembly of large subunits of ribulosebisphosphate carboxylase in isolated pea chloroplasts. J Cell Biol 94:20–27
Saluja AK, McFadden BA (1982) Modification of active site histidine in ribulosebisphosphate carboxylase/oxygenase. Biochemistry 21:89–95
Sasaki Y, Ishiye M, Sakihama T, Kamikubo T (1981) Light-induced increase of mRNA activity coding for the small subunit of ribulose-1,5-bisphosphate carboxylase. J Biol Chem 256:2315–2320
Sasaki Y, Sakihama T, Kamikubo T, Shinozaki K (1983) Phytochrome-mediated regulation of two mRNAs, encoded by nuclei and chloroplasts of ribulose 1,5-bisphosphate carboxylase/oxygenase. Eur J Biochem 133:617–620
Saver JV, Knowles JR (1982) Ribulose-1,5-bisphosphate carboxylase: enzyme-catalyzed appearance of solvent tritium at carbon 3 of ribulose 1,5-bisphosphate reisolated after partial reaction. Biochemistry 21:5398–5403
Schirch L, Mozzarelli A, Ottonello S, Rossi GL (1981) Microspectrophotometric measurements on single crystals of mitochondrial serine hydroxymethyltransferase. J Biol Chem 256:3776–3780
Schloss JV (1983) Primary deuterium isotope effects with (3-^{2}H)-RuBP in the carboxylase reaction of ribulosebisphosphate carboxylase: effect of pH and CO_2. Fed Proc, Fed Am Soc Exp Biol 42:1923
Schloss JV, Lorimer GH (1982) The stereochemical course of ribulose-bisphosphate carboxylase. J Biol Chem 257:4691–4694

Schloss JV, Norton IL, Stringer CD, Hartman FC (1978) Inactivation of ribulosebisphosphate carboxylase by modification of arginyl residues with phenylglyoxal. Biochemistry 17:5626–5631
Schloss JV, Phares EF, Long MV, Norton IL, Stringer SD, Hartman FC (1979) Isolation, characterization, and crystallization of ribulosebisphosphate carboxylase from autotrophically grown *Rhodospirillum rubrum*. J Bacteriol 137:490–501
Schmidt GW, Mishkind ML (1983) Rapid degradation of unassembled ribulose-1,5-bisphosphate carboxylase small subunits in chloroplasts. Proc Natl Acad Sci USA 80:2632–2636
Schneider G, Brändén CI, Lorimer GH (1984) Preliminary x-ray diffraction study of ribulose-1,5-bisphosphate carboxylase from *Rhodospirillum rubrum*. J Mol Biol 175:99–102
Servaites JC (1985) Crystalline ribulose biphosphate carboxylase/oxygenase of high integrity and catalytic activity from *Nicotiana tabacum*. Arch Biochem Biophys 238:154–160
Shedlovsky T, MacInnes DA (1935) The first ionization constant of carbonic acid, 0 to 38 °C, from conductance measurements. J Am Chem Soc 57:1705–1710
Shinozaki K, Sugiura M (1982) The nucleotide sequence of the tobacco chloroplast gene for the large subunit of ribulose-1,5-bisphosphate carboxylase/oxygenase. Gene 20:91–102
Siegel MJ, Lane MD (1973) Chemical and enzymatic evidence for the participation of 2-carboxy-3-ketoribitol-1,5-diphosphate intermediate in carboxylation of ribulose diphosphate. J Biol Chem 248:5486–5498
Siegelman HW, Hind G (eds) (1978) Photosynthetic carbon assimilation. Plenum, New York
Singer SJ, Eggman L, Campbell JM, Wildman SG (1952) The proteins of green leaves. IV. A high molecular weight protein comprising a large part of the cytoplasmic protein. J Biol Chem 197:233–239
Smith SM, Ellis RJ (1981) Light-stimulated accumulation of transcripts of nuclear and chloroplast genes for ribulosebisphosphate carboxylase. J Mol Appl Genet 1:127–137
Somerville CR, Fitchen J, Somerville S, McIntosh L, Nargang F (1983) Enhancement of net photosynthesis by genetic manipulation of photorespiration and ribulosebisphosphate carboxylase/oxygenase. Miami Winter Symp 20:295–309
Spielmann A, Ortiz W, Stutz E (1983) The soybean chloroplast genome. Mol Gen Genet 190:5–12
Sue JM, Knowles JR (1978) Retention of the oxygens at C-2 and C-3 of D-ribulose-1,5-bisphosphate in the reaction catalyzed by ribulose-1,5-bisphosphate carboxylase. Biochemistry 17:4041–4044
Sue JM, Knowles JR (1982a) Ribulose-1,5-bisphosphate carboxylase: fate of the tritium label in (3-^{3}H)ribulose 1,5-bisphosphate during the enzyme-catalyzed reaction. Biochemistry 21:5404–5410
Sue JM, Knowles JR (1982b) Ribulose-1,5-bisphosphate carboxylase: primary deuterium kinetic isotope effect using (3-^{2}H) ribulose 1,5-bisphosphate. Biochemistry 21:5410–5414
Tabita FR, McFadden BA (1974) Ribulose-1,5-bisphosphate carboxylase from *Rhodospirillum rubrum*. J Biol Chem 249:3459–3464
Takabe T, Akazawa T (1975) Further studies on the subunit structure of *Chromatium* ribulose-1,5-bisphosphate carboxylase. Biochemistry 14:46–50
Takabe T, Incharoensakdi A, Akazawa T (1984a) Essentiality of the small subunit (B) in the catalysis of RuBP carboxylase/oxygenase is not related to substrate-binding in the large subunit (A). Biochem Biophys Res Commun 122:763–769
Takabe T, Rai AK, Akazawa T (1984b) Interaction of constituent subunits in ribulose-1,5-bisphosphate carboxylase from *Aphanothèce halophytica*. Arch Biochem Biophys 229:202–211
Takruri IAH, Boulter D, Ellis RJ (1981) Amino acid sequence of the small subunit of ribulose-1,5-bisphosphate carboxylase of *Pisum sativum*. Phytochemistry 20:413–415

Tobin EM (1981) Phytochrome-mediated regulation of mRNAs for the small subunit of ribulose-1,5-bisphosphate carboxylase and the light-harvesting chlorophyll a/b-protein in *Lemna gibba*. Plant Mol Biol 1:35–51

Vater J, Gaudszun T, Lange B, Erdin N, Salnikow J (1983) Characteristic features of the regulatory functions of the ribulose-1,5-bisphosphate carboxylase/oxygenase from spinach. Z Naturforsch 38C:418–427

Viro M, Kloppstech K (1983) Gene expression in the developing barley leaf under varying light conditions. Planta 157:202–209

Vu JCV, Allen LH, Bowes G (1984) Dark/light modulation of ribulose biphosphate carboxylase activity in plants from different photosynthetic categories. Plant Physiol 76:843–845

Weaver KE, Tabita FR (1983) Isolation and partial characterization of *Rhodopseudomonas sphaeroides* mutants defective in the regulation of ribulosebisphosphate carboxylase/oxygenase. J Bacteriol 156:507–515

Weissbach A, Horecker BL, Hurwitz J (1956) The enzymatic formation of phosphoglyceric acid from ribulose diphosphate and carbon dioxide. J Biol Chem 218:795–810

Welch GR (1977) On the free energy "cost of transition" in intermediary metabolic processes and the evolution of cellular infrastructure. J Theor Biol 68:267–291

Whitfeld PR, Bottomley W (1983) Organization and structure of chloroplast genes. Annu Rev Plant Physiol 34:279–310

Wildman SG (1979) Aspects of fraction I protein. Arch Biochem Biophys 196:598–610

Wildman SG, Bonner J (1947) Proteins of green leaves. I. Isolation, enzymic properties, and auxin content of spinach cytoplasmic proteins. Arch Biochem 14:381–413

Wildner GF, Henkel J (1979) The effect of divalent metal ions on the activity of Mg^{2+} depleted ribulose-1,5-bisphosphate oxygenase. Planta 146:223–228

Wolfenden R (1972) Analog approaches to structure of transition state in enzyme reactions. Acc Chem Res 5:10–18

Wong CH, Pollak A, McCurry SD, Sue JM, Knowles JR Whitesides GM (1982) Synthesis of ribulose 1,5-bisphosphate: routes from glucose 6-phosphate (via 6-phosphogluconate) and from adenosine monophosphate (via ribose 5-phosphate). Methods Enzymol 89:108–121

Yeoh HH, Badger MR, Watson L (1980) Variations in $K_m(CO_2)$ of ribulose-1,5-bisphosphate carboxylase among grasses. Plant Physiol 66:1110–1112

Yeoh HH, Badger MR, Watson L (1981) Variations in kinetic properties of ribulose-1,5-bisphosphate carboxylases among plants. Plant Physiol 67:1151–1155

Yeoh HH, Stone NE, Watson L (1982) Taxonomic variation in the subunit amino acid compositions of RuBP carboxylases from grasses. Phytochemistry 21:71–80

Zurawski G, Perrot B, Bottomley W, Whitfeld PR (1981) The structure of the gene for the large subunit of ribulose-1,5-bisphosphate carboxylase from spinach chloroplast DNA. Nucleic Acids Res 9:3251–3270

The Chloroplast Thylakoid Membrane – Isolation, Subfractionation and Purification of Its Supramolecular Complexes

B. ANDERSSON and J. M. ANDERSON

1 Introduction

The thylakoid membrane of plant chloroplasts is the site of the primary steps in the photosynthetic conversion of light to chemically bound energy. This is a complicated process that requires several partial reactions, including light capture, charge separation, electron transport, proton translocation, and enzyme catalysis. The coordination and regulation of this process require a defined molecular organization of the thylakoid membrane. A great deal of our present understanding of the photosynthetic process has been made possible by the ability to perform structural and functional studies on isolated thylakoids, subthylakoid fractions and pure supramolecular protein complexes. The aim of this chapter is to give a methodological description of the main isolation procedures available for such thylakoid material.

2 Function and Organization of the Thylakoid Membrane

Before going into experimental details, a short summary of the current view on the structure and function of the thylakoid membrane will be given. This membrane contains all the components required for the light-driven production of reduced $NADP^+$ and ATP, which is necessary for the reduction of CO_2 to organic carbon compounds. The two photosystems, I and II, drive the electron transport from water to $NADP^+$ (Fig. 1). This process starts with the harvesting of light by the antennae pigments, mainly chlorophylls. The captured light energy migrates from the bulk chlorophyll a molecules into a special reaction center chlorophyll a species, designated P-700, which becomes excited. It loses one electron to a primary acceptor with a low standard oxidation potential which is believed to be another chlorophyllous molecule. The electron is then passed on to three low potential iron-sulfur proteins believed to work in series and where at least two have ferredoxin type Fe-S centers. The final electron transfer, from reduced ferredoxin to $NADP^+$, is catalyzed by the flavoprotein ferredoxin-$NADP^+$ oxido-reductase (FNR). The electrons required to fill the electron holes in P-700^+ are supplied by photosystem II. In this photosystem chlorophyll b is involved in the light harvesting as a complement to chlorophyll a. The energy is transferred to a reaction center chlorophyll a (P-680), which upon excitation donates one electron to a pheophytin species. The second electron transfer step is from the pheophytin to another acceptor designated Q_A, which is believed to be a quinone-iron protein.

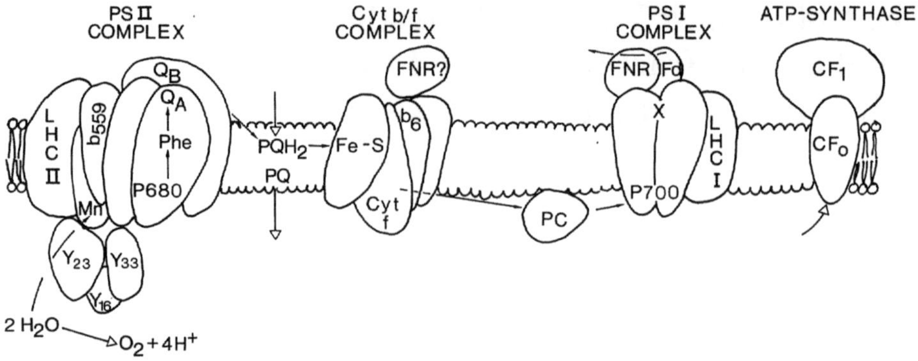

Fig. 1. Schematic and simplified model of the current view on the organization of the thylakoid proteins involved in light-harvesting, electron transport and photophosphorylation. *Arrows with filled heads* electron flow; *arrows with open heads* proton transport. *PC* plastocyanin; *PQ* plastoquinone; Y_x subunits of the oxygen evolving system; *Fd* ferredoxin; *FNR* ferredoxin-NAPP$^+$ oxido/-reductase; *Phe* pheophytin; *X* Fe-S centers at the acceptor side of PS I; Q_A and Q_B acceptors of PS II, P-680 and P-700 reaction centres of photosystem II and I respectively

The next acceptor, the Q_B-protein, acts as a two-electron gate. The bound quinone of this protein is believed to exchange with the reduced plasoquinone pool. From the reduced plastoquinones the electrons are transported to P-700$^+$ through the Rieske iron-sulfur center, cytochrome f, and the copper protein plastocyanin. The electrons donated to P-680$^+$ are derived from the photosynthetically unique water oxidation process. The identity of the catalytic components of this reaction and the precise chemical events remain to be determined. There is a general consensus that protein-bound manganese plays a central role in this reaction.

The linear electron flow from water to NADP$^+$ is coupled to phosphorylation of ADP to ATP. According to the chemiosmotic theory, the driving force for the phosphorylation is a proton gradient across the thylakoid membrane. Proton translocation sites are associated with the water oxidation reaction and the plastoquinone hydrogen shuttle (Fig. 1). The actual ATP synthesis takes place on the enzyme ATP synthetase. In addition, a cyclic electron flow around photosystem I can drive ATP formation.

For a detailed and updated view concerning photosynthetic electron transport, the reader is referred to a recent review by Haehnel (1984).

In addition to the sequential arrangement of the electron transport components discussed above, the three-dimensional organization of the components in the thylakoid membrane must be considered. In higher plants, the thylakoid membrane is differentiated into appressed and nonappressed regions (Figs. 2 and 3). The appressed membranes, which all participate in close membrane-membrane interactions, are arranged in stacks of thylakoids (Fig. 3). The nonappressed thylakoids are all in contact with the soluble stromal space. The nonappressed thylakoids can be divided into stroma lamellae, grana and membranes, and grana margins (Fig. 3). The differentiation into appressed and nonappressed

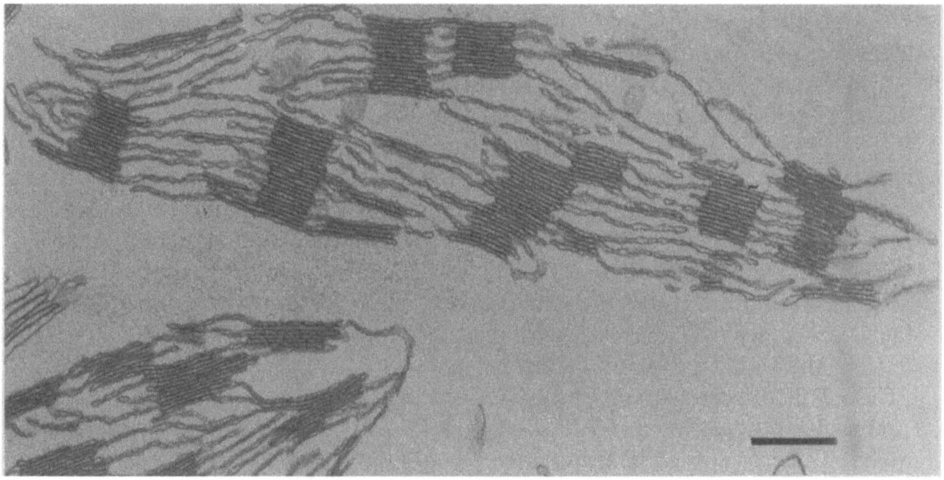

Fig. 2. Electron micrograph of isolated spinach thylakoids with maintained differentiation into appressed and non-appressed membrane regions. Bar 0.5 μm

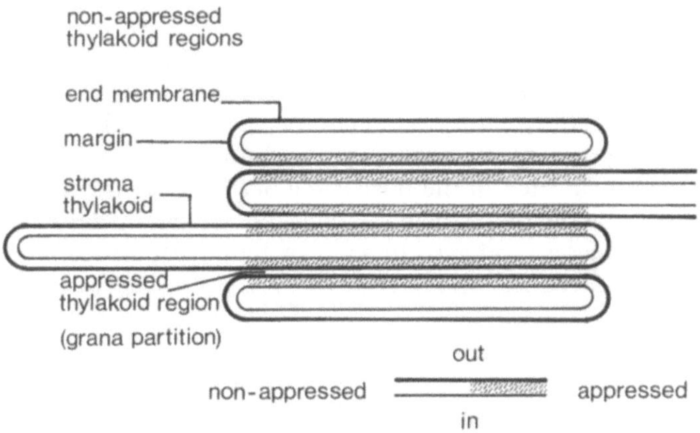

Fig. 3. Schematic illustration of a stacked thylakoid membrane indicating the various membrane regions

thylakoid regions is only maintained in the presence of cations, while in low salt buffers all the thylakoid membranes destack and appear as single nonappressed membranes.

Most data suggest the thylakoids to be arranged as a fluid lipid-protein mosaic type of membrane (Anderson 1975). About 50% of the thylakoid mass is made up by lipids. Strikingly, phospholipids comprise only a very small proportion of the total lipids, while 80% is neutral galactolipids. Moreover, most of the acyl chains of these lipids have an unusually high degree of unsaturation. Particularly dominating is the linolenic acid (18:3), which can account for up to 90% of the acyl chains. Most of the protein components are embedded in the lipid bilayer. The majority of the approximately 50 different thylakoid polypeptides do not appear as single protein species in the membrane, but are arranged in four supramolecular complexes. Three of these complexes are engaged in electron transport, the photosystem II complex, the cytochrome b/f complex and the photosystem I complex, and one involved in the synthesis of ATP, the ATP synthase. The photosystem I and II complexes both contain chlorophyll, which is not free but bound to specific protein subunits. Both these complexes are arranged with an inner core complex (CC I and CC II) containing chlorophyll a and the reaction centers. These cores are surrounded by light-harvesting complexes (LHC I and LHC II) which are made up by chlorophyll a/b proteins. Each of the supramolecular complexes contains several intrinsic and membrane-spanning proteins (Fig. 1). In addition there are extrinsic proteins, located at either membrane surface, such as ferredoxin, plastocyanin and the 33,000, 23,000 and 16,000 mw proteins of the water oxidation reaction.

Each of the components has a defined organization in the thylakoid bilayer. As in all biological membranes, the thylakoids have a defined transverse asymmetry. The extrinsic proteins are located either at the outer or inner thylakoid surfaces. Those functioning at the acceptor side of the photosystems are located at the outer surface, while those of the donor sides are located at the inner surface. The membrane-spanning complexes are arranged in asymmetrically across the membrane, thereby exposing different protein moeties at the outer and inner membrane surfaces. Finally, the thylakoid lipids show a partial transverse asymmetry.

In addition to the transverse asymmetry, the thylakoid membrane has a pronounced lateral heterogeneity. The photosystem II complex is concentrated in the appressed thylakoid regions, while the photosystem I complex and the ATP synthase are confined to the nonappressed regions. The cytochrome b/f complex seems to be evenly distributed or located in the border regions between the appressed and nonappressed regions. For a detailed overview on the organization of the thylakoid membrane the reader is referred to a review by Barber (1983).

3 Isolation of Thylakoid Membranes

The isolation of functional thylakoid membranes is not considered to be a great art within the field of organelle separation. Few significant methodological improvements have been described in the last 15 years or so. Still, as will be discussed in this section, there are some points that require attention.

Thylakoids are chloroplasts which have lost the two outer envelope membranes and the soluble components of the stroma (Fig. 2). According to the old terminology of Spencer and Wildman (1962), such preparations were referred to as broken or Class II chloroplasts. Under the light microscope they are clearly distinct from intact or Class I chloroplasts. The latter show a bright, highly reflecting appearance, while the former are quite dark with clearly visible grana. Another terminology was introduced by Hall (1972), who designated the thylakoids type C chloroplasts in contrast to intact ones which were called type A or B. Today these classifications have been mainly abandoned and in this review the term thylakoid membranes or simply thylakoids will be used.

Isolated thylakoid membranes are not capable of CO_2 fixation but have good rates of electron transport when supplied with ferredoxin, $NADP^+$, or artificial electron acceptors. Usually, they can perform photophosphorylation and even show photosynthetic control. Apart from retained function, a good thylakoid preparation should fulfill some other criteria. The membrane bilayer should be intact in order to allow the build-up of a transmembrane gradient during electron transport. Moreover, the differentiation between stacked and nonstacked regions should be maintained. During the preparation care should be taken to avoid changes in the polypeptide composition due to protein degradation or release. Also the contamination by stroma and extra-chloroplast material should be minimized. How many of these criteria need to be fulfilled is dependent on the particular use of the isolated thylakoids. In the case of a functional study a higher structural and functional integrity is required than in the case of starting material for protein purification.

Isolation of Thylakoids from Intact Chloroplasts. Just as the best intact chloroplasts are isolated by osmotic rupture of protoplasts (Leegood and Walker 1983), the best thylakoids are obtained from isolated intact chloroplasts. In this way the thylakoid membrane, during isolation, is protected by the chloroplast envelope from degrading proteolytic enzyme and lipases present in the extra chloroplast compartments of the cell. Moreover, one avoids nonspecific adsorption of nonchloroplast proteins and nucleic acids to the thylakoid membrane surface. In particular it has been shown that there is a massive binding of nuclear DNA to thylakoids which have been exposed to cytoplasmic material (Blomquist et al. 1975).

The intact chloroplasts can be isolated by any available method that gives above 95% of intactness (Leegood and Walker 1983). For the subsequent osmotic rupture of the envelope, the chloroplast pellet should be suspended in a suitable hypotonic medium. We have found that efficient breakage is obtained by 1 mM Tricine pH 6.5, 5 mM $MgCl_2$. The chloroplasts are kept in the osmotic shock medium for about 5 min and then pelleted by centrifugation at 1 000 *g* for 5 min. The pellet is washed twice in an isotonic medium such as 300 mM sucrose/sorbitol, 10 mM Tricine pH 6.5, 5 mM $MgCl_2$. The final pellet is resuspended in a suitable assay or storage medium. It is important to include the magnesium ions in the osmotic shock medium to keep the thylakoids stacked but also to preserve certain functions (Shahak et al. 1980). Alkaline pH should be avoided, since it increases the risk of release of extrinsic membrane proteins and destabilizes the water oxidation reaction. The envelope breakage only takes 1 min. Prolonged incubation in the osmotic shock medium is not necessary and should be avoided since it will eventually lead to loss of the compact thylakoid structure due to spontaneous vesiculation of the membrane. The centrifugation steps, are usually performed at too high a speed and for a longer time than is necessary. The centrifugation at only 1 000 *g* for 5 min recovers around 90% of the thylakoids (Gardeström et al. 1978). Such low speed centrifugation minimizes the sedimentation of contaminating mitochondria and other small extra-thylakoid material. Removal of the chloroplast stroma proteins requires at least two washings of the isolated thylakoids in an isotonic medium. For complete removal of the abundant ribulose-1,5-bisphosphate carboxylase/oxygenase even more washings may be required.

Direct Isolation of Thylakoids. Isolation of thylakoids from intact chloroplasts is not a realistic alternative when large quantities of material are required for subfractionation or isolation of certain thylakoid components. For such purposes a more direct isolation procedure, such as that described below, has to be used. Spinach leaves (30 g) were homogenized in 100 ml of preparation medium composed of 300 mM sucrose, 10 mM Tricine pH 6.5, 5 mM $MgCl_2$ for 5 s in a rotating-knife blendor. The homogenate was filtered through four layers of nylon net with a pore size of 20 μm. The filtrate was centrifuged in a SS-34 type of rotor for 2 min at 1 000 *g*. The pellet was resuspended and washed once in the preparation medium using a 10-min centrifugation. This pellet, containing around 50% intact chloroplasts, was suspended in an osmotic shock medium of 1 mM Tricine pH 6.5, 5 mM $MgCl_2$, as described above. The osmotic breakage was followed by two washings at 1 000 *g* for 10 min in the preparation medium. The final pellet was resuspended in a suitable assay or storage medium. This preparation gives in the case of spinach a yield of 0.2 mg chlorophyll per g of leaves. If still a higher yield is desired the first 2 min centrifugation can be increased to 5 min without much increase in extra-chloroplast contamination.

This procedure gives pure, structurally intact and functional thylakoids from most species common in laboratory practise. From more rigid types of leaves, such as in the case of grass, succulents, or pine needles, more vigorous homogenization will be required. This will inevitably lead to increased damage of structure and activity. Precutting such leaves into smaller pieces prior to the maceration usually reduces the damaging effects. To make the maceration more efficient one should try to minimize the air space above the homogenization mixture during blending. In this way the time a certain leaf piece is in contact with the rotating knives will be increased to allow for a shorter overall blending time. This can be achieved by constructing a lid that fits just above the the air–liquid interface.

The use of a nylon net with defined pore size gives a more controlled filtration than the more widely used Miracloth or cheese cloth. Most buffers can be used for thylakoid preparations, but some attention must be paid to buffers with inhibitory effects such as Tris. For a comprehensive review of buffers suitable in thylakoid research, the reader is referred to a paper by Good and Izawa (1972). If thylakoids are prepared from fresh, healthy leaves, the addition of protective agents such as ascorbate or BSA is not required. In certain plants with high content of phenolic compounds, the inclusion of substances such as polyvinylpyrrolidone or polyethylene glycol is advantageous.

If the thylakoids have to be stored prior to use, care has to be taken to preserve their structure and function. Storage in the refrigerator can preserve the photosystem I activity for several days, but photosystem II oxygen evolution and photophosphorylation will be deactivated. Freezing thylakoids in a normal deep freezer (−16 °C) should be avoided, since it leads to irreversible aggregations. Rapid freezing in liquid nitrogen in the presence of a suitable cryoprotectant is recommended. As a cryoprotectant we usually use 5% dimethylsulfoxide. To further minimize freezing injuries, it is important to keep as high a thylakoid concentration as possible and to use a storage buffer with high osmolarity and low ionic strength (Hincha et al. 1984). Still, even if all the precautions mentioned above are taken, the frozen material must always be compared with the fresh material for any particular function that is to be studied.

4 Thylakoid Membrane Subfractionation

The complexity of the reactions taking place at the thylakoid membrane makes it difficult to study one individual reaction without disturbing influences from other reactions. This problem can be partly overcome by the use of specific inhibitors in combination with artificial electron acceptors and donors (Izawa 1980). Moreover, illumination with light of different spectral quality allows some discrimination in the excitation between the two photosystems. Specific biochemical or biophysical probes such as kinetic spectroscopy or ESR may allow a functional study of a certain component in situ. However, an ultimate analysis of a reaction requires that its components are isolated and characterized with respect to structure and function. Thus, it can be regarded as equally important to subfractionate an organelle membrane into its constituents as it is to subfractionate a cell into its organelle compartments. Methods for the subfractionation of a biological membrane usually include two principal steps, a disintegration procedure followed by a method to separate the membrane fragments. The disintegrations can be divided into two main categories, those involving detergents and those based upon mechanical shearing. In contrast to the variety in the fragmentation procedures, there is very little variation in the subsequent methods for separation of the membrane particles. Mainly differential or density gradient centrifugations have been used, thereby separating the particles according to size and density differences. An alternative separation procedure is partition in aqueous polymer two-phase systems (Albertsson 1971; Albertsson et al. 1982). This separation method, which is described in detail by Larsson (this volume), separates membrane particles according to differences in surface properties such as charge and hydrophobicity. Thus, phase partition may be an alternative when different centrifugal techniques do not give satisfactory separation. In many cases phase partition has proven to be a suitable complement to centrifugation. By combining the two methods, a two-dimensional membrane separation can be obtained, separating according to size and density in the first dimension and surface properties in the second (Fig. 4).

The first subfractionations of the thylakoid membrane were reported in the mid 1960's, when the first separations of photosystem I and II were made. Initial works that should be mentioned are those of Boardman and Anderson (1964) and Wessels (1968), who used digitonin-based fractionation, while Vernon et al. (1966) used Triton X-100. Other methods relied on mechanical fragmentation, such as passage of thylakoids through a French pressure cell (Michel and Michel-Wolwertz 1970; Sane et al. 1970) or subjection to sonication (Jacobi 1971). A later

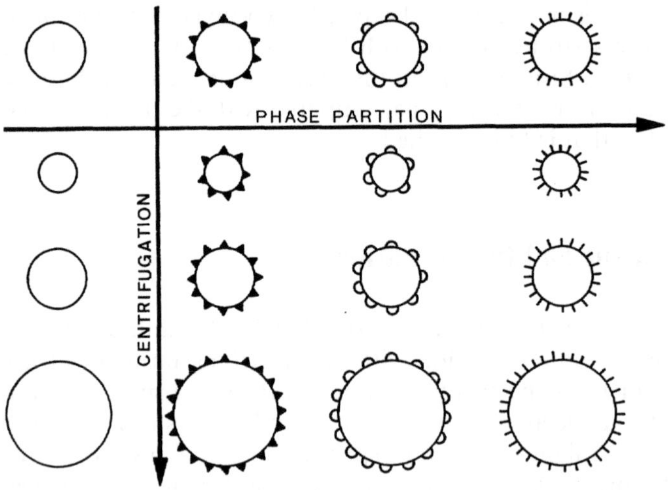

Fig. 4. Two-dimensional separation of membrane vesicles by combining centrifugation and aqueous polymer two-phase partition. Note the improved resolution when differences both in size and surface properties are used to fractionate a complicated vesicle population

development was the isolation of inside-out thylakoid vesicles using the phase partition technique (Andersson et al. 1977; Andersson and Åkerlund 1978; Andersson et al. 1980). This section will describe some procedures for subthylakoid fractionation with retained membrane structure, while the next section will deal with the isolation of supramolecular complexes from solubilized thylakoid membranes.

4.1 Photosystem I Stroma Lamellae Thylakoids

Digitonin is the detergent that has proven most useful for the isolation of photosystem I membrane particles. Low amounts of digitonin appear to be quite selective in fragmenting the nonappressed stroma lamellae vesicles (Fig. 5) while leaving the appressed grana stacks mainly intact (Wehrmeyer 1962). A standard procedure for quite some time has been that of Anderson and Boardman (1966). Their D-144 fraction, highly enriched in photosystem I, was obtained by differential centrifugation following digitonin incubation of thylakoids at a detergent: chlorophyll weight ratio around 15 (Boardman 1971).

A more recent and improved digitonin fractionation procedure has been described by Peters et al. (1981, 1983): Thylakoids were suspended in 20 mM TES-KOH buffer pH 7.8, 250 mM sorbitol, 25 mM KCl, 25 mM NaCl, 5 mM $MgCl_2$. Digitonin (Merck, twice recrystallized from ethanol) was added to give a final digitonin concentration of 0.2% (w/v) and a digitonin/chlorophyll (w/w) ratio of 1. The final chlorophyll concentration was 2 mM. The mixture was stirred for

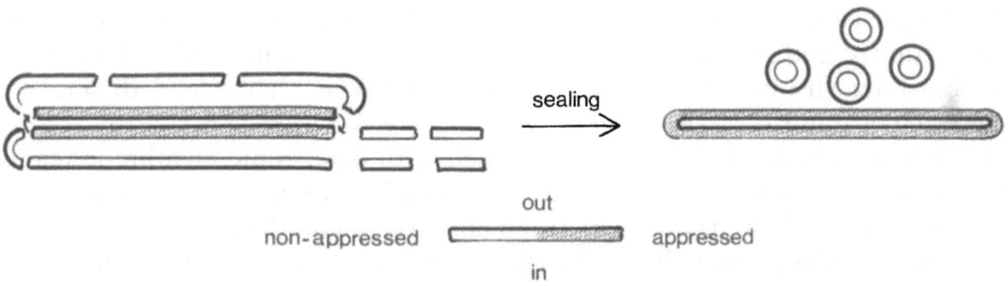

Fig. 5. Fragmentation mechanism for stacked thylakoid membranes. Breakage occurs preferentially at the grana margins. Non-appressed thylakoids reseal into right-side-out vesicles while appressed thylakoids reseal into inside-out vesicles. In the case of detergent fractionation the latter do not seal to form closed vesicles

30 min in the dark at 4 °C, followed by a threefold dilution with the suspension medium. After differential centrifugation at 10,000 *g* for 30 min and 50,000 *g* for 30 min the photosystem I material was collected from the last supernatant by centrifugation at 129,000 *g* for 60 min. This D-129 pellet was carefully resuspended with a pottering tube in a medium containing 5 mM TES-KOH pH 7.8, 2.5 mM KH_2PO_4, 20 mM NaCl, 20 mM KCl, 5 mM $MgCl_2$ at a chlorophyll concentration of 1–4 mg ml^{-1}. In electron microscopy the D-129 material appears as vesicles with an average diameter around 50 nm (Peters et al. 1981). The material is highly photosystem I enriched and depleted in photosystem II as judged from activity and compositional analysis (Table 1). In contrast, the amount of cytochrome b/f complex is about the same as in the starting thylakoid material. In comparison with previous detergent photosystem I preparations the D-129 vesicles show a quite high retention of photosystem I activity, as well as coupled cyclic photophosphorylation (Peters et al. 1983). One main reason is that the low digitonin concentration used allows extrinsic proteins such as the coupling factor (CF_1) and plastocyanin to remain membrane-bound. The drawback of the

Table 1. Properties of subthylakoid fractions[a]

	Y-100	D-129[b]	B5	BBY
PS II activity μmol O_2 mg^{-1} chl h^{-1}	12	0.5	200	275
PS I activity μmol O_2 mg^{-1} chl h^{-1}	490	760	50	21[c]
Chlorophyll a/b	6.5	4.5	2.0	1.8
PS I and PS II % chl in SDS-PAGE				
CC-I+LHC-I	69	63	7	2
CC-II+LHC-II	22	3	82	74
Free chlorophyll	9	34	11	24
Cytochrome f mol/10^3 mol chl	0.8	0.8	0.9	0.4[d]
ATP synthase mol/10^3 mol chl	3.3	2.4	0.3	

[a] Unless otherwise stated observations from the authors
[b] Peters et al. (1983)
[c] Dunahay et al. (1984)
[d] Lamm et al. (1983)

method is the yield, which is as low as 2% on a chlorophyll basis. Thus, if a higher yield is required and some deactivations can be accepted, the method of Anderson and Boardman, using a higher amount of detergent, is recommended, since it gives a yield around 12% (Boardman 1971).

Even if detergent-based methods are used with care, some membrane rearrangements such as delipidation cannot be excluded. Thus, if an intact membrane structure is desired, a mechanical fragmentation method should be chosen. Highly pure photosystem I stroma lamellae vesicles can be obtained by differential centrifugation following disruption of thylakoids by French press (Sane et al. 1970; Arntzen et al. 1972; Henry and Lindberg-Møller 1981) or Yeda press (Åkerlund et al. 1976; Andersson and Anderson 1980). As illustrated in Fig. 5, the shearing forces during such press treatment have shown to disrupt the thylakoid membrane in the borders between the appressed and nonappressed regions (Sane et al. 1970). The following paragraph describes the isolation of stroma lamellae vesicles using a Yeda pressure cell (Åkerlund et al. 1976; Andersson and Anderson 1980). The advantage with the Yeda press compared to the French press is that it is easier to operate and that a high oxygen pressure is avoided by using an inert gas pressure.

Thylakoid material suspended in 50 mM sodium phosphate pH 7.4, 150 mM NaCl at a chlorophyll concentration around 1 mg ml^{-1} was allowed to stand for 30 min at 4 °C to assure complete stacking. The material was poured into a prechilled Yeda press (Shneyour and Avron 1970) (Yeda Scientific Instruments, Rehovot, Israel) fitted to a nitrogen gas cylinder. The thylakoid suspension was passed through the needle valve of the press at a dropwise flow rate using a pressure of 10 MPa. After all material was passed through the press, the procedure was repeated once at the same pressure. The final homogenate was centrifuged at 40,000 *g* for 30 min. The supernatant was carefully collected by pipette and transferred to an ultracentrifugation tube which was spun at 100,000 *g* for 30–60 min. The supernatant which contains virtually no chlorophyll was discarded, while the pellet (Y-100) was collected and suspended in a suitable buffer. This Y-100 material shows a pronounced photosystem I enrichment and possesses high activities (Table 1). As is the case for other stroma lamellae preparations, it shows no enrichment in the cytochrome b/f complex (Cox and Andersson 1981). The Y-100 fraction retains all of the coupling factor (Berzborn et al. 1981) and most of the plastocyanin (Haehnel et al. 1981). The yield on a chlorophyll basis is 5–7%. The yield can be increased to 10% by suspending the 40,000 *g* pellet in the 50 mM sodium phosphate pH 7.4, 150 mM NaCl medium and passing it twice more through the press followed by the two centrifugal steps.

Using consecutive passages of the 40,000 *g* pellet through a French press operated at 27.5 MPa Arntzen et al. (1972) could get a quite high yield of pure stroma lamellae vesicles. If three French press and 40,000 *g* centrifugation cycles were performed, as much as 15% yield could be obtained. Henry and Lindberg-Møller (1981) point out that an improved purity can be obtained if the lower portion of the 40,000 *g* supernatant is not collected for the final ultracentrifugation step.

Most stroma lamellae preparations contain some residual photosystem II and LHC II. This can probably partially be explained by contamination from the appressed thylakoid regions. However, this cannot be the full explanation, since the photosystem II seen in stroma lamellae vesicles appears to differ from the bulk photosystem II of the appressed grana regions. It has a higher requirement for light saturation (Armond and Arntzen 1977) and possesses properties typical of photosystem IIβ rather than the main photosystem IIα (Anderson and Melis 1983). Moreover, it has been shown that the LHC II which becomes phosphorylated migrates from the appressed thylakoid region out to the nonappressed stroma lamellae (Larsson et al. 1983; Kyle et al. 1984).

4.2 Photosystem II Oxygen Evolving Thylakoid Preparations

The isolation of pure and active photosystem II thylakoids has been a major problem in which progress has been made only in the last years. Early detergent and mechanical fragmentations of thylakoids followed by centrifugation gave a heavy fraction (Anderson and Boardman 1966; Vernon et al. 1966; Sane et al. 1970; Jacobi 1971) containing more or less intact grana stacks (Sane et al. 1970). These preparations were usually quite active in oxygen evolution, but their content of photosystem I was still quite high. Several attempts were made to fractionate such grana stacks by further detergent treatments (Vernon et al. 1971; Arntzen et al. 1972; Boardman 1972; Wessels et al. 1973). These procedures gave very pure photosystem II particles, but usually these were inactivated with respect to oxygen evolution. Two different approaches have now made it possible to obtain very pure and highly active photosystem II thylakoid preparations. These are press disruption of thylakoids followed by aqueous polymer two-phase partition (Åkerlund et al. 1976; Andersson and Åkerlund 1978; Henry and Lindberg-Møller 1981; Åkerlund and Andersson 1983; Andersson 1984) or detergent-based fractionation under controlled pH and salt conditions (Berthold et al. 1981; Yamamoto et al. 1981; Kuwabara and Murata 1982; Ford and Evans 1983; Dunahay et al. 1984).

4.2.1 Isolation by Press Treatment and Phase Partition

The original procedure for subfractionation of mechanically fragmented thylakoids by phase partition was reported by Åkerlund et al. (1976). Thylakoid fragments obtained by Yeda press treatment were separated into two distinct subpopulations, one with a high affinity for the upper phase and the other for the lower phase. The material of the lower phase was highly enriched in photosystem II and showed appreciable rates of oxygen evolution. Moreover, these thylakoids were turned inside out with respect to the original sidedness of the membrane (Andersson et al. 1977; Andersson and Åkerlund 1978). This original procedure has been improved with respect to yield (Åkerlund and Andersson 1983) and purity (Henry and Lindberg-Møller 1981; Andersson 1984).

Isolation of everted photosystem II thylakoids by phase partion according to Andersson (1984): Stock solutions of 20% (w/w) dextran T-500 and 40% (w/w) polyethylene glycol 3350 (previously named polyethylene glycol 4000) were prepared as described in detail by Larsson, this volume. Thylakoids from spinach

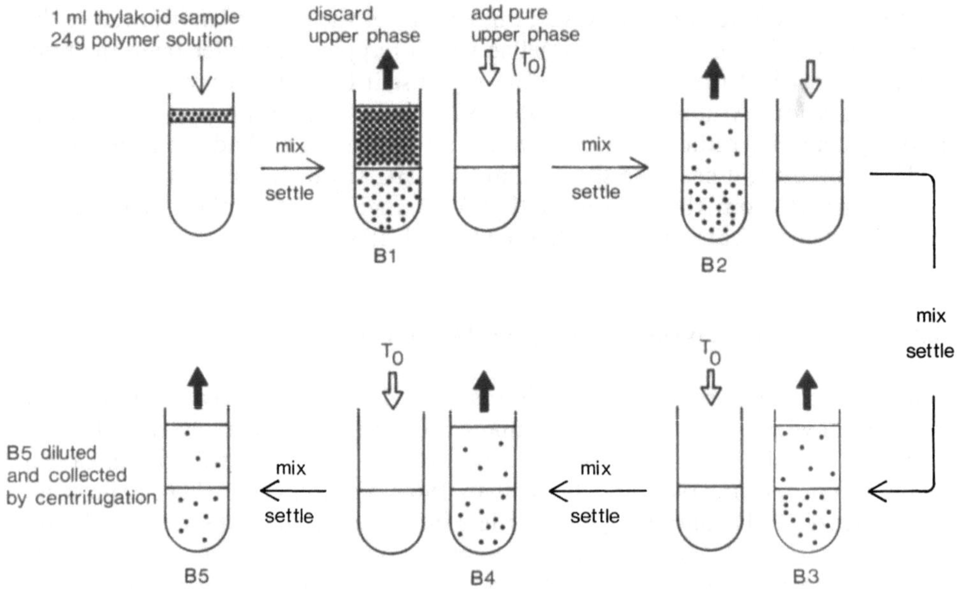

Fig. 6. Purification of photosystem II oxygen-evolving vesicles by phase partition following press disruption of stacked thylakoids. The lower phases (B_x) are extracted with pure upper phase (T_0) until the B_5 fraction is obtained

leaves were prepared as described in Section 3 and finally suspended in 100 mM sucrose, 10 mM sodium phosphate pH 6.5, 5 mM NaCl, 5 mM $MgCl_2$ to give a final chlorophyll concentration of 4 mg ml^{-1}. The suspension was passed twice through the Yeda press at a nitrogen pressure of 10 MPa in the same way as described for the preparation of the stroma lamellae vesicles (Sect. 4.1). Immediately after the press treatments, EDTA to a final concentration of 5 mM was added and two more press treatments were performed. Residual undisrupted thylakoids and starch were removed by centrifugation at 1000 *g* for 10 min. One ml of the supernatant was added to a 24-g polymer mixture to yield a phase system of the following final composition; 5.55% (w/w) dextran T-500, 5.55% (w/w) polyethylene glycol 3350, 10 mM sodium phosphate pH 7.4, 5 mM NaCl and 20 mM sucrose. This 24-g mixture was prepared by weighing into a centrifugation tube; 20.53 g of solution A and 3.47 g of the 40% (w/w) polyethylene glycol. Solution A was prepared by mixing 139.7 g 20% (w/w) dextran T-500, 24.15 g 200 mM sodium phosphate pH 7.5, 24.15 g 100 mM NaCl, 8.1 g 1 M sucrose and 217.2 g water. The final phase system containing the thylakoid sample was brought to 2 °–3 °C and then mixed by 40 inversions of the tube (Fig. 6). The subsequent phase settling was facilitated by centrifugation at 1000 *g* for 3 min. The upper phase was removed and discarded, while the lower phase (B1) was kept and supplied with 13 ml of a pure upper phase (T0) obtained from a bulk phase system with the same composition as the sample system (except that no thylakoids were incluced). After addition of the upper phase, the phase system was mixed and settled as described above yielding fraction B2 (Fig. 6). The repartitions with T0 phase was usually repeated three more times, yielding fraction B5. The B5 material was collected by twofold dilution in a suitable buffer and centrifugation at 100,000 *g* for 30–60 min. The upper phase (T0) was obtained from a 250-g bulk phase system prepared the day before by weighing into a separating funnel; 69.38 g 20% (w/w) dextran T-500, 34.69 g 40% polyethylene glycol 3350, 12.5 g 200 mM sodium phosphate pH 7.4, 12.5 g 100 mM NaCl, 5.0 g 1 M sucrose and finally water to 250 g. After the mixture had been equilibrated to 2 °–3 °C it was carefully mixed and allowed to settle overnight. The upper phase was collected and used for the partition steps as described above.

The B5 material contained inside-out vesicles with a very high photosystem II enrichement and with high rates of oxygen evolution (Table 1) (Andersson 1984). The yield on a chlorophyll basis was around 5%. If a higher yield should be required, 5.7% of both polymers should be used but this leads to a somewhat lower photosystem II purity (Åkerlund and Andersson 1983).

It should be noted that the polymer concentrations given above may not be entirely applicable in all cases, since commercial batches of the polymers differ somewhat in their molecular weight distribution and this influences the partition of membranes. A good strategy is to start by trying the polymer concentrations given above and then, if necessary, adjust the concentrations until a suitable separation is obtained. The B1 phase should contain some 10–15% of the chlorophyll. If less material than this is obtained, the polymer concentrations should be raised to around 5.7–5.9% of each polymer. On the other hand, if to much material is obtained the polymer concentrations should be lowered to around 5.2–5.5%. It should also be stressed that the temperature and salt composition of the phase system also influence the partition and should therefore be controlled to correspond to the conditions given above.

4.2.2 Isolation by Detergent Fractionation

In recent years a number of pure photosystem II preparations with high rates of oxygen evolution have been obtained from detergent fractionation of thylakoids (Berthold et al. 1981; Yamamoto et al. 1981; Kuwabara and Murata 1982; Ford and Evans 1983; Dunahay et al. 1984). The procedure that has become most widely used is that of Berthold et al. (1981) which in a modified form (Ghanotakis et al. 1984) will be described below:

Thylakoids were suspended in 50 mM Mes pH 6.0, 5 mM $MgCl_2$, 15 mM NaCl, 1 mM ascorbate. Triton X-100 (25%), made up in 50 mM Mes pH 6.0, 5 mM $MgCl_2$, 15 mM NaCl supplied with 100–200 units of catalase/100 ml, was added dropwise with stirring to give 25 mg Triton mg^{-1} chlorophyll. The conditions were choosen so that the final chlorophyll concentration became 2 mg ml^{-1}. After 30 min of stirring in the dark at 4 °C, the suspension was centrifuged at 40,000 *g* for 30 min and the pellet was resuspended in the Mes, NaCl, $MgCl_2$ buffer. The chlorophyll concentration was determined and Triton X-100 was added from the 25% stock solution to give a final concentration of 5 mg mg^{-1} chlorophyll. The final chlorophyll concentration was once again 2 mg ml^{-1}. After 5 min dark incubation at 4 °C the suspension was again centrifuged at 40,000 *g* for 30 min. The pellet, which consists of photosystem II particles, was resuspended in 400 mM sucrose, 50 mM Mes pH 6.0, 5 mM $MgCl_2$, 15 mM NaCl to a final chlorophyll concentration of 3 mg ml^{-1} and frozen in liquid nitrogen. The second Triton treatment can be omitted, in which case a somewhat lower photosystem II purity is obtained.

4.2.3 Choice of Preparation

The oxygen-evolving photosystem II thylakoid preparations show several structural and compositional similarities. However, there are also some differences which should be considered when choosing the preparation method. A comprehensive comparative study on the various detergent preparations has been made by Dunahay et al. (1984). They can all be regarded as essentially equivalent even if the purity and activity rates vary somewhat. Obviously the preparation of Yamamoto et al. (1981) contains more contaminating photosystem I than the other preparations. The preparations obtained by press disruption and phase partition show about the same purity as those obtained from the detergent fractionations as revealed by various SDS-PAGE, chlorophyll, and activity analysis (Table I) (Henry and Lindberg-Møller 1981; Lindberg-Møller et al. 1984; Andersson 1984). A clear advantage with the photosystem II particles obtained by the detergent-based methods is that their oxygen-evolving activity is more stable than that of the particles obtained by phase partition. Another advantage with the detergent-based methods is that they give a yield around 20–50%, while the yield from the phase partition is not more that 5%. The yield of the latter can be increased up to 40% by raising the polymer concentrations and applying fewer partition steps (Åkerlund and Andersson 1983), but this leads to increased photosystem I contamination. The photosystem II preparations obtained from mechanical or detergent fragmentation are derived from the appressed grana regions (Andersson et al. 1980; Dunahay et al. 1984; Lindberg-Møller et al. 1984) and formed mainly according to the mechanism outlined in Fig. 5. Thus, they all show an everted orientation, i.e., the inner thylakoid surface is exposed to the surrounding medium. Some of the preparations (Henry and Lindberg-Møller 1981; Berthold et al. 1981; Kuwabara and Murata 1982) appear as rod-like appressed double membranes with no internal volume, due to complete internal stacking. The other preparations are more curved and some show a small but visible internal space (Andersson et al. 1978; Dunahay et al. 1984). One important difference is that the fragments obtained with mechanical disruption and phase partition are sealed vesicles (Andersson et al. 1978) in contrast to the detergent derived ones which are open at their edges (Dunahay et al. 1984; Lindberg-Møller et al. 1984). Another difference is that the content of cytochrome b/f complex differs in the various preparations. Whether this is due to selective removal of the complex during isolation, or to its location in the border region between appressed and non-appressed regions as suggested by Barber (1983) and Ghirardi and Melis (1983), remains to be established. The detergent-derived photosystem II particles obtained according to Berthold et al. (1981) show a substantial loss of lipids (Gounaris et al. 1983). Such a delipidation is likely to occur in all the detergent procedures, since one action of detergents on biological membranes is the displacement of lipids from the bilayer (Helenius and Simons 1975).

In conclusion, if large quantities of photosystem II thylakoids are required and there is no demand for a completely intact bilayer or sealed membranes, the detergent-based methods seem to the best choice. They have proven particularly advantageous for studies on polypeptides located at the inner thylakoid surface and with regulatory roles in oxygen evolution (Andersson 1985). On the other hand, if a more native or intact system is required, i.e., for studies on membrane organization, the press treatment and phase partition procedures are recommended. If a combined procedure for pure photosystem I and photosystem II thylakoids is desired, the nondetergent procedure of Henry and Lindberg-Møller (1981) is recommended.

4.3 Separation of Inside-Out and Right-Side-Out Thylakoid Vesicles with the Same Composition

For studies on the transverse asymmetry of thylakoid components the everted preparations described in Section 4.2 have certain drawbacks; (1) it is not easy to obtain the corresponding right-side-out thylakoids even if the inside-out vesicles turn right-side-out upon sonication (Sundby et al. 1982); (2) their high photosystem II enrichment excludes studies on the transverse organization of photosystem I components; (3) in the case of detergent-derived everted thylakoids they are not sealed and therefore not fully impermeable to probes such as proteolytic enzymes, chemical modifiers or antibodies. The ideal preparation for studies on the transverse asymmetry should give sealed inside-out and right-side-out vesicles with the same overall composition. Such vesicles can be obtained by Yeda press fragmentation of destacked and acid-treated thylakoids, followed by phase partition (Andersson et al. 1980):

Thylakoids were suspended in 10 mM Tricine pH 7.4, 100 mM sucrose and centrifuged at 1 000 *g* for 10 min. The well-drained pellet was resuspended to a chlorophyll concentration around 1 mg ml^{-1} in the same buffer and allowed to stand for 1.5 h at 4 °C. This leads to destacking and lateral randomization of the thylakoid membrane (Staehelin 1976). Then the pH of the suspension was dropped to pH 4.7 by the use of 0.1 M HCl. At this pH all the thylakoid membranes become paired, which is a prerequisite for the formation of inside-out vesicles upon fragmentation (Andersson et al. 1980). These low pH treated thylakoids are passed dropwise through the Yeda press twice at a nitrogen pressure of 10 MPa. Immediately after this treatment the pH of the suspension was brought back to 7.4 using 0.1 M NaOH. After centrifugation of the homogenate at 40,000 *g* for 30 min, the pellet was resuspended in 10 mM sodium phosphate pH 7.4, 5 mM NaCl, 100 mM sucrose and passed twice more through the press. This material was centrifuged at 1 000 *g* for 10 min to remove starch before being subjected to phase partition. 5 ml of the suspension (about 1 mg chlorophyll ml^{-1}) was added to a 20-g polymer mixture to yield a phase system of the following final composition; 5.7% (w/w) dextran T-500, 5.7% polyethylene glycol 3350, 10 mM sodium phosphate pH 7.4, 5 mM NaCl, 20 mM sucrose. The phase mixture was prepared by weighing into a centrifugation tube 16.44 g of a solution B and 3.56 g of the 40% (w/w) polyethylene glycol 3350. The solution B was prepared by mixing

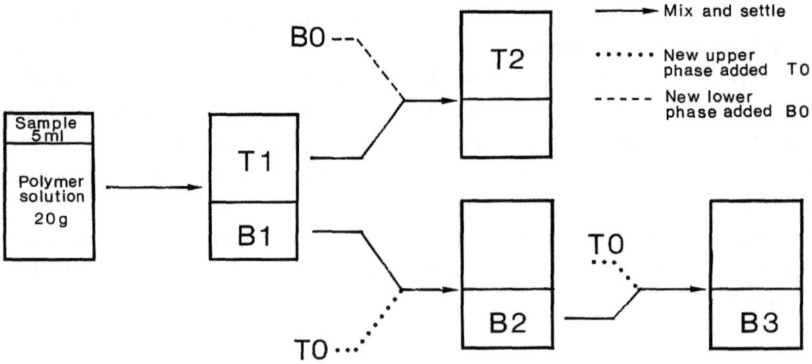

Fig. 7. Phase partition procedure for separation of right-side-out and inside-out thylakoid vesicles of the same overall composition. The T2 fraction contained the right-side-out vesicles while the B3 fraction contained the inside-out ones

71.25 g 20% (w/w) dextran T-500, 10.0 g 200 mM sodium phosphate pH 7.4, 10.0 g 100 mM NaCl and 73.13 g of water. After addition of the 5 ml sample the phase mixture was brought to 2 °–3 °C and mixed by 40 inversions of the tube. Phase settling was facilitated by a 1 000 *g* centrifugation for 3 min. The upper (T1) phase was removed from the lower (B1) phase (Fig. 7) and added to 9 ml of pure lower phase (B0). The B1 phase was supplied with 13 ml of pure upper phase (T0). The two tubes were shaken and the phase systems were allowed to settle, as described above yielding fractions T2 and B2 (Fig. 7). The B2 phase was repartitioned once more with 13 ml of the T0 phase yielding the B3 fraction. The T2 and B3 fractions were collected and diluted twofold with a suitable buffer, and centrifuged at 100,000 *g* for 30–60 min. The pure upper (T0) and lower (B0) phases were obtained from a 250 g bulk phase system. This was prepared by weighing into a separating funnel; 71.25 g 20% (w/w) dextran T-500, 35.63 g 40% (w/w) polyethylene glycol 3350, 12.5 g 200 mM sodium phosphate pH 7.4, 12.5 g 100 mM NaCl, 5.0 g 1 M sucrose and water to a final weight of 250 g. The solution was mixed at 2 °–3 °C and allowed to settle overnight. The upper and lower phases were collected and stored separately until use. As discussed in Section 4.2.1 the optimal polymer concentrations may vary from those given above, due to variations in the average molecular weight of different commercial polymer batches. Proton transport measurements revealed that the T2 material consisted of right-sided vesicles, while the B3 material consisted of inside-out thylakoid vesicles (Andersson et al. 1980). The yield of T2 material was around 80% while the yield of the B3 material was 5–10%. As judged from activity and compositional studies the two membrane populations had the same composition. Thus, the T2 and B3 vesicles differed only in sidedness. This preparation procedure has been used for studies on the transverse location of plastocyanin (Haehnel et al. 1981), LHC II (Andersson et al. 1982), the photosystem I complex (Anderson 1984) and the thylakoid galactolipids (Sundby and Larsson 1985).

5 Isolation of Thylakoid Supramolecular Complexes

This last section will be devoted to a brief description of the isolation of each of the four supramolecular complexes; the photosystem I, photosystem II, cytochrome b/f and ATP synthase complexes (Fig. 1). For the photosystem I and II complexes the isolation of their respective light-harvesting assemblies, LHC I and LHC II, will also be described.

5.1 The Photosystem I Complex and the Light-Harvesting Complex of Photosystem II (LHC II)

The first isolation and detailed characterization of a supramolecular photosystem I complex was reported by Bengis and Nelson (1975). Later Burke et al. (1978) from Triton X-100 solubilized pea thylakoids could obtain both the photosystem I and LHC II complexes. When the Triton solubilization and the subsequent isolation was done by sucrose density centrifugation in the absence of cations, the photosystem I complex aggregated and was located in the lower part of the gradient, while LHC II was nonaggregated and located at the top of the gradient. The location of the complexes in the gradient was reversed if the cations were included. The method of Burke et al. (1978) was slightly modified later by Mullet et al. (1980) as a means of simple and rapid purification of "native" photosystem I complex which has all its antenna chlorophyll still associated. This Triton X-100 photosystem I complex had a chlorophyll/P-700 ratio of 110. It includes the P-700-chlorophyll a-proteins of the inner core complex of photosystem I and a peripheral light-harvesting antenna which has specific chlorophyll a/b-proteins (LHC I) which are distinct from the main chlorophyll a/b proteins (LHC II). In addition, it contains several low molecular weight polypeptides. The LHC II contains at least two main polypeptides in the 23,000–28,000 mw region. It binds up to half of the total chlorophyll a and most of the chlorophyll b. The method of Ryrie (1983) described below is a modification of the procedures of Arntzens laboratory (1978, 1980) which enables both the photosystem I complex and LHC II to be isolated by a single-step sucrose density gradient centrifugation.

Thylakoids, prepared from 50 g of spinach leaves, were washed twice with 50 mM sorbitol, 0.75 mM EDTA pH 7.8 and once in 50 mM sorbitol, 5 mM Tricine pH 7.8 in order to remove residual cations from the thylakoid membrane. The EDTA-treated pellet was resuspended in water at a chlorophyll concentration of 0.8 mg ml^{-1} and Triton X-100 [20% solution (w/v), pH 7.5] was added to 0.75%. The mixture was stirred for 30 min on ice and centrifuged at 45,000 *g* for 15 min. The extract was then loaded onto 0.1–1.5 M linear sucrose gradients containing 1 mM Tricine pH 7.5, 0.04% Triton X-100 and centrifuged in a Beckman SW-27 rotor at 27,000 rpm for 16–20 h at 4 °C. Part of the upper band showing intense red fluorescence was LHC II and the lower, very weak fluorescent zone was the photosystem I complex.

The photosystem I complex was either diluted with extra 20 mM Tricine buffer pH 7.5 (10 vol buffer) and collected by centrifugation at high speed (e.g., 220,000 *g* for 30 min) in order to remove excess Triton, or dialyzed against

20 mM Tricine pH 7.5 for several hours at 4 °C. The photosystem I complex was stored in liquid nitrogen in 20 mM Tricine pH 7.5 containing 5% glycerol.

LHC II was precipitated from the upper zone by addition of 1 M $MgCl_2$ such that the final concentration of $MgCl_2$ was 10 mM. The mixture was stirred at room temperature for 15 min, cooled and placed above an equal volume of 20 mM Tricine pH 7.5, containing 0.5 M sucrose in 15 or 30 ml corex tubes. Following centrifugation at 10,000 *g* for 15 min, the pellet was resuspended in a small volume of 20 mM Tricine pH 7.5 to a chlorophyll concentration of 1 mg ml^{-1} for immediate use or storage in liquid nitrogen. If less aggregated LHC II is required, the pellet should be subjected to several washes with 5 mM EDTA pH 7.5 to remove Mg^{2+} ions.

An alternative way of isolating LHC II is by detergent-based phase partition (Albertsson and Andersson 1981). It has been shown by Albertsson (1973) that micelles of the detergent Triton X-100 partition to the upper phase of an aqueous dextran/polyethylene glycol two phase system. The basis for protein separation by such a Triton-containing phase system is the selective partition to the polyethyleneglycol-rich upper phase of solubilized proteins with bound detergent, while unsolubilized materials stay in the lower phase or at the interface (Fig. 8). Below follows a description of how this approach can be used for the purification of LHC II from the other thylakoid proteins due to the fact that it is more resistant to extraction into the upper phase (Albertsson and Andersson 1981).

A thylakoid pellet (2.5 mg chlorophyll) obtained after washings with 50 mM sodium phosphate pH 7.4, 5 mM $MgCl_2$, 50 mM sucrose was suspended in 6 ml of distilled water immediately before addition to the polymer solution. The whole sample was added to a mixture of 8.4 g 20% (w/w) dextran T-500, 2.64 g 40% (w/w) polyethylene glycol 8000 (earlier designated 6000), 1.20 g 200 mM sodium phosphate pH 6.8, 2.40 g 1 M NaCl, and 3.36 g of water. This yields a 24-g phase system of the following final composition; 7% (w/w) dextran T-500, 4.4% (w/w) polyethylene glycol 8000, 10 mM sodium phosphate pH 6.8, 100 mM NaCl and around 0.1 mg chlorophyll ml^{-1}. After mixing the solution at room temperature it was centrifuged at 3000 *g* for 5 min to facilitate phase settling. The clear upper phase was removed and replaced by 12 ml of an upper phase solution composed of 7% (w/w) polyethylene glycol 8000, 1% (w/w) Triton X-100, 100 mM NaCl, 10 mM sodium phosphate pH 6.8. After mixing and settling, the upper phase containing material enriched in photosystem I was removed. The partition with the Triton-containing upper phase solution was repeated six times. After the

Fig. 8. Principal for the isolation of hydrophobic membrane proteins in an aqueous polymer two-phase system containing Triton X-100. Micelles of this detergent partition to the polyethylene glycol rich upper phase. Thus proteins interacting with the detergent, such as a solubilized membrane protein, will get a high affinity for the upper phase

seventh extraction 12 g of upper phase solution without Triton X-100 was added. After mixing and settling the upper phase was removed and the same extraction repeated once. These two final steps were performed in order to remove excess unbound Triton from the LHC II material which was located at the interface as a thick layer. The upper and lower bulk polymer phases were removed by pipette and discarded. The interfacial material was diluted with 6 ml water and centrifuged at 10,000 *g* for 10 min. The supernatant was discarded and the pellet was washed with another 6 ml of water. The pellet was suspended in 0.03–0.1% Triton X-100 and centrifuged at 10,000 *g* for 10 min. The supernatant was collected and made 0.1 M with respect to $MgCl_2$. This precipitated most of the green material which was collected by centrifugation. The pellet contained essentially pure LHC II. This phase partition procedure is quite rapid and can be completed within 3 h including the removal of excess detergent. The method can also easily be scaled up for bulk purification of LHC II.

5.2 The Light-Harvesting Complex of Photosystem I (LHC I)

In 1983, Haworth et al. isolated the supramolecular chlorophyll a/b-protein complex associated with photosystem I, by dissociating the native Triton X-100 photosystem I complex with a detergent mixture of Zwittergent-16 and dodecyl-β-D-maltoside followed by sucrose gradient centrifugation. This complex has a chlorophyll a/b ratio around 3.6 and a fluorescence emission maximum at 735 nm at 77 K. On SDS-PAGE, LHC I contains 25,000, 23,500, 21,500, and 20,000 mw polypeptides. The Triton X-100 photosystem I complex of Mullet et al. (1980) was isolated as described above. The zone of photosystem I complex collected from the sucrose gradient was either dialyzed against 50 mM sorbitol overnight at 4 °C or immediately diluted with 40 mM Tricine pH 7.8 and centrifuged at 360,000 *g* for 1 h to remove the Triton X-100 and sucrose. In the former case the photosystem I complex was then collected by centrifugation at 40,000 *g* for 10 min. The collected photosystem I complex was then resuspended in 20 mM Tricine pH 7.8 to a chlorophyll concentration of 1.0 mg ml^{-1} and treated with an equal volume of the detergent mixture (Zwittergent-16; 4 mg ml^{-1} and dodecyl-β-D-maltoside, 3 mg ml^{-1}) (Calbiochem.) for 60 min with stirring at 4 °C. The detergent extract (3 ml) was loaded onto a 0.1–1 M sucrose density gradient (13 ml) containing 20 mM Tricine pH 7.8, 1% dodecyl-β-D-maltoside placed over a 2 M sucrose cushion (2 ml) and centrifuged at 100,000 *g* for 15–18 h in a Beckman SW-28 rotor at 4 °C. The green zone at the top of the gradient contained the LHC I and those further down the tube corresponded to partly dissociated photosystem I complex.

5.3 The Inner Core Complex of Photosystem II (CC II)

The first CC II complex to be fully characterized was that isolated by Satoh and Butler (1978), Satoh (1979), and Satoh et al. (1983). This procedure involves several detergent treatments, sucrose density gradient centrifugation, two ion exchange chromatographic steps and preparative isoelectric focusing. The method is time-consuming, costly, and not easy to perform. The rapid and simplified method of Westhoff et al. (1983) yields highly purified CC II. The procedure outlined below is based on this method as modified by Bricker et al. (1985). The isolated complex contains polypeptides of 53,000, 49,000, 45,000, 34–33,000, and 9,000 mw and was highly depleted in the LHC II 23–28,000 apopolypeptides. The 49,000 polypeptide may by the P-680-chlorophyll a-protein of photosystem II and the 43,000 polypeptide is the inner antenna chlorophyll a-protein of photosystem II. The 53,000 polypeptide appears to be a slower migrating form of the 49,000 protein. Intrinsic proteins in the 31–34,000 range include the herbicide-binding protein, the D_2 protein and the 9,000 polypeptide which is cytochrome b-559. The room temperature absorbance maximum of the CC II was at 674 nm and its fluorescence emission at 77 K was at 685 and 695 nm (Bricker et al. 1985).

The initial steps of the procedure of Bricker et al. (1985) follows the isolation of photosystem II particles according to Kuwabara and Murata (1982). Thylakoid membranes were suspended in 50 mM Mes-HCl pH 6.0, 400 mM sucrose, 5 mM $MgCl_2$, 15 mM NaCl together with the protease inhibitors PMSF, ε-amino caproic acid and benzamidine at 2–3 mg chlorophyll ml^{-1}. An aliquot of Triton X-100 (20%) was added to give a detergent/chlorophyll (w/w) ratio of 20:1. After mixing, the suspension was immediately centrifuged at 2000 *g* for 2 min and the supernate was recovered and centrifuged at 34,000 *g* for 10 min. The pellet from the latter centrifugation was resuspended in 50 mM Mes pH 6.0, 5 mM $MgSO_4$, 50 mM NaCl and centrifuged at 2000 *g* for 2 min, and the supernate centrifuged at 34,000 *g* for 10 min. This photosystem II preparation had high rates of oxygen evolution with exogenous electron acceptors. This preparation was then treated twice with 1.0 M Tris-HCl pH 9.25 at a chlorophyll concentration of 0.5 mg ml^{-1} for 20 min at 4 °C to remove the extrinsic 33,000, 23,000, and 16,000 mw polypeptides of oxygen evolution. This polypeptide-depleted preparation was resuspended in 50 mM Mes pH 6.0, 200 mM sucrose, 30 mM NaCl, 5 mM $MgCl_2$ plus the three protease inhibitors at a chlorophyll concentration of 1 mg ml^{-1}. The detergent, dodecyl-β-D-maltoside (Calbiochem) was added from a 20% stock solution to give a detergent/chlorophyll (w/w) ratio of 10:1. This mixture was incubated for 20 min at 4 °C with stirring, and then centrifuged at 34,000 *g* (little or no pelletable chlorophyll should be observed). The dodecyl maltoside extract (1.5–2.0 ml aliquots) was loaded onto 10–30% linear sucrose gradient containing 20 mM Mes-NaOH pH 6.5, 0.05% Triton X-100 and centrifuged in a Beckman SW-27 rotor at 24,000 rpm for 40 h. The upper dark zone of LHC II, which was highly fluorescent under an ultraviolet lamp, was separated from the lower, paler, weakly fluorescent zone of CC II.

5.4 The Cytochrome b/f Complex

Following the first isolation of a 125,000 mw cytochrome b/f complex without appreciable PQH_2:PC oxidoreductase activity by Nelson and Neumann (1972), Hurt and Hauska (1981) isolated a cytochrome b/f complex from spinach thylakoids that catalyzed electron transport from plastoquinol to plastocyanin. This complex is functionally analogous to the cytochrome b/c complex of mitochondria and photosynthetic bacteria. Several modifications of the original Hurt and Hauska preparation have been described, the most significant being the method of Clark and Hind (1983). The spinach cytochrome b/f complex of Hurt and Hauska contained five polypeptides on SDS-PAGE; 34,000 and 33,000 (cytochrome f), 23,500 (cytochrome b-563), 20,000 (Rieske Fe-S centre), and 17,400 (subunit IV). The molar ratios are 1 cytochrome f:2 cytochrome b-563:1 Fe-S center. The principal difference between this and the cytochrome b/f complex of Clark and Hind is that the latter also contains a 37,000 polypeptide which may corresponds to ferredoxin-NADP$^+$ oxido-reductase (FNR). Since the FNR copurified with the cytochrome b/f complex and comigrates on sucrose density gradients FNR has been suggested to be an integral component of the cytochrome b/f complex (Clark et al. 1984).

Isolation procedure according to Hurt and Hauska (1982): Spinach leaves (500 g) were homogenized in 600 ml of 50 mM Tricine pH 8.0, 200 mM sucrose, 10 mM NaCl (STN buffer). The homogenate was filtered and intact chloroplasts isolated by centrifugation at 1000 *g* for 2 min. The pellet was resuspended in different buffers and centrifuged repeatedly as follows; (1) 0.15 M NaCl (700 ml) and centrifuged at 13,000 *g* for 10 min. (2) STN buffer containing 2 M NaBr (chlorophyll concentration 1 mg ml^{-1}) and left for 30 min at 4 °C, after which an equal volume of cold water was added and the suspension was centrifuged at 13,000 *g* for 20 min. (3) as (2). (4) STN buffer (one l) and centrifuged at 13,000 *g* for 10 min. The washed thylakoid pellet was resuspended in STN buffer at a chlorophyll concentration of 3 mg ml^{-1} and diluted with an equal volume of STN buffer containing 60 mM octyl-β-D-glucoside (Sigma), 1% sodium cholate, 10% ammonium sulphate. The suspension was incubated with stirring at 4 °C for 30 min and then centrifuged at 300,000 *g* for 1 h in a Beckman Ti 60 rotor. Saturated ammonium sulphate at 4 °C was added to the supernatant to give 45% saturation, making allowance for the 10% ammonium sulphate initially present. The suspension was stirred at 4 °C for 10 min, and the precipitate was removed by centrifugation at 13,000 *g* for 10 min. Saturated ammonium sulphate solution was then added to 55% saturation, the mixture was stirred for 10 min and centrifuged at 13,000 *g* for 10 min. The supernatant was discarded while the pellet was dissolved in minimal volume of 30 mM Tris-succinate pH 6.5 containing 0.5% sodium cholate. This suspension was dialyzed against one l of the same buffer at 4 °C for no longer than 45 min.

Octyl-β-D-glucoside was then added to the dialyzed solution to a final concentration of 30 mM. The extract was then applied to a 7–30% (w/v) linear sucrose gradient containing 30 mM Tris-succinate pH 6.5, 0.5% sodium cholate, 30 mM octyl-β-D-glucoside, 0.1% soybean lecithin (w/v) and centrifuged in a Beckman SW-41 rotor at 40,000 rpm for 20 h. The reddish-brown bands of the cytochrome b/f complex were collected from the gradient and frozen in liquid nitrogen.

Isolation procedure according to Clark and Hind (1983): Spinach leaves (250 g) were homogenized in one l of 50 mM Tricine pH 7.8, 0.4 M sucrose. The homogenate was filtered through two layers of Miracloth or eight layers of wet cheesecloth, then centrifuged at 3000 *g* for 10 min. The pellet was resuspended by homogenization in 250 ml of 5 mM EDTA, 2 mM Tricine pH 7.8 and centrifuged at 15,000 *g* for 20 min. The pelleted thylakoids were resuspended in 0.5 mM EDTA, 0.2 mM Tricine pH 7.8 and stirred at 4 °C for 15 min, then centrifuged at 25,000 *g* for 20 min. The pellet was resuspended in 10 mM Tricine pH 8.2, 10 mM $MgCl_2$, at a chlorophyll concentration of 1.4–1.6 mg ml^{-1}. The EDTA-washed thylakoids were then diluted 1:1 with 60 mM octyl-β-D-glucoside, 1% sodium cholate, 100 mM Tris-HCl pH 8.2. The mixture was stirred at 0 °C for 1 h and centrifuged at 300,000 *g* for 1.5 h. The pellet was discarded and the protease inhibitor, PMSF, added to the supernatant to give a nominal concentration of 0.5 mM. The extract was then concentrated two- to threefold using an Amicon concentrator fitted with an XM-100A membrane. MES buffer was added to yield a final concentration of 20 mM using a 0.5 M MES pH 6.5 stock solution. The concentrated extract was passed through an ascorbate-reduced equine heart cytochrome c-Sepharose column (bed volume 8–10 ml) which had been pre-equilibrated with 20 mM sodium MES pH 6.5, 0.5% sodium cholate. Equine heart cytochrome c (Sigma) was coupled to cyanogen-bromide-activated Sepharose 4B-CP (Sigma) (Godinot and Gautheron 1979) or prepared according to Cuatrecasas (1980). After the gel had been poured, the column bed was washed with 0.2 M Tris-HCl ph 8.0, 0.4 M NaCl, 1% Triton X-100 until no more colour was eluted and then washed with water. The bed was resuspended in water and the gel was washed batchwise with 80 mM Tricine pH 7.5, 0.5% cholate, then in water and finally in 20 mM MES ph 6.5, 0.5% cholate. Sodium ascorbate was included in the last batchwise wash, and the gel was then poured and washed free of ascorbate with the MES/cholate buffer before use. The same washing procedure was used to regenerate the gel after each use, and the cytochrome c-Sepharose was stored in buffer containing 0.04% NaN_3.

Ten to twenty ml aliquots of the concentrated extract were run through the column at a flow rate of 1–2 ml min^{-1}. The column was held at room temperature and the eluent collected on ice so the extract was not allowed to warm for more than 20 min. The cytochrome b/f complex was eluted from the affinity column with 10 ml 20 mM MES ph 6.5, 0.5% sodium cholate. The eluant was brought to 40% saturation by the addition of solid ammonium sulphate. After stirring for 20 min at 0 °C, the suspension was centrifuged at 11,000 *g* for 20 min and the

Table 2. Purification of cytochrome b/f complex[a]

Purification step	Vol (ml)	Cyt f (μM)	Cyt f Chl	Cyt f protein (nmol mg^{-1})	Cyt b-563 Cyt f (mol mol^{-1})	Yield (Cyt f) %	Activity[b]
Chloroplasts	116	2.4	597	0.41	2.8	100	7.5
Detergent extract	210	0.63	7.6	0.98	2.1	48	7.5
Cyt c-sepharose eluate	128	1.05	6.5	1.38	2.0	48	7.5
60% ammonium sulphate	43	2.4	1.1	2.70	1.9	37	8.9
Bio-gel P300 eluate	69	13.2	1.1	8.25	1.9	33	9.0

[a] Clark and Hind 1983

[b] Plastoquinol-plastocyanin oxidoreductase activity (μmol h^{-1} nmol cyt f^{-1})

supernatant was brought to 60% saturation with ammonium sulphate. After stirring for 20 min, the precipitated protein was collected by centrifugation at 12,000 *g* for 20 min. The pellet was resuspended in a minimum volume (4 ml) of 30 mM octyl-β-D-glucoside, 0.5% sodium cholate, 30 mM Tris-succinate pH 6.5. The suspension, together with a grain of PMSF, was kept overnight at 0 °C. It was then reduced with sodium ascorbate and applied to a Bio-gel P-300 column (1.8 × 17 cm) which had been pre-equilibrated with 20 mM Tricine pH 7.5, 0.5% cholate. The cytochrome b/f complex was eluted with the same buffer as a dark brown band from which leading and tailing portions were discarded. The isolated complex was stored in liquid nitrogen in the presence of 5% (v/v) glycerol and a grain of PMSF. The yield was 200–360 nmol of cytochrome f per kg of leaves. It is stable for at least 6 months at 77 K, or 4 days at 4 °C. The yield and purity of the cytochrome b/f complex through the various preparations steps are summarized in Table 2.

5.5 The ATP Synthase (CF_0–CF_1)

The first successful attempt to isolate an active CF_0–CF_1 complex from thylakoid membranes was that of Carmeli and Racker (1973). Subsequently it was found that ribulose-1,5-bisphosphate carboxylase/oxygenase, a major contaminant in the preparation, could be removed prior to the main isolation procedure by washing the thylakoids with 0.15 M NaCl (Winget et al. 1977). The isolated ATP synthase has an apparent molecular weight of 435,000. The coupling factor portion (CF_1) is composed of five subunits of mw 60,000, 56,000, 37,000, 19,000, and 14,000 in a molar ratio of 2:2:1:1:2 while the intrinsic CF_0 portion contains three subunits of 15,000, 12,500, and 8,000 (Nelson 1982). Below follows a description of the procedure described by Pick and Racker (1979).

Thylakoids were washed twice in 0.15 M NaCl, 10 mM Tricine pH 8.0. Prior to detergent solubilization, the thylakoids were incubated with 50 mM DTT for 10 min at 4 °C. Thylakoid membranes at a chlorophyll concentration of 2 mg ml^{-1} were then stirred with a final concentration of 0.5% sodium cholate and 1% octyl-β-D-glucoside (Calbiochem) for 15 min at 0 °C. The suspension was then centrifuged at 230,000 *g* for 1 h and saturated ammonium sulphate (room temp.) added to 37% saturation. After stirring for 20 min at 0 °C, the suspension was centrifuged at 10,000 *g* for 10 min, the pellet discarded and the supernatant brought to 44% saturation with the saturated ammonium sulphate solution. After stirring for 20 min, the suspension was centrifuged as above, the supernatant carefully removed and the pellet dissolved in 30 mM Tris-succinate pH 6.5, 0.2% Triton X-100, 0.1 mM ATP, 0.5 mM EDTA, 0.1% sonicated soybean phospholipids so that the protein/Triton X-100 ratio was around 5:1. The extract was then applied to linear 7–30% sucrose gradients containing the same components as above, and centrifuged at 35,000 rpm for 15–18 h in a Beckman SW-40 rotor. Alternatively, the final purification step could be performed with a linear 7–40% sucrose gradient containing 0.4% sodium cholate in the 30 mM Tris-succinate pH 6.0, ATP, EDTA, and 0.1% soybean lipids as described above. A purification of five- to sixfold was obtained with a recovery of about 35% of the original 32 P_i-ATP exchange activity (Pick and Racker 1979). The extrinsic portion of the ATP synthase (CF_1) can be isolated from thylakoids by the EDTA extraction procedure of Lien and Racker (1971) as modified by Nelson (1980) to include a prior 0.15 M NaCl wash to remove contaminating ribulose-1,5-bisphosphate carboxylase/oxygenase.

References

Åkerlund HE, Andersson B (1983) Quantitative separation of spinach thylakoids into photosystem II-enriched inside-out vesicles and photosystem I-enriched right-side out vesicles. Biochim Biophys Acta 725:34–40

Åkerlund HE, Andersson B, Albertsson PÅ (1976) Isolation of photosystem II enriched membrane vesicles from spinach chloroplasts by phase partition. Biochim Biophys Acta 449:525–535

Albertsson PÅ (1971) Partition of cell particles and macromolecules, 2nd edn. John Wiley & Sons, New York

Albertsson PÅ (1973) Application of the phase partition method to a hydrophobic membrane protein, phospholipase A1 from *Escherichia coli.* Biochemistry 12:2525–2530

Albertsson PÅ, Andersson B (1981) Separation of membrane components in detergent containing polymer phase systems. Isolation of the light-harvesting chlorophyll a/b protein. J Chromatogr 215:131–141

Albertsson PÅ, Andersson B, Larsson C, Åkerlund HE (1982) Phase partition – A method for purification and analysis of cell organelles and membrane vesicles. In: Glick D (ed) Methods of biochemical analysis, vol 28. Wiley & Sons, New York, pp 115–150

Anderson JM (1975) The molecular organization of chloroplast thylakoids. Biochim Biophys Acta 416:191–235

Anderson JM (1984) Molecular organization of the chloroplast thylakoid membrane. In: Sybesma C (ed) Advances in photosynthesis research, vol 3. Nijhoff/Dr W Junk, The Hague, pp 1–10

Anderson JM, Boardman NK (1966) Fractionation of the photochemical systems of photosynthesis. Chlorophyll contents and photochemical activities of particles isolated from spinach chloroplasts. Biochim Biophys Acta 112:403–421

Anderson JM, Melis A (1983) Localization of different photosystems in separate regions of chloroplast membranes. Proc Natl Acad Sci USA 80:745–749

Andersson B (1984) Isolation of inside-out thylakoid vesicles with increased photosystem II purity – Lateral index of thylakoid components. In: Sybesma C (ed) Advances in photosynthesis research, vol 3. Nijhoff/Dr W Junk, The Hague, pp 223–226

Andersson B (1985) Proteins participating in photosynthetic water oxidation. In: Staehelin LA, Arntzen CJ (eds) Photosynthesis III. Encyclopedia of plant physiology, vol 19. Springer, Berlin Heidelberg New York (in press)

Andersson B, Åkerlund HE (1978) Inside-out membrane vesicles isolated from spinach thylakoids. Biochim Biophys Acta 503:462–472

Andersson B, Anderson JM (1980) Lateral heterogeneity in the distribution of chlorophyll-protein complexes of the thylakoid membranes of spinach chloroplasts. Biochim Biophys Acta 593:427–440

Andersson B, Åkerlund HE, Albertsson PÅ (1977) Light-induced reversible proton extrusion by spinach-chloroplast photosystem II vesicles by phase partition. FEBS Lett 77:141–145

Andersson B, Simpson DJ, Høyer-Hansen G (1978) Freeze-fracture evidence for the isolation of inside-out spinach thylakoid vesicles. Carlsberg Res Commun 43:77–89

Andersson B, Sundby C, Albertsson PÅ (1980) A mechanism for the formation of inside-out membrane vesicles. Preparation of inside-out vesicles from randomized chloroplast lamellae. Biochim Biophys Acta 599:391–402

Andersson B, Anderson JM, Ryrie IJ (1982) Transbilayer organization of the chlorophyll-proteins of spinach thylakoids. Eur J Biochem 123:465–472

Armond PA, Arntzen CJ (1977) Localization and characterization of photosystem II in grana and stroma lamellae. Plant Physiol 59:398–404

Arntzen CJ, Dilley RA, Peters GA, Shaw ER (1972) Photochemical activity and structural studies of photosystems derived from chloroplast grana and stroma lamellae. Biochim Biophys Acta 256:85–107

Barber J (1983) Photosynthetic electron transport in relation to thylakoid membrane composition and organization. Plant Cell Environ 6:311–322

Bengis C, Nelson N (1975) Purification and properties of the photosystem I reaction center from chloroplasts. J Biol Chem 250:2783–2788

Berthold DA, Babcock GT, Yocum CF (1981) A highly resolved oxygen-evolving photosystem II preparation from spinach thylakoid membranes. FEBS Lett 134:231–234

Berzborn RJ, Müller D, Roos P, Andersson B (1981) Significance of different quantitative determinations of photosynthetic ATP-synthase CF_1 for heterogeneous CF_1 distribution and grana formation. In: Akoyunoglou G (ed) Photosynthesis vol 3. Balaban, Philadelphia, pp 55–66

Blomquist G, Larsson C, Albertsson PÅ (1975) A study of DNA from chloroplasts separated by counter-current distribution. Acta Chem Scand Ser B 29:838–842

Boardman NK (1971) Subchloroplast fragments: Digitonin method. In: San Pietro A (ed) Methods in enzymology, vol 23. Academic Press, New York, pp 268–276

Boardman NK (1972) Photochemical properties of a photosystem II subchloroplast fragment. Biochim Biophys Acta 283:469–482

Boardman NK, Anderson JM (1964) Isolation of spinach chloroplast particles containing different proportions of chlorophyll a and chlorophyll b and their possible role in the light reactions of photosynthesis. Nature 203:166–167

Bricker TM, Pakrasi HB, Sherman LA (1985) Characterization of a spinach PS II core preparation isolated by a simplified method. Arch Biochem Biophys 237:170–176

Burke JJ, Ditto CL, Arntzen CJ (1978) Involvement of the light-harvesting complex in cation regulation of exitation energy distribution in chloroplasts. Arch Biochem Biophys 187:252–263

Carmeli C, Racker E (1973) Partial resolution of the enzymes catalyzing photophosphorylation. J Biol Chem 248:8281–8287

Clark RD, Hind G (1983) Isolation of a five-polypeptide cytochrome b–f complex from spinach chloroplasts. J Biol Chem 258:10348–10354

Clark RD, Hawkesford MJ, Coughlan SJ, Bennett J, Hind G (1984) Association of ferredoxin-$NADP^+$ oxidoreductase with the chloroplast cytochrome b–f complex. FEBS Lett 174:137–142

Cox RP, Andersson B (1981) Lateral and transverse organization of cytochromes in the chloroplast thylakoid membrane. Biochem Biophys Res Commun 103:1336–1342

Cuatrecasas P (1980) Protein purification by affinity chromatography: Derivatization of agarose and polyacrylamide beads. J Biol Chem 254:3059–3065

Dunahay TG, Staehelin LA, Seibert M, Olijivie PD, Berg SP (1984) Structural, biochemical and biophysical characterization of four oxygen evolving photosystem II preparations from spinach. Biochim Biophys Acta 764:179–193

Ford RC, Evans MCW (1983) Isolation of a photosystem II preparation from higher plants with highly enriched oxygen evolution activity. FEBS Lett 160:159–164

Gardeström P, Ericsson I, Larsson C (1978) Preparation of mitochondria from green leaves of spinach by differential centrifugation and phase partition. Plant Sci Lett 13:231–239

Ghanotakis DF, Babcock GT, Yocum CF (1984) Structural and catalytic properties of the oxygen-evolving complex. Correlation of polypeptide and manganese release with the behaviour of Z^+ in chloroplasts and a highly resolved preparation of the photosystem II complex. Biochim Biophys Acta 765:388–398

Ghirardi ML, Melis A (1983) Localization of photosynthetic electron transport components in mesophyll and bundle sheath chloroplasts in *Zea mays*. Arch Biochem Biophys 224:19–28

Godinot C, Gautheron DC (1979) Separation of right-side-out and inside-out submitochondrial particles by affinity chromatography on Sepharose-cytochrome c. In: Fleischer S, Packer L (eds) Methods in enzymology, vol 55. Academic Press, New York, pp 112–114

Good NE, Izawa S (1972) Hydrogen ion buffers. In: San Pietro A (ed) Methods in enzymology, vol 24. Academic Press, New York, pp 53–74

Gounaris K, Whitford D, Barber J (1983) The effect of thylakoid lipids on an oxygen-evolving photosystem II preparation. FEBS Lett 163:230–234

Haehnel W (1984) Photosynthetic electron transport in higher plants. Annu Rev Plant Physiol 35:659–693

Haehnel W, Berzborn RJ, Andersson B (1981) Localization of the reaction side of plastocyanin from immunological and kinetic studies with inside-out thylakoid vesicles. Biochim Biophys Acta 637:389–399

Hall DO (1972) Nomenclature for isolated chloroplasts. Nature New Biol 235:125–126

Haworth P, Watson JL, Arntzen CJ (1983) The detection, isolation and characterization of a light-harvesting complex which is specifically associated with photosystem I. Biochim Biophys Acta 724:151–163

Helenius A, Simons K (1975) Solubilization of membranes by detergents. Biochim Biophys Acta 415:29–79

Henry LEA, Lindberg-Møller B (1981) Polypeptide composition of an oxygen evolving photosystem II vesicle from spinach chloroplasts. Carlsberg Res Commun 46:227–242

Hincha DK, Schmidt JE, Heber U, Schmitt JM (1984) Two mechanisms of freeze-thaw inactivation of thylakoid membranes. In: Sybesma C (ed) Advances in photosynthesis research, vol 3. Nijhoff/Dr W Junk, The Hague, pp 47–50

Hurt E, Hauska G (1981) A cytochrome b/f complex of five polypeptides with plastoquinol-plastocyanin-oxidoreductase activity from spinach chloroplasts. Eur J Biochem 117:591–599

Hurt E, Hauska G (1982) Identification of the polypeptides in the cytochrome b/f complex from spinach chloroplasts with redox-center-carrying subunits. J Bioenerg Biomembr 14:405–424

Izawa S (1980) Acceptors and donors for chloroplast electron transport. In: San Pietro A (ed) Methods of enzymology, vol 69. Academic Press, New York, pp 413–434

Jacobi G (1971) Subchloroplast fragments: Sonication method. In: San Pietro A (ed) Methods in enzymology, vol 23. Academic Press, New York, pp 289–296

Kuwabara T, Murata N (1982) Inactivation of photosynthetic oxygen evolution and concomitant release of three polypeptides in the photosystem II particles of spinach chloroplasts. Plant Cell Physiol 23:533–539

Kyle DJ, Kuang TY, Watson JL, Arntzen CJ (1984) Movement of a sub-population of the light-harvesting complex (LHC-II) from grana to stroma lamellae as a consequence of its phosphorylation. Biochim Biophys Acta 765:89–96

Lamm E, Baltimore B, Ortiz W, Chollar S, Melis A, Malkin R (1983) Characterization of a resolved oxygen-evolving photosystem II preparation from spinach thylakoids. Biochim Biopyhs Acta 724:201–211

Larsson UK, Jergil B, Andersson B (1983) Changes in the lateral distribution of the light-harvesting chlorophyll a/b protein complex induced by its phosphorylation. Eur J Biochem 136:25–29

Leegood RC, Walker DA (1983) Chloroplasts. In: Hall JL, Moore AL (eds) Isolation of membranes and organelles from plant cells. Academic Press, London, pp 185–210

Lien S, Racker E (1971) Preparation and assay of chloroplast coupling factor CF_1. In: San Pietro A (ed) Methods in enzymology, vol 23. Academic Press, New York, pp 547–555

Lindberg-Møller B, Høj PB, Henry LEA (1984) Electron microscopic characteristics of photosystem II preparations and their inactivation and reactivation with respect to oxygen evolution. In: Sybesma C (ed) Advances in photosynthesis research, vol 3. Nijhoff/Dr W Junk, The Hague, pp 219–222

Michel JM, Michel-Wolwertz MR (1970) Fractionation and photochemical activities of photosystems isolated by broken chloroplasts by sucrose-density gradient centrifugation. Photosynthetica 4:146–155

Mullet JE, Burke JJ, Arntzen CJ (1980) Chlorophyll-proteins of photosystem I. Plant Physiol 65:814–822

Nelson N (1980) Coupling factors from higher plants. In: San Pietro A (ed) Methods of enzymology, vol 69. Academic Press, New York, pp 301–313

Nelson N (1982) Structure and function of the higher plant coupling factor in electron transport and photophosphorylation. In: Barber J (ed) Topic in photosynthesis, vol 4. Elsevier Biomedical Press, Amsterdam, pp 81–104

Nelson N, Neumann J (1972) Isolation of a cytochrome b_6-f particle from chloroplasts. J Biol Chem 247:1817–1824

Peters FALJ, Dokter P, Kooij T, Kraayenhof R (1981) Studies on stable energy-conserving photosystem I-enriched thylakoid vesicles. In: Akoyunoglou A (ed) Photosynthesis, vol 1. Balaban, Philadelphia, pp 691–700

Peters FALJ, Van Wielink JE, Wong Fong Sang HW, De Vries S, Kraayenhof R (1983) Studies on well coupled photosystem I-enriched subchloroplast vesicles. Content and redox properties of electron-transfer components. Biochim Biophys Acta 722:460–470

Pick U, Racker E (1979) Purification and reconstitution of the N,N-dicyclohexylcarbodiimide-sensitive ATPase complex from spinach chloroplasts. J Biol Chem 254:2793–2799

Ryrie IJ (1983) Freeze-fracture analysis of membrane appression and protein segregation in model membranes containing the chlorophyll-protein complexes from chloroplasts. Eur J Biochem 137:205–213

Sane PV, Goodchild DJ, Park RB (1970) Characterization of chloroplast photosystems 1 and 2 separated by a non-detergent method. Biochim Biophys Acta 216:162–178

Satoh (1979) Polypeptide composition of the purified photosystem II pigment-protein complex from spinach. Biochim Biophys Acta 546:84–92

Satoh K, Butler WL (1978) Low temperature spectral properties of subchloroplast fractions purified from spinach. Plant Physiol 61:373–379

Satoh K, Nakatani HY, Steinback KE, Watson J, Arntzen CJ (1983) Polypeptide composition of a photosystem II core complex – presence of a herbicide-binding protein. Biochim Biophys Acta 724:142–150

Shahak Y, Crowther D, Hind G (1980) Endogenous electron transport in broken chloroplasts. FEBS Lett 114:73–78

Shneyour A, Avron M (1970) High biological activity in chloroplasts from *Euglena gracilis* prepared with a new gas pressure device. FEBS Lett 8:164–166

Spencer DL, Wildman SG (1962) Observations on the structure of grana containing chloroplasts and a proposed model of chloroplast structure. Aust J Biol Sci 15:199–610

Staehelin LA (1976) Reversible particle movements associated with unstacking and restacking of chloroplast membranes in vitro. J Cell Biol 71:27–56

Sundby C, Larsson C (1985) Transbilayer organization of the thylakoid galactolipids. Biochim Biophys Acta 813:61–67

Sundby C, Andersson B, Albertsson PÅ (1982) Conversion of everted thylakoids into vesicles of normal sidedness exposing the outer grana partition membrane surface. Biochim Biophys Acta 688:709–719

Vernon LP, Shaw ER, Ke B (1966) A photochemical active particle derived from chloroplast by the action of the detergent Triton X-100. J Biol Chem 241:4101–4109

Vernon LP, Shaw ER, Ogawa T, Raveed D (1971) Structure of photosystem I and II of plant chloroplasts. Photochem Photobiol 14:343–357

Wehrmeyer W (1962) Elektronenmikroskopische Untersuchungen zur präparativen Gewinnung einer Granafraktion aus isolierten Chloroplasten. Z Naturforsch 17B:54–57

Wessels JSC (1968) Isolation and properties of two digitonin soluble pigment protein complexes from spinach. Biochim Biophys Acta 153:497–500

Wessels JSC, Van Alphen-van Waveren O, Voorn G (1973) Isolation and properties of particles containing the reaction centre complex of photosystem II from spinach chloroplasts. Biochim Biophys Acta 292:741–752

Westhoff P, Alt J, Herrmann RG (1983) Localization of the genes of the two chlorophyll a-conjugated polypeptides (mol. wt. 51 and 44 kD) of the photosystem II reaction centre on the spinach plastid chromosome. EMBO J 2:2229–2237

Winget D, Kanner N, Racker E (1977) Formation of ATP by the adenosine triphosphate complex from spinach chloroplasts reconstituted together with bacteriorhodopsin. Biochim Biophys Acta 460:490–499

Yamamoto Y, Ueda T, Shinkai H, Nishimura M (1981) Preparations of O_2-evolving photosystem-II subchloroplasts from spinach. Biochim Biophys Acta 679:347–350

The Isolation and Characterization on Nongreen Plastids

J. A. Miernyk

1 Introduction

Most studies of plastids have dealt with green plastids from photosynthetic tissues. This emphasis has been fully deserved considering the roles of chloroplasts in the photosynthetic assimilation of carbon, nitrogen, and sulfur (Jensen 1980), amino acid metabolism (Bryan 1976), fatty acid (Stumpf 1980), terpene (Kreuz and Kleinig 1981) and complex lipid (Mudd and Dezacks 1981) synthesis. Perhaps because of the accumulation of a large body of knowledge derived from these chloroplast studies, there has in recent years been an increasing interest in nongreen, nonphotosynthetic plastids.

The recent interest in nongreen plastids is manifold. Etioplasts are thought to represent an intermediate stage in the conversion of proplastids to chloroplasts, and as such have served as a model system for the study of this process. Recent reports on the biochemical nature and role of the prolamellar bodies (Dahlin et al. 1983; Lutz and Nordmann 1983; Ryberg et al. 1983) illustrate the usefulness of the model. Similarly, there has been considerable interest in the roles of nongreen plastids in purine and ureide metabolism in leguminous plants (Boland and Schubert 1983; Shelp et al. 1983), and in fatty acid synthesis and accumulation in developing oilseeds (Dennis and Miernyk 1982). Interest in acetate-derived compounds has led to examination of nongreen plastids in relation to pigment (Camara et al. 1982; Kreuz et al. 1982), terpene (Green et al. 1975), and hydrocarbon (Simcox et al. 1975) synthesis. Finally, studies of nongreen plastids and of the conversion of nongreen to green plastids (Leech and Leese 1982) have contributed to a beginning in the understanding of plastid genesis.

The purpose of this chapter is to point out some of the peculiarities of organelle isolation from nongreen tissues and hopefully to convey the author's enthusiasm for the study of these organelles. As the plastids from developing castor oil seeds are the nongreen plastids which have been studied to the greatest extent, some details of their isolation and characterization are presented. I hope that the information gained from these studies will stimulate others to take a greater interest in these and other nongreen plastids. Plant biology as a whole cannot help but be enriched by the results.

2 The Terminology of Nongreen Plastids

Plastid terminology was described in considerable detail by Kirk and Tilney-Bassett (1978), and has recently been briefly reviewed (Dennis and Miernyk 1982). It is nonetheless appropriate to define the usages applied herein.

2.1 Proplastids

Proplastids are small, colorless, undifferentiated plastids found primarily in meristematic tissues. Proplastids are the progenitors of the various types of mature or specialized plastids. When viewed in situ by transmission electron microscopy, a few small starch grains are occasionally present, but little internal membrane organization is visible.

Unfortunately, the term proplastid has frequently been used incorrectly. There has been a tendency to describe any nonpigmented plastids as proplastids, without consideration of the degree of maturation or specialization. A review of plant cell fractionation (Quail 1979) has even attempted to legitimize the misuse of proplastid as descriptive of "plastids derived from tissue that is not normally photosynthetic." The most prevalent misuse has been in the literature of the plastids from developing of germinating castor oil seeds (e.g., Wong and Benedict 1983). As noted by Garland and Dennis (1980), the generic plastid is an acceptable usage for castor oil seed plastids with leucoplast (Miernyk and Dennis 1983), being somewhat more descriptive. Of the extensive literature referring to proplastids, it is possible that only in the case of *Euglena* (Dockerty and Merrett 1979) have true proplastids been isolated and characterized.

2.2 Etioplasts

Plants grown in the complete absence of light are elongate, pale yellow, and are said to be etiolated. The plastids found in the leaf cells of such plants are etioplasts. The lipid composition of etioplasts resembles that of chloroplasts and in addition to the colorless polar lipids, carotenoids, protochlorophyll, and quinones are present (Dahlin et al. 1983; Ryberg et al. 1983). Viewed in situ etioplasts are irregularly elipsoid in shape (Lutz and Nordmann 1983) but when isolated are roughly spherical (Wellburn and Wellburn 1971). The most prominent morphological feature of etioplasts is the occurrence of one to four prolamellar bodies. These bodies appear to be made up of interconnecting tubules, presumably composed of membranous material. The term prolamellar body originated from the observation that during the etioplast to chloroplast transition there is an apparent conversion of these bodies into the thylakoids (Dahlin et al. 1983; Lutz and Nordmann 1983; Ryberg et al. 1983). While etioplasts may be regarded as a developmental state in chloroplast formation, they most probably pesist only during the abnormal interruption of this process. Interestingly, the cotyledons of large seeded plants, such as peas and beans, which serve only a storage function, do not develop etioplasts during dark germination.

2.3 Chromoplasts

Chromoplasts are plastids specialized for the accumulation of lipid-soluble pigments. Most fruits are yellow, orange, or red due to the accumulation of high concentrations of carotenoids in the chromoplasts. Chromoplasts are also responsible for the colors of many flower petals and specialized underground storage organs such as those of carrots and sweet potatoes. Chromoplasts are often very lipid-rich, for example Straus (1954) reported that isolated carrot root chromoplasts were 72% lipid by weight. Chromoplasts are approximately the size of chloroplasts, but range in shape from long and thin to near spherical. Electron-dense, osmiophillic inclusions can generally be observed in chromoplasts. Sitte (1977) has proposed a classification scheme based upon the inclusion ultrastructure; globular, fibrillar, membranous, reticulo-tubular, or crystalline. The functional significance of chromoplasts is not yet well understood.

2.4 Amyloplasts

The amyloplast is a mature nongreen plastid specialized as a storage depot for carbohydrate in the form of starch. Morphologically, amyloplasts are characterized by the lack of a well-developed internal membrane system, and the occurrence of starch grains. The number of starch grains within amyloplasts ranges from a single large grain, in potatoes, to many smaller grains in the plastids found in root cap cells. When photosynthetically produced carbohydrate is in excess of metabolic demands, it is transported into the amyloplasts and incorporated into the storage polymers. Later, when demand exceeds the supply available from photosynthesis, the starch is depolymerized and the carbohydrate monomers exported. Amyloplasts are common in specialized storage tissues such as cotyledons, endosperm, and tubers. Amyloplasts are additionally found in differentiated root tissues, where it has been proposed that they play a role in geotropic responses (Thomson and Whatley 1980).

2.5 Leucoplasts

Leucoplast is a general descriptive term indicating a lack of plastid pigmentation. No functional significance is implicit in this term. In contrast to proplastids, leucoplasts are larger, mature, and metabolically specialized. While proplastids are found in meristematic regions, leucoplasts are usually found in specialized tissues. In situ leucoplasts are generally ovoid but they are spherical when isolated under iso-osmotic conditions. Leucoplasts typically have a uniformly granular matrix containing numerous ribosomes, but with few osmiophillic globuli and lacking substantial internal membrane development.

An example of the sometimes confused state of plastid terminology is a recent report by Gleizes et al. (1983) on monoterpene hydrocarbon biosynthesis by isolated leucoplasts of *Citrofortunella*. While noting that they had actually isolated plastid-derived vesicles, the authors suggested that their work represented, "...

the first successful attempt in obtaining a leucoplast-enriched fraction." In fact, the isolation of intact leucoplasts from developing castor oil seeds occurred more than 10 years before (although at the time they were incorrectly described as proplastids), and there has been substantial study of their role in terpene metabolism (Green et al. 1975; Simcox et al. 1975).

2.6 Other Nongreen Plastids

Elaioplasts and proteinoplasts are nongreen plastids containing large quantities of lipid and protein respectively. Most observations of these plastids have been at the ultrastructural level (Kirk and Tilney-Bassett). Both types of plastid have a relatively limited distribution and nothing is known about their physiological or metabolic significance.

Plastids are extensively interconvertable and the reader should recognize that the previously described terminology is an attempt at a static definition of a dynamic system. Terms such as etiochloroplast and amylochloroplast are not uncommon in the literature, describing forms intermediate in plastid development. It is not always possible, or even desirable, to satisfy the demands of a rigid terminology, and the generic term plastid is often the most appropriate usage.

Thomson and Whatley (1980) have proposed a plastid developmental sequence which emphasizes the dynamic nature of these organelles. Their model uses the term eoplast as descriptive of the early stages of plastid development. Pregranal plastids (including etioplasts as a possible variant) represent possible stages prior to the development of mature chloroplasts. In aging plant leaves, chloroplasts or chromoplasts pass into a senescent stage, the gerontoplast. The developmental model certainly serves to emphasize plastid interconversion and is additionally useful in consideration of plastid evolution. It appears, however, to be based heavily on ultrastructural observations and may not adequately take into account metabolic considerations.

3 Basics of Plastid Isolation and Separation

3.1 Experimental Design

The reason for plastid isolation plays a major role in experimental design. Preparation of a crude, plastid-enriched fraction by differential pelleting may be an appropriate starting point from which to purify a plastid enzyme, but it is certainly not appropriate for detailed analysis of enzyme localization! Conversely, the low capacity of analytical gradients hardly makes them appropriate as a first step in the purification and analysis of any biochemical. The best method must be chosen based upon the specific goals of the experimenter.

Analytical studies are typically done by density gradient centrifugation, and this method has proven to be a powerful tool for plant scientists. There are, however, instances in which density gradient centrifugation is not very effective. Some chromoplasts are so lipid-rich that they will not sediment in aqueous solutions, and it is impossible to isolate intact amyloplasts by any centrifugal method. In such instances it is necessary to adopt specialized methodology such as flotation gradients, gel filtration, phase partition, etc.

Another major consideration in experimental design is the nature of the plant material to be studied. In many instances a biochemical process of limited distribution (e.g., purine/ureide synthesis) is the object of study and the experimenter has little choice in plant material. Even in these instances, tissues of a particular age or developmental stage may provide certain advantages. If, on the other hand, a widely occurring metabolic process is of interest, such as the development of the photosynthetic apparatus, it may be best to choose a plant such as maize, where a large amount of material is available, and from which etioplasts can be easily isolated (Leech and Leese 1982).

By careful choice of plant material and methodology it is possible to obtain intact plastids, from almost any source, in quantities and of sufficient purity for meaningful biochemical analysis.

3.2 Isolation Medium

The primary components of any isolation medium are a buffer for pH control, and some sort of osmotic support. When making a choice of the buffer system to be used it is, as always, necessary to consider the nature of the isolation. If one is interested in quantitation of plastid metabolites, or in in vitro metabolism by isolated, intact plastids, then phosphate- or pyrophosphate-based buffers should be avoided, as they promote loss of plastid metabolites (triose-phosphates, adenylates, etc.). If the localization of macromolecules is an experimental goal, then phosphate buffers have the advantages of good pH control, that pH is unaffected by temperature, and that they are inexpensive.

If a phosphate buffer is inappropriate, imidazole-based buffers are quite good. If commercial imidazole is to be used, it should first be purified by recrystallization to remove UV-absorbing and occasionally enzyme-inhibiting contaminants. Imidazole buffers can also be useful in control of metal ion concentrations (Perrin and Dempsey 1971).

Tricine buffers are extensively used in plant organelle isolations, but in the author's personal experience they have always resulted in lower yields of organelles and enzyme activities. The Good buffers (Good et al. 1966) can be very useful, but are substantially more expensive than phosphate- or imidazole-based systems. While the choice of pH is up to the individual investigator, values around neutrality give consistently good results.

Most nongreen plastids are more fragile than chloroplasts, perhaps due to the lack of a well-developed internal membrane system. For this reason greater osmotic support is required to maintain plastid integrity, 0.4 to 0.7 M osmoticum giving uniformly good results. Generally, sucrose, sorbitol, and mannitol are equally useful, however, commercial sucrose is contaminated with variable amounts of glucose and fructose and in some instances its use should be avoided (e.g., Miernyk and Dennis 1983). Potassium chloride is occasionally used as an osmoticum, primarily in work originating in the laboratory of R. J. Ellis (1981). This technique seems not to have been used in isolation of nongreen plastids, but could possibly be advantageous in some instances.

As a rule, the simplest isolation medium which *works* is the best one to use! For example, the rationale for inclusion of equimolar EDTA and Mg^{2+} is not obvious, although there are claims of benefit (Walker 1980). Divalent cations can affect the surface properties and aggregation of plastids (Nakatani and Barber 1977). These properties have occasionally been exploited for improved separations (Scott-Burden and Canvin 1975), however, in most instances the aggregation impedes separations and not only should divalent cations be excluded from the isolation medium, but chelators should be included.

There are reports of plastid membrane stabilization by KCl, and low concentrations of Ficoll (2–2.5% w/v) may also help stabilize isolated organelles.

When plant tissues contain large quantities of phenolic compounds, addition of either soluble or insoluble PVP is generally beneficial (Walker 1980). Ascorbate or isoascorbate (2 mM) and 2-mercaptoethanol (up to 14 mM) will additionally help to prevent phenol oxidation and subsequent organelle damage, and it may be beneficial to lower the pH of the isolation medium as phenolics bind more tightly to PVP at acidic pH values (Loomis 1974).

Generally, proteases are not a problem in isolation of intact organelles. If proteolysis results in reduced yields, inclusion of BSA (up to 5% w/v) in the isolation medium will help alleviate the problem. Specific inhibitors of each of the major classes of proteases are readily available from commercial sources, and their addition to the isolation medium may be of benefit. Particularly good inhibitors of serine proteases are available (leupeptin, 1 μM; p-amidinophenyl methylsulfonyl fluoride, 0.2 mM).

As a preliminary for developing a suitable plastid isolation medium I recommend; 50 mM imidazole – HCl, pH 6.9, 500 mM sucrose, 2% (W/V) Ficoll 400 and 1 μM leupeptin. Additions or deletions, as described above, can be made according to the demands of a particular plant material.

3.3 Tissue Disruption

The method of tissue disruption is, of course, dependent upon the plant tissue examined. Most commonly used methods rely upon the adaptation of various kitchen utensils (e.g., Waring-type blenders, onion grinders, electric knives, etc.). Blenders, even those adapted to use razor blades, generally result in relatively low yields. Similarly, short pulses with a Polytron-type tissue disrupter will release intact plastids, but a large proportion of the total organelles is damaged due to shear forces.

Some plant tissues, such as developing oilseeds, are relatively soft, and may be efficiently homogenized with a mortar and pestle. In certain instances it may be beneficial to include an abrasive, such as acid-washed silica, during homogenization.

High yields of intact organelles can be obtained by chopping the tissue with razor blades, but hand chopping with other than very small samples is exceedingly tedious. An excellent solution involves the adaptation of an electric knife to use two single-edged razor blades, allowing large amounts of material to be rapidly and efficiently processed.

Nongreen tissues are generally structurally less rigid than green tissues, and it is usually possible to obtain high yields of intact organelles. Ultimate yields will, however, result from the gentle disruption of protoplasts, and with difficult to isolate plastids such as amyloplasts protoplasts may be the only way to achieve any practical yield at all. Preparation of protoplasts, their purification and subsequent use in organelle isolation have been recently reviewed (Galun 1981). As with all procedures, a small temporal investment in optimization of methodology will yield increases in experimental return.

Regardless of the method of homogenization, some sort of filtration step should be included prior to centrifugation. Nylon mesh, cheesecloth, and Miracloth are commonly used, but regardless of the material of choice it should be moistened with isolation medium before use, otherwise recoveries will be reduced.

3.4 General Methods of Chloroplast Isolation

Chloroplasts have been separated from other organelles and cellular components by a variety of methods, including, phase partition (Larsson and Andersson 1979), gel permeation (Cobb 1977), free-flow electrophoresis (Dubacq and Kader 1968) and nonaqueous methods (Bowman and Dyer 1982). The most widely used method, however, has been density gradient centrifugation (Miflin and Beevers 1974). Centrifugal methods can be easily conducted on either a preparative or analytical scale, and may be based upon rate-zonal sedimentation, isopycnic banding, or a combination of the two. While sucrose gradients are the most flexible and widely used, colloidal silica is gaining increasing attention, and certain modified carbohydrates (Metrizamide, Nycodenz, etc.), or polymers (Ficoll) offer advantages in terms of density, viscosity, etc. In virtually all instances, methodology devised for the isolation and purification of chloroplasts can be successfully adapted for use with nongreen plastids. Specific examples of the separation and purification of plastids from developing endosperm of castor oil seeds are presented below.

4 Isolation of Nongreen Plastids from Developing *Ricinus* Endosperm

4.1 Rate-Zonal Sedimentation

The different organelles have characteristic sedimentation coefficients, and by manipulating centrifugation time and force it is possible to prepare fractions enriched in various components. This method is primarily useful as a preliminary step prior to further analysis by methods having greater resolving power but lesser capacity. In some instances, but only after a system is well understood through analytical evaluation, differential pelleting may yield organelle preparations sufficiently enriched that further purification is not required.

Chloroplast-enriched fractions can be prepared by brief low speed pelleting (Walker 1980). Nongreen plastids often have a smaller sedimentation coefficient, and longer centrifugation at higher g-forces is usually required to prepare fractions enriched in these fractions. Storage tissues from developing oil seeds contain protein bodies, very dense organelles which complicate isolation of plastid-enriched fractions.

Solution. Grinding medium, 50 mM imidazole, pH 6.9, 500 mM sucrose, 1 μM leupeptin, 2.5% (w/v) Ficoll.

4.1.1 Protocol

Ricinus endosperm, selected at the appropriate developmental stage (Ireland and Dennis 1981), is gently ground in a chilled mortar and pestle with two to two and a half volumes of grinding medium. As this tissue is soft and easily homogenized, more rigorous methods are not required. While yields from this simple method are generally acceptable (≧50%), they can be increased by chopping the tissue with an electric knife adapted to use single-edged razor blades. Very high yields are possible by first preparing protoplasts, but the amounts of hydrolytic enzymes required preclude this method when large quantities of material are to be processed. Homogenates are poured through eight layers of cheese-cloth moistened with grinding medium plus a single layer of Miracloth into precooled centrifuge tubes. Typically, the initial centrifugation is 500 g for 5 min. The resulting pellet ($P_{0.5}$) is primarily debris, unbroken cells, and protein bodies. The supernatant ($S_{0.5}$) is transferred to a clean tube and centrifuged at 10,000 *g* for 10 min. The supernatant plus floating lipid body fraction (S_{10}) is poured off and the inside of the tube wiped with absorbant tissue. The pellet (P_{10}) may be gently resuspended in a small volume of grinding medium using a fine paintbrush or a small cotton swab, followed by a brief period of gentle swirling. Uniform resuspension of the plastid pellet is critical for consistent results. Aggregates of organelles can lead to significant difficulties in interpretation of data. When a plastid is examined by density gradient centrifugation, it is advantageous to allow the resuspended plastids to sit in a conical tube in an icebath for 20 min prior to application of the sample to a gradient. In some instances it may be desirable to dilute the P_{10} with a larger volume of grinding medium and to repeat the 10,000 *g* centrifugation. The resulting washed pellet (WP_{10}) is substantially enriched in plastids (Table 1).

Centrifugation for removal of debris, preliminary preparation of plastid-enriched fraction, or isolation of crude plastids by rate-zonal sedimentation (WP_{10}), is described for 30-ml Corex tubes and a Sorvall HB-4 rotor used in a Sorvall preparative centrifuge. For use of other equipment, appropriate conversions for g-force (r_{max}), time, or volume can be easily made. Use of a swing-out rotor is recommended, to minimize wall effects, but acceptable results may be obtained with fixed-angle rotors (Sorvall SS-34, etc.).

Table 1. Isolation of a plastid-enriched fraction from developing *Ricinus* endosperm and purification by rate-zonal sedimentation

Marker enzyme	Total activity	Percent distribution				Re-cov-ery
		$S_{0.5}$	S_{10}	P_{10}	WP_{10}	
Alcohol dehydrogenase	26	100	98	2	0	100
Succinate dehydrogenase	24	100	80	18	6	106
Catalase	93	100	45	55	40	118
Choline phosphotransferase	9	100	88	2	2	90
Ribulose bisphosphate carboxylase	12	100	39	61	48	104
Triose-phosphate isomerase (plastid isozyme)	106	100	41	59	53	176

Total activity measurements are μmol min^{-1} g^{-1} fr. wt. except for catalase, which is in Lück units. Recoveries are calculated by summing each of the fractions. In several instances recoveries are in excess of 100%, reflecting the occurrence of an organelle-associated non-specific inhibitor (J. Miernyk and D. Dennis, in preparation)

Unless otherwise noted, all of the gradient centrifugations described here are routinely conducted using 25 × 89 mm Ultra-Clear tubes, a Beckman SW-28 rotor and a Beckman L8-55 preparative ultracentrifuge. If other rotors are to be used, the appropriate conversions for g-force (r_{av}) or volume can be easily made.

4.1.2 Analysis

The use of discrete marker characteristics is absolutely necessary for any valid interpretation of cell fractionation. Both morphological and biochemical markers may be used, but the greatest confidence is inspired when both types of marker are simultaneously monitored. Agreement between results obtained by cell fractionation with in situ ultrastructural or cytochemical data provides the ultimate support of experimental hypotheses. Evaluation of the purity of an organelle fraction requires the use of both positive and negative markers. In other words, not only must the occurrence of a particular fraction be demonstrated, but the lack of contamination by other fractions also. Quail (1979) has provided a particularly lucid treatment of the marker concept, and also a compendium of markers for use in evaluating plant cell fractionation.

Extensive analysis of the organelles from endosperm of germinating castor oil seeds by Beevers and associaties (Beevers 1979) has resulted in the general acceptance of the following organelle marker enzymes (details of the assays are provided in the references cited), cytosol, alcohol dehydrogenase (EC 1.1.1.1) (Donaldson et al. 1980); mitochondria, succinate dehydrogenase (EC 1.3.99.1) (Singer et al. 1973); peroxisomes, catalase (EC 1.11.1.6) (Lück 1965); endoplasmic reticulum, the phospholipid biosynthetic enzymes (Moore et al. 1973); plastids, ribulose 1,5-bisphosphate carboxylase (EC 4.1.1.39) (Benedict 1973). No suitable marker enzymes exist for the lipid and protein storage organelles from developing endosperm, but they can be evaluated using the chemical markers triacylglycerol and lectins or other storage protein subunits. Other enzyme activities may be useful for localization of the plastids on gradients (e.g., any of the glycolytic enzymes) but as these enzymes are present as isozymes in multiple compartments (Miernyk and Dennis 1982), they are difficult to use in yield evaluation. Acetyl-CoA carboxylase (EC 6.4.1.2) is located exclusively in the plastids (Scott-Burden and Canvin 1975), but appears at least in part membrane-associated (Finlayson and Dennis 1983) and may overestimate recoveries relative to stromal markers. *Ricinus* plastids are not significantly pigmented, precluding use of this as a useful marker characteristic. Reporting marker enzyme data must be in accord with the concept of quantitative cell fractionation (DeDuve 1964) and should include the activities of the total original homogenate and all derivative fractions.

The structural integrity of chloroplasts and nongreen plastids can be evaluated by phase-contrast light microscopy. All plastids are bounded by inner and outer envelope membranes and this characteristic can be useful in evaluation of integrity of the organelles by electron microscopy. While the functional integrity of chloroplasts may be determined by ferricyanide-dependent O_2 evolution (Lilley et al. 1975), no comparable assay exists for nongreen plastids. The use of plastid enzyme latency can be a suitable alternative (e.g., Echeverria et al. 1984).

In evaluating the isolation of plastid-enriched fractions protein must be quantitated. While measurement of absorbance at 280 nm is a simple method of analysis, it should not be used to the exclusion of standard chemical methods (Bradford 1976; Lowry et al. 1951). Refractometry can be used to monitor gradient fractions. It is completely inadequate to simply assume that what is prepared as a linear gradient is in fact perfectly linear.

4.1.3 Comments

Isolation of *Ricinus* leucoplasts by rate-zonal sedimentation has been used primarily as a preparative step and as such has been very valuable as a preliminary to isozyme evaluation (Simcox and Dennis 1978) and enzyme purification and characterization (Finlayson and Dennis 1983). More elaborate washing procedures (e.g., Pardo et al. 1980) do not significantly increase the purity of the plastid fractions isolated by rate-zonal sedimentation, but do substantially reduce yields.

It is perhaps notable that plastids isolated from developing castor oil seeds have a smaller sedimentation coefficient than plastids isolated from germinating endosperm. The basis of this difference remains obscure. The occurrence of ribulose 1,5-bisphosphate carboxylase in *Ricinus* leucoplasts is another interesting point. These plastids are not, and cannot become photosynthetically competent, and the enzyme does not seem to be active in vivo (Wong and Benedict 1983). While ribulose bisphosphate carboxylase serves as a useful marker for *Ricinus* endosperm plastids, it does not occur universally in nongreen plastids.

4.2 Isopycnic Banding on Linear Sucrose Gradients

Isopycnic banding is based upon centrifugation through a linear gradient until an equilibrium density is reached, determined by the chemical composition of the or-

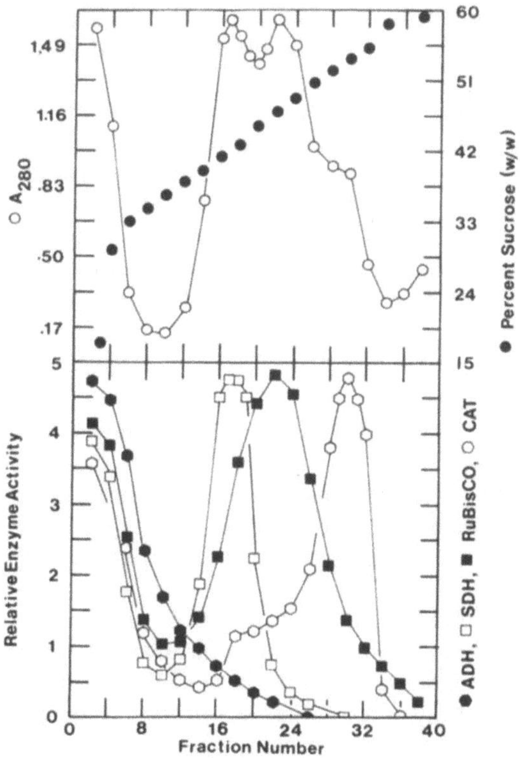

Fig. 1. Isolation of organelles from developing castor oil seeds by isopycnic gradient centrifugation. Forty g of developing endosperm were homogenized with a mortar and pestle, clarified by low-speed centrifugation, and a plastid-enriched fraction prepared by rate-zonal sedimentation at 10,000 *g* for 10 min. The resuspended plastids were loaded onto a 30-ml linear 35 to 55% (w/w) sucrose gradient and centrifuged at 100,000 *g* for 4 h using a Beckman SW-28 rotor. To calculate actual enzyme activities as units fraction^{-1}, multiply ordinate values by 0.77 for alcohol dehydrogenase (ADH), 0.66 for succinate dehydrogenase (SDH), 40 for ribulose-1,5-bisphosphate carboxylase (RuBisCO), and 0.98 for catalase (CAT). Activity of RuBisCO in fractions *24* through *34* represents 59% of that applied. Sucrose concentrations were determined by refractometry

ganelles and their permeability to the gradient media. By manipulation of the slope and extent of the gradient it is possible to separate all the major organelles.

Stock Solutions. 60% (w/w) sucrose in deionized water; 200 mM TES buffer, pH 7.5.

4.2.1 Protocol

Prepare 15 ml of 55% (w/w) sucrose containing 20 mM TES, pH 7.5. Prepare 15 ml of 35% sucrose containing 20 mM TES, pH 7.5. Place a 6 ml cushion of 60% sucrose in the bottom of a 39 ml centrifuge tube, overlayered by a 30 ml linear gradient of 35 to 55% sucrose. Up to 2.5 ml of sample (up to 40 mg protein) in homogenization buffer can be loaded onto a gradient. The gradient is then centrifuged for four hours at 100,000 *g*, followed by fractionation and analysis.

A typical separation of the organelles from a crude plastid-enriched fraction prepared from developing *Ricinus* endosperm is presented in Fig. 1.

4.2.2 Comments

In contrast to results obtained by methods based upon sedimentation coefficient, the behavior of *Ricinus* endosperm plastids on equilibrium sucrose gradients appears identical to that of green plastids (Miflin and Beevers 1974). The equilibrium position on these gradients corresponds to a buoyant density of 1.21 gm ml^{-1}.

Advantages of the equilibrium centrifugation method are its simplicity, and the ability to resolve all of the major organelles. Chloroplasts have been extensively studied using this method and it is thus possible to make certain assumptions and comparisons with the *Ricinus* leucoplasts.

Disadvantages include the relatively long centrifugation times required, the low capacity of linear gradients, and possible damage to plastid membranes from dehydration due to the high sucrose concentrations. Additionally, some sort of gradient-forming apparatus is required.

4.3 Rate-Zonal Sedimentation on Linear Sucrose-Magnesium Co-Gradients

Organelles have distinct sedimentation coefficients, based in part upon size, mass, and permeability. Relatively brief centrifugation can often be used to separate organelles which have different sedimentation coefficients when equilibrium centrifugation is ineffective.

While it is possible to purify the plastids from developing castor oils seeds by isopycnic banding on linear sucrose gradients, Scott-Burden and Canvin (1975) chose to consider the often reported (e.g., Nakatani and Barber 1977) but not well understood observations that divalent cations alter the surface properties, aggregation state and equilibrium position of plastids on sucrose gradients. Rate-zonal sedimentation on the resultant linear sucrose-magnesium co-gradients resulted in an excellent separation of the plastids from the other major organelles present in developing endosperm.

Stock Solutions. 60% (w/w) sucrose in deionized water; 200 mM TES buffer, pH 7.5; 1 M $MgCl_2$.

4.3.1 Protocol

Prepare 15 ml of 55% (w/w) sucrose containing 20 mM TES, pH 7.5 and 1 mM $MgCl_2$. Prepare 15 ml of 35% sucrose containing 20 mM TES, pH 7.5. Place a 6-ml cushion of 60% sucrose in the bottom of a 39-ml centrifuge tube, overlayered by a 30-ml linear gradient of 35–55% sucrose, 0–1 mM $MgCl_2$. Up to 2.5 ml of sample (up to 40 mg protein) in homogenization buffer (approximately 17% sucrose) can be loaded onto the gradient. The gradient is then centrifuged for 20 min at 100,000 *g*, followed by fractionation and analysis.

A typical separation of the organelles from a crude plastid-enriched fraction prepared from developing *Ricinus* endosperm is present in Fig. 2.

4.3.2 Comments

A prominent advantage of the linear sucrose-magnesium co-gradients is the ability to rapidly resolve all the major organelles. This power of resolution is particularly useful for analysis of the subcellular localization of enzymes.

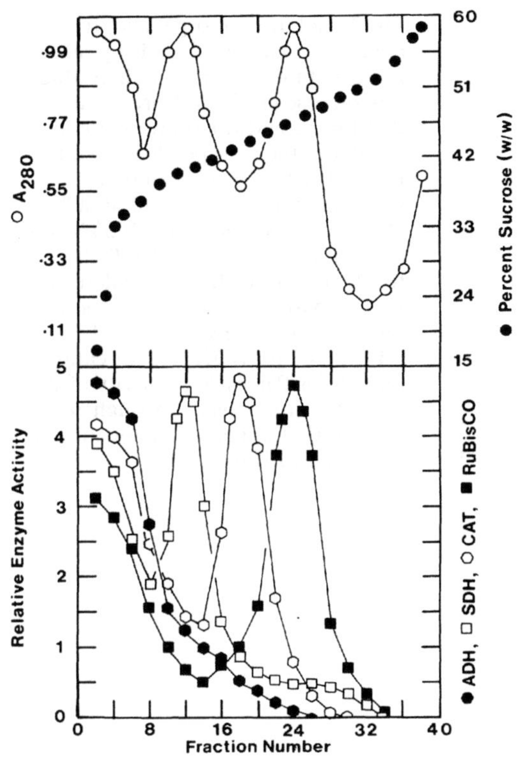

Fig. 2. Isolation of organelles from developing castor oil seeds by rate-zonal sedimentation on a linear sucrose-magnesium co-gradient. Thirty two g of developing endosperm were homogenized and the resuspended plastid-enriched fraction layered onto a 30 ml linear 35 to 55% sucrose, 0 to 1 mM $MgCl_2$ gradient and centrifuged at 100,000 *g* for 20 min. To calculate actual activities of marker enzymes in units $fraction^{-1}$ multiply ordinate values by 0.61 for ADH, 0.58 for SDH, 0.88 for CAT and 33 for RuBisCO. Activity of RuBisCO in fractions 18 to 30 represents 65% of that applied. Other details as described for Fig. 1

Disadvantages include the relatively low capacity of the linear gradients (overloading results in band broadening and decreased resolution), and that the final purified plastids are very dehydrated from the concentrated sucrose solutions. Additionally, some sort of apparatus for generation of the linear gradients is required.

4.4 Rate-Zonal Sedimentation on Discontinuous Sucrose Gradients

As previously discussed, the various organelles have distinct sedimentation coefficients and this characteristic may be used to separate the organelles on linear density gradients. In some instances separations can be improved by the introduction of discontinuities in the gradients. Organelles with smaller sedimentation coefficients tend to stop as they reach the discontinuities or "steps" in the gradients.

Stock Solutions. 60% (w/w) sucrose in deionized water; 200 mM TES buffer, pH 7.5, 1 M $MgCl_2$.

4.4.1 Protocol

For a 39-ml gradient, prepare 18 ml of 45% (w/w) sucrose containing 20 mM TES buffer, pH 7.5, and 1 mM $MgCl_2$, and 11 ml of 35% sucrose, 20 mM TES,

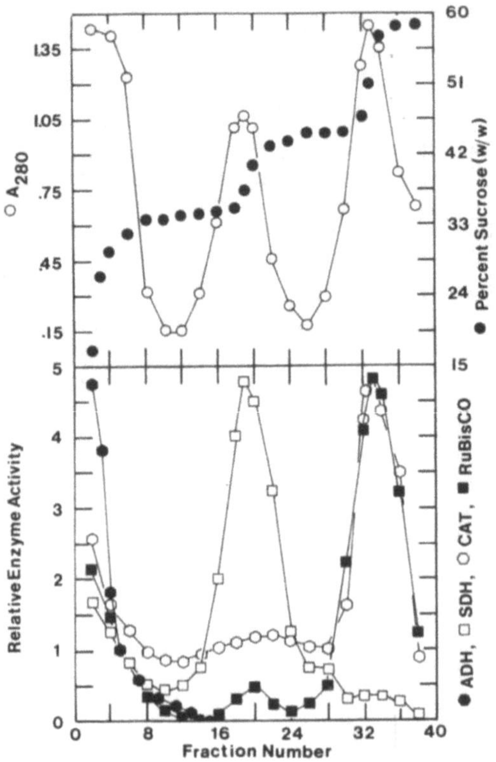

Fig. 3. Isolation of organelles from developing castor oil seeds by rate-zonal sedimentation on a discontinuous sucrose gradient. Forty-four g of developing endosperm were homogenized and the resuspended plastid-enriched fraction layered onto a gradient consisting of 6 ml of 60%, 18 ml of 45%, and 11 ml of 35% sucrose, and centrifuged at 100,000 *g* for 20 min. In order to calculate actual activities of marker enzymes, multiply ordinate values by 0.85 for ADH, 0.61 for SDH, 1.0 for catalase and 55 for RuBisCO. Activity of RuBisCO in fractions *28* through *38* is 81% of that applied. Other details as described for Fig. 1

1 mM $MgCl_2$. Place a 6-ml cushion of 60% sucrose in the bottom of the tube followed by layers of 45% and 35% sucrose solutions. Use of a thin cylinder of cork, of a diameter slightly smaller than that of the centrifuge tube, can result in very sharp boundaries between sucrose layers. The sucrose solutions should be gently pipetted on the floating cork which helps prevent mixing of the layers. A segment of wire forced through the cork cylinder before use will facilitate removal of the cork after the sample has been applied to the gradient. Up to 2.5 ml of sample (up to 90 mg protein) in homogenization buffer can be loaded onto a gradient. The gradient is then centrifuged for 20 min at 100,000 *g*, followed by fractionation and analysis.

A typical separation of the organelles present in a crude plastid-enriched fraction prepared from developing *Ricinus* endosperm is present in Fig. 3.

4.4.2 Comments

Advantages of rate-zonal sedimentation on discontinuous sucrose gradients are that separations are much more rapid than isopycnic banding, and that the capacity of the gradients is as least double that of any of the linear gradients. It is relatively simple to prepare reproducible gradients using the floating cork method.

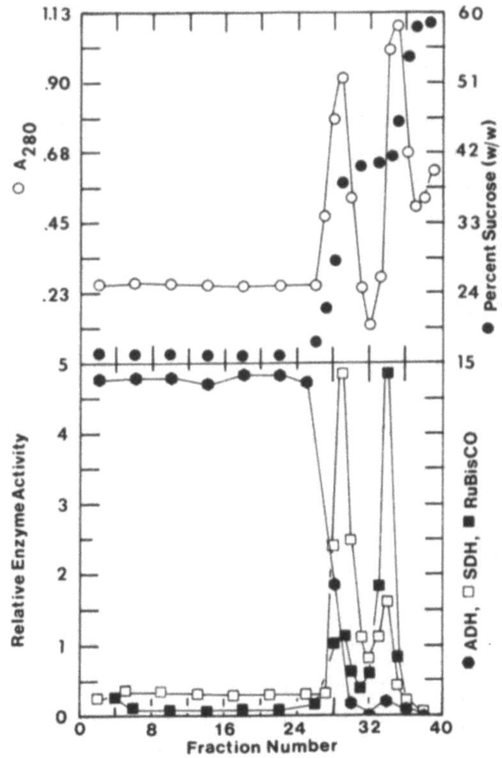

Fig. 4. Large-scale preparative isolation of plastids from developing castor oil seeds by rate-zonal sedimentation on a discontinuous sucrose gradient. One hundred and seventeen g of developing endosperm were homogenized and the resuspended plastid-enriched fraction layered onto a gradient consisting of 4 ml 60% and 7 ml 42.5% sucrose then centrifuged at 100,000 *g* for 20 min. To calculate actual activities of marker enzymes, multiply ordinate values by 0.39 for ADH, 0.33 for SDH and 114 for RuBisCO. For the sake of clarity, CAT activity is not plotted, but peak activity was coincident with peak RuBisCO activity

Inclusion of $MgCl_2$ in the sucrose solutions is necessary for consistent, sharp separations.

Disadvantages are that this method does not resolve plastids and peroxisomes, and in common with all sucrose gradients, damage to membranes due to dehydration is possible.

While rate-zonal sedimentation on discontinuous sucrose gradients is a useful analytical tool, the great capacity of these gradients allows them to be easily adapted for preparative use. By forming a gradient consisting of only 4 ml of 60% (w/w) sucrose overlayered with 7 ml of 42.5% sucrose, it is possible to load as much as 26.5 ml (as much as 300 mg protein) of a crude plastid-enriched fraction. The protocol is otherwise identical to that described for the analytical gradients. There is, of course, some loss of resolution when a gradient is so heavily loaded, but it is possible to obtain a fraction free of cytosolic contamination and with relatively low mitochondrial contamination (Fig. 4).

The state of the art in large-scale preparation of organelles involves use of a zonal rotor. It should be possible to load as much as 200 ml of a crude plastid-enriched fraction onto a sucrose gradient formed in a Beckman JCF-Z, or equivalent, rotor. While zonal rotors have been used for isolation of glyoxysomes (Bortman et al. 1981; Miernyk and Trelease 1981) and other plant organelles (Sautter et al. 1981), they seem not yet to have been used for plastid isolation.

4.5 Rate-Zonal Sedimentation on Discontinuous Percoll Gradients

The principals of separation are the same as for rate-zonal sedimentation on discontinuous sucrose gradients. The viscosity of the nonpenetrating, colloidal silica solutions is relatively low, however, allowing equivalent separations with significantly shorter centrifugation times. Additionally, the low osmolality of Percoll solutions allows gradients to be nearly iso-osmotic throughout.

Stock Solutions. Two hundred mM EDTA, pH 7.0; 1 M TES buffer, pH 7.5; 2 M sucrose; PB-Percoll solution [3% (w/v) polyethyleneglycol (PEG) 4000, 1% (w/v) bovine serum albumin, in 100% Percoll].

4.5.1 Protocol

From stock solutions, prepare gradient solutions which are 10, 22, 35, and 80% (all w/v) PB-Percoll. Each solution should include the following components at the indicated final concentrations, TES, 20 mM; EDTA, 2 mM; sucrose, 400 mM. For the 80% PB-Percoll solution it will be necessary to add solid sucrose, rather than the stock solution, in order to obtain the desired final concentration.

In a 30-ml Corex tube pipette a 5-ml cushion of the 80% PB-Percoll solution. This is then overlayered by 7.5 ml each of 35, 22, and 10% PB-Percoll solutions. The floating cork method, described for discontinuous sucrose gradients, should be used to form sharp boundaries between the PB-Percoll solutions.

Up to 2.5 ml of sample (up to 40 mg protein) may be layered onto a gradient. Centrifugation should be for exactly 3 min at 9500 rpm in a Sorvall HB-4 rotor, exclusive of acceleration or deceleration (without brake) time. After centrifugation the gradients in the glass Corex tubes can be fractionated by inserting a thin segment of tubing nearly to the bottom of the centrifuge tube and either pumping the gradient out into a fraction collector, or pumping a very dense solution (e.g., Maxidens, Nyegaard & Co.) into the tube and collecting the displaced fractions from the top. It is possible to construct a calibration curve to convert the refractive index of the Percoll to buoyant densities, however, it would seem equally useful to evaluate these gradients simply by the refractive index values. If the Percoll itself is thought to interfere with subsequent analyses, it is easily removed by centrifugation after diluting the fractions with several volumes of isolation medium.

Analysis of a typical separation of the organelles from a crude plastid-enriched fraction by rate-zonal sedimentation on a discontinuous Percoll gradient is presented in Fig. 5.

4.5.2 Comments

This protocol is based upon the method of Mullet and Chua (1983) for the isolation of pea chloroplasts, but has been optimized for use with developing *Ricinus* endosperm. The position of the peroxisomes on Percoll gradients is very sensitive to PEG in the gradient solutions. In gradients lacking PEG, peroxisomes migrate much further and are difficult to separate from intact plastids. The peroxisomes

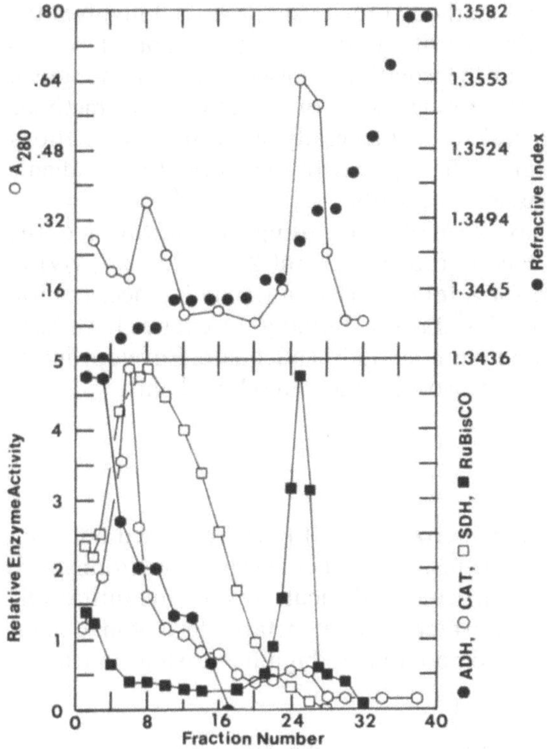

Fig. 5. Isolation of organelles from developing castor oil seeds by rate-zonal sedimentation on a discontinuous Percoll gradient. Seventeen g of developing endosperm were homogenized and the resuspended plastid-enriched raction layered onto a 27.5-ml gradient composed of 5 ml 80%, 7.5 ml 35%, 7.5 ml 22% and 7.5 ml 10% Percoll solutions, then centrifuged at 9500 rpm for 3 min using a Sorvall HB-4 rotor. To calculate marker enzyme activities in units fraction^{-1}, mutiply ordinate values by 0.29 for ADH, 0.21 for SDH and 0.53 for CAT. RuBisCO is in arbitrary units as determined by rocket immunoelectrophoresis. RuBisCO in fractions *21* through *26* is 81% of that applied to the gradient. Refractive index measurements were made with a Bausch and Lomb refractometer. Other details as described for Fig. 1

migrate less far into PEG-containing gradients and are cleanly separated from *Ricinus* leucoplasts.

Advantages of plastid isolation on discontinuous Percoll gradients are the very short centrifugation times required and the iso-osmotic conditions used throughout. The protocol can also be accomplished using a preparative centrifuge rather than the ultracentrifuge typically required for organelle separations. Additionally, organelles isolated on Percoll gradients appear somewhat more pure than with comparable sucrose gradients. There seems to be a "scrubbing" effect from passage through the colloidal silica.

Disadvantages of Percoll gradients are that they are relatively expensive, and have a lower capacity than sucrose gradients. Additionally, there are reports of inhibitors of enzyme activity or factors which adversely effect organelle integrity (e.g., Schmitt and Edwards 1982) even in the purified, PVP-coated colloidal silica. In some instances it may be necessary to dialyze or treat with activated charcoal before use.

Mills and Joy (1980) presented a simple, one-step method for the isolation of ntact plastids from pea leaf tissue by pelleting through a layer of 40% Percoll. While this method can be adapted for the isolation of *Ricinus* leucoplasts, the yields were lower than reported for chloroplasts and the purity of the isolated fraction was substantially less, primarily due to contamination with peroxisomes and protein bodies.

A modification of the centrifugal methods previously described involves the use of a vertical rotor and reorienting density gradient. Initial reports have appeared on the use of vertical rotors for the isolation of organelles from nongreen plant (Donaldson 1982; Gregor 1977) tissues. Any of the centrifugal methods may be adapted for use with a vertical rotor. Generally, equivalent separations can be obtained in much shorter times. The capacity of the reorienting gradients is slightly less than that of more classical gradients.

Promising preliminary results have been obtained using a modification of the method of Stocco (1983). Linear Ficoll gradients, 39 ml 2.5 to 22.5% (w/v) in 400 mM sucrose, were centrifuged for a total of 12 min (including acceleration time) at 10,000 r.p.m. in a Beckman Vti-50 rotor. Initially, recoveries have been good and separation of leucoplasts from mitochondria adequate. Envelope integrity appeared equivalent to that with the iso-osmotic Percoll gradients.

4.6 Nonaqueous Methods

Nearly all organelle isolation and characterization methods are carried out in an aqueous milieu. There can, however, be instances where such an environment is disadvantageous. For example, it is particularly difficult to study quantitatively the subcellular localization of small water-soluble metabolites. Nonaqueous methods can sometimes be used to avoid potential ambiguities arising from aqueous methods.

4.6.1 Isopycnic Banding on Linear Hexane-CCl Gradients

Generally, tissues are frozen in liquid N_2 and then freeze-dried. The dry powder is then homogenized in hexane, mixtures of hexane and CCl_4, or in some instances anhydrous glycerol. After filtration to remove debris, the extracts can be layered onto hexane/CCl_4 gradients and centrifuged as with conventional aqueous gradients. After a brief centrifugation, chloroplasts band at a specific gravity of approximately 1.05 in these gradients (Keys 1968).

The successful use of nonaqueous gradients for the isolation of chloroplast DNA has been recently reported (Bowman and Dyer 1982). The requirement for complete organelle intactness in order to obtain adequate yields is not as great with nonaqueous methods, because contaminating nucleases are inactive. *Ricinus* leucoplasts have been isolated (J. A. Miernyk, unpublished) by a method identical to that described for chloroplasts (Bowman and Dyer 1982). It has thus far been difficult to isolate uncontaminated, high molecular weight leucoplast DNA, and it might be beneficial to examine the nonaqueous method toward this end.

A significant disadvantages of nonaqueous methods is the volatility of the solvents, and the toxicity of the vapors.

A second method, one which is not truly nonaqueous but which does involve centrifugation through a nonaqueous phase, is silicon oil centrifugal filtration. This method is primarily used to rapidly separate organelles from an aqueous incubation medium, although a recent adaptation (Lilley et al. 1982) allows the separation of various organelles one from another as well as from the incubation medium.

4.6.2 Silicon Oil Centrifuged Filtration

Silicon oil filtration has been polarized by Heldt and his associates (reviewed, Heber and Heldt 1981), for the study of carrier-mediated transport and for quantitative analysis of organelle metabolite pool sizes. Relatively small samples are examined by this method, but by conducting parallel replicate experiments, it is possible to use marker enzymes as a measure of cross-contamination of fractions and to calculate recoveries.

In essence, a small volume of quenching solution is placed in a microcentrifuge tube, then overlayered by silicon oil of the appropriate density. Plastids in their aqueous incubation medium are then layered on top of the silicon oil. To terminate incubations, the tubes are centrifuged and the organelles pelleted through the silicon oil. Results may be evaluated either by product appearance in the pellet or substrate disappearance from the incubation medium. Nonspecific effects may be evaluated using ^{3}HOH and ^{14}C-sucrose, sorbitol, or dextrin.

The quenching solution can be 1.4 M perchloric acid for metabolite studies, or buffer solutions for measurement of enzyme activities. In either case, the tubes may be frozen in liquid N_2 after centrifugation and analyzed when convenient.

The choice of silicon oil to be used is best left to individual experimenters, as a wide variety of brands and mixtures have been reported in the literature. A few hours of trial and error will generally result in a protocol optimized for the particular organelles to be examined.

Results of metabolite uptake and incorporation experiments with isolated intact *Ricinus* leucoplasts were suggestive of carrier-mediated transport (Miernyk and Dennis 1983). While it was possible to measure the volume of the ^{3}HOH permeable, ^{14}C-sorbitol impermeable space by silicon oil centrifugal filtration (approximately 3 µl mg protein^{-1}, J. A. Miernyk and B. J. Shelp, unpublished) it was difficult to convincingly demonstrate carrier-mediated uptake of ^{14}C-3-phosphoglycerate or ^{14}C-glucose. Further studies will be necessary to better understand this important aspect of plastid metabolism. It will also be of interest to examine the apparent requirement for transport of 6-phosphogluconate by these plastids (Simcox et al. 1977).

The method of silicon oil centrifugal filtration is relatively simple and has been extensively used to study chloroplasts. The equipment and materials used are relatively inexpensive and readily available.

4.7 Noncentrifugal Methods

The centrifugal methods previously described are broadly useful in studies of isolated plant organelles, and have been successfully used for the isolation of both green and nongreen plastids from many different plant sources. In some cases, intact plastids can be isolated only by methods which avoid any centrifugal steps (e.g., amyloplasts). The great density of starch grains causes them to break through the plastid envelope during sedimentation. While the envelope membranes may reseal it is, especially in the case of low molecular weight compounds, difficult to assess losses accurately.

4.7.1 Gel Permeation

One separation method which does not require sedimentation is gel filtration. Although plastids from developing *Ricinus* endosperm do not contain starch grains, gel filtration was evaluated in comparison with the centrifugal methods.

Solution. Four hundred mM sucrose containing 20 mM TES buffer, pH 7.5, and 1 mM $MgCl_2$.

4.7.1.1 Materials

A chromatography tube (2.6 × 70 cm) with the appropriate flow adaptors, etc., a peristaltic pump, Sephadex G-50, Sephacryl S-1000.

4.7.1.2 Protocol

If Sephadex is to be used, it should be hydrated in the buffered sucrose solution overnight. Sephacryl is supplied hydrated and can be solution exchanged by two to three cycles of centrifugation followed by resuspension in buffered sucrose. Degassing of the slurry and pouring of the column should be done essentially as described by the manufacturer. When a suitable column of either Sephadex or Sephacryl has been prepared, two bed volumes of buffered sucrose should be passed through the column, loosely packing it. Up to 5 ml of a crude plastid-enriched fraction may be applied to the column, followed by several column volumes of buffered sucrose. Flow rates should be maintained at 1 ml min^{-1} for Sephadex columns and up to 3 ml min^{-1} when using Sephacryl.

Representative analyses of the separations of organelles in a crude plastid-enriched fraction prepared from developing *Ricinus* endosperm on columns of Sephadex G-50 (A) or Sephacryl S-1000 (B) are presented in Fig. 6.

4.7.1.3 Comments

The use of gel filtration to isolate intact chloroplasts was first reported by Wellburn and Wellburn (1971). The method was subsequently used by Cobb (1977) to isolate algal chloroplasts, and by other workers for other organelles (Andersson and Abrahamsson 1983; Harley et al. 1978). In each instance, the isolation of intact organelles was reported. Unfortunately, in no instance was the purity of the isolated organelles rigorously defined, nor were recoveries reported.

While it was possible to isolate intact plastids by Sephadex gel filtration, separation from other organelles was not as good as with the centrifugal methods. Similar experiments using Sephacryl S-1000 resulted in separations equal to those obtained with most density gradient methods.

Advantages of gel filtration are the ability to maintain iso-osmotic conditions throughout. Additionally, it is possible to isolate intact organelles without centrifugation. While Sephadex gel filtration did not adequately resolve plastids from other organelles, it was possible to isolate intact plastids free of cytosolic contamination.

Disadvantages of gel filtration are the amount of time required and the low recoveries. Additionally, once poured, a column may only be used two to three

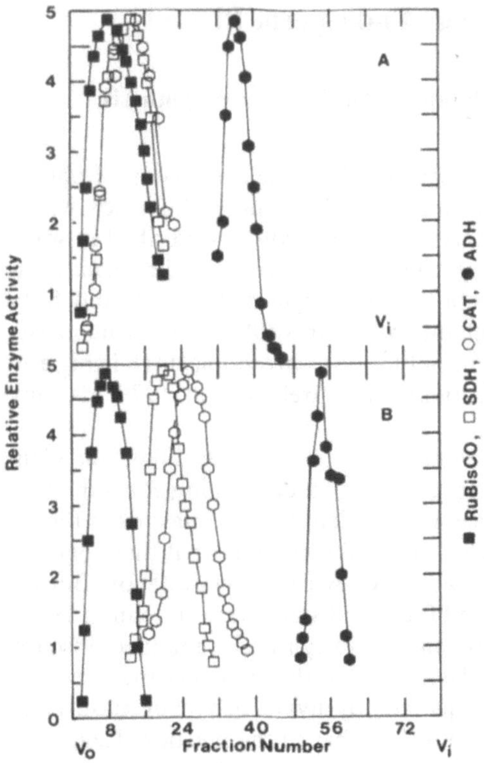

Fig. 6. Separation of organelles from developing castor oil seeds by gel filtration using Sephadex G-50 (**A**) or Sephacryl S-1000 (**B**). Recoveries of RuBisCO in fractions *2* through *20* (**A**) and *4* through *18* (**B**) were 29% and 37% of that applied to the columns, respectively. For clarity, only ADH activity is presented in the soluble protein regions of the elution profiles. V_o void volume; V_i included volume

times before it becomes too tightly packed and must be disassembled and repoured.

4.7.2 Phase Partition

Other potentially useful noncentrifugal methods include phase partition (Larsson and Andersson 1979), successfully used by B. d. Moore and R. N. Trelease (unpublished) to isolate plastids from developing cotton embryos when no centrifugal method had succeeded.

4.7.3 Unit-Gravity Sedimentation

Novel 1 g sedimentation methods have been developed by Fishwick and Wright (1980) to isolate amyloplasts from potato tubers, and Echeverria et al. (1984) to isolate amyloplasts from developing maize endosperm.

While it is possible to isolate intact plastids by all of the methods described herein, none is without flaws. No method should be considered absolute to the exclusion of further attempts at improvement. Continued progress by centrifuge and rotor manufacturers and the companies supplying density gradient media, and more importantly the ongoing ingenuity of the investigators, will result in a new generation of methodology which may render obsolete the methods which seem so useful today.

5 Metabolic Capabilities of *Ricinus* Endosperm Plastids

5.1 Glycolysis, the Pentose-Phosphate Pathway and Fatty Acid Synthesis

These plastids, highly differentiated and specialized for the conversion of photosynthetically produced carbohydrate to lipid, are the nongreen plastids which have been studied to the greatest extent. As such, they will be described in some detail as a model system. Available survey data support this use of the *Ricinus* leucoplasts as a qualitative model for plastid carbohydrate metabolism and fatty acid synthesis in developing oilseeds (Ireland and Dennis 1980), other leucoplasts (cauliflower floret, J. A. Miernyk, 1984, unpublished), chromoplasts (Kleinig and Liedvogel 1980), and probably green plastids as well (Ireland et al. 1979; Murphy and Leech 1977; Stitt and apRees 1980).

D. T. Canvin and his coworkers (Drennan and Canvin 1969; Zilkey and Canvin 1972) observed that fatty acid synthetic activity in developing *Ricinus* endosperm was particulate in nature and subsequently identified the plastids as the organelles involved. Further examination led to the observations that these plastids were the sole subcellular localization of acetyl-CoA carboxylase (Scott-Burden and Canvin 1975), acyl carrier protein and fatty acid synthetase (Drennan 1971). To this point the source of acetyl-CoA for fatty acid synthesis remained obscure, but further studies by the group of D. T. Dennis led to the discovery that developing *Ricinus* endosperm contained two distinct pyruvate dehydrogenase complexes, one located in the mitochondria and the other in the plastids (Reid et al. 1975). *Ricinus* plastid pyruvate dehydrogenase complex was subsequently characterized and its regulatory properties examined (Reid et al. 1975; Thompson et al. 1977a, b). The search for the source of pyruvate led to the then surprising discovery of a separate glycolytic pathway, distinct from typical cytosolic glycolysis, located in the plastids (Dennis and Green 1975). Further studies in Dennis's laboratory at Queen's University demonstrated that the *Ricinus* leucoplasts contain a complete glycolytic sequence from hexokinase through pyruvate kinase, and in each case the plastid and cytosolic glycolytic activities have been assigned to distinct isozymes (DeLuca and Dennis 1978; Miernyk and Dennis 1982; Simcox and Dennis 1978; Simcox et al. 1975) (Table 2). The *Ricinus* plastid and cytosolic glycolytic isozymes have all been purified and characterized to some degree (Garland and Dennis 1980; Ireland et al. 1980; Miernyk and Dennis 1983, 1984). Isolated, intact *Ricinus* leucoplasts were fully capable of the in vitro synthesis of fatty acids from free hexoses or glycolytic intermediates, supporting the enzyme localization data (Miernyk and Dennis 1983).

Similar results, with one significant exception, were obtained for the pentose-phosphate cycle enzymes, the exception being the apparent absence from these plastids of the enzyme catalyzing the first reaction of the pathway, glucose-6-phosphate dehydrogenase (Simcox and Dennis 1978). An isozyme of glucose-6-phosphate dehydrogenase was present in chloroplasts (Schnarrenberger et al. 1973; Srivastava and Anderson 1983), but all attempts to measure this activity in vitro from isolated *Ricinus* plastids have failed (Simcox and Dennis 1978). Recently, Yamada and his associates (Satoh et al. 1983) presented data which they

Table 2. Subcellular distribution of the *Ricinus* endosperm glycolytic enzyme activities

Enzyme	Activity (nmol min^{-1} $seed^{-1}$)	Percent distribution		
		Cytosol	Plastids	Mito-chondria
Invertase	50	100	0	0
Sucrose synthase	5	100	0	0
UDPG-pyrophosphorylase	90	88	12	0
Hexokinase	170	69	21	10
Glucosephosphate isomerase	1550	77	23	0
Phosphofructo-1-kinase	91	42	58	0
Inorganic pyrophosphate: fructose-6-phosphate 1-phosphotransferase	386	100	0	0
Aldolase	550	79	21	0
Glyceraldehyde-3-phosphate dehydrogenase	3655	86	14	0
Triose-phosphate isomerase	11675	91	9	0
3-Phosphoglycerate kinase	5215	86	14	0
Phosphoglycerate mutase	2375	80	20	0
Enolase	4185	69	31	0
Pyruvate kinase	2050	61	39	0
Pyruvate dehydrogenase complex	195	0	58	42

interpreted as evidence for glucose-6-phosphate dehydrogenase activity from the *Ricinus* plastids, suggesting that it had been previously overlooked due to nonoptimal assay conditions. A careful re-examination of the *Ricinus* leucoplasts using both the conditions of Simcox and Dennis (1978) and of Satoh et al. (1983) failed to detect any significant glucose-6-phosphate dehydrogenase activity. These observations were further extended using the very sensitive Enzyme Linked Immunosorbent Assay and, consistent with the original observations of Simcox and Dennis (1978), we were unable to detect any glucose-6-phosphate dehydrogenase protein in the isolated plastids (J. A. Miernyk, L. Peleato, D. T. Dennis, in preparation). It is unclear at this time whether the apparent lack of glucose-6-phosphate dehydrogenase activity in *Ricinus* leucoplasts is of metabolic or regulatory significance. It would, however, be of considerable interest to examine plastids of various types, from a variety of different plants, for glucose-6-phosphate dehydrogenase activity. It has been reported that plastids from germinating castor oil (Nishimura and Beevers 1981) and hazel seeds (Gosling and Ross 1979) also lack glucose-6-phosphate dehydrogenase activity.

Apart from glucose-6-phosphate dehydrogenase activity, *Ricinus* leucoplasts contained activities of each of the steps of the classical F-type pentose phosphate pathway. As with glycolysis, each of the plastid activities has been attributed to a distinct isozyme (e.g., Simcox and Dennis 1978). The function of the plastid pentose-phosphate pathway in developing oilseeds is specifically to provide NADPH for fatty acid biosynthesis (Agrawal and Canvin 1971).

5.2 The Calvin Cycle

Although they are not, and cannot become, photosynthetically competent, the plastids from developing *Ricinus* endosperm contain ribulose 1,5-bisphosphate carboxylase (Benedict 1973; Wong and Benedict 1983). While present in plastids from germinating castor oil seeds (Youle and Huang 1976), the plastids from developing endosperm lacked fructose 1,6-bisphosphatase activity (Simcox et al. 1975). Plastids from both sources lacked NADP-glyceraldehyde-3-phosphate dehydrogenase activity (Wong and Benedict 1983). The purpose of a partial Calvin cycle in a nonphotosynthetic tissue is not immediately apparent. It was originally thought that ribulose 1,5-bisphosphate carboxylase might be involved in the refixation of respiratory CO_2 (Benedict 1973), but Wong and Benedict recently concluded that this is not the case (1983). The most likely explanation would seem to be that in *Ricinus* endosperm plastid development is unavoidably linked to ribulose 1,5-bisphosphate carboxylase synthesis. It has been suggested that ribulose bisphosphate carboxylase in leaf tissue may serve as a storage protein (Kleinkopf et al. 1970). In the case of developing castor oil seeds, with their massive accumulation of well-defined storage proteins (Tully and Beevers 1976; Youle and Huang 1976), a reserve function for ribulose bisphosphate carboxylase seems unlikely. At any rate, the carboxylase serves as an excellent marker for plastid development and for the identification of isolated plastids. Ribulose bisphosphate carboxylase was absent from the nongreen plastids of developing sunflower seeds, cauliflower florets, and maize endosperm (J. A. Miernyk, unpublished), making the developing *Ricinus* endosperm plastids atypical, in this regard, among nongreen plastids.

5.3 Nitrogen Metabolism

Chloroplasts are in part specialized for nitrogen assimilation and subsequent metabolism (Jensen 1980), and Fowler and his associates (Emes and Fowler 1979a, b, 1983) have examined in some detail the participation of pea root leucoplasts in nitrite assimilation. In contrast, the developing endosperm of castor oil seeds receives reduced nitrogen in the form of amino acids from the leaves of the parent plant to be used for the synthesis of both catalytic and storage proteins. Under these conditions there is no need for the plastids to participate in the initial steps of nitrogen assimilation. Consistent with this, Lees and Dennis (1981) found that while *Ricinus* leaf chloroplasts contained a specific isozyme of glutamate dehydrogenase, developing seed plastids were devoid of this activity. All glutamate dehydrogenase activity in the developing endosperm was located within the mitochondria. Similarly, Hirel, and Miernyk (in preparation) found that *Ricinus* leucoplasts contained neither the activity nor the protein (tested immunochemically) of the typical chloroplast glutamine synthetase isozyme, although green tissue of castor oil plants did have the characteristic chloroplast isozyme. All glutamine synthetase activity in the developing endosperm could be attributed to the cytosolic isozyme. Chloroplasts and the developing *Ricinus* endosperm leucoplasts probably represent extremes of the spectrum of plastid involvement in nitrogen metabolism. While MacDonald and apRees (1983) recently reported the occurrence of nitrite reductase in amyloplasts isolated from suspension cultures of *Glycine max* cells, little other comparative information is available concerning nitrogen metabolism in nongreen plastids.

5.4 Terpenoid Metabolism

Plastids play a central role in the synthesis and metabolism of isoprene-derived products including pigments, sterols, hydrocarbons and plant hormones (Camara 1984; Sitton and West 1975; Spurgeon et al. 1984). West and his associates have examined several aspects of terpenoid metabolism in developing *Ricinus* endosperm. Green et al. (1975) noted the localization of isopentenyl pyrophosphate isomerase and several prenyl transferases in the plastids from developing endosperm. One of the prenyl transferases was a farnesyl transferase unique to the plastids. While studying gibberellin biosynthesis, Simcox et al. (1975) found that essentially all of the kaurene synthetase B activity was located in the *Ricinus* leucoplasts, implicating these organelles in hormone synthesis. Much of the recent work in West's lab has involved the diterpene phytoalexin casbene (Sitton and West 1975). It has been observed (P. D. Simcox and C. A. West, unpublished) that casbene synthetase is largely if not entirely localized in the plastids from developing *Ricinus* endosperm. It would seem that although unpigmented themselves, the developing castor oil seed plastids are nonetheless quite active in terpenoid metabolism.

6 Composition and Biochemical Properties

6.1 Structure

While one generally thinks of plastids as having a characteristic lens shape, this is really true only for chloroplasts (cf. Cobb 1977) and is probably due to their extensive internal membrane organization. Transmission electron micrographs of developing *Ricinus* endosperm revealed that while they have a typical plastid double membrane envelope, the leucoplasts have very few internal membranes (Fig. 7). In situ the plastids are pleiomorphic, generally appressed among the lipid and protein storage organelles. When the leucoplasts are isolated under iso-osmotic conditions, they appear completely spherical.

6.2 Protein Composition

From the section on metabolic capabilities it is obvious that the developing *Ricinus* endosperm plastids are quite active. The noted enzymatic activities account for approximately 50 polypeptide subunits. Electrophoretic analysis of leucoplast

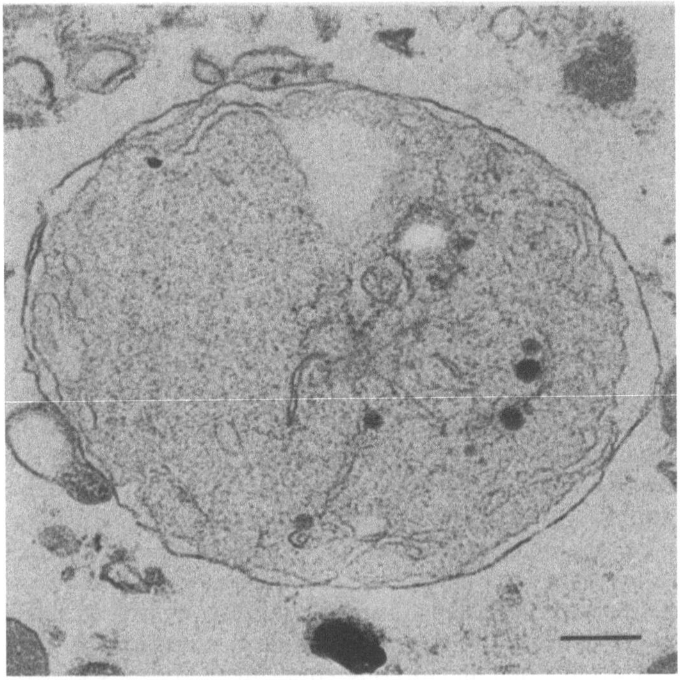

Fig. 7. Transmission electron micrograph of *Ricinus* leucoplast isolated by rate-zonal sedimentation on a discontinuous sucrose gradient. Bar = 1 μm. (B. J. Rapp, unpublished)

stromal preparations indicates a complexity greater than can be attributed to known metabolic activities. Two-dimensional gel analysis of stromal proteins (isoelectric focusing in the first dimension followed by SDS-polyacrylamide electrophoresis in the second) suggests > 100 discreet subunits (S. A. Boyle, 1983, unpublished). While many of the unidentified proteins are no doubt ribosomal proteins and other components of the plastid transcriptional and translational systems, there remains the possibility of the existence of as yet unidentified metabolic pathways.

In agreement with the results of Keegstra and Cline (1982) with pea chloroplasts, there is no evidence for the occurrence of glycoproteins in the *Ricinus* leucoplasts. Similarly, there is no evidence for leucoplast lipoproteins.

6.3 Membranes

Chloroplasts are unique among plant organelles in terms of membrane lipid composition. Chloroplast membranes are typified by a large proportion of galactosyl diacylglycerols (Roughan and Slack 1982), and both envelopes and internal membranes are largely composed of mono- and digalactosyl diacylglycerols. Developing *Ricinus* endosperm plastids lack a substantial internal membrane development, and the total membrane glycerolipid composition has some characteristics

Table 3. Acyl-lipid composition of leucoplasts isolated from developing *Ricinus* endosperm

Lipid class	Species	Percent fatty acid composition						Proportion of total lipids
		16:0	16:3	18:0	18:1	18:2	18:3	
Phospholipids								70.6
	PS	33.0	0	0	48.8	12.0	5.8	1.4+1.2[a]
	PI	52.0	0	0.2	6.0	28.8	12.0	12.0+1.5
	PC	41.2	0	6.2	2.0	39.5	10.1	23.7+1.3
	PE	39.8	0	4.1	4.9	40.5	9.8	25.1+1.5
	PG	54.0	0	4.0	8.8	28.0	4.4	4.7+0.3
	PA	28.9	0	0.2	62.0	4.5	2.8	3.5+0.7
Glycolipids								23.7
	MGDG	12.0	1.2	4.9	0.2	1.6	79.4	
	DGDG	9.9	0.9	3.1	0.4	1.2	84.1	
Neutral lipids								5.7
	DG	33.0	0	1.3	32.1	32.0	1.1	
	SE	36.4	0	54.3	9.0	0.3	0	

Abbreviations, PS, phosphatidylserine; PI, phosphatidylinositol; PC, phosphatidylcholine; PE, phosphatidylethanolamine; PG, phosphatidylglycerol; PA phosphatidic acid; MGDG, monogalactosyldiacylglycerol; DGDG, digalactosyldiacylglycerol; DG, diglyceride; SE, sterol ester

[a] Mean ± SD, n = 4

typical of chloroplasts, as well as of other organelles (Table 3). The fatty acid composition of the galactolipid fraction from developing *Ricinus* leucoplasts is similar to that of chloroplasts, containing primarily 18:3. The fatty acid composition of the phospholipids is mainly 16:0 and 18:2 (Table 3), typical of membrane phospholipids (Roughan and Slack 1982). The phospholipid composition of the developing endosperm plastids was nearly identical to that reported by Donaldson and Beevers (1977) for plastids isolated from germinating castor oil seeds.

There has been considerable recent interest in chloroplast membrane proteins, both in terms of function (Cline et al. 1981) and of plastid biogenesis (Flugge and Heldt 1984). Comparative studies were recently initiated on the developing *Ricinus* endosperm leucoplasts (J. A. Miernyk, S. A. Boyle, D. T. Dennis, 1983, unpublished). Plastids isolated by rate-zonal sedimentation on discontinuous sucrose gradients were subfractionated according to Cline et al. (1981) and analyzed by SDS polyacrylamide gel electrophoresis (Fig. 8). Preliminary results indicated a complexity as great as found with chloroplasts (Cline et al. 1981). Independent studies suggested the occurrence of a phosphate-ester transporter (Miernyk and Dennis 1983) and of galactosyl transferase activity (J. A. Miernyk 1983, unpublished) associated with the *Ricinus* leucoplast envelope, but the subunit peptides of neither has yet been unequivocally identified. Acetyl-CoA carboxylase activity seemed to be in part membrane-associated in developing *Ricinus* plastids (Finlayson and Dennis 1983), most likely via the biotin carboxyl carrier protein component. There was no acetyl-CoA carboxylase activity associated with isolated envelope vesicles (J. A. Miernyk and S. A. Finlayson 1982, unpublished), suggesting that the association may be with the rudimentary internal membrane system.

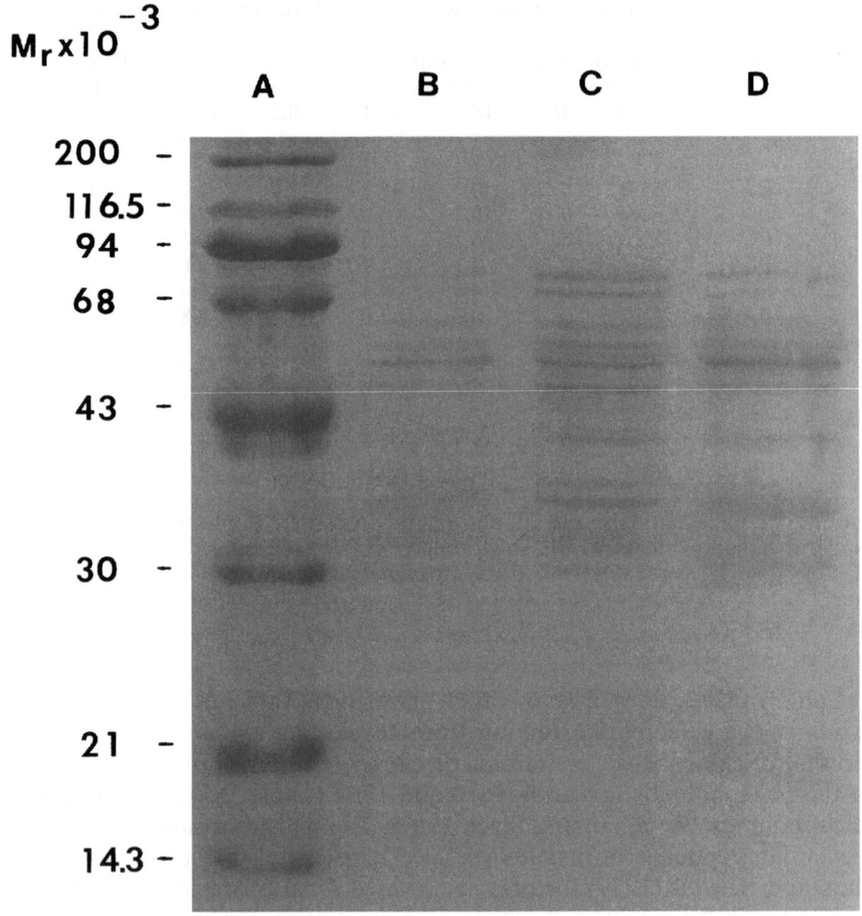

Fig. 8. Analysis by SDS-PAGE of *Ricinus* leucoplasts isolated by rate-zonal sedimentation on a discontinuous sucrose gradient, then subfractionated by the method of Cline et al. (1981). Track *A* molecular mass standards; Track *B* isolated plastid envelopes; Track *C* intact plastids; Track *D* membrane-free stromal fraction. The 10% acrylamide gel was run using the buffer system of Laemmli, stained with Coomassie R-250, then destained with 5% acetic acid

Various glycolytic enzymes from mammalian tissues have been reported to be membrane associated (Felgner et al. 1979; Font et al. 1975; Wilson 1978). The instances studied in the greatest detail have been the association of glyceraldehyde-3-phosphate dehydrogenase with red cell (Gilles 1982) or lense (Lenstra and van Raaig Bluemendal 1982) membranes, and the association of hexokinase with the mitochondrial outer membrane (Felgner et al. 1979; Wilson 1978). Preliminary reports have suggested the membrane association of certain chloroplast glycolytic isozymes (Rudd and Anderson 1982). Possible membrane association was examined in detail with *Ricinus* leucoplasts and pea seedling etioplasts (J. A. Miernyk

Table 4. Subfractionation of plastids isolated from etiolated pea seedlings and developing castor oil seed endosperm

	Ricinus leucoplasts		Pea etioplasts	
	Soluble	Membrane-associated	Soluble	Membrane-associated
Hexokinase	7.1	0.1	–	–
Glucose-phosphate isomerase	171.3	0.0	–	–
Phosphofructokinase	72.0	0.2	68.4	0.3
Aldolase	19.8	0.2	–	–
Triose-phosphate isomerase	180.3	0.0	172.3	0.7
Glyceraldehyde-3-phosphate dehydrogenase	15.0	0.1	6.4	0.2
Phosphoglycerate kinase	13.5	0.0	1.5	0.0
Phosphoglycerate mutase	8.8	0.1	0.3	0.0
Enolase	15.6	0.1	3.0	0.0
Pyruvate kinase	8.8	0.0	0.7	0.0

Unpublished observations of J. Miernyk and W. Hekman, 1983. Enzymes were assayed according to (Miernyk and Dennis 1982, Simcox et al. 1977). Activities are nmol/min/isolated plastid fraction.

and W. E. Hekman 1983, unpublished). No evidence of membrane association of any glycolytic isozyme was found with either type of nongreen plastid (Table 4).

While metabolism within *Ricinus* leucoplasts has been extensively studied, little is known about the transport capabilities of these organelles. Future studies of the function of the plastid membrane proteins will greatly increase our understanding of the overall metabolic capabilities of these organelles, and of their interrelationships with other organelles.

6.4 Nucleic Acids

The knowledge of chloroplast DNA, and of the structure and expression of chloroplast genes has advanced tremendously in recent years (reviewed Whitfield and Bottomley 1983). In contrast, virtually nothing is known about the DNA of nongreen plastids. As the developing *Ricinus* endosperm plastids contain ribulose bisphosphate carboxylase, it has been assumed that, in analogy to chloroplasts, they also contain DNA, including at least the large subunit gene. The isolation and characterization of *Ricinus* leucoplast DNA has recently begun in D. T. Dennis's laboratory at Queen's University (Thomson et al. 1983). While *Ricinus* leaf chloroplasts and seed leucoplasts are differentiated for distinct metabolic specialties, preliminary results suggest that the two plastid genomes are very similar, if not identical. Comparisons thus far are based upon restriction analysis of DNA isolated and purified from each type of plastid (A. Thomson and D. T. Dennis, unpublished), and size comparisons based on contour lengths (Fig. 9). The size of the *Ricinus* plastid genome (chloroplast and leucoplast) is approximately 136 kilobase pairs (A. Thomson and D. T. Dennis, in preparation). A coarse restric-

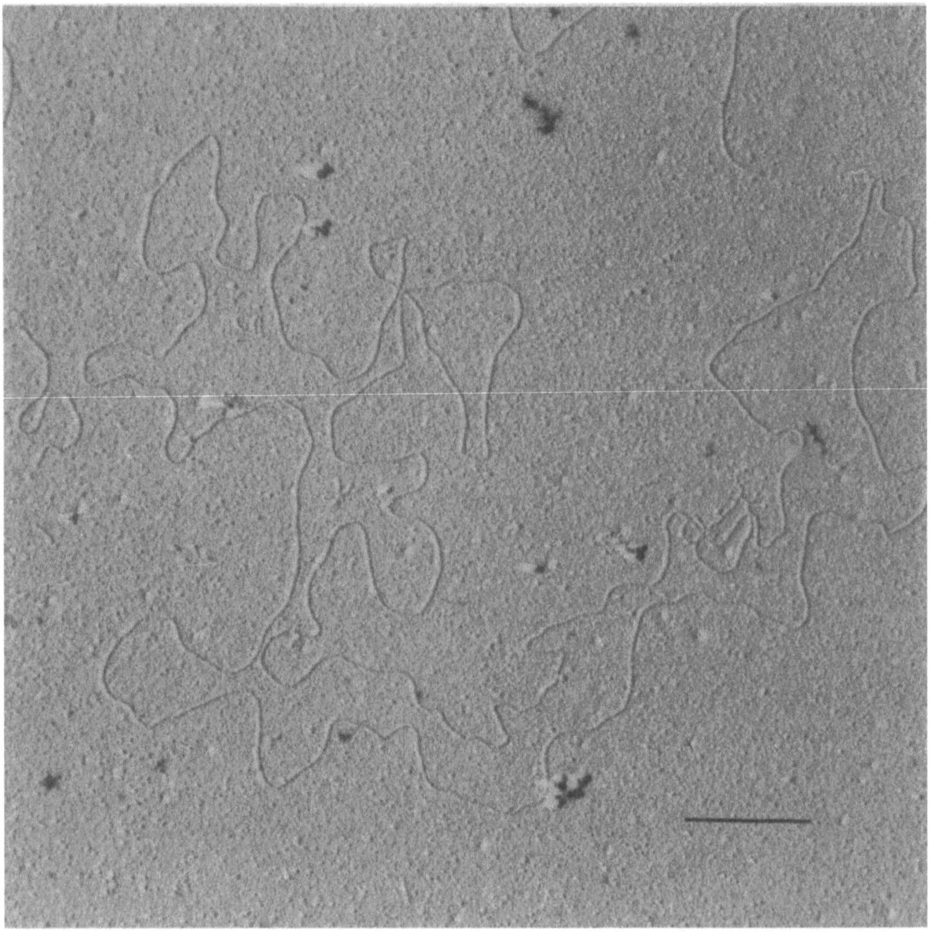

Fig. 9. Electron microscopy of a *Ricinus* leucoplast DNA-cytochrome c mixture. Plastids were isolated by rate-zonal sedimentation on a discontinuous sucrose gradient, disrupted, and the nucleic acids extracted and purified. Bar = 1 μm

tion map of *Ricinus* leucoplast DNA has been prepared, and several restriction fragments cloned into the plasmid pAT153. In progress is the development of a fine restriction map and the location and characterization of various plastid genes.

As the differences between *Ricinus* chloroplasts and leucoplasts cannot be attributed to plastid chromosome structure, the differentiation must be due to differential regulation of plastid gene expression, or to the contribution of the nuclear genome. Evidence is accumulating which suggests that the plastid glycolytic and pentose-phosphate cycle isozymes are encoded by nuclear DNA, synthesized in the cytosol and taken up post-translationally by plastids (Feierabend and Gringel 1983). Before this suggestion can be accepted as a truism, it will be necessary to locate directly the genes coding for these proteins. As with proteins and lipids, comparison of nucleic acids from green and nongreen plastids will certainly yield valuable information on the response of plant cells to external stimuli, and on the regulation of gene expression.

7 Future Prospects

Methodology for the isolation of nongreen plastids has now reached the stage where it is possible to prepare these organelles in quantities and of sufficient purity, for detailed biochemical analysis. Areas of the biology of these plastids which deserve additional attention include metabolic capabilities, genesis at all biochemical levels, and communication with other organelles.

While carbohydrate and lipid metabolism by nongreen plastids have received the greatest attention, much remains to be examined even in these areas. The metabolism of amino acids and other nitrogenous compounds by chloroplasts has been studied in some detail (Bryan 1976; Jensen 1980), but very little is known about these aspects of metabolism in nongreen plastids. It seems likely that carbohydrate, lipid, and amino acid metabolism will be similar in all types of plastids and that pecularities will be a matter of degree rather than extent.

Despite some recent attention, as related to ureide synthesis (Boland and Schubert 1983; Shelp et al. 1983), almost nothing is known about the role of plastids in the synthesis of purines and pyrimidines. With the recent expansion of interest in plastid molecular biology, this is one area which requires examination.

While the basic tenents of plastid biogenesis have been established, none of the details has yet been elucidated. Further study is especially needed related to lipid and membrane synthesis, and protein uptake and processing, activities which appear to be light-driven in green tissues (Mullet and Chua 1983). Either these processes are substantially different in nongreen plastids, or our apparent understanding is deceptively incomplete.

The role of the nuclear and plastid genomes in plastid genesis is becoming clearer, but most details have come from the study of green tissues. Virtually nothing is known about gene expression by nongreen plastids, and no information is yet available concerning regulation of gene expression or of coordination of expression of the two genomes.

It seems likely that one of the next major advances in plant cell biology will involve communication among organelles. Recent results from the in vitro reconstitution of several organelles involved in complex lipid synthesis (Drapier et al. 1982) are a move in the direction of understanding organelle communication, but very much remains to be done in this area and with other cellular processes.

As well as signal transduction, communication involves the specific transport of substrates and products among organelles. Through the pioneering studies of Heldt and associates (Heber and Heldt 1981), considerable progress has been made toward understanding of plastid transport, but with a very few exceptions (e.g., Kleinig and Liedvogel 1980) nothing has been reported on transport by nongreen plastids.

Many of the above described processes can be advantageously studied with nongreen plastids. As a greater knowledge of these organelles accumulates, it will become increasingly clear that nongreen plastids deserve at least as much study as their green counterparts.

Acknowledgments. The author wishes to acknowledge the free exchange of information with the *Ricinus* leucoplast group at Queen's University, D. T. Dennis, S. A. Boyle, W. Hekman, S. Finlayson, A. Thomson, and P. S. MacDougall. Thanks are due J. B. Ohlrogge for his critical evaluation of this manuscript, and J. Pruneau and L. Winfry for help in its preparation. Financial support in the form of a fellowship from Monsanto is gratefully acknowledged.

References

Agrawal PK, Canvin DT (1971) The pentose phosphate pathway in relation to fat synthesis in the developing castor oil seed. Plant Physiol 47:672–675

Andersson T, Abrahamsson H (1983) Purification of mitochondria and secretory granules isolated from pancreatic β-cell using Percoll and Sephacryl S-1000 superfine. Anal Biochem 132:82–88

Beevers H (1979) Microbodies in higher plants. Annu Rev Plant Physiol 30:159–193

Benedict CR (1973) The presence of ribulose-1,5-bisphosphate carboxylase in the nonphotosynthetic endosperm of germinating castor beans. Plant Physiol 51:755–759

Boland MJ, Schubert KR (1983) Biosynthesis of purines by a proplastid fraction from soybean nodules. Arch Biochem Biophys 220:179–187

Bortman SJ, Trelease RN, Miernyk JA (1981) Enzyme development and glyoxysome characterization in cotyledons of cotton seeds. Plant Physiol 68:82–87

Bowman CM, Dyer TA (1982) Purification and analysis of DNA from wheat chloroplasts isolated in nonaqueous media. Anal Biochem 122:108–118

Bradford MM (1976) A rapid and sensitive method for the quantitation of microgram quantities of protein utilizing the principle of protein-dye binding. Anal Biochem 72:248–254

Bryan JK (1976) Amino acid biosynthesis and its regulation. In: Bonner J, Varner JE (eds) Plant biochemistry, 3rd edn. Academic Press, New York, p 525

Camara B (1984) Terpenoid metabolism in plastids. Sites of phytoene synthetase activity and synthesis in plant cells. Plant Physiol 74:112–116

Camara B, Bardat F, Moneger R (1982) Sites of biosynthesis of carotenoids in *Capsicum* chromoplasts. Eur J Biochem 127:255–258

Cline K, Andrews J, Mersey B, Newcomb EH, Keegstra K (1981) Separation and characterization of inner and outer envelope membranes of pea chloroplasts. Proc Natl Acad Sci USA 78:3595–3599

Cobb AH (1977) The relationship of purity to photosynthetic activity in preparation of *Codium fragile* chloroplasts. Protoplasma 92:137–146

Dahlin C, Ryberg M, Axelsson L (1983) A possible structural role for carotenoid precursors in etioplasts. Physiol Plant 59:562–566

DeDuve C (1964) Principles of tissue fractionation. J Theor Biol 6:33–59

DeLuca V, Dennis DT (1978) Isoenzymes of pyruvate kinase in proplastids from developing castor bean endosperm. Plant Physiol 61:1037–1039

Dennis DT, Green TR (1975) Soluble and particulate glycolysis in developing castor bean endosperm. Biochem Biophys Res Commun 64:970–975

Dennis DT, Miernyk JA (1982) Compartmentation of nonphotosynthetic carbohydrate metabolism. Annu Rev Plant Physiol 33:27–50

Dockerty A, Merrett MJ (1979) Isolation and enzymic characterization of *Euglena* proplastids. Plant Physiol 63:468–473
Donaldson RP (1982) Nicotinamide cofactors (NAD and NADP) in glyoxysomes, mitochondria, and plastids isolated from castor bean endosperm. Arch Biochem Biophys 215:274–279
Donaldson RP, Beevers (1977) Lipid composition of organelles from germinating castor bean endosperm. Plant Physiol 59:259–263
Donaldson RP, Glashofer LM, Stopak SS, Zaras AG (1980) Alcohol dehydrogenase in germinating castor bean. Plant Physiol 65:S773
Drapier D, Dubacq JP, Tremolieres A, Mazliak P (1982) Cooperative pathway for lipid biosynthesis in young pea leaves: oleate exportation from chloroplasts and subsequent integration into complex lipids of added microsomes. Plant Cell Physiol 23:125–135
Drennan CH (1971) The subcellular localization, purification and partial characterization of acyl carrier protein from developing castor oil seeds. Dissertation, Queen's Univ, Kingston, Ontario
Drennan CH, Canvin DT (1969) Oleic acid synthesis by a particulate preparation from developing castor oil seeds. Biochim Biophys Acta 187:193–200
Dubacq J-P, Kader J-C (1968) Free flow electrophoresis of chloroplasts. Plant Physiol 61:465–468
Echeverria E, Boyer C, Liu K-C, Shannon J (1984) Isolation of amyloplasts from developing maize endosperm. Plant Physiol 77:513–519
Ellis RJ (1981) Chloroplast proteins: Synthesis, transport and assembly. Annu Rev Plant Physiol 32:111–137
Emes MJ, Fowler MW (1979 a) The intracellular location of the enzymes of nitrate assimilation in the apices of seedling pea roots. Planta 144:249–253
Emes MJ, Fowler MW (1979 b) Intracellular interactions between the pathways of carbohydrate oxidation and nitrate assimilation in plant roots. Planta 145:287–292
Emes MJ, Fowler MW (1983) The supply of reducing power for nitrite reduction in plastids of seedling pea roots (*Pisum sativum L.*). Planta 158:97–102
Feierabend J, Gringel G (1983) Plant transketolase: subcellular distribution, search for multiple forms, site of synthesis. Z Pflanzenphysiol 110:247–258
Felgner PL, Messer JL, Wilson JE (1979) Purification of hexokinase-binding protein from the outer mitochondrial membrane. J Biol Chem 254:4946–4949
Finlayson SA, Dennis DT (1983) Acetyl-Coenzyme carboxylase from the developing endosperm of *Ricinus communis*. I. Isolation and characterization. Arch Biochem Biophys 225:576–585
Fishwick MJ, Wright AJ (1980) Isolation and characterization of amyloplast envelope membranes from *Solanum tuberosum*. Photochemistry 19:55–59
Flugge UI, Heldt HW (1984) The phosphate-triose phosphate-phosphoglycerate translocator of the chloroplast. Trends Biochem Sci 9:530–533
Font B, Vial C, Gautheron DC (1975) Intracellular and submitochondrial localization of pig heart hexokinase. FEBS Lett 56:24–29
Galun E (1981) Plant protoplasts as physiological tools. Annu Rev Plant Physiol 32:237–266
Garland WJ, Dennis DT (1980) Plastid and cytosolic phosphofructokinases from the developing endosperm of *Ricinus communis*. I. Separation, purification, and initial characterization of the isozymes. Arch Biochem Biophys 204:302–309
Garland WJ, Dennis DT (1980) Plastid and cytosolic phosphofructokinases from the developing endosperm of *Ricinus communis* II. Comparison of the kinetic and regulatory properties of the isozymes. Arch Biochem Biophys 204:310–317
Gilles RJ (1982) The binding site for aldolase and G3PDH in erythrocyte membranes. Trends Biochem Sci 7:41–42
Gleizes M, Pauly G, Carde J-P, Marpeau A, Bernard-Dagan C (1983) Monoterpene hydrocarbon biosynthesis by isolated leucoplasts of *Citrofortunella mitis*. Planta 259:373–381
Good NE, Winget GD, Winter W, Connolly TN, Izawa S, Singh RMM (1966) Hydrogen ion buffers for biological research. Biochemistry 5:467–477

Gosling PG, Ross JD (1979) Characterization of glucose-6-phosphate dehydrogenase and 6-phosphogluconate dehydrogenase from hazel cotyledons. Phytochemistry 18:1441–1445
Green TR, Dennis DT, West CA (1975) Compartmentation of isopentenyl pyrophosphate isomerase and prenyl transferase in developing castor bean endosperm. Biochem Biophys Res Commun 64:976–982
Gregor HD (1977) A new method for the rapid separation of cell organelles. Anal Biochem 82:255–257
Harley HL, Black MK, Wedding RT (1978) Preparation of plant mitochondria using gel filtration columns. New Phytol 81:223–231
Heber U, Heldt HW (1981) The chloroplast envelope: structure, function and role in leaf metabolism. Annu Rev Plant Physiol 32:139–168
Ireland RJ, Dennis DT (1980) Isozymes of the glycolytic and pentose phosphate pathways in storage tissues of different oil seeds. Planta 149:476–479
Ireland RJ, Dennis DT (1981) Isozymes of the glycolytic and pentose-phosphate pathways during the development of the castor oil seed. Canad J Bot 59:1423–1425
Ireland RJ, DeLuca V, Dennis DT (1979) Isozymes of pyruvate kinase in etioplasts and chloroplasts. Plant Physiol 63:903–907
Ireland RJ, DeLuca V, Dennis DT (1980) Characterization and kinetics of isozymes of pyruvate kinase from developing castor bean endosperm. Plant Physiol 65:1188–1193
Jensen RG (1980) Biochemistry of the chloroplast. In: Tolbert NE (ed) The plant cell. Academic Press, New York, pp 274–313 (The biochemistry of plants, vol 1)
Keys AJ (1968) The intracellular distribution of free nucleotides in the tobacco leaf. Biochem J 108:118
Keegstra K, Cline K (1982) Evidence that envelope and thylakoid membranes from pea chloroplasts lack glycoproteins. Plant Physiol 70:232–237
Kirk JTO, Tilney-Bassett RAE (1978) The plastids: their chemistry, structure, growth and inheritance. Elsevier Biomedical Press, Amsterdam
Kleinig H, Liedvogel B (1980) Fatty acid synthesis by isolated chromoplasts from the daffodil. Energy sources and distribution patterns of the acids. Planta 150:166–169
Kleinkopf GE, Huffaker R, Matheson A (1970) Light-induced *de novo* synthesis of ribulose-1,5-bisphosphate carboxylase in greening leaves of barley. Plant Physiol 46:416–418
Kreuz K, Kleinig H (1981) On the compartmentation of isopentenyl diphosphate synthesis and utilization in plant cells. Planta 153:578–581
Kreuz K, Beyer P, Kleinig H (1982) The site of carotenogenic enzymes in chromoplasts from *Narcissus pseudonarcissus* L. Planta 154:66–69
Larsson C, Andersson B (1979) Two-phase methods for chloroplasts, chloroplast elements, and mitochondria. In: Reid E (ed) Plant organelles. Ellis Horwood, Chichester, pp 35–46 (Methodological surveys in biochemistry, vol 9)
Leech RM, Leese BM (1982) Isolation of etioplasts from maize. In: Edelman M, Hallick RB, Chua N-H (eds) Methods in chloroplast molecular biology. Elsevier Biomedical Press, Amsterdam, p 221
Lees EM, Dennis DT (1981) Glutamate dehydrogenase in developing endosperm, chloroplasts, and roots of castor bean. Plant Physiol 68:827–830
Lenstra JA, van Raaig Bluemendal AJM (1982) One of the protein components of lens fiber membranes is glyceraldehyde 3-phosphate dehydrogenase. FEBS Lett 148:263–266
Lilley R McC, Fitzgerald MP, Rienits KG, Walker DA (1975) Criteria of intactness and the photosynthetic activity of spinach chloroplast preparations. New Phytol 75:1–10
Lilley R McC, Stitt M, Mader G, Heldt HW (1982) Rapid fractionation of wheat leaf protoplasts using membrane filtration. The determination of metabolite levels in the chloroplasts, cytosol, and mitochondria. Plant Physiol 70:965–970
Loomis WD (1974) Overcoming problems of phenolics and quinones in the isolation of plant enzymes and organelles. Methods Enzymol 31:528–544
Lowry OH, Rosebrough NJ, Farr AL, Randall RJ (1951) Protein measurement with the Folin phenol reagent. J Biol Chem 193:265–275

Lück H (1965) Catalase. In: Bergmeyer H (ed) Methods of enzymatic analysis. Academic Press, New York, pp 885–894

Lutz C, Nordmann U (1983) The localization of saponins in prolamellar bodies mainly depends on the isolation of etioplasts. Z Pflanzenphysiol 110:201–210

MacDonald FD, apRees T (1983) Enzymic properties of amyloplasts from suspension cultures of soybean. Biochim Biophys Acta 755:81–89

Miernyk JA, Dennis DT (1982) Isozymes of the glycolytic enzymes in endosperm from developing castor oil seeds. Plant Physiol 69:825–828

Miernyk JA, Dennis DT (1983) Mitochondrial, plastid, and cytosolic isozymes of hexokinase from developing endosperm of *Ricinus communis*. Arch Biochem Biophys 226:456–463

Miernyk JA, Dennis DT (1983) The incorporation of glycolytic intermediates into lipids by plastids isolated from the developing endosperm of castor oil seeds (*Ricinus communis* L.). J Exp Bot 34:712–718

Miernyk JA, Dennis DT (1984) Enolase isozymes from *Ricinus communis*: partial purification and characterization of the isozymes. Arch Biochem Biophys 233:643–651

Miernyk JA, Trelease RN (1981) Control of enzyme activities in cotton cotyledons during maturation and germination. IV β-oxidation. Plant Physiol 67:341–346

Miflin BJ, Beevers H (1974) Isolation of intact plastids from a range of plant tissues. Plant Physiol 53:870–874

Mills WR, Joy KW (1980) A rapid method for isolation of purified physiologically active chloroplasts, used to study the intracellular distribution of amino acids in pea leaves. Planta 148:75–83

Moore TS, Lord JM, Kagawa T, Beevers H (1973) Enzymes of phospholipid metabolism in the endoplasmic reticulum of castor bean endosperm. Plant Physiol 52:50–53

Mudd JB, Dezacks R (1981) Synthesis of phosphatidylglycerol by chloroplasts from leaves of *Spinacia oleracea* L. Arch Biochem Biophys 209:584–591

Mullet JE, Chua N-H (1983) In vitro reconstitution of synthesis, uptake and assembly of cytoplasmically synthesized chloroplast proteins. Methods Enzymol 97:502–509

Murphy DJ, Leech RM (1977) Lipid biosynthesis from (^{14}C) bicarbonate, (2-^{14}C) pyruvate and (1-^{14}C) acetate during photosynthesis by isolated chloroplasts. FEBS Lett 77:104–168

Nakatani HY, Barber J (1977) An improved method for isolating chloroplasts retaining their outer membranes. Biochim Biophys Acta 461510–512

Nishimura M, Beevers H (1981) Isozymes of sugar phosphate metabolism in endosperm of germinating castor beans. Plant Physiol 64:31–37

Pardo AD, Chereskin BM, Castelfranco P, Franceschi VR, Wezelman BE (1980) ATP requirement for Mg chelatase in developing chloroplasts. Plant Physiol 65:956–960

Perrin DD, Dempsey B (1971) Buffers for pH and metal ion control. Chapman and Hall, London

Quail PH (1979) Plant cell fractionation. Annu Rev Plant Physiol 30:425–484

Reid EE, Lyttle CR, Canvin DT, Dennis DT (1975) Pyruvate dehydrogenase complex activity in proplastids and mitochondria of developing castor bean endosperm. Biochem Biophys Res Commun 62:42–47

Reid EE, Thompson P, Lyttle CR, Dennis DT (1977) Pyruvate dehydrogenase complex from higher plant mitochondria and proplastids. Plant Physiol 59:842–848

Roughan PG, Slack CR (1982) Cellular organization of glycerolipid metabolism. Annu Rev Plant Physiol 33:97–132

Rudd TP, Anderson LE (1982) Evidence for a membrane binding protein for NADP-linked glyceraldehyde-3-phosphate dehydrogenase. Plant Physiol 69:S506

Ryberg M, Sandelius AS, Seistram E (1983) Lipid composition of prolamellar bodies and thylakoids of wheat etioplasts. Physiol Plant 57:555–560

Satoh Y, Usami Q, Yamada M (1983) Glucose-6-phosphate dehydrogenase in plastids from developing castor bean seeds. Plant Cell Physiol 24:527–532

Sautter C, Bartscherer HC, Hock B (1981) Separation of plant cell organelles by zonal centrifugation in reorienting density gradients. Anal Biochem 113:179–184

Schmitt MR, Edwards GE (1982) Isolation and purification of intact peroxisomes from green leaf tissue. Plant Physiol 70:1213–1217

Schnarrenberger CA, Oeser A, Tolbert NE (1973) Two isozymes each of glucose-6-phosphate dehydrogenase and 6-phosphogluconate dehydrogenase in spinach leaves. Arch Biochem Biophys 154:438–448

Scott-Burden T, Canvin DT (1975) The effect of Mg^{+2} on density of proplastids from the developing castor bean and the use of acetyl-CoA carboxylase activity for their cytochemical identification. Can J Bot 53:1371–1376

Shelp BJ, Atkins CA, Storer PJ, Canvin DT (1983) Cellular and subcellular organization of pathways of ammonia assimilation and ureide synthesis in nodules of cowpea (*Viona unguiculata* L. Walp). Arch Biochem Biophys 224:429–441

Simcox PD, Dennis DT (1978) Isoenzymes of the glycolytic and pentose phosphate pathways in proplastids from the developing endosperm of *Ricinus communis* L. Plant Physiol 61:871–877

Simcox PD, Dennis DT, West CA (1975) Kaurene synthetase from plastids of developing plant tissues. Biochem Biophys Res Commun 66:166–172

Simcox PD, Reid EF, Canvin DT, Dennis DT (1977) Enzymes of the glycolytic and pentose pathways in proplastids from the developing endosperm of *Ricinus communis* L. Plant Physiol 59:1128–1132

Singer TP, Oestreicher G, Hogue P, Contreiras J, Brandao I (1973) Regulation of succinate dehydrogenase in higher plants. I. Some general characteristics of the membrane bound enzyme. Plant Physiol 52:616–621

Sitte P (1977) Chromoplasten-bunte Objekte der modernen Zellbiologie. Biol Unserer Zeit, pp 65–74

Sitton D, West CA (1975) Casbene, an antifungal diterpene produced in cell-free extracts of *Ricinus Communis* seedlings. Phytochemistry 14:1921–1925

Spurgeon SL, Sathyamoorthy N, Porter JW (1984) Isopentenyl pyrophosphate isomerase and prenyl transferase from tomato fruit plastids. Arch Biochem Biophys 230:446–454

Srivastava DK, Anderson LE (1983) Isolation and characterization of light and dithiothreitol-modulatable glucose-6-phosphate dehydrogenase from pea chloroplasts. Biochim Biophys Acta 724:359–369

Stitt M, apRees T (1980) Capacities of pea chloroplasts to catalyse the oxidative pentose phosphate pathway and glycolysis. Phytochemistry 18:1905–1911

Stocco DM (1983) Rapid, quantitative isolation of mitochondria from rat liver using Ficoll gradients in vertical rotors. Analyt Biochem 131:453–457

Straus W (1954) Chromoplasts from carrots. Exp Cell Res 11:289–296

Stumpf PK (1980) Biosynthesis of saturated and unsaturated fatty acids. In: Stumpf PK (ed) Lipids: structure and function. Academic Press, New York, pp 177–204 (The biochemistry of plants, vol 4)

Thompson P, Reid EE, Lyttle CR, Dennis DT (1977a) Pyruvate dehydrogenase complex from higher plant mitochondria and proplastids: Kinetics. Plant Physiol 59:849–853

Thompson P, Reid EE, Lyttle CR, Dennis DT (1977b) Pyruvate dehydrogenase complex from higher plant mitochondria and proplastids: Regulation. Plant Physiol 59:854–858

Thomson A, Miernyk JA, Dennis DT (1983) Plastid DNA from *Ricinus communis*. Proc Annu Meet Can Soc Plant Physiol

Thomson WW, Whatley JM (1980) Development of nongreen plastids. Annu Rev Plant Physiol 31:375–394

Tully RE, Beevers H (1976) Protein bodies of castor bean endosperm. Isolation, fractionation, and characterization of the protein components. Plant Physiol 58:710–716

Walker DA (1980) Preparation of higher plant chloroplasts. Methods Enzymol 69:94–104

Walker JRL (1980) Enzyme isolation from plants and the phenolic problem. What's New Plant Physiol 11:33–36

Wellburn AR, Wellburn FAM (1971) A new method for the isolation of etioplasts with intact envelopes. J Exp Bot 22:972–979

Whitfield PR, Bottomley W (1983) Organization and structure of chloroplast genes. Annu Rev Plant Physiol 34:279–310
Wilson JE (1978) Ambiquitous enzymes: variation in intracellular distribution as a regulatory mechanism. Trends Biochem Sci 3:124–125
Wong JH, Benedict CR (1983) Development and characterization of ribulose-1,5-bisphosphate carboxylase. Evidence indicating a lack of carboxylase function in CO_2 fixation in endosperm of germinating castor bean seedlings. Plant Physiol 72:37–43
Youle RJ, Huang AHC (1976) Development and properties of fructose 1,6-bisphosphate in the endosperm of castor bean seedlings. Biochem J 154:647–652
Youle RJ, Huang AHC (1976) Protein bodies from the endosperm of castor bean. Subfractionation, protein components, lectins, and changes during germination. Plant Physiol 58:703–709
Zilkey B, Canvin DT (1972) Subcellular localization of oleic acid biosynthesis enzymes in the developing castor bean endosperm. Biochem Biophys Res Commun 34:646–653

Mitochondria

J. F. Jackson

1 Introduction

Among the eukaryotes, the function of mitochondria in carrying out oxidative phosphorylation appears highly conserved, and yet the size, organization, and structure of the mitochondrial genome varies enormously (Sederoff 1984). Although most animal mitochondria have a uniformly monomeric circular DNA of about 16 kb, it is the fungal, and even more, the mitochondria of higher plants, which have the bewildering array of size distribution in contained DNA. Thus the size, of mitochondrial genomes in higher plants can vary from approximately 100 kb to as much as 2400 kb, and can be a mixture of linear and circular molecules as well. Advances in this field have been so rapid that some of the properties of mitochondrial DNA can now be used as an analytical tool in detecting, for example, certain types of cytoplasmic male sterility in maize (Pring and Levings 1978). This trait is of major economic importance, being used in the commercial production of hybrids (Duvick 1965). It is maternally inherited and is characterized by easily recognized changes in mitochondrial DNA structure (Schardl et al. 1984).

Many other relevations concerning mitochondrial DNA structure and function have followed, so that it now seems that some chloroplast genes are contained also within the corn mitochondrial genome (Lonsdale et al. 1983), and that in yeast at least mitochondrial genes can be found in the nucleus of the cell (Farelly and Butow 1983). In some cases the nuclear counterpart is silent (corn, yeast) or, in the case of *Neurospora* (Van den Boogaart et al. 1982), is active. All these instances show the use to which a knowledge of DNA species present in mitochondria can increase our understanding of plant mitochondria and its function in relation to the rest of the cell. For this reason we include here essentials of the preparation of mitochondrial DNA and a brief description of its analysis, together with recent techniques and tests for integrity used in the preparation of mitochondria for the study of the better-understood oxidative phosphorylation functions of mitochondria. In bringing the two areas together in one chapter, we hope both areas may benefit, and that those setting out on plant biochemical studies may be made aware of the possibilities and essentials of the techniques involved in both disciplines.

Mitochondrial preparation for DNA analysis need not be quite so involved as that for oxidative studies; we begin therefore with a description of these methods, together with an explanation of our present understanding of cytoplasmic male sterility, which we use as an example of the use to which DNA analysis can be put.

2 Preparation for DNA Analysis

2.1 Cytoplasmic Male Sterility and Structure of Mitochondrial DNA

Of the many complexities of the plant mitochondrial genome, perhaps one of the better understood and certainly one with acknowledged commercial implications is that associated with cytoplasmic male sterility (CMS). Four types of CMS have been identified in *Zea mays* so far; S (USDA), T (Texas), and C (Charrva) all caŭse pollen abortion, giving male sterility, and were identified by their patterns of fertility restoration by nuclear restorer genes (Duvick 1965). Recognition of these characteristics have enabled commercial breeders to prevent self-pollination and bring about the production of hybrids with associated "hybrid vigor" through cross-pollination. The N (normal) cytoplasm allows normal pollen development, even in the absence of restorer genes. The restriction analysis of mitochondrial DNA can unambiguously distinguish these types, and it has been suggested that the mitochondrial genome is the cytoplasmic determinant of male fertility in maize (Schardl et al. 1984).

The best-studied type is S, where the presence of S-1 (6.4 kb) and S-2 (5.4 kb) episomes (both linear) characterizes S mitochondrial genomes. Much of the S-mitochondrial DNA is in the form of linear mitochondrial chromosomes with copies of S-1 and S-2 at the ends of linear molecules, in addition to occurring free as S-1 and S-2 episomes (Schardl et al. 1984). When S-type cytoplasms are reverted to fertile forms, the free S-1 and S-2 episomes are no longer detectable, and is thought that rearrangement of the S-episomes are involved in fertility reversion (Fig. 1).

Circularization of the mitochondrial genome appears to play a role in fertility reversion, while current evidence suggests that this major alteration of genome structure is due to a specific mutation of internally located S-2 sequences (Schardl et al. 1984). In any case, restriction analysis of fertile revertants gives a result very different from S-type male steriles, and forms the basis of the example used in this section.

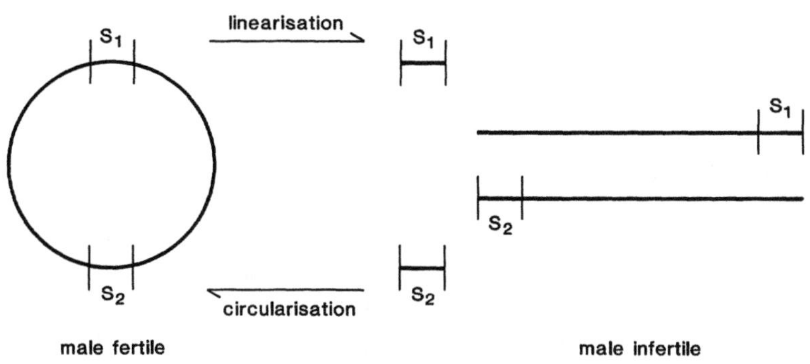

Fig. 1. Representation of mitochondrial genome in fertile and sterile S-type maize

2.2 Isolation of Mitochondria for DNA Preparation

This method is essentially that of Day and Hanson (1977), with modifications as described by Pring and Levings (1978) and by Kemble et al. (1980). It involves homogenization of maize tissue in mannitol-buffer solution, collection of mitochondria by differential centrifugation, and finally removal of extra-mitochondrial DNA by deoxyribunuclease treatment.

Seeds of *Zea mays* are surface-sterilized with 2% w/v sodium hypochlorite solution for 15 min, washed with water, imbibed in running tap water overnight prior to germination on water-moistened towels for 4 days at 27 °C in the dark. Approximately 5 g of etiolated shoots (about 30 seedlings) are homogenized in a mortar and pestle with 15 ml 0.5 M mannitol – 0.01 M N-Tris (hydroxymethyl) methyl-2-aminoethane sulfonic acid (TES) pH 7.2 – 0.001 M EGTA-0.2% BSA – 0.05% cyseine, for 30 s, the temperature being kept at 0 ° to 4 °C. The homogenate is filtered through four layers of buffer muslin and one layer of Miracloth (Calbiochem.), and the filtrate centrifuged at 1000 *g* for 10 min. The supernatant is then recentrifuged at 12,000 *g* for 10 min, and the mitochondrial pellet resuspended in homogenizing buffer (minimal amount ca. 0.4 ml) and again centrifuged at 1000 *g* for 10 min. To the supernatant add $MgCl_2$ and DNAase I (Worthington, DPFF) to give a concentration of 0.01 M and 10 μg g^{-1} fresh weight starting tissue respectively. Incubate at 4 °C for 1 h to hydrolyse extra-mitochondrial DNA, then centrifuge this preparation through a layer of 0.6 M sucrose – 0.01 M TES, pH 7.2 – 0.02 M EDTA at 10,000 *g* for 20 min. The pellet containing the mitochondria, now free of extramitochondrial DNA, is washed twice with the sucrose-buffer-EDTA solution to eliminate the deoxyribosonuclease. A temperature of 0 ° to 4 °C was maintained through all the above procedures.

2.3 Preparation of Mitochondrial DNA

Mitochondria can be lysed in 0.05 M Tris-HCl, pH 8 – 0.01 M EDTA – 2% sarkosyl NL 97 – 0.012% autodigested pronase for 1 h at 37 °C. Lysates are than made 2 M with ammonium acetate and the DNA purified by three successive phenol-chloroform extractions, precipitation with three volumes of ethanol and ethanol washes, before finally suspending in 0.05 ml 0.005 M Tris-HCl, pH 8 – 0.00025 M EDTA, or in the appropriate restriction buffer (see below).

2.4 Electrophoresis of Mitochondrial DNA

The DNA solution is made 5% glycerol and 0.005% bromophenol blue before electrophoresis at room temperature in 1.5% agarose horizontal slab gels 21 × 18 × 0.5 cm (14 sample wells), at 30 mA overnight. Electrophoresis buffer is 0.04 M Tris, pH 7.8 – 0.005 M sodium acetate – 0.001 M EDTA (Bedbrook and Bogorad 1976). The gel is then stained in 0.5 μg ml^{-1} ethidium bromide for 30 min, before vizualization of DNA bands on gels by UV illumination. A Polaroid type 47 high speed Land film or Kodak P × P film van be used to photograph the gels, using a Wratten 23A filter. Four different patterns of low molecular weight bands can be observed with mitochondrial DNA from maize, corresponding to N, T, S, and C cytoplasms (Kemble et al. 1980).

2.5 Restriction Analysis of Mitochondrial DNA

The DNA preparation is added to sterile plastic tubes containing the appropriate restriction enzyme and restriction buffer. Approximately 1 μg of DNA is used in each digestion, the DNA being dissolved in the necessary restriction buffer, before being added to the enzyme solution. Hind III restriction endonuclease (Miles

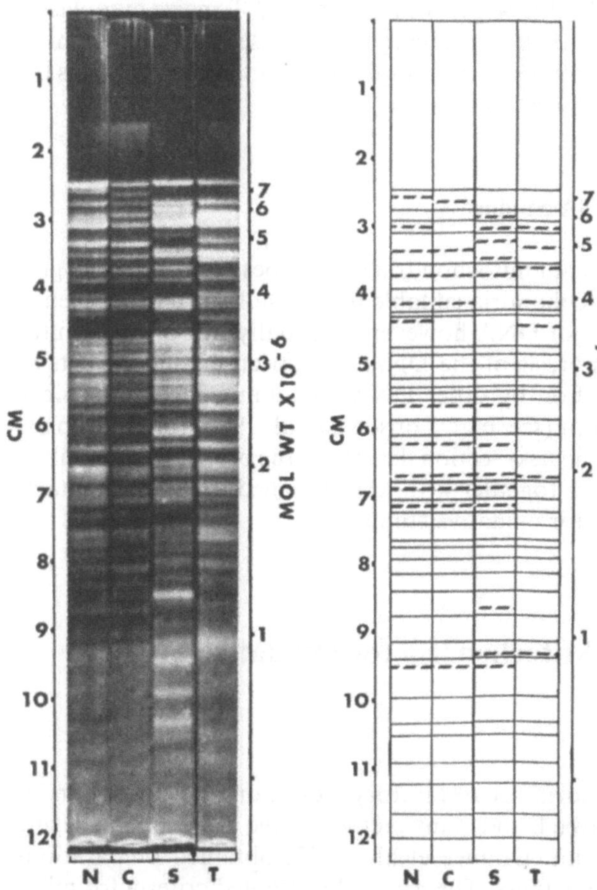

Fig. 2. Restriction endonuclease Hind III digestion of mitochondrial DNA from N (normal) cytoplasm, and from C, S, and T male-sterile cytoplasm. Agarose gel electrophoresis patterns are shown on the left and a schematic on the right, where dashed lines indicate fragments not common to all cytoplasms, and solid lines those common to all tested. Molecular weight of fragments are indicated on both diagrams. Reproduced by kind permission from Pring and Levings (1978)

Laboratories, Inc.) is assayed in 50 μl 50 mM NaCl – 6 mM $MgCl_2$-6 mM Tris-HCl, pH 7.5, with 5 μg BSA; EcoRI restriction endonuclease in 25 μl 50 mM NaCl – 5 mM $MgCl_2$-100 mM Tris-HCl, pH 7.5; Sal I in 50 μl 6 mM Tris-HCl, pH 7.9-6 mM mercaptoethanol – 100 mM NaCl; Bam I in 50 μl 100 mM Tris-HCl, pH 7.5 – 10 mM $MgCl_2$. Digestions should be carried out with sufficient enzyme to achieve limit digestion, and terminated by the addition of sodium dodecyl sulfate to 1% and heating at 60 °C for 10 min. As before, bromophenol blue and glycerol are added to each digestion before electrophoresis.

Agarose gel electrophoresis is conducted as described above, except that 0.5, 0.7, 0.9, or 1% agarose and two buffer systems (0.03 M Tris HCl-0.036 M NaH_2PO_4 – 0.002 M Na_2EDTA, pH 7.8; or 0.09 M H_3BO_3 – 0.09 M Tris HCl – 0.002 M Na_2EDTA, pH 7.5) are necessary to resolve all bands. Electrophoresis was for 15 to 17 h at 1.5 to 1.9 v cm^{-1}. An example is given of the Hind III restriction pattern of mitochondrial DNA from N, C, S, and T maize cytoplasms (Fig. 2; after Pring and Levings 1978). Each cytoplasm possesses a unique mitochondrial DNA, the restriction patterns are complex with more than 50 bands being observed with the normal (N) cytoplasm.

2.6 Notes on Mitochondrial DNA Studies

The method of preparation of plant mitochondria above does not necessarily give a high percentage of intact mitochondria suitable for physiological experiments. It does yield good preparations of DNA however, essentially free of extra-mitochondrial DNA. Gel electrophoresis of the DNA or restriction analysis acts as a built-in check on contamination by other DNA-containing organelles, since they may give a different pattern (see, e.g., Lonsdale et al. 1983, for chloroplast and mitochondrial patterns). The next section deals with the preparation of intact mitochondria for other purposes together with tests designed to establish intactness of obtained mitochondria for oxidative studies.

3 Preparation of Intact Mitochondria for Oxidative Studies

3.1 Introduction

Preparation of plant mitochondria for respiratory oxidation and phosphorylation studies has been well described, and many methods have been applied. Using a discontinuous sucrose gradient, Dounce et al. (1972) obtained preparations with good homogeneity and intactness. The generally accepted criteria for mitochondrial integrity include strong respiratory control by ADP (intact mitochondria but not submitochondrial particles shut off respiration when ADP is exhausted), high P:O ratios (the accepted upper limiting value of the P:O ratio is theoretically 3, a value never achieved in practice. This implies that the passage of one pair of electrons traversing the electron transfer chain from NADH to oxygen is coupled to the generation of three molecules of ATP), and general appearance on electron micrographs. While these criteria are important, studies for other activities of mitochondria may require mitochondria free from other subcellular structures, and perhaps, for example, greater attention needs to be paid to the intactness or otherwise of the outer of the two membranes of mitochondria. This may be measured by estimation of succinate : cytochrome c reductase activity before and after osmotic shock with 0.3 M sucrose (Dounce et al. 1972). This enzyme is located on the inner mitochondrial membrane, so that integrity or otherwise of the outer membrane can be estimated because it is only after rupture of the outer membrane that the substrate can get to the active site of the enzyme (Yamaya et al. 1984). For these reasons the Percoll discontinuous gradient method is described here, as it gives good quality mitochondrial preparations on all these counts.

3.2 Mitochondrial Preparation and Purification

The method described below is essentially a modification of that of Nishimura et al. (1982) and Yamaya et al. (1984). All steps are carried out at 4 °C. Maize shoots (5 days old, prepared as described above) are washed in water, dried on filter paper, weighed. Approximately 50 g fresh weight shoots are chopped with razor blades in an extraction buffer (250 ml) consisting of 0.4 M mannitol – 0.1 M Hepes-KOH, pH 7.5 – 1 mM EDTA – 0.1% (w/v) BSA – 0.6% (w/v) insoluble PVP, and the chopped material homogenized in the mortar and pestle. The homogenate is passed through muslin and Miracloth as described above, and centrifuged for 10 min at 1000 *g* to sediment debris, starch, chloroplasts and aggregated organelles. The supernatant is now centrifuged at 12,000 *g* for 10 min to give the crude mitochondrial pellet.

To purify the mitochondria a discontinuous Percoll gradient is employed. The pellet is gently dispersed in 1 ml of suspending buffer composed of 0.3 M mannitol – 20 mM Hepes-KOH, pH 7.5 – 0.1% w/v fatty acid free BSA, and layered on top of a discontinuous Percoll gradient. The latter was formed by adding first 3 ml 60% v/v Percoll, then 4 ml 4.5% Percoll, 4 ml 28% Percoll, and finally 4 ml 5% Percoll. In each case the Percoll solution contained 0.25 M sucrose – 20 mM Hepes-KOH, pH 7.5 0.1% w/v fatty acid free BSA. Centrifugation is carried out at 30,000 *g* for 30 min (SW-27.1 rotor in Beckmann L5-75 ultracentrifuge). The mitochondrial fraction collects at the interface between the 45% and 28% Percoll solutions, and is removed with a Pasteur pipette. To remove the Percoll, the mitochondrial fraction is diluted tenfold with suspending buffer and centrifuged for 10 min at 10,000 *g*. The mitochondrial pellet is again suspended in the buffer solution, recentrifuged at 10,000 *g* for 10 min. The final mitochondrial pellet is carefully dispersed in 0.5 ml of suspension buffer.

3.3 Tests for Integrity of Mitochondria

3.3.1 Succinate: Cytochrome c Reductase

Mitochondria containing 0.1 to 1.0 mg protein (Lowry et al. 1951) is added to a solution containing 5 mM phosphate, pH 7.2 – 0.05 mM cytochrome C – 1 mM KCN, and reaction initiated with 10 mM succinate, final volume 3 ml. One test is done in 0.3 M sucrose and another in the absence of sucrose, when the outer membrane is disrupted. Activity is followed with a spectrophotometer, measuring reduction of cytochrome c at 551 nm. For a preparation from light-grown corn, Yamaya et al. (1984) report 242 and 19 nmol cytochrome c reduced min^{-1} mg^{-1} mitochondrial protein, in the absence and presence of 0.3 M sucrose respectively, suggesting that the outer membrane was intact for approximately 92% of the mitochondria in their preparation.

3.4 Tests for Integrity

3.4.1 Measurement of Oxygen Consumption for Respiratory Control and P/O Ratios

The respiratory control ratios and P/O ratios are estimated by measuring oxygen uptake polarographically at 25 °C using a Clark-type electrode system (Nishimura et al. 1982). This electrode system is available from Hansatech Ltd (Hardwick Industrial Estate, Kings Lynn, Norfolk, U.K.). The reaction medium contains 0.3 M mannitol – 10 mM Na-phosphate, pH 7.2 – 5 mM $MgCl_2$ – 10 mM KCl – 0.1% defatted BSA and mitochondria, total volume 1.0 ml. Substrates malate (30 mM) or succinate (10 mM) are added, and oxygen electrode traces followed with time (see Yamaya et al. 1984), adding 150 μM ADP when necessary to initiate consumption and thus demonstrate respiratory control by ADP. Yamaya et al. (1984) report respiratory control ratios of about 5 (i.e., oxygen consumption in the presence of ADP: oxygen consumption when ADP depleted) with malate substrate, for corn mitochondria prepared as described above. P/O ratios, calculated from these results were approximately 2.5. These figures are representative of those expected for good quality mitochondrial preparations.

The work of Dounce et al. (1972) should be consulted for methods relating to preparation and observation on mitochondria by electron microscopy.

3.5 Notes on the Methods

The method and tests given above are those often used for current mitochondrial studies for example, on transport of ions and metabolites (Yamaya et al. 1984) and similar studies. However, it should be remembered that specialist experiments may have different needs. If, for example, mitochondrial studies are needed where chloroplasts are especially to be excluded then the method of Bergman et al. (1980) should be consulted. Studies designed to show that β-oxidation of fatty acids in plant cells does not reside in the mitochondria would need to exclude or take account of peroxisomes which may be the site of such β-oxidation (Gerhardt 1983).

An alternative sucrose gradient method for mitochondrial purification is described by Miller and Romani (1965) and used recently for example, for studies on DNA polymerase and RNA polymerase in potato tuber mitochondria (Apelbaum et al. 1984).

References

Apelbaum A, Vinkler C, Spakiotakis E, Dilley DR (1984) Increased mitochondrial DNA and RNA polymerase activity in ethylene-treated potato tubers. Plant Physiol 76:461–464

Bedbrook JR, Bogorad L (1976) Endonuclease recognition sites mapped on *Zea mays* chloroplast DNA. Proc Natl Acad Sci USA 73:4309–4313

Bergman A, Gardestrom P, Ericson I (1980) Method to obtain a chlorophyll free preparation of intact mitochondria from spinach leaves. Plant Physiol 66:442–445

Boogaart P van den, Samello J, Agsteribbe E (1982) Similar genes for a mitochondrial ATPase subunit in the nuclear and mitochondrial genomes of *Neurospora crassa*. Nature 298:187–189

Day DA, Hanson JB (1977) On methods for the isolation of mitochondria from etiolated corn shoots. Plant Sci Lett 11:99–104

Dounce R, Christensen EL, Bonner WD (1972) Preparation of intact mitochondria. Biochim Biophys Acta 275:148–160

Duvick DN (1965) Cytoplasmic pollen sterility in corn. Adv Genet 13:1–56

Farelly F, Butow RA (1983) Rearranged mitochondrial genes in the yeast nuclear genome. Nature 301:296–301

Gerhardt B (1983) Peroxisomes as site of β oxidation in plant cells. Plant Physiol 72S:170

Kemble RJ, Gunn RE, Flavell RB (1980) Classification of normal and male-sterile cytoplasms in maize. II. Electrophoretic analysis of DNA species in mitochondria. Genetics 95:451–458

Lonsdale DM, Hodge TP, Howe LJ, Stern DB (1983) Maize mitochondrial DNA contains a sequence homologous to the ribulose-1,5-biphosphate carboxylase large subunit gene of chloroplast DNA. Cell 34:1007–1014

Lowry OH, Rosebrough NJ, Farnn AL, Randall RJ (1951) Protein measurement with Folin phenol reagent. J Biol Chem 193:265–275

Miller LA, Romani RJ (1965) Sucrose density gradient distribution of mitochondrial protein and enzymes from preclimateric and climateric pears. Plant Physiol 70:1385–1390

Nishimura M, Dounce R, Akazawa T (1982) Isolation and characterization of metabolically competent mitochondria from spinach leaf protoplasts. Plant Physiol 69:916–920

Pring DR, Levings CS (1978) Heterogeneity of maize cytoplasmic genomes among male-sterile cytoplasms. Genetics 89:121–136

Schardl CL, Lonsdale DM, Pring DR, Rose KR (1984) Linearization of maize mitochondrial chromosomes by recombination with linear episomes. Nature 310:292–296

Sederoff RR (1984) Structural variation in mitochondrial DNA. Adv Genet 22:1–108

Yamaya T, Oaks A, Matsumoto H (1984) Stimulation of mitochondrial calcium uptake by light during growth of corn shoots. Plant Physiol 75:773–777

Endoplasmic Reticulum

R. L. JONES

1 Introduction

The endoplasmic reticulum (ER) serves as one of the principal biosynthetic sites for a wide variety of molecules ranging from small molecular weight amino acids and organic acids to proteins and other polymers. The ER also serves as one of the major centers of membrane biogenesis, and the membrane flow hypothesis infers that the ER plays an important role in the formation of other components of the endomembrane system.

In this chapter an attempt is made to review recent advances in our understanding of ER structure, especially the relationship between tubular ER (tER) and cisternal ER (cER) and the biosynthesis of ER membrane lipids and proteins. The isolation and purification of the ER are discussed in the context of these advances. The literature was reviewed up to the end of 1983, although no attempt was made to provide a comprehensive review of all areas of research pertinent to ER structure and function. Several books and articles on methods for ER research have been published (Chrispeels 1980; Hall and Moore 1983; Price 1982; Quail 1979; Reid 1979). These references were valuable in the preparation of this article and they are recommended to the reader for further study.

2 Structure and Organization of the ER

The endoplasmic reticulum was first described by light microscopists as the ergastoplasm of cells, but it is the distinctive form of this membrane system as seen by electron microscopy that allows for its most accurate definition. Endoplasmic reticulum in plant cells forms an anastomosing system of smooth-surfaced (smooth ER, sER) or ribosome-studded (rough ER, rER) membranes which are folded into a series of flattened sacs (cisternal ER, cER) or appear as a network of tubules (tubular ER, tER) (Fig. 1).

There is compelling evidence that tubular and cisternal ER are structurally distinct components of the ER system. When visualized by a combination of low or high voltage electron microscopy (EM) and zinc iodide-osmium tetroxide stain, the ER of cell types as widely different as storage parenchyma of legume cotyledons (Harris 1979; Harris and Chrispeels 1980; Harris and Oparka 1983) and root cap cells of corn (Hawes et al. 1981; Juniper et al. 1982) appears as a continuous interconnected network of tER and cER (Fig. 1). A similar differentiation of the ER into tubular and cisternal forms was observed by Hepler (1980,

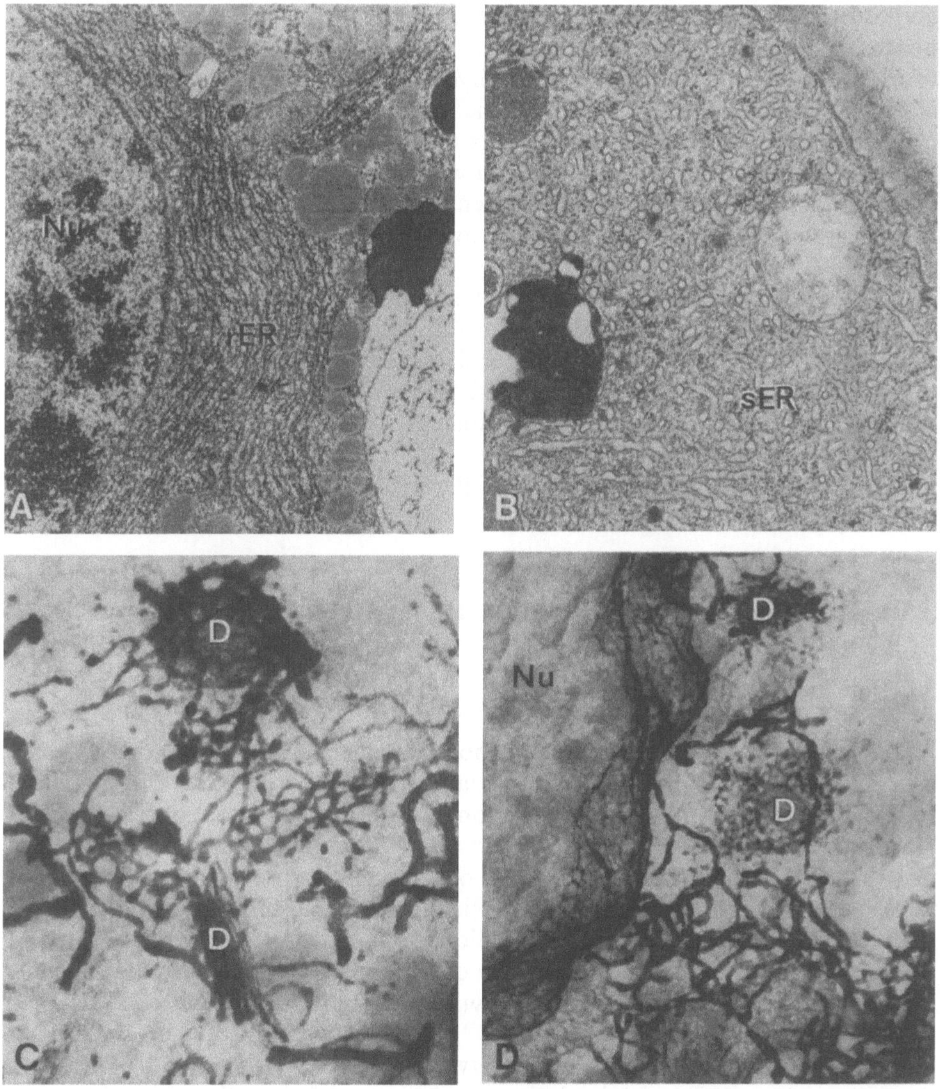

Fig. 1. Cisternal (**A**, **B**) and tubular (**C**, **D**) ER of plant cells. **A** Rough ER from barley aleurone, reproduced from Jones (1980). **B** Smooth ER from *Primula kewensis* petals, courtesy of B.E.S. Gunning. **C** Tubular ER from mung bean (*Vigna radiata*) cotyledons, reproduced with permission from Harris and Oparka (1983). **D** Tubular ER from maize root cap, reproduced with permission from Juniper et al. (1982). *D* dictyosome; *Nu* nucleus; *rER* rough ER; *sER* smooth ER

1981, 1983) in lettuce roots post-fixed with a mixture of osmium tetroxide and potassium ferricyanide and viewed at low voltage.

Although the two forms of ER can be best resolved by staining with zinc iodide or potassium ferricyanide in the presence of osmium tetroxide, the existence of cER and tER was well documented using conventional EM stains (Gunning and Steer 1975). These two parts of the ER system can be distinguished in EM images on the basis of size, the diameter of the tER being two to three times the width of the cER (Harris 1979). In aldehyde-fixed and thin-sectioned tissue, tER is generally viewed in cross-section as membrane vesicles, while cER is generally organized into lamellae that are frequently found in stacks. The density of ribosomes on the surface of ER also allows for the distinction between the two forms of this membrane, since cER is densely studded with ribosomes, while the tubular system has no or only a few ribosomes (Goosen-deRoo et al. 1983; Harris 1979, 1980).

The answer to whether all SER is of the tubular type will await an examination of those cells having a predominance of this membrane using zinc iodide in potassium ferricyanide. Harris (1979) speculates that tER may be a transition form of ER that might be involved in its synthesis or degradation. This notion would be in keeping with the membrane flow concept of membrane biogenesis proposed by Mollenhauer and Morré (1980) and would be consistent with current ideas on the synthesis of the lipid components of the ER (see below).

3 Interactions Between Tubular and Cisternal ER

3.1 Role in Protein Transport

A role for tER in the intracellular transport of proteins from rER to dictyosomes was put forward by Harris (1979) based largely on the observation that transport vesicles were not observed in thick-sectioned material. The recent observations of Harris and Oparka (1983) and Juniper et al. (1982) lend direct support to the notion that tER is involved in the process of intracellular protein transport. These workers have shown direct connections between ER and the dictyosome that are of three types: (1) The cis-face of the dictyosome is connected directly to tER, (2) the trans-face of the dictyosome is connected by fine tubules to ER, and (3) dictyosomes and ER are connected via the Golgi-ER-lysosome (GERL) network (Harris and Oparka 1983).

These numerous interconnections between plant dictyosomes and ER suggest that protein transport can take different routes according to the types of protein being sequestered. Harris and Oparka (1983) propose that acid hydrolases may take the path from cER to dictyosomes via GERL, while others may pass directly from cER to the trans-face of the dictyosome. In this way the interaction between plant dictyosomes and ER resembles that in animals, which until recently was considered a more complex system for the sorting of transported proteins (Rothman 1981).

3.2 Role in Cell Division

Other functions have been proposed for tER and cER based on the distribution of these membranes during mitosis and cytokinesis (Hawes et al. 1981; Hepler

1980, 1983). During mitosis in corn (*Zea mays* L.) and barley (*Hordeum vulgare* L.) roots, disappearance of the nuclear envelope is coupled with the appearance of a network of cER that surrounds the spindle apparatus. This arrangement of ER seems to isolate the spindle from the cytoplasm and thus exclude cytoplasmic organelles. Elements of tER are also found interspersed among the dividing chromosomes of barley and corn. During anaphase and telophase, tER is concentrated in the regions of the spindle poles as well as in the plane of cell plate formation. Hepler (1980) has proposed that the ER that is initially associated with elements of the mitotic spindle might be involved in the regulation of chromosome movement by regulating local Ca^{2+} concentrations. He also speculated that these elements of the ER might serve to anchor the mitotic spindle during movement of chromosomes to the pole.

Tubular ER forms a complex network with vesicles during cell plate formation (Hawes et al. 1981; Hepler 1980, 1983). These tubular elements are different from the vesicles that coalesce to form the cell plate, and it has been proposed that the tER acts as a template for the organization of the new cell wall. Hepler (1983) has suggested that, by playing a role in Ca^{2+} sequestration, tER might also influence the fusion of vesicles in the cell plate and therefore regulate cell-wall synthesis.

4 Synthesis and Degradation of ER

4.1 Membrane Proteins

Like other components of the endomembrane system, the ER of plant cells is composed largely of protein (about 60–65% of total membrane mass) and lipid (30%) (Phillip et al. 1976). As expected, the protein complement of ER is diverse and consists of integral and peripheral proteins (Fig. 2), as well as those proteins sequestered in the lumen of the ER. Separation of the integral proteins of castor bean (*Ricinus communis*) endosperm ER by two-dimensional polyacrylamide gel electrophoresis shows the presence of at least 30 polypeptides (Goldberg et al. 1982). This study also demonstrated that ER proteins from endosperm of maturing seeds were qualitatively different from those isolated from the endosperm of germinating seedlings. Proteins from the ER of maturing seeds had molecular weights in the 45,000–68,000 mol.wt. range, while polypeptides from the ER of seedlings were of considerably lower molecular weights. The authors inferred a functional difference between the ER of mature seeds and seedlings based on this qualitative difference in polypeptide composition.

Although there is a large body of evidence that the ER of plant cells is involved in the synthesis of protein destined for transport to storage compartments or the cell exterior (see Chrispeels 1980; Chrispeels and Jones 1980), the site of synthesis of ER membrane proteins in plant cells is not well understood. The proteins of the ER membrane can be classified according to their position in the membrane (Fig. 2), and evidence from work with animal cells suggests that the site of protein

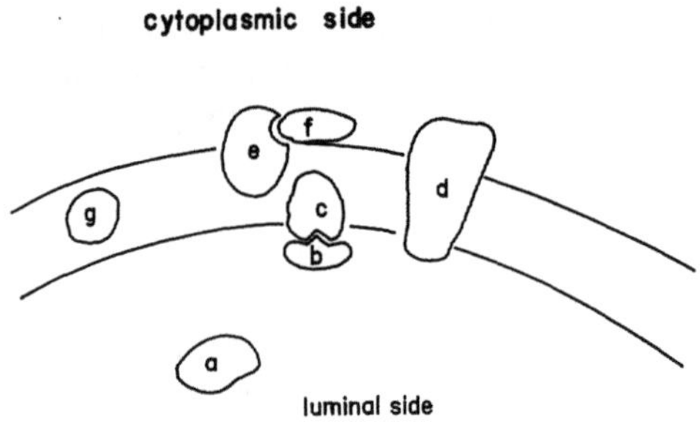

Fig. 2. Possible dispositions of ER membrane proteins. *a* Luminal protein; *b* luminal-face peripheral membrane protein; *c* luminal-face integral protein; *d* transmembrane protein; *e* cytoplasmic-face integral membrane protein; *f* cytoplasmic-face peripheral membrane protein; *g* intramembranous protein (hypothetical) (Sabatini and Kreibich 1976)

synthesis, whether on membrane-bound or free polysomes, and the mechanism for insertion of the protein into the membrane may vary according to the disposition of the protein in the membrane (Svardal and Pryme 1981).

In the case of cytochrome b-5, which is an integral protein of the ER, the hydrophobic sequences at the carboxy terminus of the protein are embedded within the membrane, while the hydrophilic portion of the protein faces the cytoplasmic side of the ER (Rachubinski et al. 1980). The enzyme NADH-cytochrome b-5 reductase is also an integral ER protein inserted into the membrane with its hydrophilic region exposed to the cell cytoplasm (Meldolesi and Borgese 1981). Both cytochrome b-5 and the cytochrome b-5 reductase are synthesized on free polyribosomes and become integrated into the ER posttranslationally.

The synthesis of other intrinsic ER membrane proteins occurs on membrane-bound polyribosomes (Fujikuriyama et al. 1979; Lodish et al. 1981). The synthesis of viral coat protein G of vesicular stomatitis virus (VSV G) occurs on ER-bound polyribosomes as does the synthesis of cytochrome P-450. The VSV G protein is synthesized on ER and its transport across the ER membrane involves the participation of a signal sequence (Lodish et al. 1981) in a manner analogous to the transmembrane transport of secreted polypeptides (Walter et al. 1982). The signal sequence is cleaved from the growing VSV G protein in the lumen of the the ER and the polypeptide chain is glycosylated as it elongates by the addition of two carbohydrate chains. When synthesis of this protein is complete, the VSV G molecule spans the membrane with the glycosylated N-terminal end folded in the lumen of the ER and the carboxy-terminal end, consisting of some 20 amino acids, projecting from the cytoplasmic face of the ER (Lodish et al. 1981).

Comparable information on the synthesis of intrinsic proteins of plant membranes is not available. However, the presence of hydrophilic and hydrophobic domains within the proteins might dictate sites for synthesis and mechanisms of transport similar to those reported above for animals.

4.2 Membrane Lipids

Much more is known about the synthesis of the lipid components of plant ER. Phospholipid represents the major class of lipids of the ER membranes in plant tissues. For example, in the endosperm of *Ricinis communis*, 74% of total ER lipid is accounted for by phosphatidyl choline (PC, 33%), phosphatidyl ethanolamine

(PE, 27%) and phosphatidyl inositol (PI, 14%) (Donaldson and Beevers 1977). The distribution of phospholipids in the ER membrane is assymetric. For example, 52% of ER PC is found in the outer leaflet of the lipid bilayer and 20% is in the inner leaflet (Moore 1982). In the glyoxysome of *R. communis* endosperm on the other hand, 78% of the PC is associated with the inner leaflet of the membrane and only 22% is found on the outer part of the lipid bilayer (Moore 1982).

The ER can be regarded as one of the major sites of phospholipid biosynthesis in plant cells (Moore 1982). In *R. communis* endosperm the enzymes responsible for the synthesis of PC, PE, and PI, as well as phosphatidyl glycerol and phosphatidylserine (PS) are all localized in the ER (Moore 1982). Indeed, in this tissue the synthesis of PC, PE, PI, and PS may be confined to the membranes of the ER.

Because enzymes of phospholipid metabolism are concentrated in the ER, this membrane system may play a pivotal role in the biogenesis of other organelles (Chrispeels 1980). The complex interrelationships between cER and tER and elements of the dictyosome and GERL suggest that the role of ER is more complex than is inferred from biosynthetic data. Evidence from Chrispeels' laboratory indicates that the ER is a dynamic system where synthesis and turnover of membrane may involve an interplay between its tubular and cisternal components (Gilkes and Chrispeels 1980a, b; Harris and Chrispeels 1980). During germination and early seedling growth in *Phaseolus aureus* there is a preferential loss of tER followed by an increase in cER prior to mobilization of stored protein. It is not known whether these data reflect the synthesis of new ER from old ER. It is easy to speculate, however, that tER could provide the substrates for the biogenesis of membranes either by membrane flow (Morré 1975) or by the participation of phospholipid transfer proteins (PTPs) (Moore 1982). The activities of several PTPs have been documented in several plant tissues, in particular PTPs for PC and PE have been reported from *R. communis* endosperm (Moore 1982).

5 Isolation and Characterization of ER

Although the ER has been described in detail at the ultrastructural level based largely on its characteristic morphology, a characterization of this membrane system at the biochemical level has proved more elusive. The challenge that is presented to the biochemist is twofold. In order to provide a chemical characterization of ER the membrane must first be isolated, but isolation of the membrane requires some knowledge of its biochemical features. It is not surprising, therefore, that the characterization of ER and other organelles has progressed as methods for the isolation of these membranes have improved.

5.1 Isolation Media

Since ER is isolated from the same cellular milieu as other organelles, the details of pH, osmolarity, and temperature are well understood and have been cataloged in detail in several recent accounts of organelle isolation from plant tissues (Hall and Moore 1983; Quail 1979; Reid 1979). When isolating ER, however, particular attention must be paid to the cation concentration and composition of the isolation medium. Plant ER (Lord 1983), like its animal counterparts (Dallner and Azzi 1982), has a high negative surface charge, and addition of cations, particularly divalent cations, to the isolation medium promotes aggregation of these membranes. Investigators using high concentrations of ionic buffers, for example 100 mM sodium phosphate, also report the aggregation of isolated plant organelles (Jelsema et al. 1977).

The levels of Mg^{2+} and K^+ in isolation media must be rigorously controlled, since these ions influence dissociation of ribosomes and their consequent detachment from the surface of the ER (Sabatini et al. 1966). For the isolation of ER from plant tissues Lord et al. (1973) developed an isolation medium that has been widely used. Smooth-surfaced ER is isolated in a medium containing 10 mM KCl, 1 mM EDTA, 150 mM Tricine (pH 7.5), and 400 mM sucrose. Rough-surfaced ER is isolated in the same medium with the addition of 3 mM $MgCl_2$.

The preparation of tER for studies of protein synthesis has relied on methods similar to those described by Lord et al. (1973) (see Adeli and Altosaar 1983; Cameron-Mills et al. 1978; Hurkman and Beevers 1982; Puchel et al. 1979) with modifications aimed at minimizing degradation of RNA. These modifications include the use of RNase-free glassware (Matthews and Mifflin 1980) and the inclusion of specific agents such as heparin (Greene 1981) in the homogenizing medium to reduce polysome degradation.

5.2 Tissue Homogenization

Because of the rigidity of the cell wall, considerable force must be applied to break open plant cells. Numerous methods have been used to overcome this mechanical barrier (Price 1983), but the methods of choice for the isolation of intact ER in high yield generate low shear forces. The simplest and one of the more successful methods reguires that the tissue be chopped by hand with a sharp razor blade (Lord et al. 1973; Ray et al. 1969b). While yielding intact ER, this method limits the amount of tissue that can be homogenized. Mechanical devices that incorporate the low shear forces of razor blade chopping and offer increased yields include devices to mimic manual chopping (Beevers and Mense 1977; Morré 1971), as well as a more sophisticated apparatus that can process several kilograms of tissue in a few minutes (Leigh et al. 1979). Perhaps the most suitable and inexpensive device for routine homogenization of plant tissue is the one adapted from a commercial carving knife where the reciprocating knife blades are modified by removing the honed blade and brazing a razor blade holder to the shank of each blade (Beevers and Mense 1977). The chopper is operated by holding it vertically.

5.3 Organelle Isolation

Centrifugation is almost always used at some stage in the purification of plant organelles, and excellent reviews of this technique are provided by Price (1982,

1983). Homogenates of plant tissue are generally subjected to differential centrifugation as the first step in the purification of ER. Centrifugation at low speed (1000 *g*, 10 min) or filtration through cheese cloth, or a combination of these methods, is invariably used to remove cell debris and cell-wall material, and ER can be prepared from this low speed supernatant by density gradient methods or by further differential centrifugation.

5.3.1 Molecular Sieve Chromatography

Several investigators have used molecular sieve chromatography of low speed supernatants as a step in ER isolation prior to further purification by centrifugation. Gel filtration using Sephadex G-200 and Sepharose 4-B has been used for the separation of organelles for the purification of ER from barley aleurone (Firn 1975; Jones 1980; Locy and Kende 1978) as well as from storage tissues of other cereals (Adeli and Altosaar 1983) or legumes (Gilkes and Chrispeels 1980a; Van Der Wilden et al. 1980). This method has two particular advantages. It separates organelles from soluble components of the homogenate and at the same time suspends them in a medium of known composition (Fig. 3; Jones 1980). This method has particular utility for the isolation of organelles from storage tissues of plants since these tissues are particularly rich in cations and hydrolytic enzymes that could interfere with the isolation of ER (Jones 1980).

5.3.2 Differential Centrifugation

Although differential centrifugation has been used for the preparation of rough or smooth microsomes, it is clear that these preparations contain variable amounts of membrane derived from ER, dictyosomes, plasmalemma and tonoplast (Dallner 1974; Quail 1979). In a study of auxin (naphthalene acetic acid, NAA) binding in corn coleoptiles, Ray (1977) showed that the ER was distributed in pellets following centrifugation at 2000, 10,000, 41,000, and 130,000 *g* (Table 1), although the highest specific activity of NAA binding was in the 130,000 *g* pellet. By varying the time of each step of differential centrifugation,

Table 1. Sedimentation of NAA- and NPA-binding activities in differential centrifugation Binding assays were performed on one-tenth of the particulate material of each centrifugal fraction, derived from 10.8 g fresh weight of coleoptile tissue; 20,000 cpm of (^{14}C) NAA or 5,000 cpm of ^{3}H-NPA were supplied in each assay. Centrifugal force is given for r_{av} (Ray 1977)

Centrifugation step		Particulate	^{3}H-NPA binding		^{14}C-NAA binding		Specific ^{14}C-NAA
Force	Time	protein per assay	Non-specific	Specific	Non specific	Specific	binding mg^{-1} protein
g	min	mg	cpm	cpm	cpm	cpm	cpm mg^{-1}
2,000	10	0.49	85	405	155	100	204
10,000	10	0.99	105	555	260	270	270
41,000	15	1.06	120	560	280	590	550
123,000	20	0.53	95	335	175	610	1,150

Table 2. Comparison of two centrifugation procedures used to separate initially mitochondria and microsomes from corn root homogenates. (Nagahashi and Hiraike)

Fraction	Cyt c Oxidase			pH 6.5, K^+-Stimulated ATPase			Antimycin A-intensive NADH cyt c reductase		
	Specific activity	Total activity		Specific activity	Total activity		Specific activity	Total activity	
	μmol mg^{-1} protein min^{-1}	μmol $fraction^{-1}$ min^{-1}	%	μmol mg^{-1} protein h^{-1}	μmol $fraction^{-1}$ h^{-1}	%	μmol mg^{-1} protein min^{-1}	μmol $fraction^{-1}$ min^{-1}	%
1,000 g– 6,000 g	0.881	4.60	92.7[a]	5.40	28.2	34.5[a]	0.123	0.64	27.6[a]
6,000 g–80,000 g	0.107	0.36	7.3	5.43	53.5	65.5	0.503	1.68	72.4
1,000 g–13,000	0.800	4.82	93.9	5.70	47.8	58.8	0.163	0.98	38.7
13,000 g–80,000 g	0.072	6.1	0.31	3.18	33.5	41.2	0.366	1.55	61.3

[a] Expressed as percentage of total activity recovered in fractions obtained by differential centrifugation up to 80.000 *g*

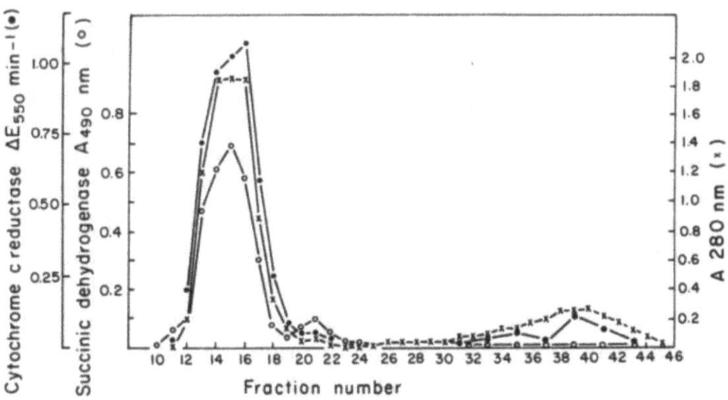

Fig. 3. Separation of particulate and soluble fractions of a barley aleurone homogenate by gel filtration on Sepharose 4B (Jones 1980)

it is possible to achieve a substantial enrichment of either pellet or supernatant fractions for a particular organelle (Nagahashi and Hiraike 1982). In a homogenate from corn roots Nagahashi and Hiraike (1982) used differential centrifugation to prepare a pellet containing 92.7% of the mitochondria present following centrifugation at 6,000 *g* for 20 min and a pellet containing 72.4% of the ER after centrifugation of the post-mitochondrial supernatant at 80,000 *g* for 35 min (Table 2).

When differential centrifugation is used as a step in the purification of ER, experience has shown that the method should be used only as a means of removing contaminating organelles such as mitochondria or plasmamembrane from the supernatant fractions (Table 2). If microsomes are first concentrated from the supernatant by pelleting, then these organelles do not separate well when subsequently purified by density gradient centrifugation (Ray 1977). Ray (1977) and others have concentrated microsomes by centrifugation of supernatant fractions onto a dense sucrose cushion prior to further purification. On the other hand, I have achieved good separation of barley aleurone organelles by loading the fraction excluded from a Sepharose 4-B molecular sieve column directly onto the density gradient (Fig. 3 and Jones 1980). This latter procedure has the disadvantage that less organelle protein can be applied to each gradient, while it avoids the potentially harmful effects of diluting an osmotically fragile organelle fraction collected on a sucrose cushion (Ray 1977).

5.3.3 Density Gradient Centrifugation

Density gradient centrifugation is the method of choice for the purification of ER membranes, and sucrose is the medium that has been favored for the gradient. Other gradient materials have been used for the isolation of plant organelles, and for the isolation of chloroplasts, mitochondria, and nuclei from green tissues these media may be superior to sucrose (Price 1983). These media include iodinated derivatives of benzoic acid (Metrizamide, Renograffin, Urograffin) that make solutions of high density and low viscosity. Metrizamide, for example, possesses no ionizable groups, is stable at room temperature, and at a concentration of 500 g l^{-1} has a density of 1.28 g cm^{-3} and a viscosity of 6 cp at 20 °C (Rickwood and Birnie 1975). By contrast, at a concentration of 500 g l^{-1}, sucrose has a density of only 1.19 g cm^{-3} but a viscosity of 8.5 cp. Although Metrizamide has not been widely used for the isolation of ER, it has found particular application for separation of the plasma membrane (Jacobs and Hertel 1978). Since plasma membrane appears to be less permeable to Metrizamide than other components of the microsomal fraction, this density gradient medium may prove useful in separating ER from fractions contaminated with plasmalemma (Hertel 1979).

Both rate-zonal and isopycnic density gradients have been utilized for the separation of ER from plant cells, and in cases where organelles have the same bouyant density but different size, a combination of both methods has been used to achieve purification of ER (Jones 1980; Koller and Kindl 1982; Ray 1977).

Rate-zonal gradients separate organelles based on differences in their sedimentation coefficients, and this type of centrifugation is generally used to separate organelles that are of similar densities but different sizes. For example, rough microsomes of barley aleurone have densities in the range of 1.18–1.19 g cm^{-3}, while the mitochondria of this tissue also band at this density (Jones 1980). Centrifugation of a fraction from an isopycnic gradient containing rER and mitochondria on a rate-zonal gradient separates the ER from the mitochondria.

Rate-zonal centrifugation, on the other hand, does not separate ER from other microsomal components. These membranes can be separated by isopycnic centrifugation, based on the principle that particles will come to equilibrium with the medium when particle bouyant density is equal to the density of the medium. Isopycnic centrifugation has been used successfully by Ray (1977) to separate microsomal membranes of corn coleoptiles that could not be separated by rate-zonal centrifugation (Fig. 4). When these microsomes were subjected to isopycnic centrifugation, ER and plasmamembrane fractions were readily separated (Fig. 5)

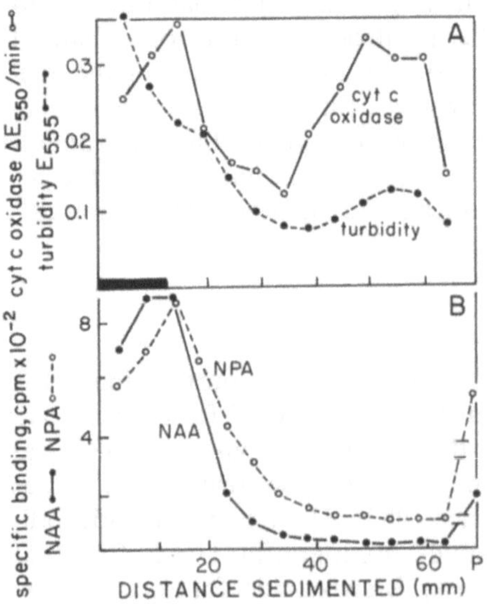

Fig. 4. Rate zonal sedimentation of NAA- and NPA-binding particles from a maize coleoptile homogenate (Ray 1977)

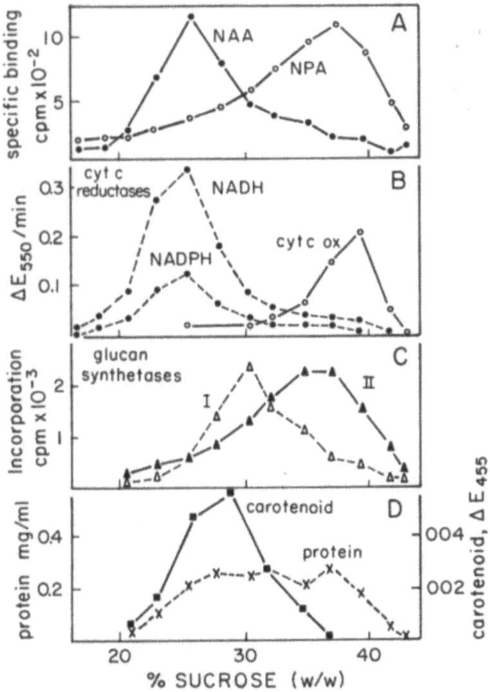

Fig. 5. Isopycnic density gradient fractionation of NAA- and NPA-binding activities and comparison enzymes associated with particles in the 6780 *g* supernatant of a maize coleoptile homogenate (Ray 1977)

and were determined to have bouyant densities of 1.15 g cm^{-3} and 1.10 g cm^{-3} respectively (Ray 1977).

A combination of rate zonal and isopycnic centrifugation was used to separate ER from malate synthetase-containing particles in homogenates from cucumber (*Cucumis sativus*) cotyledons (Koller and Kindl 1978). These investigators also employed floatation centrifugation to achieve separation of organelles on isopycnic gradients. A cucumber cotyledon homogenate was first subjected to rate-zonal centrifugation and two zones from this gradient containing ER at 23–26% (w/w) sucrose (light fraction) and 27–31% (w/w) sucrose (heavy fraction) were brought to 35% sucrose and recentrifuged isopycnically on a 10–60% (w/w) gradient. Recentrifugation gave clear separation of particles and established that the ER from both light and heavy fractions possessed an equilibrium density of 1.115 g cm^{-3} (Koller and Kindl 1978).

5.4 Identification of ER Membranes

5.4.1 Magnesium Shift

The criterion that has received the widest acceptance for the identification of ER concerns the so-called Mg shift (Lord 1983; Quail 1979). Rough ER undergoes a shift in its bouyant density when the Mg^{2+} level of the homogenate is changed (Quail 1979). In the absence of Mg^{2+}, ribosomes dissociate from the membrane

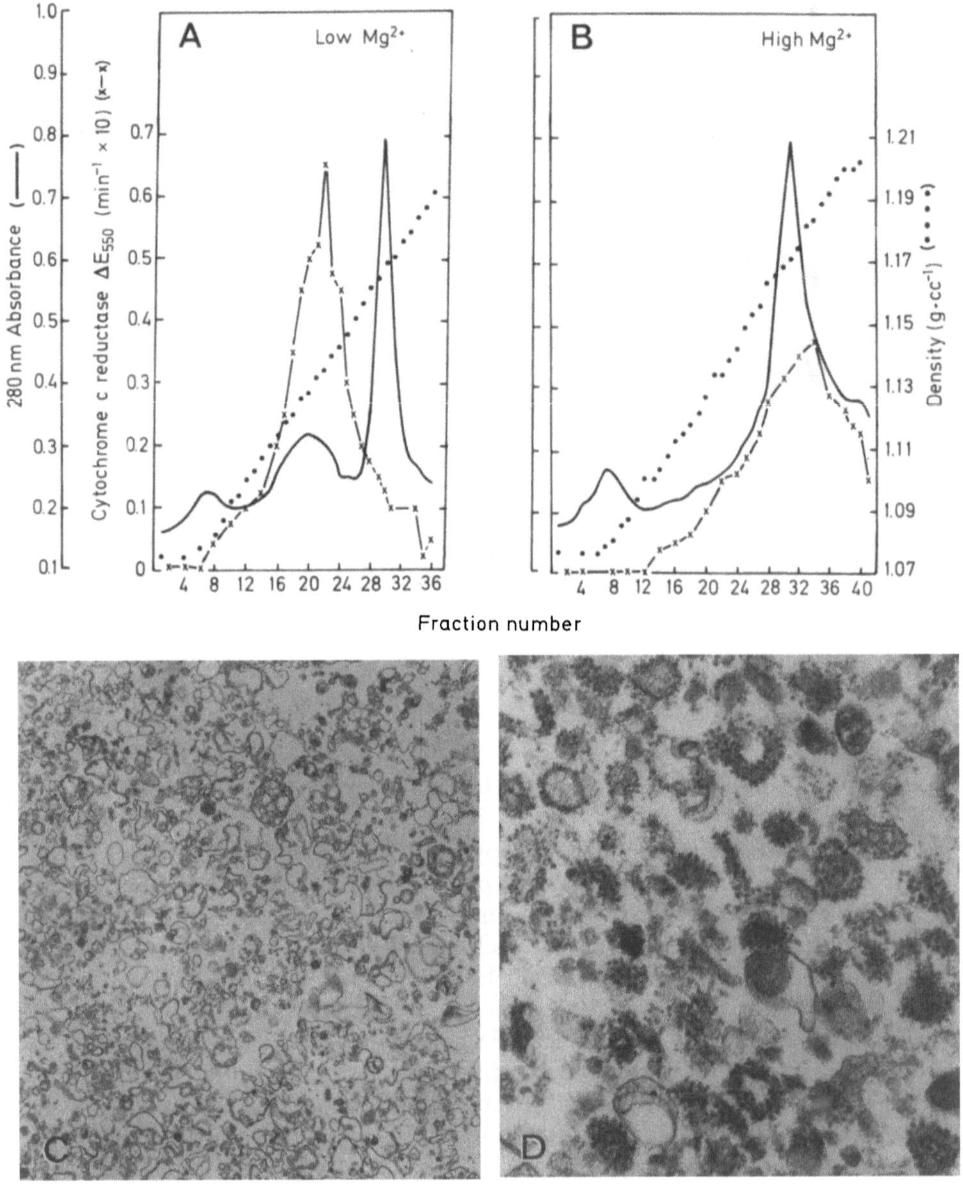

Fig. 6. A, B Separation of barley aleurone organelles by isopycnic centrifugation in low (**A**) and high (**B**) magnesium following filtration on Sepharose 4B. **C, D** Smooth (**C**) and rough (**D**) ER isolated from barley aleurone in the presence of low (**C**) or high (**D**) magnesium (All from Jones 1980). **C** × 7700; **D** × 37,000

resulting in a decrease in its bouyant density from approximately 1.15–1.19 g cm^{-3} to 1.10–1.13 g cm^{-3} (Fig. 6). There are several precautions that must be adhered to when this procedure is used to identify ER, including the use of low Mg^{2+} concentrations and the careful monitoring of the density of other organelles. High concentrations of Mg^{2+} and other divalent cations (>20 mM) cause plant organelles to aggregate and consequently change their bouyant densities. When using the Mg^{2+} shift as a criterion for identifying ER, it is essential to monitor the bouyant densities of other organelles to establish that the shift is due to ribosome removal from ER rather than the clumping of organelles. From Fig. 6 it is evident that homogenization of aleurone tissue in the presence of 3 mM Mg^{2+} causes no change in the bouyant density of mitochondria but causes a change in position of ER from 1.13 g cm^{-3} to 1.18 g cm^{-3}.

Proof that removal of Mg^{2+} causes rER to lose ribosomes and appear as smooth-surfaced membrane can be obtained by electron microscopy (Lord 1983; Quail 1979). For barley aleurone ER, the membranes isolated in the presence of 3 mM Mg^{2+} are rough-surfaced and possess those features characteristic of rER, while those membranes isolated in the absence of Mg^{2+} are devoid of ribosomes (Fig. 6).

5.4.2 Marker Enzymes

Enzymes and other proteins have also been used to identify the ER of plant and animal cells. Controversy exists as to what constitutes a valid marker enzyme for ER, since many enzymes that were at one time regarded as characteristic of ER have been found to be associated with other cytoplasmic compartments (Table 3). A good example of such an enzyme is NADH-linked cytochrome c reductase, which at one time was believed to be useful for distinguishing among the components of the microsomal fraction. This enzyme is present in mitochondria and its activity is sensitive to antimycin-A in contrast to the ER-localized enzyme whose activity is little affected by this drug (Nagahashi and Beevers 1978). Morré and Buckhout (1979) found that NADH cytochrome c reductase is present in other fractions of the microsomal pellet including the dictyosome, and they suggested that this enzyme activity is distributed ubiquitously among organelles.

The NADPH-linked enzyme (NADPH cytochrome c reductase), on the other hand, is a more reliable marker for ER. This enzyme, although present in homogenates at only a fraction (10% or less) of the NADH-linked activity, is absent from mitochondria and is present in other components of the microsomal fraction at lower specific activities (Quail 1979).

Enzymes of phospholipid metabolism are also found in the ER, but as with other so-called marker enzymes, their activities are not exclusively confined to ER (Lord 1983; Moore 1982; Moreau and Stumpf 1982; Quail 1979). Choline phosphotransferase, for example, has been used by many investigators as an ER marker enzyme; however, there can be no doubt that this activity is also located in membranes of the dictyosome (Montague and Ray 1977; Morré and Buckhout 1979).

One of the features which make enzymes of phospholipid metabolism particularly useful marker enzymes for ER is their location within the membrane itself.

Table 3. Enzymes, proteins and specialized functions that have been localized in the ER and the other organelles of plant cells

Enzyme, protein or function	ER	Other organelles	Reference
"Mg shift"	Yes	No	Lord (1983), Quail (1979)
Ribosomal RNA (16S, 26S RNA)	Yes	Yes	Dobberstein et al. (1974), Gressel and Quail (1976), Kagawa et al. (1973), Lord (1983)
NADH cytochrome c reductase	Yes	Yes	Bowles et al. (1979), Lord (1983), Morré and Buckhout (1979), Quail (1979), Ray (1977)
NADPH cytochrome c reductase	Yes	Yes	Bowles et al. (1979), Lord (1983), Morré and Buckhout (1979), Quail (1979), Ray (1977)
o-Lysophospholipid acyl transferase	Yes	?	Moreau and Stumpf (1982), Stymne and Glad (1981)
Phosphatidate phosphatase	Yes	?	Herman and Chrispeels (1980), Moore (1982), Moore et al. 1973), Moore and Sexton (1978)
Choline phosphotransferase	Yes	?	Montague and Ray (1977), Moore (1982)
Ethanolamine phosphotransferase	Yes	Yes	Montague and Ray (1977), Moore (1982)
Phosphatidyl ethanolamine methyl transferase	Yes	Yes	Moore (1976, 1982)
Phosphatidyl serine base exchange	Yes	?	Moore (1975, 1982)
Phosphatidyl serine decarboxylase	Yes	?	Moore (1975, 1982)
Phosphatidate cytidyl transferase	Yes	Yes	Moore (1982)
CDP diglyceride: inositol phosphatidyl transferase	Yes	?	Moore et al. (1973), Sexton and Moore (1978)
Phosphatidyl inositol base exchange	Yes	?	Moore (1982), Sexton and Moore (1981)
CDP diacyl glycerol: glycerol-P-phosphatidyl transferase/PGP phosphatase	Yes	Yes	Moore (1974, 1982)
Phosphatidyl choline transfer protein	Yes	Yes	Mazliak and Kader (1980), Moore (1982), Yamada et al. (1978)
Phospatidyl ethanolamine transfer protein	Yes	Yes	Mazliak and Kader (1980), Moore (1982), Yamada et al. (1978)
SAM △ 24 cycloartenol methyl transferase	Yes	?	Hartmann-Bouillon and Benveniste (1978)
Cycloeucalenol obtusifoliol isomerase	Yes	?	Hartmann-Bouillon and Benveniste (1978)
Farnesyl pyrophosphate squalene synthetase	Yes	?	Hartmann et al. (1977)
Kaurene oxidation to GA_{12}-aldehyde	Yes	?	Graebe (1982), Hasson and West (1976)
Cinnamate 4-hydroxylase	Yes	?	Benveniste et al. (1977), Potts et al. (1974), Saunders et al. (1977), Young and Beevers (1976)

Table 3 (continued)

Enzyme, protein or function	ER	Other organelles	Reference
Tyrosine N-hydroxylase	Yes	?	Saunders et al. (1977)
Nitrile hydroxylase	Yes	?	Saunders et al. (1977)
p-Chloro-N-methyl aniline N-methlase	Yes	?	Young and Beevers (1976)
Prolyl hydroxylase	Yes	Yes	Cohen et al. (1983), Wienecke et al. (1982)
Indoleacetic acid oxidase	Yes	?	Waldrum and Davies (1981)
Signal peptide processing	Yes	?	Adeli and Altosaar (1983), Bollini and Chrispeels (1979), Bollini et al. (1983), Burr and Burr (1981), Cameron-Mills et al. (1978), Greene (1981), Higgins and Spencer (1981), Hurkman and Beevers (1982), Larkins and Hurkman (1978), Matthews and Mifflin (1980), Puchel et al. (1979), Sengupta et al. (1981)
Glycoprotein mannosyl transferase	Yes	?	Beevers and Mense (1977), Bollini et al. (1982), Bollini et al. (1983), Conder and Lord (1983), Lehle et al. (1978), Marriott and Tanner (1979), Mellor and Lord (1979), Mellor et al. (1979), Nagahashi and Beevers (1978), Nagahashi et al. (1980)
Glycoprotein N-acetyl glucosamine transferase	Yes	?	Beevers and Mense (1977), Bergner and Tanner (1981), Bollini et al. (1982, 1983),Chrispeels et al. (1982), Conder and Lord (1983), Davies and Delmer (1981), Marriott and Tanner (1979), Nagahashi and Beevers (1978), Nagahashi et al. (1980), Sengupta et al. (1981)
Glycoprotein galactosyl transferase	Yes	?	Lang (1982), Meldolesi and Borgese (1981), Mellor and Lord (1978)
Glycoprotein fucosyl transferase	Yes	Yes	Roberts et al. (1980), Sturm and Kindl (1983)
Polyprenyl-N-acetyl glucosamine transferase	Yes	?	Beevers and Mense (1977), Bergner and Tanner (1981), Davies and Delmer (1981), Kara and Kindl (1982), Marriott and Tanner (1979), Sengupta et al. (1981)
Polyprenyl mannosyl transferase	Yes	?	Beevers and Mense (1977), Davies and Delmer (1981), Lehle et al. (1978), Marriott and Tanner (1979), Mellor et al. (1979), Nagahashi and Beevers (1978), Nagahashi et al. (1980)

Table 3 (continued)

Enzyme, protein or function	ER	Other organelles	Reference
Polyprenyl galactosyl transferase	Yes	?	Mellor and Lord (1978, 1979)
Polysaccharide fucosyl transferase	Yes	?	James and Jones (1979)
Protein disulfide isomerase	Yes	?	Freedman (1979), Roden et al. (1982)
Auxin binding	Yes	Yes	Hertel (1979), Normand et al. (1975), Oostrom et al. (1980), Quail (1979), Ray (1977), Rothman (1981), Vreugdenhil et al. (1980)
ATP-dependent Ca^{2+} transport	Yes	Yes	Buckhout (1983), Dieter and Marmé (1980, 1982), Salimath and Marmé (1983)
Cytochrome b-5	Yes	?	Lord et al. (1973), Phillip et al. (1976)
Cytochrome P-450	Yes	?	Hasson and West (1976), Lord et al. (1973), Markham et al. (1972), Potts et al. (1974), Rich and Lamb (1977)
Alkaline lipase	Yes	?	Muto and Beevers (1974)
α-Mannosidase	Yes	Yes	Van Der Wilden and Chrispeels (1983)
Vicillin peptidohydrolase	Yes	Yes	Van Der Wilden et al. (1980)
Ribonuclease	Yes	Yes	Van Der Wilden et al. (1980)
α-Amylase	Yes	?	Jones (1980), Jones and Jacobsen (1982), Locy and Kende (1978)
Allantoinase	Yes	?	Hanks et al. (1981)
Other miscellaneous storage proteins and peptides including lectins	Yes	Yes	Beevers and Mense (1977), Bollini and Chrispeels (1979), Bollini et al. (1982, 1983), Chrispeels (1983), Chrispeels et al. (1982), Greene (1981), Higgins and Spencer (1981), Hurkman and Beevers (1982), Larkins and Hurkman (1978), Matthews and Mifflin (1980), Nagahashi and Beevers (1978), Puchel et al. (1979), Sengupta et al. (1981)

These proteins appear to be integral membrane proteins in both animals and plants and they can be solubilized from the ER only after extraction with detergents. Phosphatidyl inositol synthetase from soybean (*Glycine max* L.) seedlings, for example, is localized in the ER (Carman and Dougherty 1980) and can be solubilized from the ER with the nonionic detergent Brij W-1 (Robinson and Carman 1982).

Numerous steps in terpenoid metabolism in plants also involve reactions localized on the ER (Graebe 1982; Hartmann et al. 1973, 1977; Hartmann-Bouillon and Beveniste 1978; Hasson and West 1976). In corn coleoptiles several sterol biosynthetic enzymes are localized on microsomal membranes (Hartmann-Bouillon and Beveniste 1978). S-Adenosyl-L-methionine-Δ24 cycloartenol methyl transferase and cycloeucalenol-obtusifoliol isomerase are both localized on ER, while UDP-glucose sterol glycosyltransferase is localized on the plasma membrane.

The biosynthesis of the gibberellins in higher plants involves several steps which are localized on microsomal membranes (Graebe 1982; Hasson and West 1976). These steps are catalyzed by monooxygenases, require O_2 and NADPH, and depend on cytochrome P-450 for electron transport. Results from Graebe's laboratory also indicate that these activities are confined to the ER membranes in pumpkin (*Cucurbita maxima*) (Graebe 1982).

Microsomes of plant and animal origin are known to possess other monooxygenase (hydroxylase) activities, and evidence suggests that these may be found primarily on ER. Cinnamic acid 4-monooxygenase catalyzes the hydroxylation of trans cinnamic acid and utilizes cytochrome P-450 as the electron donor. In both plants (Beveniste et al. 1977; Saunders et al. 1977; Young and Beevers 1976) and animals this enzyme has been localized on the ER. In addition to the presence of cinnamate hydroxylase, the ER of sorghum seedlings also contains hydroxylases which convert tyrosine to hydroxytyrosine and p-hydroxyphenyl acetonitrile to form p-hydroxy(s)mandelonitrile (Saunders et al. 1977).

The formation of hydroxypyroline-rich proteins from peptidyl prolyl in plants and animals involves the participation of a membrane-bound peptidyl prolyl hydroxylase which is dependent on O_2 and Fe^{2+} (Cohen et al. 1983; Grant and Jackson 1976). In animal tissue this activity seems to be associated exclusively with the ER, but it has been found in both ER and dictyosome membranes of rye grass and carrot tissue (Cohen et al. 1983; Wienecke et al. 1982).

The synthesis of proteins and their concomitant signal processing is one of the caracteristics of rER in animal cells. The capacity of ER to carry out the cleavage of the signal sequence cotranslationally is a diagnostic feature of this membrane system, and therefore signal-processing may be used to characterize the ER membrane (see Walter et al. 1982).

Evidence is now accumulating that ER of plant origin may also participate in the cleavage of an amino acid sequence from the N-terminus of polypeptide chains translated on membrane-bound polysomes. The mRNA's for a wide variety of plant proteins are translated on the rER and the newly synthesized polypeptides are translocated into the lumen of the ER (Adeli and Altosaar 1983; Bollini and Chrispeels 1979; Bollini et al. 1983; Burr and Burr 1981; Cameron-Mills et al. 1978; Greene 1981; Higgins and Spencer 1981; Hurkman and Beevers 1982; Larkins and Hurkman 1978; Matthews and Mifflin 1980; Puchel et al. 1979; Sengupta et al. 1981).

Although direct evidence supporting the removal of a signal sequence from plant proteins during co-translational processing is lacking, several observations support the notion that this processing is of the signal type. First, plant proteins that are secreted or transported from the ER to other organelles are synthesized and sequestered in the lumen of the ER have a molecular weight which is 1500 to 2000 less than proteins synthesized in vitro on free polysomes or purified mRNA. Second, addition of microsomal membranes to in vitro protein-synthesizing systems achieves co-translational processing (Higgins and Spencer 1981; Burr and Burr 1981). Co-translational processing of pea seed vicillin subunits was observed when mRNA was translated in the presence of microsomal membranes from pea or dog pancreas (Higgins and Spencer 1981). Finally, mRNAs from plant sources are translated and processed when injected into *Xenopus* oocytes (Boston et al. 1982; Hurkman et al. 1981; Larkins et al. 1979).

Plant proteins are frequently further modified by the addition of sugar residues (Elbein 1979). Many of these glycosylations have been localized on the ER although glycosylation of proteins by dictyosomes has also been established. The additions of mannose (Beevers and Mense 1977; Bollini et al. 1982, 1983; Conder and Lord 1983; Marriott and Tanner 1979; Mellor and Lord 1979b; Mellor et al. 1979; Nagahashi and Beevers 1978; Nagahashi et al. 1980) N-acetylglucosamine (Beevers and Mense 1977; Bergner and Tanner 1981; Bollini et al. 1982, 1983; Chrispeels et al. 1982; Conder and Lord 1983; Davies and Delmer 1981; Marriott and Tanner 1979; Nagahashi and Beevers 1978; Nagahashi et al. 1980; Sengupta et al. 1981) and fucose (Roberts et al. 1980) to proteins have been reported to occur in the ER. In the case of fucose, however, there is dispute on whether glycoprotein fucosyl transferase occurs on ER, since Sturm and Kindl (1983) report that in cucumber cotyledons this activity is confined almost exclusively to the dictyosome fraction.

Chrispeels and colleagues (Bollini et al. 1982, 1983; Chrispeels 1983; Chrispeels et al. 1982) have made a particularly thorough study of the post-translational processing of the storage proteins of *Phaseolus vulgaris* cotyledons. In addition to demonstrating that these proteins are co-translationally processed to remove a signal sequence, they showed that glycosylation occurred in two steps. The first glycosylation step occurs co-translationally and independent of the ER, since glycosylated polypeptides are formed in vitro by polysome run-off. The second glycosylation step, on the other hand, adds sugar residues to only the most abundant polypeptides and requires the participation of the ER (Bollini et al. 1983).

It is clear that the synthesis and post-translational modification of many plant proteins occur on the ER. These functions might therefore prove invaluable as diagnostic tools to establish the identity of this membrane system.

5.4.3 Auxin Binding

The binding of auxins to the ER is also a highly reliable feature which helps characterize this membrane system (Normand et al. 1975; Ray 1977; Ray et al. 1969a; Rubery 1982). Ray (1977) has provided convincing evidence that NAA is bound specifically to the ER of *Zea mays* coleoptiles and that the K_D for this binding is of the order of $5–7 \times 10^{-7}$ (Ray 1977). Although auxin binding to *Z. mays* ER can be readily demonstrated, auxin binding to this membrane fraction may not be ubiquitous. Thus, the ER-binding site for auxin is absent in homogenates of zucchini hypocotyls (Jacobs and Hertel 1978), but present in tobacco pith callus (Oostrom et al. 1980) and tobacco leaf homogenates (Vreugdenhil et al. 1980).

5.4.4 Calcium Transport

In animal cells, Ca^{2+} uptake sites have been localized on the ER and the ability to transport Ca^{2+} is dependent on ATP and Mg^{2+} (Bruns 1976; Colca et al. 1983). A similar site for Ca^{2+} uptake on the ER has been described for various plant tissues (Buckhout 1983; Dieter and Marmé 1980, 1982; Salimath and Marmé 1983). Buckhout (1983) has shown that in roots of *Lepidium sativum* L. ATP-dependent Ca^{2+} transport is indeed localized in the ER. In his preparations of garden cress, Buckhout showed that rER represented 78.0% of the total membrane fraction with mitochondria representing only 8% of this fraction. Using inhibitors of mitochondrial Ca^{2+} transport together with the morphometric analysis of this membrane fraction, Buckhout (1983) argued that Ca^{2+} transport into the ER was responsible for the observed ATP-dependent Ca^{2+} transport. As with other activities of the ER, Ca^{2+} transport and its possible stimulation by calmodulin (see Dieter and Marmé 1980, 1982) may be a valuable indicator of this membrane system.

5.4.5 Structural Proteins

Several structural, nonenzymatic proteins of the ER may prove to be among the best markers for identifying this membrane system. The presence of cytochromes P-450 and b-5 in microsomal membranes has been known for more than 12 years (Markham et al. 1972) and the location of these proteins as components of the ER was established by Lord et al. (1973). These proteins can be identified by their characteristic spectra and cytochrome P-450 can be distinguished from cytochrome b-5 on the basis of difference spectra when reduced with dithionite.

Although it is not possible at this time to identify a single membrane protein which is exclusively localized on the ER (Table 3), a combination of several criteria can provide positive identification of this membrane. These criteria, however, should always include a combination of the Mg-shift together with the presence of several well established marker enzymes and preferably evidence for the presence of structural and integral membrane proteins, e.g., cytochromes b-5 and P-450.

5.5 Concluding Remarks

Pitfalls await the investigator who attempts to ascertain the role of ER in biosynthetic activities of the cell based on experiments with purified membrane fractions alone. The work of Kindl and his colleagues (Kindl 1982) emphasizes the dangers inherent in drawing conclusions based only on experiments with membranes purified on sucrose density gradients. These workers have shown that the widely held belief that microbody (glyoxysome and peroxisome) proteins are synthesized on rER and transported via the endomembrane system (Beevers 1979) is not valid.

The hypothesis that microbody proteins were synthesized on the rER was predicated on the observations that enzymes such as malate synthetase and isocitrate lyase are associated with ER membranes on density gradients; that these proteins are glycosylated and are therefore synthesized on the ER; that ER and glyoxysomes frequently appear connected in electron micrographs; and that the polypeptide composition of the microbody and ER membrane are similar. The first two of these predicates are now clearly incorrect. It has been shown that malate synthetase is not associated with the ER, and that its apparent localization in this membrane system is fortuitous, since this enzyme possesses sedimentation characteristics similar to ER (Kindl 1982). Furthermore, malate synthetase has been shown not to be a glycoprotein. Kindl and his associates have gone on to show that microbody enzymes are synthesized on free polysomes. Thus isocitrate lyase and malate synthetase are synthesized in vitro in a form indistinguishable from the mature enzyme located within the glyoxysome. It is not yet known, however, how glyoxosomal enzymes are recruited from the cytosolic pool into the developing glyoxysome (Kindl 1982).

In those investigations involving synthesis of proteins on rER, the fortuitous association of newly synthesized protein with the surface of the ER membrane can be tested by incubation of membranes with an appropriate protease (Bollini et al. 1982; Hurkman and Beevers 1982; Jones and Jacobsen 1982). If labeled protein is fortuitously associated with the surface of the ER or is a peripheral protein on the cytoplasmic side of the membrane (see Fig. 2), then it will be degraded by the protease. On the other hand, if the newly synthesized protein is a luminal protein or in some other way protected by the membrane, it will not be susceptible to proteolytic attack. As an added control in this type of experiment, detergent treatment of ER vesicles should render newly synthesized proteins, previously resistant to proteolytic attack, accessible to the proteolytic activity of the enzyme.

References

Adeli K, Altosaar I (1983) Role of endoplasmic reticulum in biosynthesis of oat globulin precursors. Plant Physiol 73:949–955

Beevers H (1979) Microbodies in higher plants. Annu Rev Plant Physiol 30:159–193

Beevers L, Mense R (1977) Glycoprotein biosynthesis in cotyledons of *Pisum sativum* L. Plant Physiol 60:703–708

Benveniste I, Salaun JP, Durst F (1977) Phytochrome-mediated regulation of a monooxygenase hydroxylating cinnamic acid in etiolated pea seedlings. Phytochemistry 17:359–363

Bergner U, Tanner W (1981) Occurrence of several gylcoproteins in glyoxysomal membranes of castor bean. FEBS Lett 131:68–72

Bollini R, Chrispeels MJ (1979) The rough endoplasmic reticulum is the site of reserve-protein synthesis in developing *Phaseolus vulgaris* cotyledons. Planta 146:487–501

Bollini R, Van der Wilden W, Chrispeels MJ (1982) A precursor of the reserve-protein, phaseolin, is transiently associated with the endoplasmic reticulum of developing *Phaseolus vulgaris* cotyledons. Physiol Plant 55:82–92

Bollini R, Vitale A, Chrispeels MJ (1983) In vivo and in vitro processing of seed reserve protein in the endoplasmic reticulum – evidence for 2 glycosylating steps. J Cell Biol 96:999–1007

Boston R, Miller TJ, Burgess RR (1982) In vitro synthesis and processing of wheat α-amylase. Plant Physiol 69:150–154
Bowles DJ, Quail PH, Morré DJ, Hartmann C (1979) Use of markers in plant cell fractionation. In: Reid E (ed) Plant Organelles: methodological surveys B. Biochemistry, vol 9. Harwood, Chichester, pp 207–223
Bruns DE (1976) Energy-dependent calcium transport in endoplasmic reticulum of adipocytes. J Biol Chem 251:7191–7197
Buckhout TJ (1983) ATP-dependent Ca^{2+} transport in endoplasmic reticulum isolated from roots of *Ledpidium sativum*. Planta 159:84–90
Burr FA, Burr B (1981) In vitro processing and uptake of prezein and other maize proteins by maize membranes. J Cell Biol 90:427–434
Cameron-Mills V, Ingversen L, Brandt A (1978) Transfer of in vitro synthesized barley endosperm proteins into the lumen of the endoplasmic reticulum. Carlsberg Res Commun 43:91–102
Carman GM, Dougherty M (1980) Subcellular localization of inositol synthetases from germinating soybeans. J Food Biochem 4:153–158
Chrispeels MJ (1980) The endoplasmic reticulum. In: Stumpf PK, Conn EE (eds) The biochemistry of plants. A comprehensive treatise, vol 1. Academic Press, New York, pp 389–435
Chrispeels MJ (1983) Biosynthesis, processing and transport of storage proteins and lectins in cotyledons of developing legume seeds. Proc Phil Trans Soc Lond B 304:309–322
Chrispeels MJ, Jones RL (1980) The role of the endoplasmic reticulum in the mobilization of reserve macromolecules during seedling growth. Isr J Bot 29:225–245
Chrispeels MJ, Higgins TJV, Spencer D (1982) Assembly of storage protein oligomers in the endoplasmic reticulum and processing of the polypeptides in protein bodies of developing pea cotyldons. J Cell Biol 93:306–313
Cohen PB, Schibeci A, Fincher GB (1983) Biosynthesis of arabinogalactan-protein in *Lolium multiflorum* (Ryegrass) endosperm cells III. Subcellular distribution of prolyl hydroxylase. Plant Physiol 72:754–758
Colca JR, Kotagal PE, Lacy PE, McDaniel ML (1983) Modulation of active Ca^{2+} uptake by islet cell endoplasmic reticulum. Biochem J 212:113–121
Conder MJ, Lord JM (1983) Heterogenous distribution of glycosyltransferases in the endoplasmic reticulum of castor bean endosperm. Plant Physiol 72:547–552
Dallner G (1974) Isolation of rough and smooth microsomes – general. Methods Enzymol 31:191–201
Dallner G, Azzi A (1972) Structural properties of rough and smooth microsomal membranes. A study with fluorescence probes. Biochim Biophys Acta 255:589–601
Davies HM, Delmer D (1981) Two kinds of protein glycosylation in a cell-free preparation from developing cotyledons of *P. vulgaris*. Plant Physiol 68:284–291
Dieter P, Marmé D (1980) Calmodulin activation of plant microsomal Ca^{2+} uptake. Proc Natl Acad Sci USA 77:7311–7314
Dieter P, Marmé D (1982) A calmodulin-dependent, microsomal ATPase from corn (*Zea mays* L.) FEBS Lett 125:245–248
Dobberstein B, Volkmann D, Klambt D (1974) The attachment of polyribosomes to membranes of the hypocotyl of *Phaseolus vulgaris*. Biochim Biophys Acta 374:187–196
Donaldson RP, Beevers H (1977) Lipid composition of organelles from germinating castor bean endosperm. Plant Physiol 59:259–263
Elbein AD (1979) The role of lipid-linked saccharides in the biosynthesis of complex carbohydrates. Annu Rev Plant Physiol 30:239–272
Firn RD (1975) On the secretion of α-amylase by barley aleurone layers after incubation in gibberellic acid. Planta 125:227–233
Freedman RB (1979) How many distinct enzymes are responsible for the several cellular processes involving thiol: protein disulfide interchange? FEBS Lett 97:201–210
Fujikuriyama Y, Negishi M, Mikawa R, Tashiro Y (1979) Biosynthesis of cytochrome P-450 on membrane-bound ribosomes and its subsequent incorporation into rough and smooth microsomes in rat hepatocytes. J Cell Biol 81:510–519

Gilkes NR, Chrispeels MJ (1980a) Endoplasmic reticulum of mung bean cotyledons. Accumulation during seed maturation and catabolism during seedling growth. Plant Physiol 65:600–604
Gilkes NR, Chrispeels MJ (1980b) The endoplasmic reticulum of mung bean cotyledons. Biosynthesis during seedling growth. Planta 149:361–369
Goldberg D, Al-Marayati S, Gonzales E (1982) A comparison of intrinsic endoplasmic reticulum membrane protein in maturing seeds and germinated seedlings of castor bean. Plant Physiol 68:280–282
Goosen-deRoo L, Burggraaf PD, Libbenga KR (1983) Microfilament bundles associated with tubular endoplasmic reticulum in fusiform cells in the active cambial zone of *Fraxinus excelsior* L. Protoplasma 116:204–208
Graebe J (1982) Gibberellin biosynthesis in cell-free systems from higher plants. In: Wareing PF (ed) Plant growth substances. Academic Press, New York, pp 71–80
Grant ME, Jackson DS (1976) The biosynthesis of procollagen. Essays Biochem 12:77–113
Greene FC (1981) In vitro synthesis of wheat (*Tritium aestivum* L.) storage proteins. Plant Physiol 68:778–783
Gressel J, Quail P (1976) Particle-bound phytochrome: differential pigment release by surfactant, ribonuclease, and phospholipase C. Plant Cell Physiol 17:771–776
Gunning BES, Steer MW (1975) Ultrastructure and the biology of plant cells. Arnold, London, p 312
Hall JL, Moore AL (1983) Isolation of membranes and organelles from plant cells. Academic Press, London, p 317
Hanks JF, Tolbert KR, Schuubert KR (1981) Localization of enzymes of ureide biosynthesis in peroxisomes and microbodies of nodules. Plant Physiol 68:65–69
Harris N (1979) Endoplasmic reticulum of mung-bean cotyledons: a high voltage electron microscope study. Planta 146:63–69
Harris N, Chrispeels MJ (1980) The endoplasmic reticulum of mung bean cotyledons: quantitative morphology of cisternal and tubular ER during seeding growth. Planta 148:293–303
Harris N, Oparka KJ (1983) Connections between dictyosomes, ER and GERL in cotyledons of mung bean (*Vigna radiata* L.). Protoplasma 114:93–102
Hartmann MA, Ferne M, Gigot E, Brandt R, Benveniste P (1973) Isolement, caracterisation et composition en sterols de fractions subcellulaires de feuilles etioles de Haricot. Physiol Veg 11:209–230
Hartmann M-A, Fonteneau P, Benveniste P (1977) Subcellular localization of sterol synthesizing enzymes in maize coleoptiles. Plant Sci Lett 45:45–51
Hartmann-Bouillon M-A, Benveniste P (1978) Sterol biosynthetic capacity of purified membrane fractions from maize coleoptiles. Phytochemistry 17:1037–1042
Hasson EP, West CA (1976) Properties of the systems for the mixed function oxidation of kaurene and kaurene derivatives in microsomes of the immature seed of *Marah macrocarpus*. Plant Physiol 58:479–484
Hawes CR, Juniper BE, Horne JC (1981) Low and high voltage electron microscopy of mitosis and cytokinesis in maize roots. Planta 152:397–407
Hepler PK (1980) Membranes in the mitotic apparatus of barley cells. J Cell Biol 86:490–499
Hepler PK (1981) The structure of the endoplasmic reticulum revealed by osmium tetroxide-potassium ferricyanide staining. Eur J Cell Biol 26:102–110
Hepler PK (1983) Endoplasmic reticulum in the formation of cell plate and plasmodesmata. Protoplasma 111:121–133
Herman EM, Chrispeels MJ (1980) Characteristics and subcellular localization of phospholipase D and phosphatidic acid phosphatase in mung bean cotyledons. Plant Physiol 66:1001–1007
Hertel R (1979) Auxin binding sites: subcellular fractionation and specific binding assay. In: Reid E (ed) Plant organelles. Methodological surveys B. Biochemistry, vol 9. Ellis Horwood, Chichester, pp 173–183
Higgins TJV, Spencer D (1981) Precursor forms of pea vicillin subunits. Modification by microsomal membranes during cell-free formation. Plant Physiol 67:205–211

Hurkman WJ, Beevers L (1982) Sequestration of pea reserve proteins by rough microsomes. Plant Physiol 69:1414–1417

Hurkman WJ, Smith L, Richtel J, Larkins BA (1981) Subcellular compartmentalization of maize storage proteins in *Xenopus* oocytes injected with zein messenger RNAs. J Cell Biol 89:292–299

Jacobs M, Hertel R (1978) Auxin binding to subcellular fractions from *Curcurbita* hypocotyls: in vitro evidence for an auxin transport carrier. Planta 142:1–10

James DW, Jones RL (1979) Intracellular localization of GDP fucose polysaccharide fucosyl transferase in corn roots (*Zea mays*). Plant Physiol 64:914–918

Jelsema CJ, Morré DJ, Ruddat M, Turner C (1977) Isolation and characterization of the lipid reserve bodies, spherosomes, from aleurone layers of wheat. Bot Gaz 138:138–149

Jones RL (1980) The isolation of ER from barley aleurone layers. Planta 150:58–69

Jones RL, Jacobsen JV (1982) Localization of α-amylase synthesis and transport in barley aleurone layers. Planta 156:421–432

Juniper BE, Hawes CR, Horne JC (1982) The relationship between the dictyosomes and the forms of endoplasmic reticulum in plant cells with different export programs. Bot Gaz 143:135–145

Kagawa T, Lord JM, Beevers H (1973) The origin and turnover of organelle membranes in castor bean endosperm. Plant Physiol 51:61–65

Kara VAK, Kindl H (1982) Membranes of protein bodies II. Detection and partial characterization of membrane glycoproteins. Eur J Biochem 121:539–544

Kindl H (1982) The biosynthesis of microbodies (peroxisomes, glyoxysomes). Int Rev Cytol 80:193–229

Koller W, Kindl H (1978) The appearance of several malate synthetase-containing cell structures during the stage of glyoxysome biosynthesis. FEBS Lett 88:83–86

Lang WC (1982) Glycoprotein synthesis in *Chlamydomonas* I. In vitro incorporation of galactose from UDP-[^{14}C]galactose into membrane-bound protein. Plant Physiol 69:678–681

Larkins BA, Hurkman WJ (1978) Synthesis and deposition of zein in protein bodies of maize endosperm. Plant Physiol 62:256–268

Larkins BA, Pedersen K, Handa AK, Hurkman WJ, Smith LD (1979) Synthesis and processing of maize storage proteins in *Xenopus laevis* oocytes. Proc Natl Acad Sci USA 76:6448–6452

Lehle L, Bowles DJ, Tanner W (1978) Subcellular site of mannosyl transfer to dolichyl phosphate in *Phaseolus aureus*. Plant Sci Lett 11:27–34

Leigh RA, Branton D, Marty F (1979) Methods for the isolation of intact vacuoles and fragments of tonoplast. In: Reid E (ed) Plant organelles: Methodological surveys B. Biochemistry, vol 9. Ellis Horwood, Chichester, pp 69–80

Locy R, Kende H (1978) The mode of secretion of α-amylase in barley aleurone layers. Planta 143:89–99

Lodish HF, Silberstein A, Porter M (1981) Synthesis and assembly of transmembrane viral and cellular glycoproteins. In: Hand AR, Oliver C (eds) Methods in cell biol, vol 23. Academic Press, New York, pp 5–25

Lord JM (1983) Endoplasmic reticulum and ribosomes. In: Hall JL, Moore AL (eds) Isolation of membranes and organelles from plant cells. Academic Press, London, pp 119–134

Lord JM, Kagawa T, Moore TS, Beevers H (1973) Endoplasmic reticulum as the site of lecithin synthesis in castor bean endosperm. J Cell Biol 57:659–667

Markham A, Hartman GC, Parke DV (1972) Spectral evidence for the presence of cytochrome P-450 in microsomal fractions from some higher plants. Biochem J 130:90P

Marriott KM, Tanner W (1979) Dolichylphosphate-dependent glycosyl transfer reactions in the endoplasmic reticulum of castor bean endosperm. Plant Physiol 64:445–449

Matthews JA, Mifflin BJ (1980) In vitro synthesis of barley storage proteins. Planta 149:262–268

Mazliak P, Kader JC (1980) Phospholipid-exchange synthesis. In: Stumpf PK (ed) The biochemistry of plants, vol 4. Lipids: structure and function. Academic Press, New York, pp 283–300

Meldolesi J, Borgese N (1981) Membrane synthesis and turnover in secretory cell systems. In: Hand AR, Oliver C (eds) Methods in cell biology, vol 23. Basic mechanisms in cellular secretion. Academic Press, New York, pp 445–460

Mellor RB, Lord JM (1978) Incorporation of D-[^{14}C]galactose into organelle glycoprotein in castor bean endosperm. Planta 141:329–332

Mellor RB, Lord JM (1979a) Involvement of lipid-linked intermediates in the transfer of galactose from UDP ^{14}C galactose to exogenous protein in castor bean endosperm homogenates. Planta 147:89–96

Mellor RB, Lord JM (1979b) Subcellular localization of mannosyl transferase and glycoprotein biosynthesis in castor bean endosperm. Planta 146:147–153

Mellor RB, Roberts LM, Lord JM (1979) Glycosylation of exogenous proteins by endoplasmic reticulum membranes from castor bean (*Ricinus communis*) endosperm. Biochem J 182:629–631

Mollenhauer HH, Morré DJ (1980) The Golgi Apparatus. In: Tolbert NE (ed) The biochemistry of plants, vol 1. Academic, New York, pp 438–488

Montague MJ, Ray PM (1977) Phospholipid-synthesizing enzymes associated with golgi dictyosomes from pea tissue. Plant Physiol 59:225–230

Moore TS (1974) Phosphatidyl glycerol synthesis in castor bean endosperm. Kinetics, requirements and intracellular localization. Plant Physiol 54:164–168

Moore TS (1975) Phosphatidyl serine synthesis in castor bean endosperm. Plant Physiol 56:177–180

Moore TS (1976) Phosphatidylcholine synthesis in castor bean endosperm. Plant Physiol 57:383–386

Moore TS (1982) Phospholipid biosynthesis. Annu Rev Plant Physiol 33:235–259

Moore TS, Sexton JC (1978) Phosphatidate phosphatase of castor bean endosperm. Plant Physiol 61S:80

Moore TS, Lord JM, Beevers H (1973) Enzymes of phospholipid metabolism in the endoplasmic reticulum of castor bean endosperm. Plant Physiol 52:50–53

Moreau RA, Stumpf PK (1982) Solubilization and characterization of an acyl-coenzyme a o-lysophospholipid acyltransferase from the microsomes of developing safflower seeds. Plant Physiol 69:1293–1297

Morré DJ (1971) Isolation of golgi apparatus. Methods Enzymol 22:130–148

Morré DJ (1975) Membrane biogenesis. Annu Rev Plant Physiol 26:441–481

Morré DJ, Buckhout TJ (1979) Isolation of golgi apparatus. In: Reid E (ed) Plant organelles: methodological surveys. Biochemistry, vol 9. Ellis Horwood, Chichester, pp 117–134

Muto S, Beevers H (1974) Lipase activity in castor bean endosperm during germination. Plant Physiol 54:23–28

Nagahashi J, Beevers L (1978) Subcellular localization of glycosyltransferase involved in glycoprotein biosynthesis in the cotyledon of *Pisum sativum* L. Plant Physiol 61:451–459

Nagahashi J, Hiraike K (1982) Effects of centrifugal force and centrifugation time on the sedimentation of plant organelles. Plant Physiol 69:546–548

Nagahashi J, Browder SK, Beevers L (1980) Glycosylation of pea cotyledon membranes. Plant Physiol 65:648–657

Normand G, Hartmann NA, Schuber F, Benveniste P (1975) Characterisation de membranes de coleoptiles de mais fixant l'auxine N-naphthyl phythalamique. Physiol Veg 13:743–761

Oostrom H, Kulescha Z, Van Vliet TB, Libbenga KR (1980) Characterization of a cytoplasmic auxin receptor from tobacco pith callus. Planta 149:44–47

Phillip E-I, Franke WW, Keenan TW, Stadler J, Jarasch E-D (1976) Characterization of nuclear membranes and endoplasmic reticulum isolated from plant tissue. J Cell Biol 68:11–29

Potts JRM, Weklych R, Conn EE (1974) The 4 hydroxylation of cinnamic acid by sorghum microsomes and the requirement for cytochrome P_{450}. J Biol Chem 249:5019–5026

Price CA (1982) Centrifugation in density gradients. Academic Press, p 429

Price CA (1983) General principles of cell fractionation. In: Hall JL, Moore AL (eds) Isolation of membranes and organelles from plant cells. Academic Press, London, pp 1–24

Puchel M, Kunz K, Parthier B, Aurich O, Bassumer R, Manteuffel R (1979) RNA metabolism and membrane bound polysomes in relation to globulin biosynthesis in cotyledons of developing field beans (*Vicia faba* L.). Eur J Biochem 96:321–329

Quail PH (1979) Plant cell fractionation. Annu Rev Plant Physiol 30:425–484

Rachubinski RA, Verma DPS, Bergeron JJM (1980) Synthesis of rat liver microsomal cytochrome b_5 by free ribosomes. J Cell Biol 84:705–716

Ray PM (1977) Auxin binding sites of maize coleoptiles are localized on membranes of the endoplasmic reticulum. Plant Physiol 59:594–599

Ray PM, Dohrmann U, Hertel R (1969a) Characterization of naphthalene acetic acid binding to receptor sites on cellular membranes of maize coleoptile tissue. Plant Physiol 59:357–364

Ray PM, Shininger TL, Ray MM (1969b) Isolation of β-glucan synthetase particles from plant cells and identification with Golgi membranes. Proc Natl Acad Sci USA 64:605–612

Reid E (1979) Plant organelles: methodological surveys B. Biochemistry, vol 9. Ellis Horwood, Chichester, p 232

Rich PR, Lamb CJ (1977) Biophysical and enzymological studies upon the interaction of trans cinnamic acid with higher plant microsomal cytochrome P_{450}. Eur J Biochem 72:353–360

Rickwood D, Birnie GD (1975) Metrizamide, a new density-gradient medium. FEBS Lett 50:102–110

Roberts LM, Mellor RB, Lord JM (1980) Glycoprotein fucosyl transferase in the endoplasmic reticulum of castor bean endosperm cells. FEBS Lett 113:90–94

Robinson ML, Carman GM (1982) Solubilization of microsomalassociated phosphatidylinositol synthetase from germinating soybeans. Plant Physiol 69:146–149

Roden LT, Miflin BJ, Freedman RB (1982) Protein disulfide-isomerase is located in the endoplasmic reticulum of developing wheat endosperm. FEBS Lett 138:121–124

Rothman JE (1981) The Golgi apparatus: two organelles in tandem. Science 213:1212–1219

Rubery PH (1982) Auxin receptors. Annu Rev Plant Physiol 32:569–596

Sabatini DD, Kreibach G (1976) Functional specialization of membrane-bound ribosomes in eucaryotic cells. In: Martonosi A (ed) Enzymes of biological membranes, vol 2. Plenum, New York, pp 531–579

Sabatini DD, Tashiro Y, Palade GE (1966) On the attachment of ribosomes to microsomal membranes. J Mol Biol 19:503–524

Salimath BP, Marmé D (1983) Protein phosphorylation and its regulation by calcium and calmodulin in membrane fractions from zucchini hypocotyls. Planta 158:560–568

Saunders JA, Conn EE, Shimada M (1977) Localization of cinnamic acid 4 monooxygenase and the membrane-bound enzyme system for Dhurrin biosynthesis in sorghum seedlings. Plant Physiol 60:629–634

Sengupta C, Deluca V, Bailey DS, Verma DPS (1981) Post translational processing of 7S and 11S components of soybean storage protein. Plant Mol Biol 1:19–34

Sexton JC, Moore TS (1978) Phosphatidyl inositol synthesis in castor bean endosperm cytidine diphosphate diglyceride: inositol transferase. Plant Physiol 62:978–980

Sexton JC, Moore TS (1981) Phosphatidyl inositol synthesis by a Mg^{2+} dependent exchange enzyme in castor bean endosperm. Plant Physiol 68:18–22

Sturm A, Kindl H (1983) Fucosyl transferase activity and fucose incorporation in vivo as markers for subfractionating cucumber microsomes. FEBS Lett 160:165–168

Stymne S, Glad G (1981) Acyl exchange between oleoyl-CoA and phosphatidyl choline in microsomes of developing soya bean cotyledons and its role in fatty acid desaturation. Lipids 16:298–305

Svardal AM, Pryme IF (1981) Aspects of the role of the endoplasmic reticulum in protein synthesis. Subcell Biochem 7:117–170

Van Der Wilden W, Chrispeels MJ (1983) Characterization of the isoenzymes of α-mannosidase located in the cell wall, protein bodies, and endoplasmic reticulum of *Phaseolus vulgaris* cotyledons. Plant Physiol 71:82–87
Van Der Wilden W, Gilkes NR, Chrispeels MJ (1980) The endoplasmic reticulum of mung bean cotyledons. Role in the accumulation of hydrolases in protein bodies during seedling growth. Plant Physiol 66:390–394
Vreugdenhil D, Harkes PAA, Libbenga KR (1980) Auxin binding by particulate fractions from tobacco leaf protoplasts. Planta 150:9–12
Waldrum JD, Davies E (1981) Subcellular localization of IAA oxidase in peas. Plant Physiol 68:1303–1307
Walter P, Gilmore R, Muller M, Blobel G (1982) The protein translocation machinery of the endoplasmic reticulum. Phil Trans R Soc Lond B Biol Sci 300:225–228
Wienecke K, Glas R, Robinson DG (1982) Organelles involved in the synthesis and transport of hydroxyproline-containing glycoprotein in carrot root discs. Planta 155:58–63
Yamada M, Tanaka T, Kader JC, Mazliak P (1978) Transfer of phospholipids from microsomes to mitochondria in germinating castor bean endosperm. Plant Cell Physiol 19:173–176
Young O, Beevers H (1976) Mixed function oxidases from castor bean endosperm. Phytochemistry 15:379–385

Polyribosomes

B. A. LARKINS

1 Introduction

Ribosomes convert the information encoded in the nucleotide sequence of mRNA into the amino acid sequence of a protein. The translational mechanism by which this occurs involves a number of biochemical reactions (Weeks 1981). Initially, ribosomal subunits and methionyl-tRNA bind to the mRNA (initiation). This is followed by sequential "reading" of nucleotide triplets and addition of the corresponding amino acids into a nascent polypeptide (elongation). After the ribosome traverses the protein-coding region of the mRNA, it reaches a "stop" codon, which directs release of the ribosome and discharge of the completed polypeptide (termination).

Because mRNA's are long linear molecules, they can be translated simultaneously by several ribosomes, and this leads to the formation of structures called *polyribosomes* or *polysomes*. Electron microscopy of plant cells reveals ribosomes and polyribosomes, free in the cytoplasm and bound to endoplasmic reticulum (ER) (Fig. 1). Their segregation is related to the types of protein synthesized (Shore and Tata 1977; Davis and Larkins 1980). Proteins that remain within the cytoplasm or are transported to the chloroplast or mitochondria are made on free polyribosomes, while those that are secreted out of the cell or are sequestered within the ER, vacuole, or lysosomes are made on membrane-bound polyribosomes.

Protein synthesis and hence polyribosome formation are regulated by several different mechanisms (Weeks 1981). They are controlled by the production and availability of specific mRNA's, the rate at which these mRNA's are translated, and the turnover of the mRNA's and proteins. Although it is difficult to study these reactions individually, overall changes in protein synthesis and gene expression can be monitored by analyzing alterations in the polyribosome populations. The validity of this approach requires that one be able to isolate undegraded polyribosomes. This can be a formidable problem because many plant tissues contain high levels of endogenous ribonuclease (RNase) as well as substances that precipitate and aggregate polyribosomes. However, methods are available that minimize RNase activity and allow efficient recovery of intact polyribosomes.

In this chapter I will discuss procedures that have proven useful for the isolation and analysis of polyribosomes from a variety of plant tissues. I have not attempted to provide an exhaustive literature review regarding the adaptation or slight modification of the basic methods, but I have illustrated the effectiveness of various buffer components that inhibit RNase activity and prevent polysome

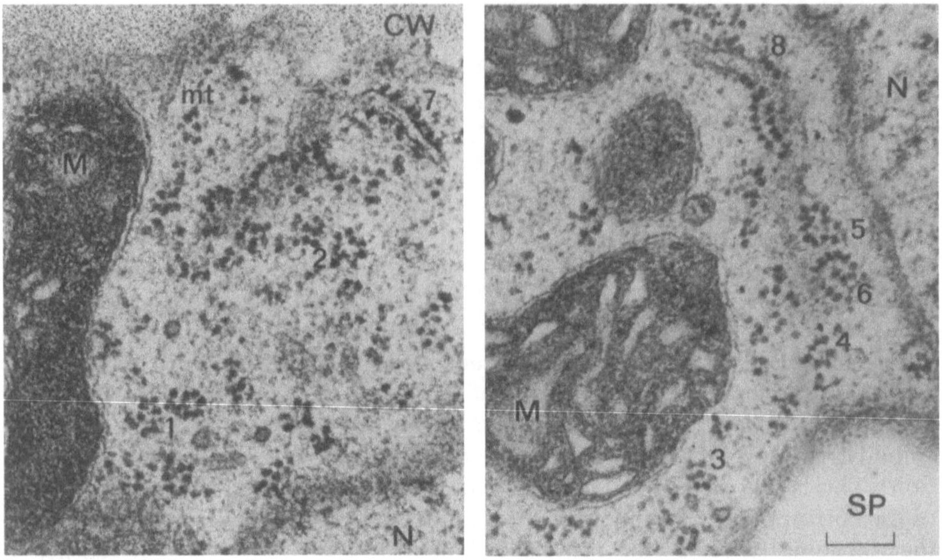

Fig. 1. Portions of cells of suspension cultures of wild carrot (*Daucus carota* L.) showing polyribosomes. *1–4* free polyribosomes sectioned to show the spiral configuration; *5, 6* polyribosomes bound to endoplasmic reticulum seen in tangential views also show the spiral configuration; *7, 8* polyribosomes bound to the endoplasmic reticulum in conventional cross-sectioned view; *M* mitochondria; *SP* starch plastid; *N* nucleus; *CW* cell wall; *mt* microtubules. (Courtesy of D. James Morré from a study with W. F. Boss)

losses. I will also discuss some of the problems associated with analysis of polyribosome profiles and some ways in which polysomes can be used to study gene expression. A more detailed description of the physical properties of polysomes and techniques for their analysis is available in the review by Noll (1969).

2 Isolation of Polysomes from Plant Cells

2.1 Factors that Affect the Stability and Recovery of Polyribosomes

Although a variety of factors can alter the recovery of intact polyribosomes from plant tissue, the single most serious problem is RNase activity. Plant cells appear to contain significantly more RNase than animal cells, so procedures that are effective for isolating intact polysomes from animal tissues are generally ineffective for plants. Many RNases are probably sequestered within vacuoles in the plant cell; however, once cell membranes are ruptured, these enzymes readily degrade RNA. This degradation is particularly apparent with polysomes because the hydrolysis of a single phosphodiester linkage in an mRNA can convert a polysome with 12 ribosomes into two polysomes with six ribosomes each. Messenger RNA isolated from such degraded polysomes is of limited value, although the polysomes themselves can be useful for studying protein synthesis (see Sect. 5.1).

In addition to RNases, divalent metal ions also affect the recovery of intact polyribosomes. Some plant tissues contain calcium in the form of calcium oxalate, calcium pectate, and calcium carbonate. Once the cells of a tissue are disrupted, calcium can cause polysome degradation and precipitation. Pea epicotyls were found to contain a calcium-activated RNase (Larkins and Davis 1973). Although it is not clear whether other plant tissues contain similar RNases, it has been demonstrated at least in leaf tissue that calcium quantitatively reduces recovery of large polysomes (Jackson and Larkins 1976). Other heavy metal ions or large, positively charged cations can have similar effects.

Plant stresses that occur prior to tissue homogenization can also affect the stability and recovery of plant polysomes. Heat stress, water stress, and anaerobiosis alter gene expression and protein synthesis (Key et al. 1982; Rhodes and Matsuda 1976; Armstrong and Jones 1973; Lin and Key 1967). Water stress has also been associated with increases in RNase activity in a variety of plant tissues (Hsiao 1973; Dhindsa and Bewley 1976). Whether or not the increased level of RNase causes polysome degradation prior to tissue homogenization, it increases the potential for polysome breakdown when the tissue is homogenized. Therefore, any steps that can be taken to prevent perturbation of the tissue prior to homogenization are beneficial.

2.2 Tissue Preparation

If polysomes are to be isolated from certain plant tissues such as meristematic regions of roots, shoots, or leaves, it is generally sufficient to simply excise the tissue and place it in chilled buffer or water until it is homogenized. If this requires only a few minutes, it is unlikely to have an effect on polysome stability. However, if tissue isolation is slow or if the tissue is easily desiccated, it is better to freeze the tissue quickly by placing it on dry ice or in liquid nitrogen. If the frozen tissue must be stored for a long period of time it should be kept in liquid nitrogen. If this is inconvenient, the tissue can be stored at −80 or −20 °C. However, after several months of storage at low temperature, tissue will lose water by lyophilization, so it should be stored in a compact, air-tight container. Tissues that contain a high percentage of water, such as fruits or tubers, may be frozen and lyophilized and polysomes prepared from the tissue powder (Drouet and Hartmann 1979).

Frozen tissue sections are difficult to disrupt. If they are added directly to cold grinding buffer, they usually form a lump of ice that is difficult to break up. It is better to freeze the tissue with dry ice or liquid nitrogen and pulverize it with a mortar and pestle or in a coffee mill or its equivalent. The frozen powder can then be slowly stirred into cold grinding buffer. The tissue homogenate should be maintained at 4 °C at all times to reduce RNase activity (see Sect. 3).

For efficient polysome recovery it is important that all cells be effectively disrupted. In many instances this can be done with a mortar and pestle. A small amount of acid-washed sand or ground glass added to the tissue can improve cell breakage. One can also use mechanical homogenizers such as a Waring blendor, Polytron homogenizer (Brinkman), or a Virtis homogenizer (Virtis). Although polysomes may be susceptible to shearing forces (Noll 1969), this is usually not a problem with these mechanical homogenizers. The time required for complete homogenization of tissue is dependent on the tissue type; however, it is generally not possible to "over-grind" a tissue sample.

2.3 Subcellular Fractionation and Polysome Isolation

Once the tissue has been thoroughly homogenized, cell-wall debris and vascular tissue should be removed by straining the brei through several layers of cheesecloth or nylon cloth. The particulate material can also be removed with the nuclei by a low speed centrifugation at 2000–2500 *g* for 5 min, but the pellet is often loose and may slide out when the supernatant is decanted.

Depending on the type of study, one may wish to analyze the total polysome population or fractionate it into free and membrane-bound polyribosomes (Fig. 2). Total polysomes can be isolated by adjusting the post-nuclear supernatant (2000 *g* supernatant) to 1% with Triton X-100 or a similar nonionic detergent to release ribosomes bound to the ER and lyse chloroplasts and mitochondria. After addition of detergent, the remaining membranous material is removed by centrifugation at 30,000 *g* for 10 min. The supernatant is then layered directly onto a sucrose gradient and the polysomes are separated by a short period of centrifugation, e.g., 120,000 *g* for 2 h (Fig. 4E). This method is rapid and provides the most accurate analysis of the actual population of polyribosomes in the tissue. However, this direct analysis may not be possible if the polysomes are too dilute in the extract. In that instance they can be concentrated by centrifugation through a 1.5–2.0 M layer of buffered sucrose (40 mM Tris-HCl, pH 8.5; 40 mM KCl; 20 mM $MgCl_2$). Although this step reduces the recovery of monosomes and small polysomes (see Sect. 3.2), it quickly separates polysomes from RNases in the extract and allows recovery of relatively pure preparations of polysomes which can be subsequently used to direct protein synthesis in vitro or for purification of mRNA.

Free and membrane-bound polyribosomes are most commonly separated by discontinuous gradient centrifugation or isopycnic gradient centrifugation (McIntosh and O'Toole 1976; Shore and Tata 1977). A standard method by which this is done is to layer a post-mitochondrial supernatant (10,000 *g* supernatant) over layers of 1.5 M and 2.0 M buffered sucrose. After a period of centrifugation the free polysomes are collected from the bottom of the 2.0 M sucrose cushion and the membrane-bound polysomes are recovered from the membrane layer at the interphase of the 1.5 M and 2.0 M sucrose layers. The major limitation with this procedure is that free ribosomes and polysomes become trapped in the membranes, so the separation may not be as effective as desired. Better separation is possible by using differential centrifugation to partially purify the membrane fraction. The membranes are removed from the tissue homogenate by centrifugation at 30,000 *g* for 15 min or are pelleted through a dilute sucrose gradient before layering on a discontinuous gradient (McIntosh and O'Toole 1976). An

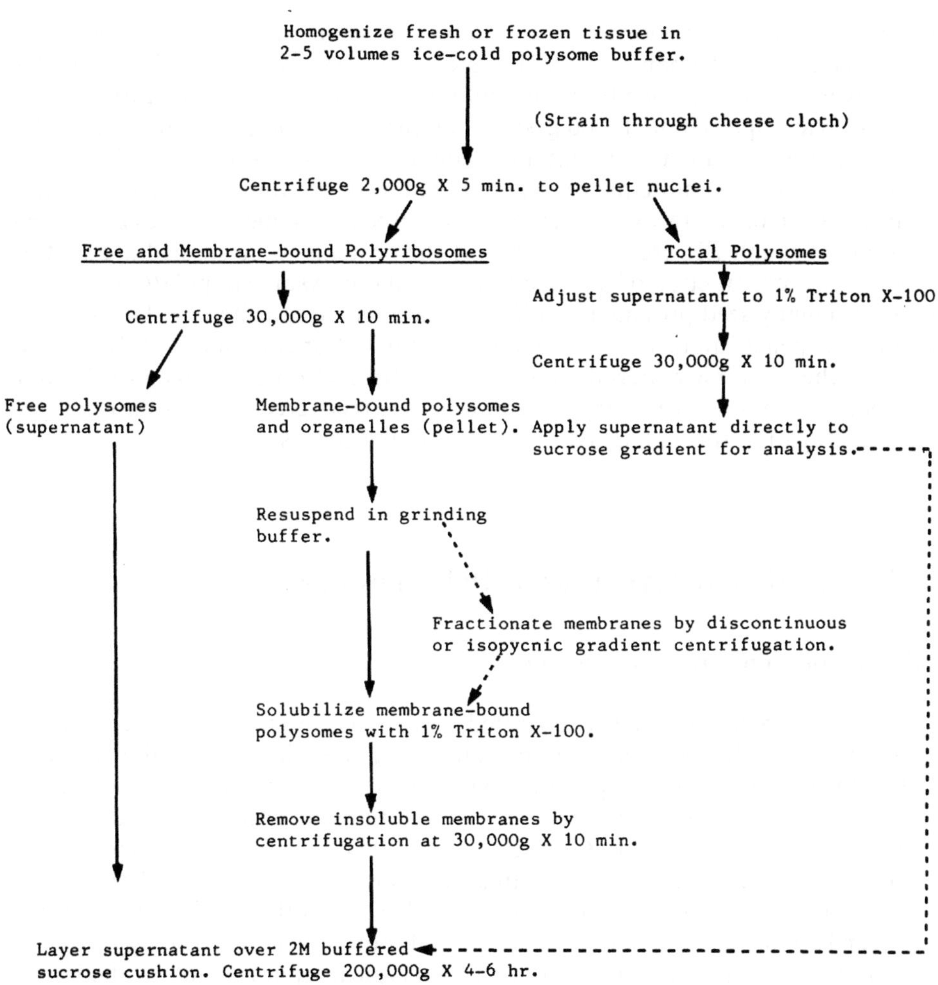

Fig. 2. Procedure for isolation of total or free and membrane-bound polyribosomes from plant tissue

alternative approach is to adjust the sucrose concentration in the initial extract to 2 M and allow the membrane to float to the surface during centrifugation while the free polyribosomes pellet. With this method there is less likelihood of trapping free polysomes in membranes, although the amount of tissue that can be processed may be limited.

Protein secretion is not a major function of most plant tissues and, with few exceptions (Armstrong and Jones 1973; Verma et al. 1975), proteins synthesized by membrane-bound polyribosomes have not been well characterized. However, several types of protein synthesized in developing seeds are made by membrane-bound polyribosomes and these have been extensively characterized (for reviews see Larkins 1981; Higgins 1984, Chrispeels 1984). Membrane-bound polysomes have been isolated from developing seeds and other plant tissues by differential centrifugation. A large portion of the rough ER sediments at relatively low gravitational force (Adelman et al. 1973; Larkins and Davies 1975), and it is possible to recover most of the membrane-bound polyribosomes by centrifugation of the post nuclear-supernatant (2000 g supernatant) at 30,000 *g* for 10–15 min. Although the membrane pellet contains a mixture of plastids as well as ER membranes, these can be subsequently fractionated by discontinuous sucrose gradient centrifugation or floatation as previously described. In some instances pure preparations of rough ER are not essential. If the proteins to be studied are synthesized by a major fraction of the membrane-bound polysome population (as is the case with many seed proteins), crude preparations are satisfactory. The 30,000 g membrane pellet can be resuspended in grinding buffer containing 1% Triton X-100 and the insoluble material removed by centrifugation at 30,000 *g* for 10 min. The polyribosomes released from membranes are recovered after they are pelleted by centrifugation through a layer of 2 M buffered sucrose.

3 Purification and Analysis of Polyribosomes

3.1 Sucrose Gradient Centrifugation

Purified polysomes are usually separated for analysis by sucrose gradient centrifugation, although electrophoretic techniques can also be used (Dahlberg et al. 1969). Sucrose gradients are prepared with an automatic mixing device or simply by diffusion of layers of buffered sucrose solutions of different concentrations (Newburn 1975; Stone 1974). Gradients of 10–60% sucrose give excellent resolution of polysomes after short centrifugation times (Allington et al. 1976). Polysome separation is based primarily on particle size, so there is better resolution between subunits, single ribosomes, and small polysomes than among the larger polysome size-classes. Figure 3 A illustrates a typical separation of purified polyribosomes. Single ribosomes and small polysomes that contain two to six ribosomes sediment more slowly in the gradient and are clearly resolved from one another. Because the ratio of size differences between polysomes containing seven, eight, and nine ribosomes is much less than that of the smaller polysomes, the larger size classes are more closely spaced and they appear less well resolved in the gradient.

Electron microscopic analysis of samples from the gradient illustrates the effective separation of these small particles. Single ribosomes (monosomes) and subunits remain at the top of the gradient (Fig. 3 B); samples from the next region of the gradient contain polysomes with clusters of primarily 2–5 ribosomes (Fig. 3 C). Although there is some contamination of small polysomes in the lower portion of the gradient, these samples contain primarily large polyribosomes (Fig. 3 D). Treatment of the polysomes with RNase causes disaggregation due to hydrolysis of the mRNA, and large polysomes are rapidly converted into smaller size classes (e.g., Fig. 5). After extensive RNase treatment, only monosomes and dimers remain.

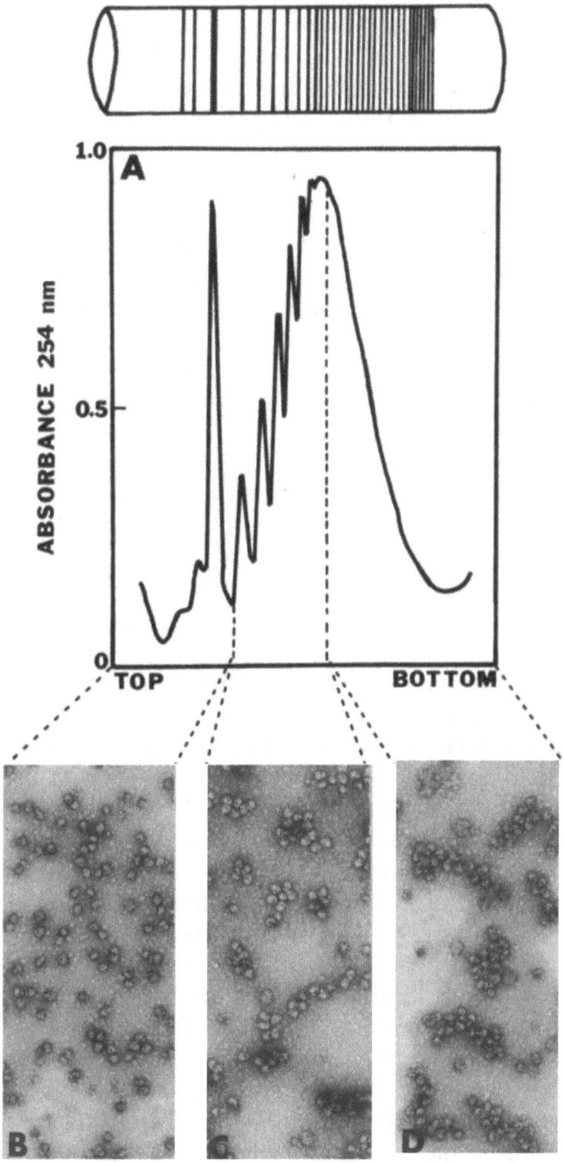

Fig. 3. Sucrose density gradient separation and analysis of polyribosomes. *Top* **A** diagrammatic representation of polyribosome separation following sucrose gradient centrifugation. *Bottom* analysis of polysome distribution by continuous UV monitoring of the gradient. **B–D** Electron microscopic analysis of gradient fractions. Samples were prepared by touching a carbon coated grid to the surface of the polysome preparation. The grid was floated on 2% (v/v) glutaraldehyde in 0.05 M sodium cacodylate, pH 7.2 for 2–5 min. Grids were touched to the surface of glass-distilled H_2O for a few seconds and negatively stained with 2.5% uranyl acetate. (×62,000)

Several types of flow cell have been developed for gradient analysis (Morton and Hirsch 1970; Allington et al. 1976). One type uses a laminar flow cell, while the other uses a larger-diameter cell designed for bulk flow. Allington et al. (1976) were able to resolve polysomes that contained up to 16 ribosomes per mRNA with a large diameter flow cell. They found that the rate at which the gradient is pumped out also affects resolution. For most biological applications such a high degree of resolution is probably not necessary, and either type of flow cell is adequate.

3.2 Purification of Polysomes with Discontinuous Sucrose Gradients

Although direct application of the post-mitochondrial supernatant to a sucrose gradient provides the most accurate representation of the polysome population, in some instances it is necessary to purify polysomes from plant extracts. For example, if the polysomes are to be used for protein synthesis in vitro or as a source of mRNA, they must be isolated in a purified form. This is usually done by pelleting the ribosomes through a 1.5–2.0 M layer of buffered sucrose. Such a discontinuous gradient, or sucrose pad as it is called, allows ribosomes and polysomes to pellet, while less dense material, including membranes and organelles, remains on top of the sucrose layer.

Optimum conditions for polysome recovery depend on the type and capacity of the rotor and the speed of centrifugation. In general a high gravitational force (e.g., 200,000 *g*) and a short centrifugation time are most effective. If a fixed angle rotor is used, a larger volume of sucrose is necessary, i.e., 40–45% of the tube capacity, because ribosomes are forced against the wall of the tube during centrifugation, and they slide down and form a pellet on the side of the tube. Without a large volume of sucrose pad the pellet becomes contaminated with microsomal membranes. If a swinging bucket rotor is used, wall effects are eliminated and the volume of sucrose can be reduced accordingly (i.e., 2–3 ml for an 11-ml tube, 4–5 ml for a 30-ml tube).

Ribosomal subunits, monosomes, and small polysomes sediment more slowly through a sucrose pad than large polysomes. When short periods of centrifugation with relatively low gravitational forces (100,000 *g*) are used, large polysomes are preferentially recovered. Figure 4 shows the recovery of polysomes with increasing periods of centrifugation time. Polysomes in Fig. 4A–D were isolated after centrifugation at 95,000 *g* for 2, 4, 72, and 90 h, respectively. The sample in Fig. 4E corresponds to an aliquot of the original post-mitochondrial supernatant that was applied directly to the sucrose gradient. Monosomes make up a smaller proportion of the ribosome population after 2 h or 4 h of centrifugation than they do in the original extract. It is not evident from these polysome profiles, but the recovery of total polysomal material is substantially reduced after short centrifugation times. Fourfold more ribosomal material was isolated after 90 h of centrifugation than 2 h, and this includes large polysomes as well as monosomes (Larkins and Davies 1975). Although recovery of polysomes improves after long centrifugation times, there is also evidence of polysome degradation in these samples. A greater proportion of monosomes and small polysomes are present in samples obtained after 72 and 90 h of centrifugation than in the original sample (Fig. 4E). The factors responsible for degradation after long periods of centrifugation are unknown, but they could involve RNases associated with microsomal membranes (Larkins and Davies 1975). Thus, with purification of polysomes through sucrose pads, there is a trade-off between the quality and the quantity of polysomes. The most serious consequence of this compromise arises when conclusions regarding changes in capacity for protein synthesis are based on polysome profiles obtained after short periods of centrifugation (Leaver and Dyer 1974; Larkins and Davies 1975). However, for purposes of in vitro protein synthesis or mRNA isolation, it is better to minimize polysome degradation even though polysome recovery is reduced.

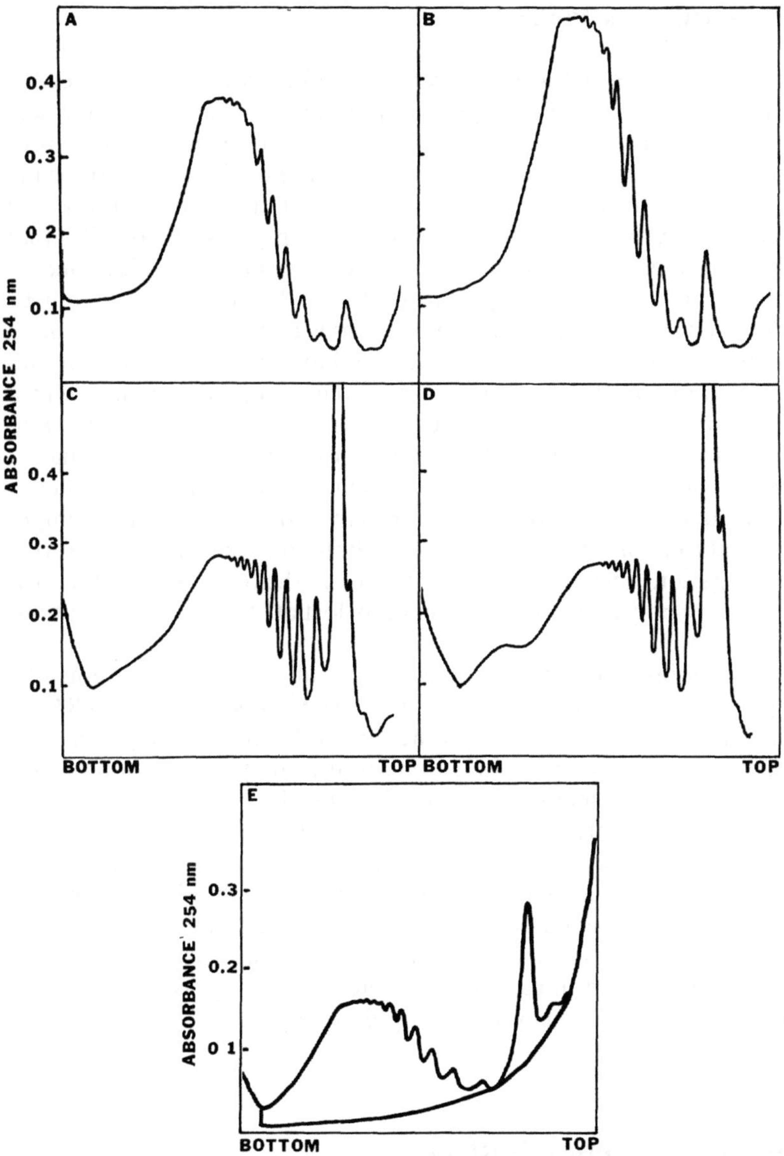

Fig. 4 A–E. Recovery of polyribosomes after centrifugation through a discontinuous sucrose gradient. Aliquots of 5 ml of the postmitochondrial supernatant from pea epicotyls were layered over 4 ml of 2 M buffered sucrose (**A–D**) or the postmitochondrial supernatant was applied directly to the gradient (**E**). The fresh weight equivalent of tissue and centrifugation time were: **A** 0.5 g, 2 h; **B** 0.5 g, 4 h; **C** 0.15 g, 72 h; **D** 0.15 g, 90 h. Sample **E** was from extract equivalent to 0.04 g of tissue. The baseline absorbance due to UV-absorbing material in the extract is indicated. (Larkins and Davies 1975)

3.3 Analysis of Polyribosome Profiles

Quantitative and qualitative changes in polysome profiles can result from genetic and physiological conditions that exist prior to polysome extraction or they may be the result of factors that alter polysome integrity during the isolation procedure. The latter factors include RNase activity, release of ribosomes following polypeptide completion (run-off), polysome aggregation and precipitation, and shearing. It is obviously important to discern changes in polysome profiles that result from these different sets of conditions. Vassart et al. (1970) developed a computer program to distinguish degradation due to RNase activity from ribosome "run-off" or shearing. RNase activity causes conversion of layer polysomes to small polysomes with a much later accumulation of monosomes. In contrast, ribosome run-off results in a rapid accumulation of monosomes with a gradual reduction in the amount of large polysomes. Shearing causes a preferential destruction of large polysomes with little effect on the level of monosomes.

Aggregation of polysomes can occur as a result of precipitation with heavy metal ions or through interactions between protein and(or) membrane components. Heavy metal ions cause polysome aggregation and result in losses during low speed centrifugation. The net effect is a decrease in the recovery of total polysomes, although large polysomes are preferentially lost at low ion concentrations (see Sect. 4.5).

Aggregation resulting from protein interactions can easily be determined by resuspending polysomes in buffer containing protease K. Since there is no RNase activity associated with protease K, polysomes are stable in relatively high concentrations of this enzyme (see Sect. 4.6). For routine analysis inclusion of protease K in the polysome resuspension buffer is beneficial.

In instances where few polysomes or apparently degraded polysomes are isolated from a tissue, it is difficult to know whether an alteration in polysomes occurred prior to or during tissue extraction. A way to approach this question is to incubate a sample of undegraded polysomes from another tissue in the post-ribosomal supernatant (Larkins et al. 1976a). If there is RNase activity, "run-off" activity, or aggregation factors, it can easily be detected after appropriate incubation and analysis of the exogenously supplied polysomes.

4 Polyribosome Extraction Buffers

Polyribosome extraction media are generally based on a TKMS buffer, i.e., they contain Tris-HCl at neutral to alkaline pH, potassium as a monovalent ion, magnesium as a divalent ion, and 0.2 M sucrose as on osmoticum. In addition to these basic components, a variety of substances can be included to inhibit RNase activity. Some inhibitors are more effective than others and certain ones have detrimental side effects.

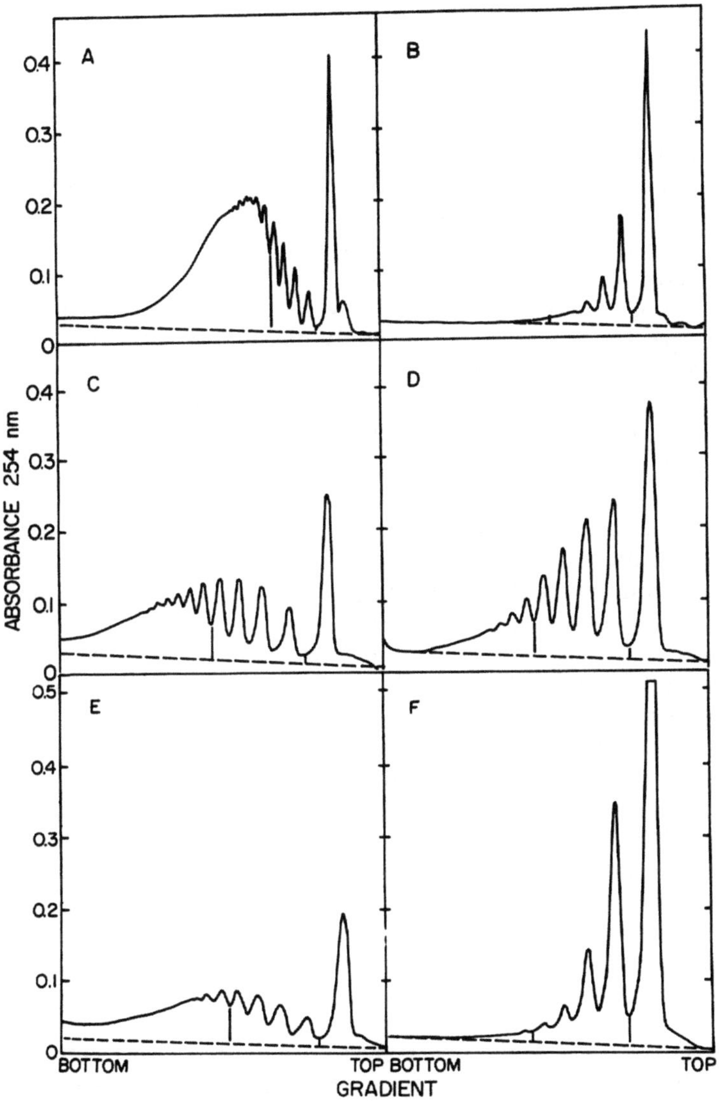

Fig. 5 A–D. Comparative resistance of polyribosomes to degradation by endogenous or exogenous ribonuclease. Polysomes were isolated from etiolated pea epicotyls; equivalent amounts of tissue were used except for **B** where only half of the sample was applied to the gradient. **A**, **B**, **C**, and **E** ground in buffer containing 200 mM Tris-HCl, pH 8.5; **D**, **F** ground in buffer containing 100 mM Tris-HCl, pH 7.5. **A** control, no RNase; **B** 0.2 μg RNase added to purified polysomes; **C**, **D** resuspended polysomes incubated for 2 h at 25 °C; **E**, **F** 0.2 μg RNase added to original grinding medium (5 ml). (Davies et al. 1972)

4.1 pH

Polysome extraction media that contain high concentrations of Tris-HCl at alkaline pH, e.g., 0.2 M, pH 8.5, are very effective at inhibiting RNase activity in both plant and animal tissue (Davies et al. 1972; Davies et al. 1977). The effectiveness of this buffer results not only from its alkaline pH, which inhibits most RNases, but also from its high ionic strength. Figure 5 A illustrates a profile of polysomes isolated from developing pea epicotyls in a buffer containing 0.2 M Tris-HCl pH 8.5. About 15% of the profile is composed of single ribosomes (monosomes), while the remainder is polysomes; the size classes with the greatest absorbance contain 7–9 ribosomes per mRNA. If the purified polysomes are treated with pancreatic RNase (Fig. 5 B) or are allowed to stand at room temperature for 2 h (Fig. 5 C), the large polysomes become degraded and the absorbance of small polysomes progressively increases; eventually only monosomes and dimers remain (Fig. 5 B). Polysomes isolated with a buffer containing 0.1 M Tris-HCl at pH 7.5 (Fig. 5 D) are more degraded than those isolated in 0.2 M Tris buffer at alkaline pH (Fig. 5 D cf. Fig. 5 A). In fact, few large polysomes can be recovered with neutral extraction buffers. The effectiveness of the high Tris concentration and the alkaline pH at inhibiting RNase activity is illustrated in Fig. 5 E and Fig. 5 F where 0.2 μg of RNase was added to these two grinding buffers prior to tissue homogenization. With 0.2 M Tris at pH 8.5 the recovery of large polysomes is slightly reduced (Fig. 5 E cf. Fig. 5 A), but with 0.1 M Tris at pH 7.5 only monosomes and dimers remain (Fig. 5 F).

Polysome buffers that contain 0.2 M Tris at pH 8.5 have proven satisfactory for isolating polysomes from a variety of plant tissues. There is evidence that slightly more alkaline pH's may improve polysome recovery in some instances (Jackson and Larkins 1976), but higher concentrations of Tris buffer do not generally improve polysome stability.

4.2 Potassium Chloride

Although pH's between 8.0 and 9.0 are important for inhibiting RNase activity, the ionic strength of the buffer is also important. Breen and co-workers (1972) showed that the recovery of large polysomes from barley tissue was improved with buffers that contained 0.4–0.6 M KCl. They attributed this to the dissociation of RNase from microsomal membranes at high ionic strength. High KCl concentration in conjunction with high Tris concentration and alkaline pH improved the yield of large polysomes from tobacco leaves (Jackson and Larkins 1976). However, high ionic strength may cause polysomes to dissociate from membranes. Therefore, if membrane-bound polysomes are to be isolated, it is necessary to reduce the KCl concentration to 0.06–0.1 M (Bol et al. 1976). Ionic strengths in excess of 0.5–0.6 M also cause dissociation of ribosomal subunits and can therefore reduce the yield of polysomal material (Breen et al. 1972). In general it is best to maintain the total ionic strength of the buffer between 0.3 and 0.5 M.

4.3 Magnesium Chloride

Magnesium is important for maintaining ribosome stability and, for this reason, most polysome buffers contain at least 5–10 mM magnesium. Some investigators prefer to use magnesium acetate rather than chloride because the acetate salt is not hygroscopic. Either form is satisfactory for most purposes.

In addition to stabilizing ribosome subunits, there is evidence that high magnesium concentrations reduce ribonuclease activity. Figure 6 shows an analysis

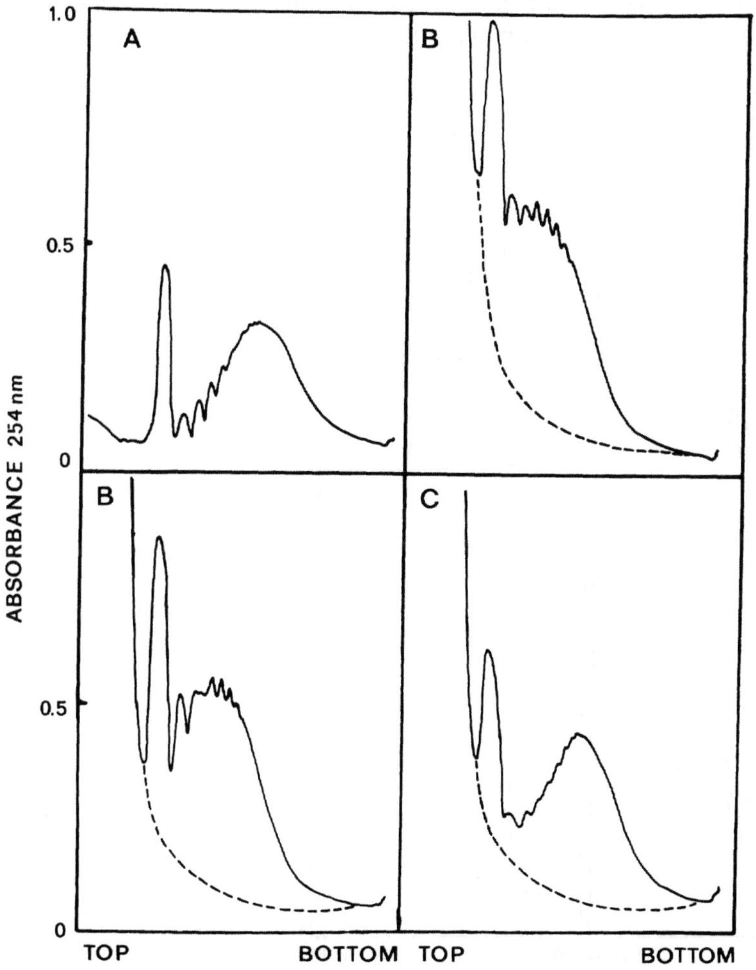

Fig. 6 A–D. Inhibition of ribonuclease by high concentration of $MgCl_2$. Purified polyribosomes were incubated in buffer at room temperature for 30 min and applied directly to a sucrose gradient (**A**) or incubated for 30 min at room temperature in post-ribosomal supernatant from maize kernels extracted in buffer (0.2 M Tris-HCl, pH 8.5; 0.2 M sucrose, 60 mM KCl; 1 mM dithiothreitol) containing varying amounts of $MgCl_2$. **B** 5 mM $MgCl_2$, **C** 10 mM $MgCl_2$, and **D** 30 mM $MgCl_2$. (Larkins and Tsai 1977)

of polysomes incubated in maize extract (post-ribosomal supernatant) prepared by grinding endosperm tissue in buffer containing 0.2 M sucrose, 0.2 M Tris-HCl, pH 8.5, 60 mM KCl and 5, 10 or 30 mM $MgCl_2$. A sample of polysomes that was not incubated in these extracts is shown in Fig. 6A. After 30 min at 25 °C a significant amount of polysome degradation occurred in the extract that contained 5 mM $MgCl_2$ (Fig. 6B) and 10 mM $MgCl_2$ (Fig. 6C) as evidenced by the reduction in large polysomes and the increased absorbance of small polysomes. The profiles of polysomes from the extract that contained 30 mM $MgCl_2$ is similar to that of the control (Fig. 6D cf. Fig. 6A). The mechanism by which magnesium stabilizes polysomes is not evident from the experiment, but it appears to prevent RNase activity. Buffers that contain 0.03 M to 0.05 M magnesium generally provide good polysome yields with minimal amounts of degradation.

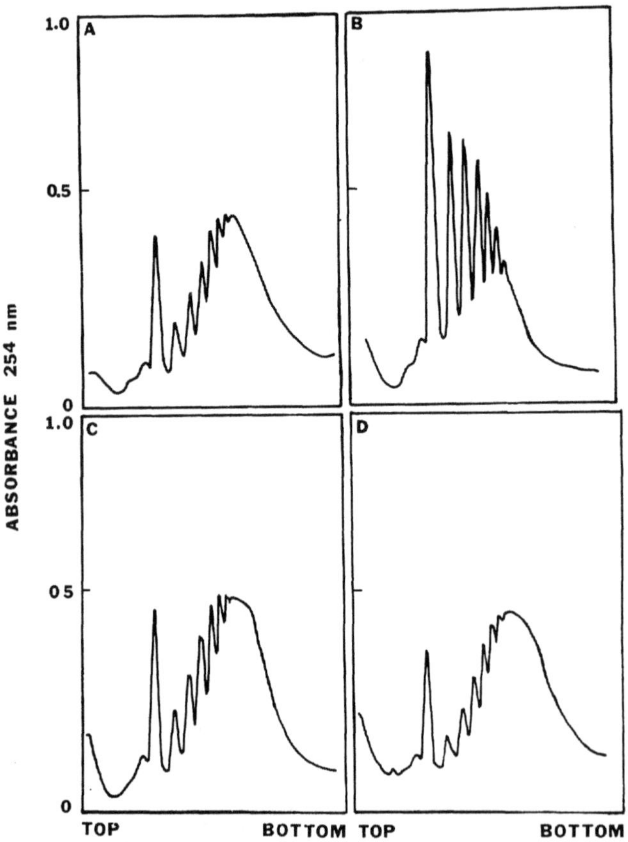

Fig. 7 A–D. Inhibition of ribonuclease by dithiothreitol. Samples of 60 maize root tips were homogenized in 20 ml of a buffer containing 50 mM Tris-HCl, pH 7.5; 20 mM KCl, 5 mM $MgCl_2$, and 0.2 M sucrose. (Weeks and Marcus 1969). Aliquots of the postmitochondrial supernatant were adjusted to contain varying amounts of dithiothreitol and incubated at 4 ° or 25 °C for 30 min. **A** 0 dithiothreitol, 4 °C, **B** 0 dithiothreitol, 25 °C, **C** 5 mM dithiothreitol, 25 °C, **D** 10 mM dithiothreitol, 25 °C

4.4 Reducing Agents

Reducing agents such as dithiothreitol (DTT) or 2-mercaptoethanol (2-ME) also help stabilize polysomes (Breen et al. 1972). Figure 7 shows an analysis of polysomes isolated from maize root tips in the presence or absence of dithiothreitol. Samples in Fig. 7 A and Fig. 7 B were isolated without DTT in the grinding buffer. The extract in Fig. 7 A was kept at 4 °C for 30 min prior to polysome isolation, while that in 7 B was held at 25 °C for the same time. A comparison of the two polysome profiles reveals that extensive degradation occurred in the extract that was incubated at 25 °C. Figure 7 C and 7 D show results of incubations of samples at 25 °C for 30 min when the grinding buffers contained 5 mM and 10 mM DTT, respectively. It is obvious that DDT protects against RNase activity, since the two polysome profiles are nearly identical to the control (Fig. 7 A). The reducing agent 2-ME has the same effect, but its concentration must be doubled to obtain equal protection.

4.5 Chelation of Divalent Metals

Heavy metal ions have been found to cause polysome aggregation (McGown et al. 1971) and significantly reduce polysome yields (Jackson and Larkins 1976).

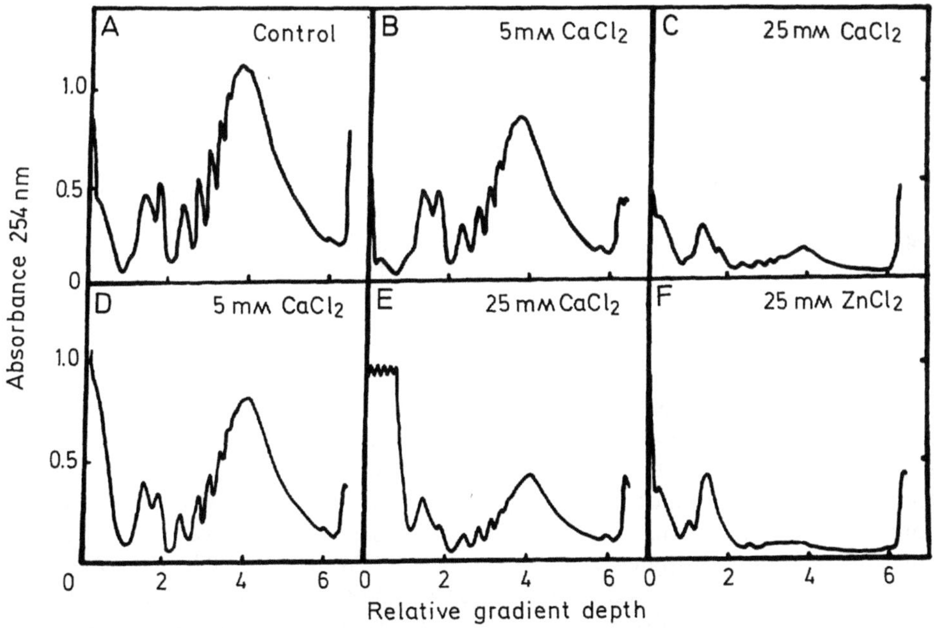

Fig. 8 A–F. Effect of divalent metal ions on polysome recovery. Unexpanded tobacco leaves (3.6 g) were ground in 36 ml of extraction buffer (0.2 M Tris-HCl; pH 9; 0.4 M KCl, 0.2 M sucrose, and 35 mM $MgCl_2$). Aliquots of 6 ml were amended with varying amounts of divalent cations. The preparations were immediately clarified and polysomes isolated. **A** control, no cation added, **B** 5 mM $CaCl_2$, **C** 25 mM $CaCl_2$, **D** 5 mM $CuCl_2$, **E** 25 mM $CuCl_2$, **F** 25 mM $ZnCl_2$. (Jackson and Larkins 1976)

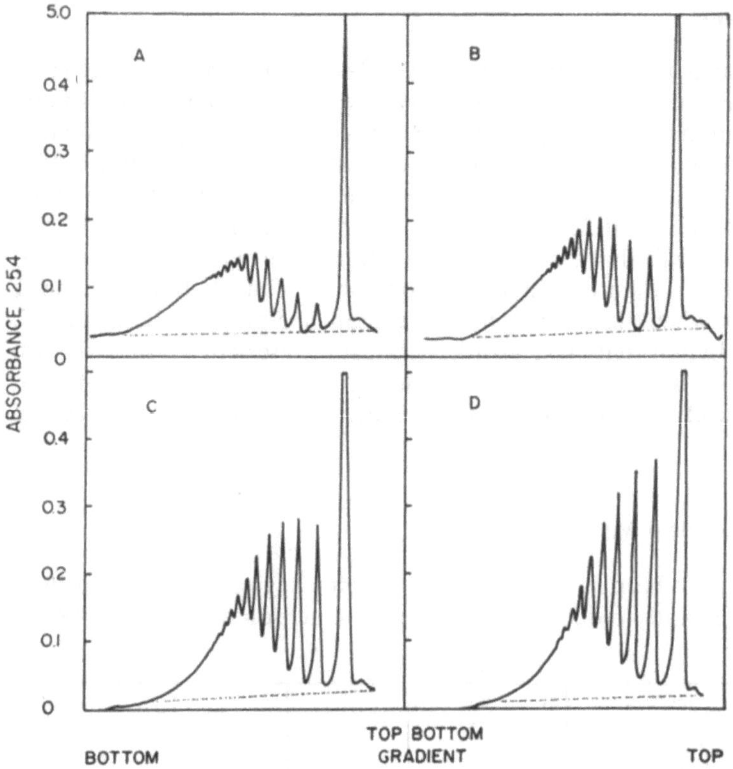

Fig. 9 A–D. Stimulation of polysome degradation by exogenous calcium. Polysomes were extracted from etiolated pea epicotyls in a buffer containing 25 mM Tris-HCl, pH 8.5; 30 mM $MgCl_2$, 60 mM KCl and 0.2 M sucrose. The postmitochondrial supernatant was incubated at 4 °C for 20 min in the presence of various concentrations of Ca^{2+} prior to polysome isolation. Samples represent polysomes recovered from 250 mg of tissue. **A** 0 Ca^{2+}, **B** 1 mM Ca^{2+}, **C** 2 mM Ca^{2+}, **D** 3 mm Ca^{2+}. (Larkins and Davies 1973)

Figure 8 illustrates the effect of Ca^{2+}, Cu^{2+}, and Zn^{2+} on the recovery of polysomes from young tobacco leaves. In this experiment a tissue homogenate was divided into six equal parts and various concentrations of metal ions were added. Even at concentrations of 5 mM (Fig. 8 B, 8 D) there is some loss of polysomes compared to the control (Fig. 8 A). Primarily the large polysomes are not recovered, although at higher ion concentrations (Fig. 8 C, 8 E, 8 F) all polysomal material is generally reduced. In this case the heavy metal ions appear to simply cause polysome precipitation, but in some tissues calcium also appears to activate RNase. Figure 9 shows the effect of exogenous calcium on polysomes in an extract from pea epicotyls. In this experiment the post mitochondrial supernatant contained 0, 1, 2, or 3 mM calcium chloride. Even with 1 mM Ca^{+2} (Fig. 9 B) there is evidence of polysome breakdown and this is markedly increased at 3 mM Ca^{2+} (Fig. 9 D).

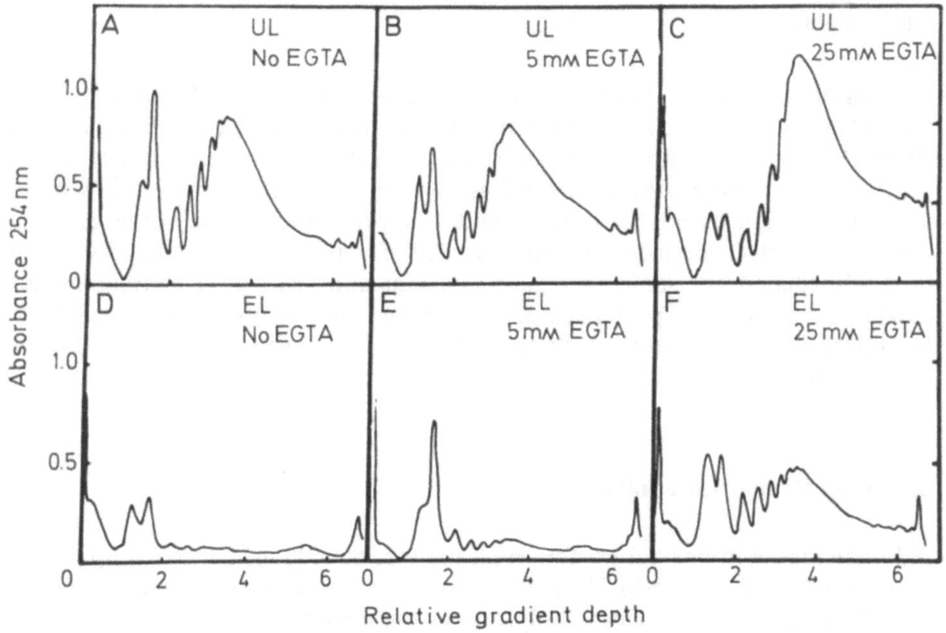

Fig. 10 A–F. Effect of EGTA on isolation of polysomes from unexpanded and expanded tobacco leaves. Polysomes extracted in buffer containing 0.2 M Tris-HCl, pH 9, 0.4 M KCl, 0.2 M sucrose, and 35 mM $MgCl_2$ with varying amounts of EGTA. **A** unexpanded leaves (*UL*), no EGTA, **B** UL 5 mM EGTA, **C** UL 25 mM EGTA, **D** expanded leaves (*EL*) no EGTA, **E** EL, 5 mM EGTA, **F** EL 25 mM EGTA. (Jackson and Larkins 1976)

The effects of heavy metal ions can be overcome by the addition of ethylene glycol-bis(2-aminoethyl ether) tetracetic acid (EGTA) to the grinding buffer. EGTA has a higher affinity for Ca^{2+} and other heavy metal ions than for Mg^{2+} (Marhol and Cheng 1970), although it can also chelate magnesium and cause ribosome dissociation. The latter problem can be avoided by using a higher concentration of Mg^{2+} than EGTA in the grinding buffer, e.g., 35 mM $MgCl_2$ and 25 mM EGTA. Figure 10 shows the effectiveness of an EGTA-containing buffer for polysome isolation from tobacco leaves. Although relatively undegraded polysomes are recovered from very young tobacco leaves in the absence of EGTA (Fig. 10 A), few if any polysomes are isolated from expanded leaf tissue (Fig. 10 D). The addition of 5 mM EGTA improves polysome recovery from both young and old leaves (Fig. 10 B, E), and recovery is enhanced even more with 25 mM EGTA (Fig. 10 C, F).

EGTA has been shown to improve polysome recovery in several leaf (Jackson and Larkins 1976; Bol et al. 1976) and fungal tissues (Gong and Lovett 1977). The effectiveness of buffers containing EGTA varies from one tissue to another (White and Murakishi 1977), and probably depends on the endogenous level of heavy metal ions.

4.6 Proteinase K

In addition to being an effective proteolytic enzyme, proteinase K has no associated RNase activity. It can therefore be used to purify mRNA (Wiegers and Hiltz 1971). Polysomes are surprisingly stable in relatively high concentrations (200 µg ml^{-1}) of proteinase K. Proteinase K has not been routinely used to isolate polysomes from plant tissue, but it was found to improve the recovery of large polysomes from radish cotyledons (Fourcroy 1980). It has also been used to dissociate protein-mediated polysome aggregates (Larkins and Tsai 1977). Polysomes isolated in the presence of proteinase K lose the capacity for protein synthesis in vitro, but they can be used to study the amount and integrity of polysomes or for recovery of mRNA.

4.7 Other Ribonuclease Inhibitors

A buffer that contains 0.2 M Tris-HCl, pH 8.5, 60–400 mM KCl (as appropriate), 35 mM $MgCl_2$, 5 mM DTT, 0.2 M sucrose, and in some instances 25 mM EGTA, is effective for isolating polysomes from most tissues. It may also be useful to include 200 µg ml^{-1} proteinase K, although this has not been tested for a variety of plant tissues. Other agents that have been used to inhibit RNase or preserve polysome integrity include bentonite, diethyl pyrocarbonate (DEP), heparin, exogenous RNA, and cycloheximide. Fine suspensions of bentonite, which is a clay mineral, have been used to adsorb RNase (Loening 1968). However, bentonite does not completely inhibit RNase activity and it also causes precipitation of ribosomes.

Several studies have shown that DEP can improve polysome isolation from plant tissues (Weeks and Marcus 1969; Anderson and Key 1971). Although DEP effectively inactivates RNases, it also acidifies the tissue homogenate, causes ribosome dissociation, and reduces ribosome efficiency for in vitro protein synthesis. DEP also causes carbethoxyethylation of purines and thereby reduces the template activity of mRNA (Denic et al. 1970). It is most effectively used to treat glassware or solutions, after which it can be destroyed by autoclaving.

Heparin, an anticoagulant from liver, provides protection from RNase activity and it has been used in some instances with plant tissue (Green 1981). However, heparin causes lysis of nuclei and for some tissues addition of heparin is not practical until after nuclei have been removed by centrifugation. This delay does not enhance recovery of intact polysomes.

In some procedures, exogenous RNA is added to the grinding buffer to provide an alternative substrate for the RNase. Although the additional RNA may afford some protection, this measure by itself does not prevent polysome degradation.

Cycloheximide (60–100 µg ml^{-1}) can also be included in the grinding buffer to preserve polysome integrity. Although not a RNAse inhibitor, cycloheximide blocks ribosomal "run-off" and leads to the recovery of fewer monoribosomes (Schlager et al. 1969).

5 Uses of Purified Polyribosomes

5.1 Changes in Protein Synthetic Activity

The ability to isolate polysomes from a plant tissue allows one to study many of its physiological and genetic features. An analysis of the percentage of monosomes versus polysomes provides an indication of protein synthetic activity, and an alteration in the proportion or size of polysomes is indicative of a metabolic change. Many studies show changes in polysome patterns resulting from hormone treatment, environmental stresses, gene activation, and tissue differentiation (Davies and Larkins 1972; Rhodes and Matsuda 1976; Armstrong and Jones 1973; Smith 1976).

5.2 In Vitro Protein Synthesis

Polysomes can be used to study the type of protein synthesized in a particular tissue. When added to a cell-free protein-synthesizing system, polysomes complete synthesis of nascent polypeptides (Marcus 1974). Even slightly degraded polysomes can be used for this purpose, since some complete polypeptides will be obtained. The synthesis of specific proteins is usually determined by reaction with antibodies (Schimke et al. 1974), and the size of the protein is determined by fluorography after SDS polyacrylamide gel electrophoresis (Laskey and Mills 1975).

Protein synthesis can also be directed by mRNA's purified from polysomes. Krystosek et al. (1975) developed a simple and direct method to obtain mRNA's from purified polysomes. Polysomes are dissolved in a buffer that contains 10 mM HEPES, pH 7.5; 0.5 M NaCl, and 0.5% SDS. It is also useful to include 200 μg ml^{-1} proteinase K (Larkins and Tsai 1977). Poly(A)-containing mRNA is obtained by passing the dissociated polysomes over oligo d(T) cellulose. At high ionic strength (0.3–0.5 M NaCl) the poly(A) tails of the mRNA bind to oligo d(T); protein and nonspecifically bound RNA are removed by repeated washing of the cellulose with buffer. Messenger RNA is eluted by washing with low ionic strength buffer (10 mM HEPES, pH 7.5). Usually several cycles of affinity chromatography are necessary to completely remove contaminating ribosomal RNA's, but relatively pure mRNA's can be obtained by this procedure (Larkins et al. 1976b).

When mRNA's are used to direct protein synthesis in vitro, proteins become labeled to a much higher specific activity than is obtained when polysomes are used. This enhances their detection in subsequent analyses. However, certain mRNA's are more efficiently translated than others (Herrlich and Schweiger 1978; Herson et al. 1979; Chroboczek et al. 1980), and the polypeptides that are produced may not reflect the in vivo situation. Furthermore, some proteins may be synthesized by mRNA's that are not retained by oligo d(T) cellulose. For this reason it is useful to compare proteins synthesized in vitro by both polysomes and purified mRNA's.

5.3 Purification of mRNA's

Sometimes it is possible to purify mRNA's encoding specific proteins by reacting polysomes with antibodies (Shapiro et al. 1974; Schell and Wilson 1979; Kraus and Rosenberg 1982). This reaction requires that the nascent polypeptide have antigenic determinants that are recognized by the antibody. Although there are some difficulties with this procedure, it has been used successfully to isolate several mRNA's encoding specific plant proteins (Vodkin 1981).

Messenger RNA's from polysomes provide efficient templates for synthesis of recombinant cDNA clones (Maniatis et al. 1982). Even though a mixed population of mRNA's are converted to cDNA's, the cloning procedure results in the isolation of individual sequences. A variety of techniques are available to identify clones that correspond to a particular mRNA. Once the clone has been identified it can be used to isolate the corresponding gene and study the regulation of its expression.

5.4 Subcellular Distribution of mRNA's

Depending on physiological and genetic conditions, mRNA's may be actively translated or they may exist in an inert form within the cell. It is generally assumed that mRNA's associated with polysomes are actively translated, and thus by comparing the distribution of mRNA's complexed with ribosomes and those free in the cytoplasm (post-ribosomal supernatant) relative expression of a particular mRNA can be assessed. Although mRNA's in polysomes may not be translated, this question may be resolved by an analysis of the proteins produced by the polysomes.

References

Aldeman MR, Blobel G, Sabatini DD (1973) An improved cell fractionation procedure for the preparation of rat liver membrane-bound ribosomes. J Cell Biol 56:191–205

Allington RW, Brakke MK, Nelson JW, Aron CG, Larkins BA (1976) Optimum conditions for high resolution gradient analysis. Anal Biochem 73:78–92

Anderson JM, Key JL (1971) The effects of diethyl pyrocarbonate on the stability and activity of plant polyribosomes. Plant Physiol 48:801–805

Armstrong JE, Jones RL (1973) Osmotic regulation of α-amylase synthesis and polyribosome formation in aleurone cells of barley. J Cell Biol 59:444–455

Bol JF, Bakhuizen CEGC, Rutgers T (1976) Composition and biosynthetic activity of polyribosomes associated with alfalfa mosaic virus infections. Virology 75:1–17

Breen MD, Whitehead EI, Kenefick DG (1972) Requirements for extraction of polyribosomes from barley tissue. Plant Physiol 49:733–739

Chrispeels MJ (1984) Biosynthesis, processing, and transport of storage proteins and lectins. Phil Trans R Soc Lond B Biol Sci 304:309–322

Chroboczek J, Witt M, Ostrowka K, Bassuner R, Puchel M, Zagorski W (1980) Seed transmissibility of plant viruses may be modulated by competition between viral and cellular messenger. A proprosal. Plant Sci Lett 19:263–270

Dahlberg AE, Dingman CW, Peacock AC (1969) Electrophoretic characterization of bacterial polyribosomes in agarose-acrylamide composite gels. J Mol Biol 41:139–147

Davies E, Larkins BA (1972) Polyribosomes from peas II. Polyribosome metabolism during normal and hormone induced growth. Plant Physiol 52:339–345
Davies E, Larkins BA, Knight RH (1972) Polyribosomes from peas. An improved method for their isolation in the absence of ribonuclease inhibitors. Plant Physiol 50:581–584
Davies E, Dumont JE, Vassart G (1977) Improved techniques for the isolation of intact thyroglobulin-synthesizing polysomes. Anal Biochem 80:289–292
Davis E, Larkins BA (1980) Ribosomes. In: Stumpf PK, Conn EE (eds) The biochemistry of plants: a comprehensive treatise, vol 1. The plant cell. Academic Press, New York, pp 413–435
Denic M, Ehrenberg L, Fedorcsak I, Solymosy F (1970) The effect of diethyl pyrocarbonate on the biological activity of messenger RNA and transfer RNA. Acta Chem Scand 24:3753–3755
Dhindsa RS, Bewley JD (1976) Water stress and protein synthesis IV. Response of a drought-tolerant plant. J Exp Bot 27:513–523
Drouet AG, Hartman CJR (1979) Requirements for extraction of polyribosomes from lyophilized peel tissue of climatic pear. Phytochemistry 18:545–547
Fourcroy P (1980) Isolation of undegraded polysomes from radish cotyledons: use of protease K and cycloheximide. Phytochemistry 19:7–10
Gong CS, Lovett JS (1977) Regulation of protein synthesis in *Blastocladiella* zoospores: factors for synthesis in nonsynthetic spores. Exp Mycol 1:138–151
Green FC (1981) In vitro synthesis of wheat (*Triticum aestivum* L.) storage proteins. Plant Physiol 68:778–783
Herrlich P, Schweiger M (1978) Discrimination of messenger RNA. FEBS Lett 87:1–6
Herson D, Schmidt A, Seal S, Marcus A (1979) Competitive mRNA translation in an in vitro system from wheat germ. J Biol Chem 254:8245–8249
Higgins TJV (1984) Synthesis and regulation of the major proteins in seeds. In: Briggs WR, Jones RL, Walbot V (eds) Annual review of plant physiology. Annual reviews, vol 35. Palo Alto, California 35:191–221
Hsiao TC (1973) Plant response to water stress. In: Briggs WR, Green PB, Jones RL (eds) Annual review of plant physiology. Annual reviews, vol 24. Palo Alto, California, pp 519–570
Jackson AO, Larkins BA (1976) Influence of ionic strength, pH, and chelation of divalent metals on isolation of polyribosomes from tobacco leaves. Plant Physiol 57:5–10
Key JL, Lin CY, Ceglarz E, Schoffl F (1982) The heat shock response in plants. In: Schlesinger M, Ashburner M, Tissieres A (eds) Heat shock: from bacteria to man. Cold Spring Harbor Lab 35:191–221
Kraus JP, Rosenberg LE (1982) Purification of low-abundance messenger RNA from rat liver by polysome immunoadsorption. Proc Natl Acad Sci USA 79:4015–4019
Krystosek A, Cawthon ML, Kabot D (1975) Improved methods for purification and assay of eukaryotic messenger ribonucleic acids and ribosomes. Quantitative analysis of their interaction in a fractionated reticulocyte cell-free system. J Biol Chem 250:6077–6084
Larkins BA (1981) Seed storage proteins: characterization and biosynthesis. In: Stumpf PK, Conn EE (eds) The biochemistry of plants: a comprehensive treatise, vol 6. Proteins and nucleic acids. Academic Press, New York, pp 449–489
Larkins BA, Davies E (1973) Polyribosomes from peas III. Stimulation of polysome degradation by exogenous and endogenous calcium. Plant Physiol 52:655–659
Larkins BA, Davies E (1975) Polyribosomes from peas V. An attempt to characterize the total free and membrane-bound polysomal population. Plant Physiol 55:749–756
Larkins BA, Tsai CY (1977) Dissociation of polysome aggregates by protease K. Plant Physiol 60:482–485
Larkins BA, Bracker CE, Tsai CY (1976a) Storage protein synthesis in maize. Isolation of zein-synthesizing polyribosomes. Plant Physiol 57:740–745
Larkins BA, Jones RA, Tsai CY (1976b) Isolation and in vitro translation of zein messenger ribonucleic acid. Biochemistry 15:5506–5510
Laskey RA, Mills AD (1975) Quantitative film detection of ^{3}H and ^{14}C in polyacrylamide gels by fluorography. Eur J Biochem 56:335–341
Leaver CJ, Dyer JA (1974) Caution in the interpretation of plant ribosome studies. Biochem J 144:165–167

Lin CY, Key JL (1967) Dissociation and reassembly of polyribosomes in relation to protein synthesis in the soybean root. J Mol Biol 26:237–247

Loening UE (1968) The occurrence and properties of polysomes in plant tissues. In: Pridham JB (ed) Plant cell organelles. Academic, New York, pp 216–227

Maniatis T, Fritsch EF, Sambrook J (1982) Molecular cloning, a laboratory manual. Cold Spring Harbor Lab

Marcus A (1974) The wheat embryo cell-free system. In: Moldave K, Grossman L (eds) Methods in enzymology, vol 30. Academic Press, New York, pp 749–761

Marhol M, Cheng KL (1970) Simple ion exchange separation of magnesium from calcium and other metal ions using ethylene glyco-bis(2-aminoethyl ether) tetraacetic acid as a complexing agent. Anal Chem 42:652–655

McGown E, Richardson A, Henderson LM, Swan PB (1971) Anomalies in polysome profiles caused by contamination of the gradients with Cu^{2+} or Zn^{2+}. Biochim Biophys Acta 247:165–169

McIntosh PR, O'Toole K (1976) The interaction of ribosomes and membranes in animal cells. Biochim Biophys Acta 457:171–212

Morton BE, Hirsch CA (1970) A high-resolution system for gradient analysis. Anal Biochem 34:544

Newburn LH (1975) Isocotables. A handbook of data for biological and physical scientists. Instrumentation Specialties Company, Lincoln, Nebraska

Noll H (1969) Polysomes: analysis of structure and function. In: Campbell PN, Sargent JR (eds) Techniques in protein biosynthesis. Academic Press, New York, pp 101–179

Rhodes PR, Matsuda K (1976) Water stress, rapid polysome reductions and growth. Plant Physiol 58:631–635

Schell MA, Wilson DB (1979) Purification of galactosidase mRNA from *Saccharomyces cerviseae* by indirect immunoprecipitation. J Biol Chem 254:3531–3536

Schimke RT, Rhoads RE, McKnight GS (1974) Assay of ovalbumin mRNA in reticulocyte lysate. In: Moldave K, Grossman L (eds) Methods in enzymology, vol 30, part F. Academic Press, New York, pp 694–701

Schlager C, Hoffman D, Hilz N (1969) Polyribosomes in tumor cells during induction of ribonuclease by cytostatic treatment. Hoppe-Seyler's Z Physiol Chem 350:1017–1022

Shapiro DJ, Taylor JM, McKnight GS, Placious R, Gonzalez C, Kieley ML, Shimke RT (1974) Isolation of hen oviduct ovalbumin and rat liver albumin polysomes by indirect immunoprecipitation. J Biol Chem 249:3665–3671

Shore GC, Tata JR (1977) Functions for polyribosome-membrane interactions in protein synthesis. Biochim Biophys Acta 472:197–236

Smith H (1976) Phytochrome-mediated assembly of polyribosomes in etiolated bean leaves. Evidence for post-translational regulation of development. Eur J Biochem 65:161–170

Stone AB (1974) A simplified method for preparing sucrose gradients. Biochem J 137:117–118

Vassart GM, Dumont JE, Cantraine FRL (1970) Simulation of polyribosome disaggregation. Biochem Biophys Acta 224:155–164

Verma DPS, Maclachlan GA, Byrne H, Ewings D (1975) Regulation and in vitro translation of messenger RNA for cellulase from auxin treated pea epicotyls. J Biol Chem 250:1019–1026

Vodkin LO (1981) Isolation and characterization of messenger RNAs for seed lectin and Kunitz trypsin inhibitor in soybeans. Plant Physiol 68:766–771

Weeks DP (1981) Protein biosynthesis: mechanisms and regulation. In: Stumpf PK, Conn EE (eds) The biochemistry of plants: a comprehensive treatise, vol 6. Proteins and nucleic acids. Academic Press, New York, pp 471–529

Weeks DP, Marcus A (1969) Polyribosome isolation in the presence of diethyl pyrocarbonate. Plant Physiol 44:1291–1294

White JL, Murakishi HH (1977) Requirement for extraction of polyribosomes from plant callus cultures. Plant Physiol 59:800–802

Wiegers U, Hiltz H (1971) A new method using "proteinase K" to prevent mRNA degradation during isolation from HeLa cells. Biochem Biophys Res Commun 44:513–519

The Nucleus – Cytological Methods and Isolation for Biochemical Studies

J. F. JACKSON

1 Introduction

Interest in the plant cell nucleus has heightened in recent times due to the rapid advances being made in the biochemistry of its DNA, RNA, and protein components. While the cytological observations on chromosome structure and behaviour have held interest in the nucleus for some time, application of recombinant DNA technology, sequencing and hybridization techniques to chromosomal DNA have focussed attention on the nucleus even more. Studies on gene expression have also seen many advances recently and much has been learnt about the events of eukaryotic mRNA biogenesis (Nevins 1983). Recognition of promoters and sites of transcriptional initiation, poly(A) addition, RNA splicing, methylation and the process of nuclear-cytoplasmic transport have been investigated and all need nuclear preparations at some stage for further investigation in the eukaryotic organism. Possession of a well-defined nucleus is one of the criteria separating eukaryotes from prokaryotes, and it is the pathway of mRNA synthesis involving this RNA processing and transport out of the nucleus which sets the eukaryotes apart from the prokaryotes as much as the other cytologically recognized nuclear differences. There is then great interest in nuclear preparation for these and other biochemical studies. This chapter will deal in the main with the larger-scale preparation of nuclei from plant cells for these biochemical studies. The properties of the nucleus will be dealt with only in so far as it explains the various techniques exploiting those properties which are used in nuclear preparation. Particular attention will be given to different procedures adopted for nuclear preparations for different purposes. The development of efficient protoplast preparation methods from plant cells has had a great effect on organelle investigations in plants (see Wagner, this vol.) including nuclear studies. These therefore find a place in this chapter. In addition, developments have taken place over recent years in the staining of chromosomal DNA by fluorescent methods for both quantitative and qualitative estimation of DNA. These are described here, since efficient staining techniques are needed during nuclear preparation and because these techniques have advanced to the level where they can be compared with the older, well-tried Feulgen procedures.

Reviews of the various methods of preparation of plant nuclei for biochemical studies, and of staining techniques for plant nuclei, are not numerous. A thorough article on some of the earlier techniques and applications is that by Mascarenhas et al. (1974), while Bonner (1976) also covers aspects of the plant nucleus. The plant nucleus and its metabolism is dealt with indirectly in an article on plant polytene chromosomes by Nagl (1981), and the nuclear envelope is well described in an article dealing with animal and plant aspects, by Franke (1974). Nuclear transplantation and nuclear properties in the alga *Acetabularia* is discussed by Schweiger and Berger (1979) and a general discussion on the structure of the nucleolus is given by Goessens (1984). The biochemistry of the cell nucleus (animal and plant) is dealt with by Garland and Mathias (1977).

2 Structure of the Plant Nucleus and Implications for Nuclear Isolation and Staining

The plant nucleus is generally spherical or disc-shaped with a diameter of about 1–10 μm depending on the plant species (Figs. 1, 2, and 3). This relatively large size means that cell breakage as a preliminary to isolation of nuclei needs to be carried out carefully so as not to disrupt the nucleus itself, and that the smallest sieve so commonly used to eliminate unbroken cells should have a pore size of 15–20 μm or greater. The major part of the cellular DNA is found in the nucleus, and even amongst the extra-nuclear DNA such as in the plastids, major homologies exist between plastid and nuclear DNA, so that many of the plastid DNA sequences are also present in the nuclear DNA (Scott and Timmis 1984). In the nucleus, DNA is organized within the chromosomes, together with basic proteins

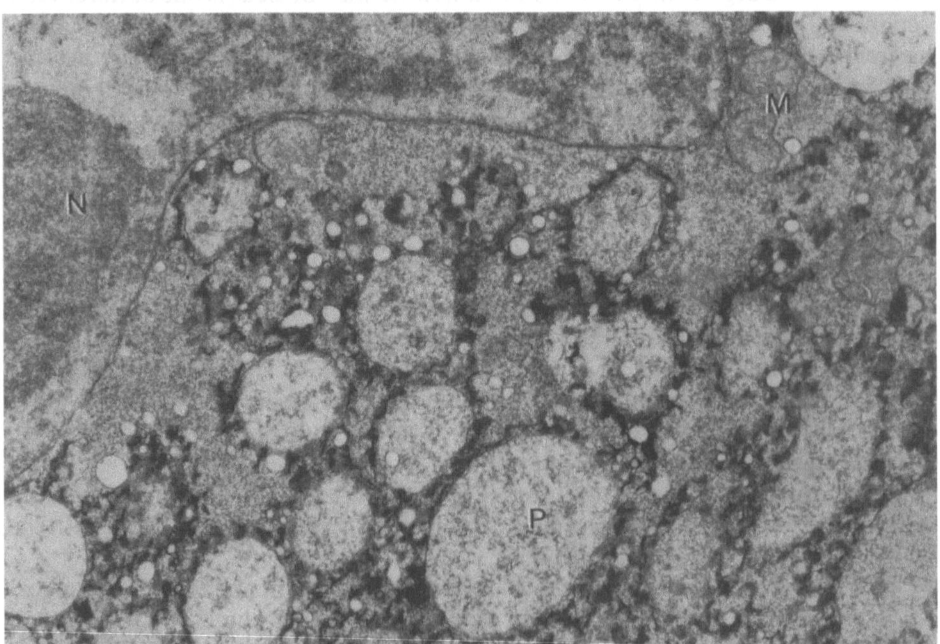

Fig. 1. Nuclear membrane of soybean seed axis. This is an electron micrograph of an axis parenchymal cell, showing the nucleus (*N*), mitochondria (*M*) and protein bodies (*P*). (×20,000) (Chabot and Leopold 1982)

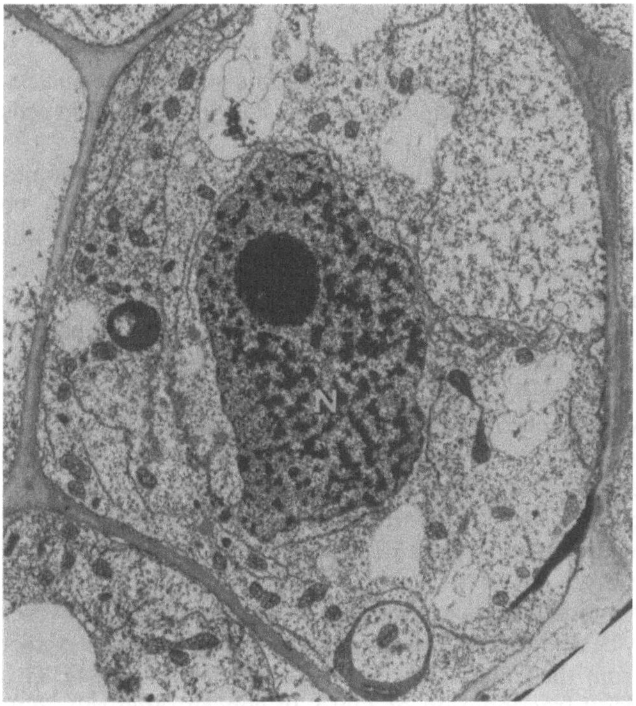

Fig. 2. Meristematic cell from the root tip of barley. The nucleus (*N*) occupies the centre of the field with a prominent nucleolus. (× 4600) (Werker et al. 1983)

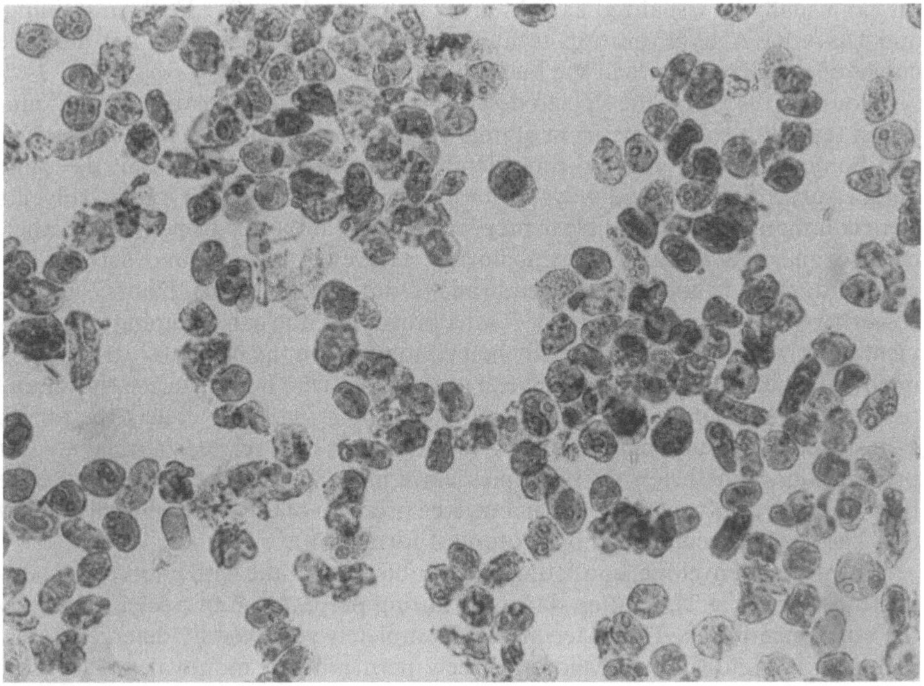

Fig. 3. Nuclei as isolated by Spiker et al. (1983). (Spiker et al. 1983)

(histones), acidic proteins and RNA. The number of chromosomes in the nucleus varies widely from species to species, but is normally constant in the diploid cells within a given species.

During the interphase segment of the mitotic cycle of the cell, chromosomes are not easily visualized and are somewhat diffuse, so that DNA appears to be spread more or less evenly throughout the nucleus. For this reason it is at this stage that much of the quantitative Feulgen staining technique has been carried out to estimate DNA content of plant cells (Harland et al. 1973). The nuclear DNA content of haploid cells from a given species is constant; a comprehensive survey of angiosperm haploid cell DNA content, using Feulgen staining techniques, is given in Bennet and Smith (1976).

During mitosis, however, the chromosomes condense and are readily identifiable at metaphase as distinct entities. Mass isolation of chromosomes is therefore usually carried out with cells blocked at metaphase. Staining of the chromosomes, possibly due to differences in degree of condensation, reveals areas of higher staining, usually called heterochromatin, as distinct from euchromatin, which stains more evenly and lightly.

DNA extraction and fractionation is not the subject of this chapter, however for certain purposes it may be desirable to isolate the nucleus first and then extract the chromosomal DNA from the isolated nuclei (Scott and Timmis 1984). This may involve buoyant density fractionation of DNA in concentrated salt solutions, which fractionates by base composition and often allows a study of "satellite" DNA. The nucleolus (Fig. 2) appears to be the site of rRNA genes, and techniques have been devised for mass preparation of this structure also. The nucleolus can usually be visualized as light areas in the nucleus after Feulgen staining, and DNA-RNA hybridization techniques have shown that there are multiple copies of the rRNA genes in the nucleolus of many organisms.

Dounce (1943) was the first to describe a procedure for isolating "clean" nuclei in reasonable amounts from animal tissue. This method, involving freezing and homogenization in a Waring blendor, yielded damaged nuclei; the same author subsequently found ways of improving his techniques. Organic solvents were commonly used for nuclear preparation with mixtures of petroleum ether, cyclohexane and/or carbon tetrachloride. However, the organic solvents increased damage to the nuclear membrane (Dounce et al. 1950). Dounce and his coworkers also observed that Ca^{2+} ions prevented nuclear clumping and fragmentation, whole Mg^{2+} apparently helped to preserve the nucleolar structure. A major advance was made by Chauveau et al. (1956) who homogenized the animal tissue in 2.2 M sucrose and centrifuged at 40,000 *g* for 1 h to collect the nuclei. Whilst it could be said that these early developments only relate to animal nuclear preparations, nonetheless the principles learnt in working with animal tissue have been found to apply to plant nuclei, once the problem of the rigid cell wall in plant cells has been overcome (e.g. by protoplast formation).

The nuclear envelope, consisting of a double membrane with intervening perinuclear space (Fig. 1), is often damaged during preparation of nuclei, especially the outer membrane. The latter may be completely removed by detergents (e.g. Triton X 100, etc.). The envelope is freely permeable to monovalent inorganic ions, amino acids, dyes, nucleosides, nucleotides, and glycolytic intermediates

(Kohen et al. 1971). While isolated nuclei take up ions readily from the media, they are not osmotic structures. There appears to be an extensive and highly specific vectorial transport of many macromolecules through the nuclear envelope, the pores providing channels for these processes. Thus mRNA appears to be transported out of the nucleus in this way, being "processed" on the way (Nagl 1981). It is, for example, possible that the poly(A) sequences are put on in the perinuclear region of the nucleus. The pore frequency may increase in cells with active RNA synthesis and there is a tendency for less pores in nuclei not so active in RNA synthesis (Mepham and Lane 1969). The pores do not appear to be "gateways" but rather associated in nonmembranous elements. These nonmembranous constituents of the pore complex tend to disappear during the various methods of nuclear preparation because of mechanical shearing and dissolution in the medium (Franke 1974). The pore complexes tend to become shrunken, and emptied of annular and internal structures particularly with high salt media.

Unlike many other cell organelles, the characterization of the plant nucleus by marker enzymes is not so precise, but neither is it so necessary due to the ease of cytological identification of nuclei by staining procedures. The animal nucleus has been shown to contain exclusively the cell's complement of ATP : NMN adenylyltransferase (Hogeboom and Schneider 1952; Branster and Morton 1956). This transferase can be purified from animal nuclear preparations (Atkinson et al. 1961). However, this enzyme does not seem to have been reported in plant cells and so considerable doubt must remain as to its usefulness in characterizing plant nuclei. It is possible that the alternative enzyme in the NAD synthetic pathway, ATP : nicotinic acid mononucleotide adenylyltransferase, which is found in *E. coli* (Preiss and Handler 1958) is utilized in plant cells. This point needs to be investigated in view of the occurrence of the animal enzyme exclusively in the nucleus.

The mitochondrion, with its own genetic system and Feulgen-positive fibres, also has a nucleus. This nuclear system has no membrane (Kurotiwa 1982), and will not be dealt with further in this chapter.

3 Cytology

3.1 Nuclei of Whole Cells

DNA. The DNA content of plant nuclei can be quantitatively determined by the well-tried microspectrophotometric Feulgen procedure (Leuchtenberger 1958; Kasten 1960, 1964; Berlyn and Cecich 1976), or by the microfluorometric methods in which fluorescence of a DNA-fluorochrome complex is excited by UV or blue-violet light (LaLoue et al. 1980; Hamilton et al. 1980; Coleman et al. 1981). Both methods have been used with success and require different instrumentation.

3.1.1 Feulgen Microspectrophotometric Method

Reagents

a) Leucobasic Fuchsin. Dissolve 1 g of basic fuchsin by pouring over it 200 ml boiling water. Shake well and cool to 50 °C. Filter. Add 30 ml N HCl to filtrate. Add 3 g potassium metabisulphite ($K_2S_2O_5$), allow to bleach for 24 h in tightly stoppered bottle in the dark. Add 0.5 g of decolorizing carbon (e.g. Norite). Shake well for 1 min and filter rapidly through course filter paper. Store in tightly stoppered bottle in dark at 4 °.

b) SO_2 – Water. To 100 ml water, add 5 ml N HCl and 5 ml 10% $K_2S_2O_5$.

Method. Fixed plant material is hydrated in 35% ethanol and then water, and incubate in 0.05 M EDTA, pH 9.0 at 45 °C for 2 h. Now rinse twice with water and hydrolyze in N HCl/60 ° for 10 min. The plant tissue is then stained in leucobasic fuchsin for 3 h. The staining time can be quite critical, and may require shorter or longer times, depending on the tissue. For example for artichoke explants, 3 h is suitable (Harland et al. 1973), while pollen from *Petunia hybrida* should be stained for a shorter time (Jackson and Linskens 1978). Pollen from *Pinus mugo* requires only 90 min staining time (Tan and Jackson 1984, personal communication). The stained tissue is then washed five times with freshly prepared SO_2-water, each wash being for 10 min before changing to fresh SO_2-water. Washing should be carried out in stoppered jars. The tissue is then rinsed in water, suspended in water and macerated if necessary with a Pasteur pipette or hypodermic syringe. Separated cells are washed several times in absolute alcohol, then a few drops are placed on a microscope slide and allowed to dry. The slide is passed through absolute ethanol, ethanol/xylene (1/1) and finally xylene. Mount in euperal.

Explanation and Comment. The Feulgen reaction consists of hydrolysing the chemically fixed DNA with a mineral acid to break the purine base-glycoside linkage, yielding "unspecified" aldehyde groups on the deoxyribase moiety. These react with the Schiffs reagent (leucobasic fuchsin) to give a DNA-specific insoluble magenta coloured compound. The instrument used to measure the amount of this coloured compound in each cell is basically a good spectrophotometer and microscope combination. The stained cells or portions thereof are not exactly equivalent to a clear, coloured, solution in a cuvette, but the system works, and several instruments are available commercially, although expensive. A careful investigation of the measurement of DNA by this method with plant cells has been carried out by Dhillon et al. (1977, 1978), and anyone contemplating using this method should refer to their work. They recommend an internal standard, such as for example, chicken red blood cells, to counteract staining variations between slides and staining batches.

The method has been used with success to investigate the haploid DNA content of various plant species (Bennet and Smith 1976), ploidy variations between cells, and for following DNA replication in plant cells which are going through a cell cycle in synchrony (Harland et al. 1973). Two-wavelength microspectrophotometry can be adopted to determine DNA content, the wavelengths chosen depending on the spectrum obtained, and typically being at approximately 490 nm and at 550 nm. Tables used to compute amounts of relative Feulgen absorption can be found in Mendelsohn (1958).

3.1.2 Microfluorometric DNA Determination

Reagents. DAPI (4′,6-diamidino-2-phenylindole), and BAO (2,5-bis[4′-aminophenyl(1′)-1,3,4-oxadiazole]) are two possible DNA-localizing fluorochromes that can be used for cytofluorometric determination of nuclear DNA. Plant material should be first fixed in ethanol-glacial acetic acid (3 : 1).

Method

a) DAPI-Staining. Prepare citrate-phosphate buffer, pH 4.0 as follows. 21.01 g citric acid monohydrate l^{-1} (0.1 M), and 28.4 g anhydrous $Na_2HPO_4l^{-1}$ (0.2 M). Mix 61.4 ml citric acid solution with 38.6 ml phosphate solution, to give pH 4.0 buffer. Dissolve 0.5 μg DAPI ml^{-1} pH 4 buffer. Then stain plant material (first washed for 10 min in distilled water) in DAPI solution for 30 min. Mount in buffer or stain solution, and examine in fluorescence microscope. DNA should fluorescence a brilliant blue-white.

b) BAO Staining. BAO *must* be freshly prepared before use. Ten mg BAO is added to 100 ml distilled water with stirring for 5 min. Add 10 ml 1 M HCl and 5 ml 10% $NaHSO_3$. Stir 10 to 20 min with slight heating and filter. Plant material is soaked in distilled water for 10 min, the hydrolysed 20 min at 37 °C in 3.5 M HCl. Rinse in 1 N HCl in cold, stain 1 to 3 h in BAO solution. Wash in sulphite water (10 ml 10% $NaHSO_3$ or anhydrous $K_2S_2O_5$, 10 ml 1 M HCl and water to 200 ml, prepared just before use). Wash in distilled water 10 min, mount in glycerine or dehydrate in ethanol and mount in xylene. Examine in fluorescence microscope as for DAPI staining. Fluorescence of the DNA is a blue-white.

Comments. A variety of equipment is available for cytofluorimetry, and for example, for visual examination a Leitz Ortholux microscope with fluorite objective is suitable, while for quantitative work a Leitz MPV photometer head equipped with EMI 6094 photomultiplier tube in combination with the first mentioned microscope gives good results. Many different fluorochrome dyes have been used to produce fluorescence in UV-irradiated DNA, including 33258 Hoechst (a bisbenzimidazole derivative used by Laloue et al. 1980), DAPI (Coleman 1978) BAO (Hull et al. 1982), mithramycin (Ward et al. 1965), propidium iodide and others. Instrumentation and technical aspects are well covered by Ruch (1973).

One problem with fluorescent techniques as applied to DNA is the relatively rapid fading of fluorescence excited by UV irradiation. For example Hull et al. (1982) report nuclear fluorescence intensity dropped to 91–96% after 3 min with BAO-stained material, while mithramycin stained nuclear DNA dropped to 33–44% within 3 min of UV excitation.

The use of DAPI in the study of nucleoids of yeast is described by Miyakawa et al. (1984).

3.2 Staining Nuclei During Isolation

Since the main thrust of this article is the preparation of nuclei for the study of the various biochemical processes carried out by nuclei, then rapid and convenient staining techniques for nuclei need to be considered. Most involve DNA staining; a large variety of stains have been used.

Nuclei from a large number of different plants has been prepared by Vlasak (1981), and this author makes the comment that the use of carmine is to be preferred over the more commonly used methyl green staining as it is simpler and more selective. Vlasak (1981) uses a 1% carmine solution in 45% acetic acid, which is mixed with an equal volume of nuclear suspension. After just 1 min, nuclei can be viewed under the microscope. There are some exceptions; Vlasak (1981) found that *Arabidopsis* nuclear preparations did not stain satisfactorily with carmine. It was found that carbolfuchsin (Kao 1975) could be used in this case, but staining times should be kept short, otherwise chloroplasts also stain with this dye.

If a microscope is available which is set up for cytofluorimetry, then a method of choice for nuclear staining is the use of fluorochrome dyes such as DAPI (see above).

3.3 Nucleolus Staining

Staining of the nucleolus is not so straightforward, but is necessary for studies where nucleolar shape and size are necessary for karyotype analyses (MacPherson and Filion 1981) in investigations concerning location of the nucleolar organizer (McClintock 1934) and in nucleolar preparation for biochemical analysis. Nucleolar staining techniques have been reviewed by Busch and Smetana (1970). Single staining methods using methylene blue or acetocarmine do not distinguish between DNA and RNA, and so chromatin and nucleoli are not differentially stained. There are methods for staining nucleoli differentially, these can involve a single stain such as acridine orange (Armstrong 1956) or multiple staining using Feulgen-methylene blue (Spicer 1961), methyl green-pyronin (Kurnick 1955) methyl green-thionin (Roque et al. 1965) and orange G-methyl green-toluidine blue (Korson 1951).

Perhaps one of the best techniques for the simple and reproducible staining of nucleoli is that described by Franklin and Filion (1981). This involves staining of nucleoli with the fluorescent dye acridine orange followed by counter-staining with methyl green, which quenches fluorescence of chromatin.

Method. Plant material is fixed in 3 : 1 absolute ethanol : glacial acetic acid at 5 °C for 24 h, and stored in 70% ethanol at 5 °C. Squashes are made in 45% acetic acid and coverslips removed by the dry ice procedure. Dip slides in 95% ethanol, dry in air and store in vacuum dessicator for several days.

To stain for nucleoli, place slides in McIlvaine's buffer (24.6 ml 0.1 M citric acid plus 15.4 ml 0.2 M Na_2HPO_4, pH 4.9) for 5 min, then stain for 20 min in 0.359 $\mu g\ ml^{-1}$ acridine orange solution made up in McIlvaines buffer (pH 4.9).

Rinse in the same buffer for 10 min, then in aqueous sodium acetate (10 mM, pH 4.9) for 5 min, and the counterstain with methyl green (0.01% in 10 mM sodium acetate, pH 4.9) for 0.5 to 5 min, depending on tissue. Rinse in sodium acetate buffer.

Nucleoli are best viewed after this procedure with a microscope equipped for cytofluorimetry (see above).

Comments. With this method, plant cells initially fluoresced a uniform bright yellow-green, and after counterstaining with methyl green (a nonfluorescent dye), the cells maintained a bright fluorescence for nucleoli and cytoplasm, but with greatly reduced fluorescence of the chromatin. Franklin and Filion (1981) consider that the quenching effect of the chromatin in acridine orange-stained cells after counterstaining with methyl green can be explained by the interaction between the dye pairs. They found the technique gave consistent and good contrast between cytoplasm, chromatin and nucleolus.

3.4 Chromosome Staining

The squash technique has been widely used for microscopic examination of plant chromosomes since its development by Belling (1921). There are problems with this technique, including difficulty in obtaining flat fields, and adequate staining. These problems can be overcome by first preparing protoplasts and air-drying these to give flattened chromosomes on the slide. The method described below is essentially that of Murata (1983), and can be applied to cells in culture, to mitotic cells of root tips and flower petals and meiotic chromosomes of pollen mother cells.

Methods. Cells grown in the presence of colchicine are centrifuged to be rid of medium and suspended in 5 ml of solution containing 2.0% cellulysin (Calbiochem) 1.0% Macerase (Calbiochem) and 0.6 M sorbitol (pH 5.5–5.6). The Macerase can be replaced by 0.1% pectolyase Y23 (Kikkoman), and may result in faster protoplast isolation (Nagata and Ishii 1979). Incubate with shaking for 2 h at 25 °. Filter cell suspension through 60 μm nylon mesh into centrifuge tube, centrifuge 65 *g* for 5 min, add fixative (3 : 1, 95% ethanol/glacial acetic acid), and store 1 h. Resuspend in fresh fixative, × 2. Use a Pasteur pipette to add 5 or 6 drops to slide, and flame or air dry.

Stain for 3–4 min with 4% Giemsa (Gurr R66, biomedical special), diluted with 1/15 M phosphate pH 6.8, rinse in buffer, then distilled water and air dry. Mount in Depex.

3.5 Other Nuclear Stains

The histones of nuclei can be stained with fast green FCF at pH 8 (Miksche 1966), while the acid nuclear proteins can be stained with the same dye at a lower pH of 5.0 (Dhar and Shah 1982). The less specific toluidine blue stain can also be used for acid nuclear proteins (Smetana and Busch 1966).

4 Isolation of Plant Nuclei – General

The isolation of nuclei for biochemical investigations from plant cells is technically much more difficult than from animal cells. The rigid and often thick cell wall calls for special methods in cell disruption, and the methods used for breakage of animal cells (homogenization with Waring blender or Potter homogenizer for example) need to be applied more vigorously if they are to be used at all for plant cells. For these reasons and because the method to be adopted for nuclear isolation may also vary according to the use the nuclei to which are to be put, a list of isolation methods for plant nuclei, together with the biochemical use which is made of these preparations and other relevant details, is given in Table 1.

Perhaps the most instructive work on disintegration techniques for plant-cell breakage in preparation for nuclear isolation is that described by Vlasak (1981). The effect of the plant-tissue source on subsequent nuclear release by various disintegration methods was investigated, using barley leaves, roots and embryos, tobacco leaves and tissue cultures, roots from *Vicia faba* and leaves of *Arabidopsis thaliana*. In all cases, the leaf was finely chopped with scissors first. Homogenization was by a type of Waring blendor with cutting propellors of different diameter, or in a Potter-Elvehjem all-glass homogenizer (clearance 0.1 mm; 1000 rpm for 10 or 15 strokes), or with a mortar and pestle, or both the latter. Additionally, homogenization by nitrogen decompression was tried in an adapted autoclave followed by forcing through a needle valve. Vlasak (1981) found that the method of homogenization was the most critical step in nuclear isolation as treatment which breaks open tough cell walls tends to disrupt the nucleus as well. On the other hand, low efficiency of cell-wall disruption limits the yield of nuclei (Kuehl 1964). Vlasak (1981) reports that the Waring-blendor-type homogenizer works best at highest speeds only and that sharpness of the blades was of no consequence. The mortar and pestle was found to break most nuclei except for those from barley embryos, while the Potter-Elvehjem homogenizer worked well for cell-suspension cultures but not at all with fibrous plant tissues. The nitrogen decompression method worked best in the hands of Vlasak (1981).

The isolation medium was also investigated by Vlasak (1981). He advises the use of slightly hypotonic medium for nuclear isolation and for further manipulation 0.6–0.9 M sucrose to prevent nuclear clumping. The use of n-octanol for younger tissue is recommended, perhaps because it acts as a weak detergent. Vlasak found that root cells have more fragile nuclei and recommends the use of up to 4% gum arabic to increase yields of intact nuclei. This work does not deal with the use of enzymes to weaken plant cell walls or to produce protoplasts prior to cell disruption. Protoplast formation allows the subsequent use of techniques already proven for animal cells, and early methods are described by Mascarenhas et al. (1974), as is another device for rupturing cells, the "pea or nuclear popper", first described by Rho and Chipchase (1962). Other, more recent descriptions of protoplast formation as a preliminary to plant nuclear isolation are to be found in the list in Table 1.

The various ways of fractionating cell organelles once cell disruption has been achieved is given in Price (1974), while some additional methods of disruption are

Table 1. Isolation of plant nuclei. This table lists in alphabetical order of authors a selection of published work on methods of preparation of plant nuclei. The table is not meant to be exhaustive, rather the emphasis has been placed on the need for variation in method of preparation to suit the biochemical parameters being measured. Where more than one publication describes plant nucleus isolation for the same purpose, preference is given to the later work unless large differences in method of preparation have been judged to warrant inclusion of earlier descriptions as well for comparative purposes

Reference	Source	Method	Purpose	Comments
Ansa et al. (1982)	Cauliflower		Study of viral replication in nucleus	Replicated virus transported to cytoplasm and there encapsulated
Arfman and Willmitzer (1982)	Tobacco		Protein kinase activity in nucleus	
Bonner et al. (1968)			Chromatin	Low ionic strength conditions
Bonner (1976)	Plants, various	Review methods	Various	Review
Brewer (1979)	*Physarum polycephalum*		Repair of DNA	Double-strand breaks
Brightwell et al. (1975)	*Physarum polycephalum*		PolyADPribose polymerase	
Buetow (1976)	Protozoa and algae	Review methods	Various	
Caboche and Lark (1981)	Soybean	Protoplast, Triton X-100, sedimentation	DNA replication	Cytokinin changes DNA amplified
Capesius and Meyer (1977)	Orchids	Protoplasts used	AT-rich DNA isolation	
Chen et al. (1983)	Soybean hypocotyl		Nucleoli	Resolves into preribosome particle and nucleolar chromatin
Dawson et al. (1982)			Total cellular DNA from single plant	
Erdmann et al (1982)	Cultured tobacco cells (leaves also)	Protoplasts, disruption sediment, percoll gradient	Protein kinase studies studies	Store −20 °C 3.5 h preparation
Franke (1974)	Onion roots and leaves	Nuclei first, fragment	Nuclear envelope	
Gallagher (1983)	Pea leaf		Study of genes for chloroplast proteins	
Gealt et al. (1976)	*Aspergillus nidulans*	Liquid N_2, centrifugation sucrose sedimentation	General	For filamentous fungi

Table 1 (continued)

Reference	Source	Method	Purpose	Comments
Gregor et al. (1974)	*Nicotiana*		Histone	
Gupta and Sen (1982)	Coconut		Uptake of metabolites by nuclei	Phytochrome effects
Hadlaczky et al. (1983)	Wheat and poppy cell cultures	Protoplasts, detergent, sucrose and centrifugation	Chromosome isolation	High purity; uses glycine-hexylene glycol buffer system
Hadlaczky et al. (1983)	Wheat and poppy cell cultures	Protoplast, hypertonic, centrifugation	Nuclei isolation	High purity
Hughes et al. (1977)	Barley, tobacco	Protoplasts used	Ultrastructure study	Triton X-100 membrane solubization
Ide (1981)	Yeast		Transcription studies	
Kuroiwa (1976)	*Physarum polycephalum*	For mitochondrial nuclei	Fine structure of mitochondrial nuclei	Histones, DNA-like fibres
Mascarenhas et al. (1974)	Whole plants	Centrifugation in sucrose	General	
Mascarenhas (1974)	Protoplasts	Centrifugation in sucrose	General	
Mascarenhas (1974)	Pollen tubes	Hypotonic rupture Sucrose layering	General	
Lohr et al. (1977)	Yeast			
Lynch and Buetow (1975)	*Euglena gracilis*	Morpholino-ethane sodium metabisulphite	General	Nuclear density
Manne and Mecke (1980)	Yeast	Protoplasts first	Nuclei and nuclear membranes	
Matsumoto et al. (1976)	*Nicotiana*		Chromatin	
Matthysse et al. (1975)	Crown gall tumor cells		T-DNA, RNA synthesis, hybridization studies	
May (1980)	Yeast		Spindle pole bodies	Used nuclease digestion
Mazliak et al. (1977)	*Helianthus annus*	Older methods, separates from starch grains	Lipid metabolism studies	^{14}C-acetate studies
Mennes et al. (1978)	Pith callus tissue	Blender, centrifugation, glycerol layer, store −20 °C	RNA polymerase cell free transcription	Activity declines at less than 10% per month

Palit et al. (1982)	Snake bean	Mechanical rupture, sucrose solution	Synth. RNA, DNA replication, transcription	
Parish et al. (1980)	*Dictyostelium discoideum*		Nucleosome isolation	Electrophoretic method
Rho and Chipchase (1962)		"Pea popper" used		
Sau et al. (1980)	Various plants	Citric acid solution, glycerol layering, sucrose solution	Nuclei from different plant organs	
Scott and Timmis (1984)	Spinach roots	Fix, HCHO	DNA for restriction analysis	
Sheridan (1973)	Lily pollen	Nonaqueous methods	Histone studies	
Sheridan (1978)	*Lilium* meiotic cells	Special methods	Synaptonemal complex study	Synaptonemal complex preserved
Simon and Becker (1976)			Chromatin	High ionic strength conditions
Spiker et al. (1983)	Dry wheat embryo	Blendor, Triton X-100 Percoll gradient	Transcription	10^8 nuclei from 20 g embryo
Takats and Wever (1972)	*Tradescantia* pollen	Sucrose solution	DNA polymerase, DNA nuclease	Prepares generative and vegetative nuclei
Talbert and Russel (1982)	*Neurospora crassa*	Percoll density gradient		
Tallman and Reeck (1980)	Leaf tissue	Protoplast first, sucrose gradient		Detergent *not* used to disrupt
Vlasak (1981)	Barley, tobacco, *Vicia faba*, *Arabidopsis*	Various disintegration, various media	Yield and appearance of nuclei	Methods found for different tissue
Willmitzer and Wagner (1981)	Tobacco cells in culture	Pectinase, cellulase, ultra-Tarrax, Percoll.	General	
Yamaguchi et al. (1977)	Dry barley seed embryo		DNA polymerase	Enzyme purified

also described. Price (1974) makes the point that centrifugation remains the principal method for separating plant particles, and that zonal (or rate-zonal) separation is superior to isopycnic (or equilibrium density) separation. Choice of gradient (zonal) materials is large, with sucrose a common one (see Table 1) but this has the complication that it is osmotically active. Various other gradient materials are listed by Price (1974), including Ficoll, glycerol and sorbitol. A more common material in use now for nuclei is Percoll (Erdmann et al. 1982), as can be seen in Table 1. Percoll is a colloidal silica coated with polyvinylpyrrolidone.

Different methods of nuclear preparation may be needed, depending on the biochemical or other use to which the preparation is put. Thus, if it is the nucleoli that is required, then the procedure of Chen et al. (1983) may well be adopted, using a Percoll gradient to separate out the nucleoli after nuclear disruption. Again, it has been suggested that nuclei prepared with organic solvents only are better able to retain DNA polymerase activity (Smith and Keir 1963); however, this is not borne out by later work, including that of Palit et al. (1982). An increasing number of publications deal with the mass isolation of chromosomes. Thus both Hadlaczky et al. (1983) and Malmberg and Griesbach (1983) have prepared chromosomes from protoplasts. The former use wheat and poppy cell cultures, and the latter tobacco, tomato, lily, onion, and pea cells. Malmberg and Griesbach (1983) were able to partially fractionate the chromosomes by size; however the in vitro size did not correspond well with in vivo size, so the fractionation has thus far not proven useful. Persual of Table 1 will show a multiplicity of other methods for nuclear preparation for different purposes.

4.1 Isolation of Plant Nuclei – Methods

Four methods of preparation of plant nuclei are dealt with here in detail, as they are either suitable for a wide range of biochemical studies, or for specialist study of contained DNA. Table 1 should be consulted for other specialist investigations using plant nuclei. Microscopically, these nuclear preparations appear as shown in Fig. 3.

4.1.1 Nuclei from Tobacco Callus Cultures (for RNA Synthesis Studies)

This method is essentially that of Bouman et al. (1981), who used it to study RNA synthesis in isolated nuclei. Both enzymatic and RNA product aspects were studied, involving cell-free transcription and RNA polymerase I and II investigation. The basic method for isolation of nuclei used by these authors was first described by Mennes et al. (1978), and the nuclei so obtained can be stored at −20 °C before use, with no more than 10% loss in contained RNA polyermase activity per month.

1. One gram of tissue is homogenized for 30 s in a Waring blendor with 5 ml of buffer A (20 mM KCl – 20 mM $MgCl_2$ – 0.6 M sucrose – 40% glycerol – 10 mM mercaptoethanol – 25 mM morpholinoethane sulphonic acid, NaOH, pH 6.0). Further disruption is achieved by homogenization in a glass/glass Potter homogenizer (1 stroke only).

2. The brei is filtered through two layers of nylon cloth (500 μm) and then through four sieves with pore size 250, 106, 45, and 10 μm.
3. Centrifugation of the filtrate is carried out at 2500 *g* for 20 min.
4. The pellet is resuspended in buffer A to which has been added 0.02% Triton X-100. After 2.5 min in this mixture, the nuclei are centrifuged at 2500 *g* for 17.5 min, and the pellet washed immediately with buffer A, and then suspended in a small volume of buffer A.
5. This nuclear suspension is layered on 24 ml buffer A containing only 30% glycerol and 1.2 M sucrose. Centrifuge at 40,000 *g* for 15 min.
6. The pellet from above is resuspended in the adjusted buffer in 5, except that the sugar concentration is changed to 1.0 M sucrose. Centrifuge at 3000 *g* for 25 min. The reduction in centrifuge speed prevents the sedimentation of starch granules and membranous material.
7. The resulting pellet is washed and concentrated finally in buffer A.
8. The nuclei can be used immediately or stored at −20 °C. RNA polymerase activity declines by only 10% per month during this storage. The DNA content of these nuclei has been determined (Mennes et al. 1977) while it was found to be free of bacterial contamination, as told by phase contrast microscopic examination and by inclusion of 5 μg/0.25 ml rifampicin in a RNA synthesis incubation medium. RNA synthesis can be investigated with these nuclear preparations using labelled ribonucleoside triphosphates in dilute buffer solutions, pH 7.9 containing BSA and dithiothreitol and monitoring TCA insoluble radioactivity afterwards (Bouman et al. 1981). RNA can be extracted for size analysis by DNase treatment, followed by SDS and proteinase K incubation, and finally extraction with phenol-chloroform-isoamyl alcohol mixtures (Bouman et al. 1981).

4.1.2 Nuclei from Tobacco Cells in Culture – for General Purpose Studies

Working with plant cells has many disadvantages for the biochemist in that if nuclei are required then the concentration of nuclei is lower in plant cells than in animal cells. For example, 100 g of rat liver has the same number of nuclei as 4 kg of plant cells in culture. Another point is that plant cells have tough cell walls, a factor which is also increased further in tissue culture of plant cells. The following method of nuclear preparation from tobacco cells in culture is also suitable for leaf cells and is basically that of Willmitzer and Wagner (1981). All steps should be carried out at 4 °C, unless otherwise stated.

1. Approximately 500 g of late lag phase cells are collected on a piece of commercial curtain cloth and washed with buffer B (0.7 M mannitol – 10 mM MES – 5 mM EDTA – 0.1% BSA – 0.2 mM phenlymethylsulphonyl fluoride – pH 5.8).
2. Suspend cells in 3 l of buffer B for 15 min at 25 °C.
3. Filter, resuspend in 3 l buffer B to which has been added 150 mg pectinase and 300 mg cellulase, and incubate 25 °C for 30 min. This time may need to be increased to 3 or 4 h for large clumps of cells. Willmitzer and Wagner (1981) purified the Serva pectinase and cellulase before use, while the pH of

buffer B was raised temporally to 11 to dissolve the enzymes and returned to pH 5.8 before adding cells.

4. Now wash cells in 2 l buffer C (0.25 M sucrose – 10 mM NaCl – 10 mM MES, pH 6.0 – 5 mM EDTA – 0.15 mM spermine – 0.5 mM spermidine – 20 mM mercaptoethanol – 0.2 mM phenylmethylsulphonyl fluoride – 0.6% Nonidet P_{40} – 0.1% BSA)
5. Resuspended in buffer C and homogenize 4 × 15 s in an ultra-Turrax (Janke and Kumkel, FRG) at full speed.
6. Repeatedly filter in sieve cloth with decreasing mesh (130, 80, and 20 μm).
7. Centrifuge 2000 *g* for 5 min.
8. Pellet (nuclei) is suspended in buffer C (volume equivalent to 60 g). Add 70 g to 100 g of buffer D (to 116 g Percoll from Pharmacia Fine Chemicals, Uppsala, add 30 g of × 5 conc. buffer C) and mix in Erlenmeyer flask and gently shake.
9. Filter the suspension through the 20 μm sieve cloth.
10. Centrifuge 1000 *g* for 5 min.
11. The pellet containing the nuclei is resuspended in 2 to 5 ml buffer C.
12. Shake in an Erlenmeyer flask.
13. Centrifuge 1000 *g* for 5 min.
    Repeat 12 and 13, three to five times until the amount of nuclei is small compared to contamination in pellet.
14. Centrifuge collected pellets at 1000 *g* for 5 min.
15. Pellet suspended in 20 ml buffer E (45 g Percoll, plus 6.0 g of × 5 concentrated buffer C and adjust to pH 6.0). This suspension should be carried out by sucking up through a Pasteur pipette. Centrifuge 5000 *g* for 5 min. The floating layer containing the nuclei is taken off and diluted with 20 ml buffer C. Centrifuge at 1000 *g* for 5 min and wash in buffer C.

The nuclei so prepared can be used for many purposes. Contained DNA can be isolated from these nuclei by proteinase K digestion followed by phenolization (Grass-Bellard et al. 1973) and restriction endonuclease analysis carried out as described by Willmitzer et al. (1980). Nuclear proteins can also be examined; histones are extracted from nuclear preparations with 0.2 M H_2SO_4 and precipitated with 5 vol acetone, while nonhistone proteins can be extracted from frozen nuclei by lysing in 8 M urea – 10 mM Tris HCl, pH 7.2 – 6% v/v 2 mercaptoethanol – 1% sodium dodecylsulphate – 2 mM EDTA – 0.2 mM phenylmethylsulphonyl fluoride – 1 μM pepstatin. The nucleic acids contained in this lysate can be pelleted by centrifugation at 100,000 *g* for 18 h and the clear supernatant applied directly to polyacrylamide gels. Both histones and nonhistones can be examined in 13 or 10.5% SDS-polyacrylamide gel electrophoresis as described by Laemmli (1970). The nucleosome structure can be examined after digestion of the nuclei with micrococcal nuclease, or the nuclei used for protein kinase, RNA polymerase or poly-ADP ribosylating activities (Willmitzer and Wagner 1981).

4.1.3 Nuclei from Soybean Cells – for DNA Studies

Attention is drawn to this method because the nuclei are obtained from an agriculturally important plant and the nuclei so prepared are used for DNA replication studies (Roman et al. 1980; Caboche and Lark 1981). In brief, the nuclei are prepared by a three-step procedure, involving the preparation of protoplasts from exponentially growing cells, the protoplasts being isolated under low osmotic stress (0.2 M KCl). The protoplasts are lysed in Triton X-100, and the nuclei collected by sedimentation. Care should be taken to keep the pH around 8.0 during isolation and subsequent handling because of a very active endonuclease in plant tissue at pH 7.5 or less, which is especially active on single-stranded DNA.

DNA can be extracted from the nuclei, after incubation with deoxyribonucleoside triphosphates etc. by first washing nuclei in 20% w/v sucrose – 3 mM $CaCl_2$ – 60 mM KCl – 25 mM Tris HCl, pH 8, then lysing in 50 mM Tris HCl, pH 7.8 – 100 mM EDTA – 150 mM NaCl – 5 mg ml^{-1} sodium dodecylsulphate – 200 μg ml^{-1} proteinase K solution at 65 °C (30 min) and the DNA purified by isopycnic centrifugation. Fractions can be dialysed against 50 mM sodium acetate – 1 mM EDTA – 40 mM Tris acetate, pH 8, and deproteinized with phenol for 1 h at 20 °C the aqueous phase extracted with ether and DNA precipitated with two volumes of ethanol. This preparation is suitable for restriction analysis (Caboche and Lark 1981).

4.1.4 Plant-Root Nuclei – for DNA Analysis

Recently plant-cell nuclei have been prepared free of chloroplasts so as to examine nuclear DNA for sequences common to chloroplast DNA (Scott and Timmis 1984). The essentials of this method are described here because it uses a different approach for a particular purpose.

1. Spinach roots are first extensively washed in dilute detergent and *fixed* in ice-cold 0.5% formaldehyde – 0.1 M Hepes, pH 7.0 for 1 to 2 h.
2. The fixed roots are homogenized 4 × 3 s in a razor blade blendor in 0.025 M Tris HCl, pH 7.5 – 0.275 M sorbitol – 0.005 M 2 mercaptoethanol –0.5% Triton X-100.
3. The homogenate is filtered through one layer of cheesecloth and three layers of Miracloth and the nuclei collected by centrifugation at 400 *g* for 5 min.
4. The nuclear pellet is resuspended in the extraction buffer, underlaid with a similar buffer containing 0.4 M sorbitol and the nuclei again collected by centrifugation 400 *g* for 5 min. This procedure is repeated three times.
5. The nuclei are lysed in 0.02 M EDTA – 0.15 M NaCl – 0.015 M trisodium citrate, pH 7.0 – 1% sarkosyl – 100 μg ml^{-1} RNAse A. After 15 min at 37 °, 5 mg ml^{-1} of predigested pronase is added, incubate a further 15 min and the nuclear DNA isolated as described by Scott and Possingham (1980).

Although DNA isolated in this way does not have as high a molecular weight as that obtained by direct extraction from root cells as described by Murray and Thompson (1980), it does give DNA, which when digested by certain restriction endonucleases gives clear restriction bands in the range 2–10 kbp.

5 Summary

The structure of the plant nucleus is considered in relation to the methodology used in the isolation of the nucleus and its cytology.

The newer fluorescence techniques for DNA complexed with fluorochrome dyes are compared to the well established Feulgen method of staining nuclei.

Methods for the mass isolation of plant nuclei are compared, stressing that the biochemical purposes to which the nuclei are to be put often determine the method to be used.

References

Ansa OA, Bowyer JW, Shepherd RJ (1982) Evidence for replication of cauliflower mosaic virus DNA in plant nuclei. Virology 121:147–156

Arfman HA, Willmitzer L (1982) Endogenous protein kinase activity of tobacco nuclei: comparison of transformed, nontransformed cell cultures and the intact plant of *Nicotiana*. Plant Sci Lett 26:31–38

Armstrong JA (1956) Histochemical differentiation of nucleic acids by means of induced fluorescence. Exp Cell Res 11:640–643

Atkinson MR, Jackson JF, Morton RK (1961) Nicotinamide mononucleotide adenylyltransferase of pig liver nuclei. Biochem J 80:318–323

Belling J (1921) On counting chromosomes in pollen-mother cells. Am Nat 55:573–574

Bennet MD, Smith JB (1976) Nuclear DNA amounts in angiosperms. Philos Trans R Soc Lond B 274:227–274

Berlyn GP, Cecich RA (1976) Optical techniques for measuring DNA quantity. In: Mikache JP (ed) Modern methods in forest genetics. Springer, Berlin Heidelberg New York

Bonner J (1976) The Nucleus. In: Bonner J, Varner JE (eds) Plant biochemistry, 3rd edn. Academic Press, New York, pp 37–64

Bonner J, Chalkley GR, Dahmus M, Farnbrough D, Fujimura F, Huang RC, Huberman J, Jensen R, Marushige K, Ohlenbusch H, Olivera B, Widholm J (1968) Isolation and characterization of chromosomal nucleoproteins. Methods Enzymol 12:3–65

Bouman H, van Paridon H, Vogelaar A, Mennes AM, Libbenga KR (1981) Size analyses of RNA synthesized in isolated tobacco. Nicotiana tabacum cultivar white-burley callus nuclei. Plant Sci Lett 22:361–368

Branster MJ, Morton RK (1956) Comparative rates of synthesis of diphosphopyridine nucleotide by normal and tumour tissue from mouse mammary gland: studies with isolated nuclei. Biochem J 63:640–646

Brewer EN (1979) Repair of radiation-induced DNA double-strand breaks in isolated nuclei of *Physarum polycephalum*. Radiat Res 79:368–376

Brightwell MD, Leech CE, O'Farrell MK, Whish WJ, Shall S (1975) Poly ADP ribose polymerase in *Polysarum polycephalum*. Biochem J 147:119–130

Buetow DE (1976) Isolation of nuclei from protozoa and algae. In: Prescott DM (ed) Methods in cell biology. Academic Press, New York, pp 283–311

Busch H, Smetana K (1970) The Nucleolus. Academic Press, New York, pp 548–575

Caboche M, Lark KG (1981) Preferential replication of repeated DNA sequences in nuclei isolated from soybean cells grown in suspension culture. Proc Nat Acad Sci USA 78:1731–1735

Capesius I, Meyer Y (1977) Isolation of nuclei from protoplasts of orchids. Cytobiologie 15:485–490

Chabot JF, Leopold AC (1982) Ultrastructural changes of membranes with hydration in soybean seeds. Am J Bot 69:623–633

Chauveau J, Moulé Y, Rouiller CH (1956) Isolation of pure and unaltered liver nuclei. Morphology and biochemical composition. Exp Cell Res 11:317–321

Chen Y-M, Huang D-H, Lin S-F, Lin C-Y, Key JL (1983) Fractionation of nucleoli from auxin-treated soybean hypocotyl into nucleolar chromatin and preribosomal particles. Plant Physiol 73:746–753

Coleman AW (1978) Visualization of chloroplast DNA with two fluorochromes. Exp Cell Res 114:95–100

Dawson JRY, Thomas K, Clegg MT (1982) Purification of total cellular DNA from a single plant. Biochem Genet 20:209–219

Dhar AC, Shah CK (1982) Cytochemical method to localize acid nuclear proteins. Stain Technol 57:151–155

Dhillon SS, Berlyn GP, Miksche JP (1977) Requirement of an internal standard for microspectrophotometric measurement of DNA. Am J Bot 64:117–121

Dhillon SS, Berlyn GP, Miksche JP (1978) Nuclear DNA content in populations of *Pinus rigida*. Am J Bot 65:192–196

Dounce AL (1943) Enzyme studies on isolated cell nuclei of rat liver. J Biol Chem 147:685–698

Dounce AL, Tishkoff GH, Barnett SR, Freer RM (1950) Free amino acids and nucleic acid content of cell nuclei isolated by a modification of Behrens technique. J Gen Physiol 33:629–642

Erdmann H, Boecher M, Wagner KG (1982) Two protein kinases from nuclei of cultured tobacco *Nicotiana tabacum* cultivar white-burley cells with properties similar to the cyclic nucleotide independent enzymes N-I and N-II from animal tissue. FEBS Lett 137:245–248

Franke WW (1974) Structure, biochemistry, and functions of the nuclear envelope. Int Rev Cytol [Suppl] 4:72–236

Franklin AL, Filion WG (1981) Acridine orange-methyl green fluorescent staining of nucleoli. Stain Technol 56:343–348

Gallagher TF (1983) Light stimulated transcription of genes for 2 chloroplast polypeptides in isolated pea leaf nuclei. Eur Mol Biol Organ J 1:1493–1498

Garland PB, Mathias AP (1977) Biochemistry of the cell nucleus. The Biochemical Society, London

Gealt MA, Sheir-Neiss G, Morris NR (1976) The isolation of nuclei from the filamentous fungus *Aspergillus nidulans*. J Gen Microbiol 94:204–210

Goessens G (1984) Nucleolar structure. Int Rev Cytol 87:107–158

Grass-Bellard M, Oudet P, Chambon P (1973) Isolation of high molecular-weight DNA from mammalian cells. Eur J Biochem 36:32–38

Gregor D, Reinert J, Matsumoto H (1974) Changes in chromosomal proteins from embryo induced carrot cells. Plant Cell Physiol 15:875–881

Gupta DNS, Sen SP (1982) Phytochrome regulation of uptake of metabolites by coconut nuclei in vitro. Plant Sci Lett 24:61–66

Hadlaczky G, Bioztray G, Praznovszky T, Dudits D (1983) Mass isolation of plant chromosomes and nuclei. Planta 157:278–285

Hamilton VT, Habberset MC, Herman CJ (1980) Flow microfluorometric analysis of cellular DNA: critical comparison of mithramycin and propidium iodide. J Histochem Cytochem 28:1125–1128

Harland J, Jackson JF, Yeoman MM (1973) Changes in some enzymes involved in DNA biosynthesis following induction of division in cultured plant cells. J Cell Sci 13:121–138

Hogeboom GH, Schneider WC (1952) Cytochemical studies VI. Synthesis of diphosphopyridine nucleotide by liver cell nuclei. J Biol Chem 197:611–620

Hughes BG, Hess WM, Smith MA (1977) Ultrastructure of nuclei isolated from plant protoplasts. Protoplasma 93:267–274

Hull HM, Howshaw RW, Wang J-Chyong (1982) Cytofluorometric determination of nuclear DNA in living and preserved algae. Stain Technol 57:273–282

Ide GJ (1981) Nucleoside 5′-γ-S-triphosphates will initiate transcription in isolated yeast nuclei. Biochemistry 20:2633–2638
Jackson JF, Linskens HF (1978) Evidence for DNA repair after ultraviolet irradiation of *Petunia hybrida* pollen. Mol Gen Genet 161:117–120
Kao KN (1975) A nuclear staining method for plant protoplasts. In: Gamborg OL, Wetter LR (eds) Plant-tissue culture methods. Natl Res Counc Can, Saskatoon, pp 60–61
Kasten FH (1960) The chemistry of Schiff's reagent. Int Rev Cytol 10:1–100
Kasten FH (1964) The Feulgen reaction, an enigma in cytochemistry. Acta Histochem 17:88–89
Kohen E, Siebert G, Kohen C (1971) Transfer of metabolites across nuclear membranes. Hoppe-Seyler's Z Physiol Chem 352:927–937
Kuehl L (1964) Isolation of plant nuclei. Z Naturforsch 19B:525–532
Korson R (1951) A differential stain for nucleic acids. Stain Technol 26:265–270
Kurnick NB (1955) Pyronin Y in the methyl green-pyronin histological stain. Stain Technol 30:213–230
Kuroiwa T (1976) Isolation of mitochondrial nuclei and their fine structures. J Electron Microsc 25:182
Kuroiwa T (1982) Mitochondrial nuclei. Int Rev Cytol 75:1–60
Laemmli UK (1970) Cleavage of structural proteins during the assembly of the head of bacteriophage T_4. Nature 227:680–685
LaLoue M, Courtois D, Manigault P (1980) Convenient and rapid fluorescent staining of plant cell nuclei with "33258" Hoechst. Plant Sci Lett 17:175–179
Leuchtenberger C (1958) Quantitative determination of DNA in cells by Feulgen microspectrophotometry. In: Danielli JE (ed) General cytochemical methods, vol 1. Academic Press, New York, pp 219–278
Lohr D, Kovaic RT, Van Holde KE (1977) Quantitative analysis of the digestion of yeast chromatin by Staphylococcal nuclease. Biochemistry 16:463–471
Lynch MJ, Buetow DE (1975) Isolation of intact nuclei from *Euglena gracilis*. Exp Cell Res 91:344–348
MacPherson P, Filion WG (1981) Karyotype analyses and the distribution of constitutive heterochromatin in five species of *Pinus*. J Hered 72:193–198
Malmberg RL, Griesbach RJ (1983) Chromosomes from protoplasts – isolation, fractionation and uptake. In: Hollaender, A (ed) Genetic engineering of plants, vol 26. Basic life sciences. Plenum, New York, pp 195–201
Manne K, Mecke D (1980) Isolation and characterization of nuclei and nuclear membranes from *Saccharomyces cerevisiae* protoplasts. FEBS Lett 122:95–99
Mascarenhas JP, Berman-Kurtz M, Kulikowski RR (1974) Isolation of plant nuclei. Methods Enzymol 31:558–564
Matsumoto H, Gregor D, Reinert J (1976) Changes in chromatin of Daucus carota cells during embryogenesis. Phytochemistry 14:41–47
Matthysse AG, Rich K, Kontak C (1975) RNA synthesis in crown gall tumour cells. Plant Physiol 56:24
May R (1980) Cytochemical characterization of spindle pole bodies of isolates nuclei from the yeast *Kluyveromyces fragils* by digestion with nucleases. Protoplasma 102:21–30
Mazliak P, Robert D, Decotte-Justin AM (1977) Lipid metabolism in isolated nuclei from germinating sunflower hypocotyls. Plant Sci Lett 9:211–234
McClintock B (1934) The relation of a particular chromosomal element to the development of the nucleoli in *Zea mays*. Z Zellforsch Mikrosk Anat 21:294–328
Mennes AM, Bouman H, van der Burg MPH, Liebenga KR (1978) RNA synthesis in isolated tobacco callus nuclei and the influence of phytohormones. Plant Sci Lett 13:329–339
Mennes AMO, Voogt E, Libbenga KR (1977) The isolation of nuclei from cultured tobacco pith explants. Plant Sci Lett 8:171–177
Mendelsohn ML (1958) The two wavelength method of microspectrophotometry II. A set of tables to facilitate the calculations. J Biophys Biochem Cytol 4:415–424
Mepham RH, Lane GR (1969) Nucleopores and polyribosome formation. Nature 221:288–289

Miksche JP (1966) DNA synthesis in primary roots of *Glycine max* during germination. Can J Bot 46:115–120
Miyakawa I, Aoi H, Sands N, Kuroiwa T (1984) Fluorescence microscopic studies of mitochondrial nucleoids during mieosis and sporulation in the yeast, *Saccharomyces cererisiae*. J Cell Sci 66:21–38
Murata M (1983) Staining air dried protoplasts for study of plant chromsomes. Stain Technol 58:101–106
Murray MG, Thompson WF (1980) Rapid isolation of high molecular weight DNA. Nucleic Acids Res 8:4321–4325
Nagata T, Ishii S (1979) A rapid method for isolation of mesophyll protoplasts. Can J Bot 57:1820–1823
Nagl W (1981) Polytene chromosomes of plants. Int Rev Cytol 73:21–54
Nevins JR (1983) The pathway of eukaryotic mRNA formation. Annu Rev Biochem 52:441–466
Palit S, Dutta R, Sarkar G, Dube DK (1982) Nucleic acid synthesis by isolated plant nuclei. Indian J Biochem Biophys 19:91–94
Parish RW, Schmidlin S, Fuhrer S, Widmer S (1980) Electrophoretic isolation of nucleosomes from *Dictyostelium descoideum* nuclei and nucleoli proteins associated with monomers and dimers. FEBS Lett 110:236–240
Preiss J, Handler P (1958) Biosynthesis of diphosphopyridine nucleotide. J Biol Chem 233:493–500
Price CA (1974) Plant cell fractionation. Methods Enzymol 31:501–519
Rho JH, Chipchase MI (1962) Incorporation of tritiated cytidine into RNA by isolated pea nuclei. J Cell Biol 14:183–192
Roman R, Caboche M, Lark KG (1980) Replication of DNA by nuclei isolated from soybean suspension cultures. Plant Physiol 66:726–730
Roque AL, Jafarey NA, Coulter P (1965) A stain for the histochemical demonstration of nucleic acids. Exp Mol Pathol 4:266–274
Ruch F (1973) Quantitative determination of DNA and protein in single cells. In: Thaer AA, Sernetz M (eds) Fluorescence techniques in cell biology. Springer, Berlin Heidelberg New York, pp 89–93
Sau H, Sharma AK, Choudhuri RK (1980) DNA, RNA and protein content of isolated nuclei from different plant organs. Indian J Exp Biol 18:1519–1522
Schweiger H-G, Berger S (1979) Nucleocytoplasmic interrelationships in *Acetabularia* and some other Dasycladaceae. Int Rev Cytol [Suppl] 9:12–44
Scott NS, Possingham JV (1980) Chloroplast DNA in expanding spinach leaves. J Exp Bot 31:1081–1092
Scott NS, Timmis JN (1984) Homologies between nuclear and plastid DNA in spinach. Theor Appl Genet 67:279–288
Sheridan WF (1973) Nonaqueous isolation of nuclei from Lily pollen and an examination of their histones. Z Pflanzenphysiol 68:450–459
Sheridan WF (1978) Preservation of synaptonemal complex structure in unfixed meiotic cells and isolated nuclei of Lilium. J Cell Biol 79:185A
Simon JH, Becker WM (1976) A polyethylene glycol/dextran procedure for the isolation of chromatin proteins (histones and nonhistones) from wheat germ. Biochim Biophys Acta 454:154–171
Smetana K, Busch H (1966) Studies on staining and localization of acidic nuclear proteins in the walker 256 carcinosarcoma. Cancer Res 26:331–337
Smith J, Keir HM (1963) DNA nucleotidyltransferase in nuclei and cytoplasm prepared from thymus tissue in non-aqueous media. Biochim Biophys Acta 68:578–588
Spicer SS (1961) Differentiation of nucleic acids by staining at controlled pH and by Schiff-methylene blue sequence. Stain Technol 36:337–340
Spiker S, Murray MG, Thompson WF (1983) DNase I sensitivity of transcriptionally active genes in intact nuclei and isolated chromatin of plants. Proc Nat Acad Sci USA 80:815–819
Takats ST, Wever GH (1972) DNA polymerase and DNA nuclease activities in S-competent and S-incompetent nuclei from *Tradescantia* pollen grains. Exp Cell Res 69:25–28

Talbert KJ, Russell PJ (1982) Nuclear buoyant density determination and the purification and characterization of wild type *Neurospora crassa* using percoll density gradients. Plant Physiol 70:704–708

Tallman G, Reeck GP (1980) Isolation of nuclei from plant protoplasts without the use of a detergent. FEBS Lett 18:271–276

Vlasak J (1981) Effect of different disintegration techniques and media on yield and appearance of isolated nuclei. Biologia Plant 23:406–413

Ward DC, Reich E, Goldberg IH (1965) Base specificity in the interaction of polynucleotides with antibiotic drugs. Science 149:1259–1263

Werker E, Lerner HR, Weinberg R, Poljajoff-Mayber A (1983) Structural changes occurring in nuclei of barley root cells in response to a combined effect of salinity and ageing. Am J Bot 70:222–225

Willmitzer L, Wagner KG (1981) The isolation of nuclei from tissue cultured plant cells. Exp Cell Res 135:69–78

Willmitzer L, De Beuckeleer M, Lemmers M, Van Montagu M, Schell J (1980) DNA from Ti plasmid present in nucleus and absent from plastids of crown gall plant cells. Nature 287:359–361

Yamaguchi H, Naito T, Tatara A (1977) Purification of DNA polymerase isolated from the nuclei in dry barley seed embryo. J Radiat Res 18:31

Microtubules

D. D. Sabnis

1 Introduction

Animal cells are endowed with an elaborate cytoskeletal network comprised of microtubules (MT's) and associated proteins, microfilaments, and one or more elements of a heterogeneous group of intermediate filaments (Roberts and Hyams 1979; Jackson 1982; Lazarides 1982): plant cells are known to possess a more limited complement of MT's and microfilaments (Lloyd 1982; Gunning and Hardham 1982; Sabnis and Hart 1982). The mechanisms whereby the cytoskeletal functions of MT's and microfilaments are expressed in plant cells are not immediately obvious. Nevertheless, accumulating cytological evidence points to an involvement of MT's at various spatial and temporal loci in cellular morphogenesis (Gunning 1982; Gunning and Hardham 1982). The involvement of regulatory proteins such as calmodulin as pleiotropic regulators of cytoskeletal organisation and mechanochemical processes in plants has been suggested by indirect evidence (Job et al. 1981; Schleicher et al. 1982; Kakiuchi and Sobue 1983). However, biochemical data on the properties, assembly and activity of plant MT's (reviewed by Sabnis and Hart 1982) are extremely sparse. Published reports on the isolation of plant tubulin are very few in number and have dealt with such disparate plant sources as yeast cells, *Chlamydomonas*, plant cell suspension cultures and excised plant epicotyls. The general applicability of any one method has never been tested. Consequently, where I have felt that it may influence choice of experimental method, I have discussed the background to, and variations of, methods applied successfully to both plant and animal tissues. In a few instances, I have cited published results that might serve as a guide in the assessment of a particular procedure.

The outstanding problems are the very low levels of MT proteins (indeed of most proteins) in plant cells, and their instability during extraction protocols that are generally successful with animal tissues. The more elaborate fractionation and concentration steps required, increase exposure of the desired proteins to degradation by hydrolytic enzymes or complex formation. Consequently, progress in the isolation of cytoskeletal proteins from plant tissues has not been remarkable.

However, several factors have engendered both an increased urgency for work in this area, as well as more optimistic prospects for success. Microheterogeneity is now a well-documented feature of dimeric tubulin, the major constituent protein of MT's (Dahl and Weibel 1979; Gozez and Sweadner 1981). Recent work has indicated that plant tubulin may possess even more distinct structural features. The α-subunit of higher plant tubulin has a marked difference in electrophoretic mobility from brain α-tubulin (Sect. 4) and conventional anti-sera raised against tubulin derived from animal sources frequently show poor cross-reactivity with higher plant MT's (S. Wick, personal communication). The responses of the plant MT cytoskeleton to treatment with anti-mitotic drugs and low temperature are also curious and presently unpredictable: compared to MT's in animal cells, a high degree of stability is displayed in many instances (Hart and Sabnis 1976c). This behaviour may be due to intrinsic differences in the tubulin molecule itself, or due to properties of microtubule-associated proteins (MAP's) that may be responsible for cross-linking the tubules to the cell membrane, to other organelles or to each other. Nothing is known concerning higher plant MAP's. A picture of tissue-specific MAP heterogeneity is emerging from work on animal cells, but there is still controversy over the identities, numbers and properties of these proteins (Berkowitz et al. 1977; Cleveland et al. 1977; Bulinsky and Borisy 1979; Runge et al. 1979; Vallee 1980; Burgoyne and Cumming 1983). An obvious need exists to isolate and characterise plant MAP's independently.

The prospects for isolating and quantitatively assaying MT proteins have been improved by several advances. The discovery of the effects of the plant drug, taxol, on stabilising the intact MT and on tilting the in vitro equilibrium between polymer and monomer in favour of the former, has been exploited for harvesting plant MT's after polymerisation in the presence of the drug (Sect. 3.4.3). Co-polymerisation of putative plant tubulin with brain tubulin has also provided a means whereby the properties of the [^{35}S]- or [^{14}C]-labelled plant co-polymer can be investigated (Sect. 3.4.4). Both polyclonal and monoclonal antibodies have been raised against tubulin from animal sources and from yeast (Sect. 6.1), permitting the development of sensitive radioimmunoassays (Sect. 6.2) and enzyme-linked immunosorbent assays (Sect. 6.3). Hybridoma technology permits the harvesting of clonal culture medium or collection of ascites fluid (Kennett 1981; Yelton and Scharff 1981), both of which provide a liberal source of antibodies which may be bound to Sepharose beads to generate an affinity matrix with very high selective and retentive properties (Sects. 6.4, 7.1). Such antibody affinity columns may permit the isolation, purification and concentration of plant tubulin in quantities adequate for further study.

2 Extraction of Microtubule Proteins

A major prerequisite for the isolation of plant tubulin is experimental material rich in MT's. Estimates in lower eukaryotes indicate that the tubulin content is less than 1% of the total protein (Fulton and Simpson 1979). In cell suspension cultures of *Nicotiana tabaccum*, tubulin may account for as much as 4% of soluble cell proteins (Yadav and Filner 1983). Among lower plants, the flagellated unicellular alga, *Chlamydomonas reinhardii*, has been heavily favoured owing to ease of culture and the ready isolation and demembranation of its MT-rich flagella (Luck et al. 1977; Witman et al. 1978). Higher plant sources have included cell suspension cultures of tobacco or rose (Morejohn and Fosket 1982; Yadav and Filner

1983), mung bean sprouts (Rubin and Cousins 1976) or azuki bean epicotyls (Mizuno et al. 1981) and maize root tips.

The plant tissue is harvested, washed and homogenised in cold buffer – usually 50–100 mM 2-(N-morpholino] ethane sulphonic acid (MES) or piperazine-N,N′-bis [2-ethane sulphonic acid] (PIPES), pH 6.8, containing 1 mM EGTA and 1 mM $MgCl_2$. 1 mM GTP is often added at this stage, but considering the rapidity with which it is hydrolysed by crude extracts, and the dubious value of its presence during extraction, it may be more economic to add the nucleotide after tissue homogenisation and initial clarification of the extract. To minimise rapid proteolysis of tubulin, the addition of protease inhibitors such as leupeptin hemisulphate (25–50 μM; Morejohn and Fosket 1982), pepstatin (5 $\mu g\ ml^{-1}$; Mizuno et al. 1981) and/or p-methyl-phenyl-sulphonyl fluoride (PMSF; 0.17 $mg\ ml^{-1}$; Kilmartin 1981) is apparently essential. Recent successful procedures have also incorporated 1–5 mM dithiothreitol (DTT) or dithioerythritol to protect free sulfhydryl groups (Morejohn and Fosket 1982; Yadav and Filner 1983). In general, recipes for extraction buffers have been based on those found most useful for MT assembly in vitro (Sect. 3.4).

The cold homogenates are filtered through Miracloth and centrifuged at 100,000 *g* for 30–60 min at 4 °C, to provide a supernatant fraction containing soluble tubulin. Subsequent treatment of this fraction has varied and includes DEAE-Sephadex ion exchange chromatography, affinity chromatography on a column of ethyl-N-(3-carboxyphenyl) carbamate-Sepharose, polymerisation of microtubules in the presence of taxol, or co-polymerisation of plant tubulin with brain tubulin. It is worth considering the background to the choice of these procedures.

3 Purification of Tubulin and MAP's

3.1 DEAE-Sephadex Ion Exchange Chromatography

The first isolation procedure applied to microtubule proteins (Weisenberg et al. 1968) employed either batch adsorption and elution from DEAE-Sephadex, or gradient elution from a column of DEAE-Sephadex A50. Colchicine-binding activity (Sect. 5) was used as an assay for tubulin and the necessity of including GTP and MG^{2+} in the medium, to preserve colchicine-binding activity, was recognised. The procedure has been modified only in minor details in subsequent publications (Weisenberg and Timasheff 1970; Frigon and Timasheff 1975; Lee et al. 1978).

The most successful application of ion exchange chromatography to the isolation of higher plant tubulin has been described by Morejohn and Fosket (1982). Rose cell suspension cultures were extracted in cold PM buffer (50 mM PIPES-KOH, pH 6.9, 1 mM EGTA, 0.5 mM $MgCl_2$, 1 mM DTT; 5 g cells per ml buffer). The filtered homogenate was made 25 μM in leupeptin hemisulphate and centrifuged at 48,200 *g* for 30 min at 4 °C. The supernatant fraction (2.4–3.3 mg

ml^{-1} protein) was mixed with DEAE-Sephadex A50 (pre-equilibrated with PM buffer) and poured into a 2.6 × 40 cm column. Weakly bound proteins were eluted with 3–5 bed volume of PM buffer containing 0.4 M KCl and 0.5 mM GTP. Tightly bound proteins were eluted from the column with 3 bed volumes of PM buffer containing 0.8 M KCl and 0.5 mM GTP. Proteins in each eluate fraction were precipitated with 50% saturated $(NH_4)_2SO_4$ and subjected to sodium dodecyl polyacrylamide gel electrophoresis (SDS-PAGE). Densitometric tracings of stained gels showed that ~63% of the polypeptides in the tightly-bound fraction had mobilities similar to the α- and β-subunits of brain tubulin. The yield of tubulin obtained by this procedure was 1–2 μg mg^{-1} of crude supernatant fraction proteins.

3.2 Phosphocellulose Chromatography

Phosphocellulose chromatography has proved to be a quick and efficient method for separating tubulin from MAP's and possible contaminants present among MT proteins isolated by cycles of MT assembly-disassembly (Sect. 3.4; Weingarten et al. 1975; Detrich and Williams 1978; Jameson et al. 1980; Sternlicht et al. 1980).

Precycled phosphocellulose (e.g. Whatman P-11, fibrous, monoammonium form) is washed with 2 × 5 vol 0.1 M $MgSO_4$ followed by 10 vol column buffer (0.1 M MES, containing 1 mM EGTA, DTT and $MgSO_4$, and 0.1 mM GTP). The phosphocellulose contracts considerably when treated with Mg^{2+}. A small column of phosphocellulose is equilibrated with 10–20 vol column buffer, till the pH and the conductivity is the same as the buffer applied. MT proteins are loaded at 3–5 mg protein ml^{-1} bed volume, and tubulin eluted with column buffer at 1 bed volume h^{-}. Bound proteins (MAP's, etc.) are then eluted with a 0 to 1 M NaCl gradient.

3.3 Affinity Chromatography

Mizuno et al. (1981) have described a procedure for the purification of tubulin from azuki bean epicotyls employing an affinity column of ethyl-N-phenyl carbamate (EPC) linked to aminoethyl Sepharose 4B. Discontinuous SDS-PAGE (Laemmli 1970) of soluble bean proteins bound to the EPC column revealed the presence of ~30 protein subunits, with enrichment of two bands migrating to the zone occupied by rabbit brain tubulin. We have substituted a DEAE-Sephacel column for the EPC "affinity" column, and have obtained very similar results with azuki bean extracts. Therefore, the value of the EPC ligand is not yet clear.

Mizuno et al. further fractionated their preparation through a column of Sephadex G-200, which resolved the proteins into two peaks, one of which represented tubulin, purified to near homogeneity as judged by SDS-PAGE or 2-D SDS-PAGE/isoelectric focussing (IEF). The remaining proteins of subunit size both smaller and larger than tubulin, were apparently (but inexplicably) separated in the second, more rapidly migrating peak.

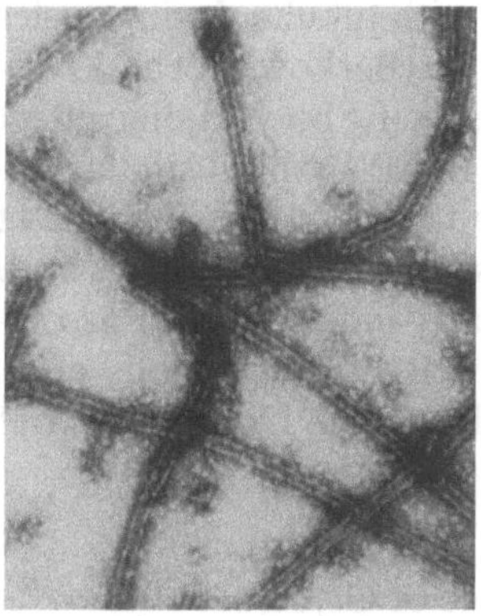

Fig. 1. Microtubules polymerised in vitro from porcine brain extracts and negatively stained with phosphotungstic acid

3.4 Cycles of Polymerisation and Depolymerisation

The conditions for in vitro assembly of MT's from homogenates of brain tissue were first defined by Weisenberg (1972). The temperature-dependent, reversible assembly of neurotubulin was shown to require the presence of GTP and Mg^{2+}, and the sequestration of calcium by chelating agents. In a suitable medium, MT's assemble at 30 °–37 °C (Fig. 1) and disassemble at 0 °–4 °C. Polymerised microtubules can be harvested by centrifugation. Most work on the assembly of MT proteins has been conducted with preparations from brain tissue (usually porcine, bovine or rat brain) since tubulin may comprise 10–20% of soluble brain protein.

While tubulin alone can assemble into microtubules (Williams and Detrich 1979), the presence of MAP's can stimulate assembly. It has been suggested that MAP's may function as ligands that modulate MT assembly and influence the structural fidelity of the polymer (Scheele and Borisy 1979). However, the uncertainty of how many of the proteins that co-polymerise with tubulin are actually associated with MT's in vivo, and the diversity of MAP's isolated from different tissues, require that this role of MAP's be subjected to much more critical assessment.

3.4.1 Pre-Conditions for Microtubule Assembly

An almost universal component of the assembly system is a zwitterionic sulphonate buffer, usually MES or PIPES (see Waxman et al. 1981). Differences in the ability of tubulin to assemble may be traced to differences in isolation procedures. Methods based on DEAE ion exchange chromatography and phosphocellulose chromatography (Weisenberg et al. 1968; Weingarten et al. 1974, 1975) remove the MAP's which are present in preparations obtained by repeated cycles of polymerisation and depolymerisation (Shelański et al. 1973; Asnes and Wilson 1979). Such tubulin, purified to homogeneity, will still reconstitute into MT's, but only at higher critical concentrations of protein (~ 1 mg ml^{-1}). It appears that all the information required for assembly in vitro is present in tubulin; the reaction requires the presence of GTP, it is enhanced by Mg^{2+} and by the sequestration of

Ca^{2+} (Lee et al. 1978). The MAP's that co-polymerise with tubulin during cycles of assembly and disassembly may do so in a constant ratio (Berkowitz et al. 1977), although the effect may be due to no more than a nonspecific polycationic effect (Lee et al. 1978). MT's reconstituted in the presence of MAP's exhibit similar lateral decorations to those on natural MT's (Murphy and Borisy 1975; Sloboda et al. 1976). More importantly, immuno-electron microscopic localisation of both high molecular weight MAP's and *tau* factors have clearly indicated their association with MT's in specific cell types or stages of the cell cycle (Connolly et al. 1977; Sherline and Schiavone 1978; Connolly and Kalnis 1980; DeBrabander et al. 1981).

3.4.2 Microtubule Assembly in the Presence or Absence of Glycerol

Shelanski et al. (1973) found that the requirement for a nucleotide triphosphate in the purification of MT proteins by cycling, could be reduced or eliminated if 4 M glycerol were included during the polymerisation steps. Reconstituted MT's are sedimented and the pellet resuspended in cold buffer without glycerol to depolymerise the MT's. The material is then clarified by cold centrifugation, incubated again with glycerol, and the procedure repeated several times. This method yields tubulin together with 10% high molecular weight MAP's. There are clear differences in the types and quantities of MAP that co-purify with tubulin in the presence and absence of glycerol. This may be related to the finding that glycerol remains very strongly bound to MT proteins (Detrich et al. 1976) even after extensive dialysis. Glycerol may promote assembly through a general thermodynamic solvent effect on the chemical potential of tubulin (Lee et al. 1978).

Consequently, purification of MT proteins by polymerisation in the absence of glycerol has been described by several workers (Olmsted and Borisy 1973; Borisy et al. 1975). Asnes and Wilson (1979) determined optimal conditions for the isolation of bovine brain tubulin, using a phosphate-glutamate buffer, without glycerol. All operations except polymerisation and warm centrifugation were done at 0 °–4 °C. Yields of MT's were optimal when assembly was conducted at 30 °C. The clarified brain homogenate was made up to a final concentration of 2.5 mM GTP, 1 mM EGTA, and 0.5 mM $MgCl_2$, incubated at 30 °C for 20 min, and then centrifuged at 39,000 *g* for 30 min. The pellet of polymerised MT's was resuspended in a small volume of buffer using a Dounce homogeniser with a loose-fitting pestle. The pellet was incubated at 0 °C for 30 min to depolymerise MT's and centrifuged at 39,000 *g* for 40 min. The supernatant fraction contains the soluble MT proteins, and the pellet contains debris and a population of cold-resistant MT's (Webb and Wilson 1980; Job et al. 1981). This sub-population of cold-stable MT's is also resistant to antimitotic drugs such as podophyllotoxin, which can cause the rapid disassembly of cold-labile MT's, but they are rapidly disassembled by millimolar concentrations of calcium in the presence of calmodulin at concentrations substoichiometric to that of tubulin (Job et al. 1981).

The supernatant fraction of Asnes and Wilson is rewarmed to 30 °C for 30 min, and the MT pellet harvested after centrifugation. The cycle can be repeated several times, although loss of MT proteins is incurred during every cycle. MT pellets may be frozen in liquid N_2 for storage.

3.4.3 The Dynamics of Polymerisation and the Use of Taxol

A dynamic equilibrium of MT's with tubulin dimers has been demonstrated in vitro (Lee and Timasheff 1975; Johnson and Borisy 1977). More recently, the treadmilling of tubulin has been firmly established (Cleveland 1982). Margolis and Wilson (1978, 1979) first demonstrated a flux of subunits through MT's at a steady state in vitro, and proposed opposite end assembly and disassembly of MT's. Bergen and Borisy (1980) showed that there was no real restriction on the relative magnitudes of the individual rate constants, and determined the four individual rate constants for subunit addition and loss at each end. These four rate constants uniquely determine both subunit flux and efficiency of flux. Several groups have now measured flux rates for subunits through MT's (Margolis and Wilson 1979; Cote and Borisy 1981) and it is clear that both the rate and efficiency of the subunit flux are sensitive functions of the precise in vitro conditions.

Taxol is a drug, isolated from the western yew (*Taxus brevifolia*), which shows antitumour activity. Schiff et al. (1979) and Schiff and Horowitz (1981) have shown that taxol acts as a promoter of calf brain MT assembly in vitro. At a protein concentration of 1 mg ml^{-1} tubulin, taxol causes a dose-dependent decrease in the lag time for assembly; 5 µM taxol eliminates the lag time. At this concentration, the drug also dramatically reduces the critical concentration of tubulin required for assembly from 0.2 mg ml^{-1} (in the absence of taxol) to 0.01 mg ml^{-1}. Furthermore, phosphocellulose-purified tubulin, which is free of MAP's, will assemble in the presence of stoichiometric concentrations of taxol. These observations have been employed in the isolation of MT proteins from both animal cells (Vallee 1982) and from plant cells (Morejohn and Fosket 1982).

In vitro assembly reactions with plant tubulin (Morejohn and Fosket 1982) were performed on fractions purified by DEAE-Sephadex chromatography (Sect. 3.1). The protein fraction, tightly bound to the ion exchange column, and eluted with 0.8 M NaCl in PM buffer, was collected by precipitation with 50% saturated $(NH_4)_2SO_4$ over 30 min. Precipitates from ~200 g of cells were dissolved in 1–1.5 ml of PM buffer containing 2 mM GTP. The solution was desalted on a Sephadex G25 column (5 ml bed volume) and the excluded volume was made 50 µM in leupeptin hemisulphate at 4 °C for 2 h. A 100-µl sample, containing 3 mg ml^{-1} protein, was made 10 µM in taxol, incubated at 37 °C for 1 h to permit MT assembly and centrifuged at 130,000 *g* for 1 h at 23 °C. Each MT pellet was resuspended in 10 µl of PM buffer containing 4 M glycerol, and prepared for electron microscopy, SDS-PAGE, or protein determinations.

In enhancing the rate and yield of brain MT assembly, and in stabilizing the polymers formed, taxol binds to the MT's and cannot readily be removed (Parness and Horowitz 1981). Hence, taxol-induced MT's cannot be purified further by successive round of polymerisation. Nevertheless, because it has a more specific affinity for MT proteins, the initial enrichment of tubulin during in vitro assembly is substantially greater than that induced by glycerol. The lower critical concentration of tubulin for assembly in the presence of taxol is also of potential importance in using plant homogenates, since fractions with tubulin levels lower than those used by Morejohn and Fosket by 1 or 2 orders of magnitude would still permit MT assembly.

3.4.4 Co-Polymerisation

Tubulins are highly conserved proteins (see reviews by Luduena 1979; Sabnis and Hart 1982; Hyams 1982) and conformational similarities permit the co-assembly of tubulins from widely disparate sources. Thus, axonemal proteins from *Chlamydomonas* co-polymerise with, and display immunochemical and electrophoretic similarities to, brain tubulin (Snell et al. 1974; Rosenbaum et al. 1975). Co-polymerisation with brain tubulin has also been used to isolate tubulins from yeast (Water and Kleinsmith 1976; Baum et al. 1978; Clayton et al. 1979; Kilmartin 1981), *Aspergillus* (Sheir-Neiss et al. 1976) and higher plants (Slabas et al. 1980; Yadav and Filner 1983).

Kilmartin (1981) grew yeast in the minimal culture medium described by Baum et al. (1978), containing 10–100 μCi/200 ml of $^{35}SO_4^{2-}$. The cells were resuspended in buffer (0.1 M PIPES-NaOH, pH 6.9, containing 1 mM EGTA, DTT, and GTP, 0.1 mM $MgCl_2$, and 17 μg ml^{-1} PMSF) and lysed by vortexing with glass beads. The extracts were clarified at 4 °C by centrifugation at 20,000 *g* for 20 min, followed by 100,000 *g* for 1 h. An equal volume of pig brain tubulin (7 mg ml^{-1}), prepared as described by Borisy et al. (1975), was added and 2–3 cycles of temperature-dependent assembly-disassembly carried out. Molecular species that co-polymerise with brain tubulin can be detected by autograms or fluorograms of polyacrylamide gels run in one- or two dimensions.

Similar procedures have been employed for the co-polymerisation of brain tubulin with putative tubulin in extracts of higher plant tissues (Yadav and Filner 1983). Tobacco cells in suspension culture, labelled with ^{35}S, were extracted using a procedure similar to that employed for yeast cells by Kilmartin (1981). The soluble ^{35}S-labelled proteins were fractionated by co-polymerisation with purified cow brain tubulin. A 2-ml aliquot of the supernatant fraction of the plant extract, containing 2 mg ml^{-1} protein, was mixed with 2 ml of ice-cold buffer containing 10 mg of brain tubulin and 4 ml of glycerol. The sample was carried through two cycles of assembly-disassembly to yield a co-polymer containing ^{35}S-labelled plant proteins.

4 Fractionation and Identification of Tubulin by SDS-PAGE

Fine (1971) first subjected brain tubulin to discontinuous electrophoresis on 10% Tris-chloride-acrylamide gels (after Laemmli 1970), containing 1 M urea, 0.1% (w/v) SDS and a running buffer of Tris-glycine, pH 8.3. Under these conditions, tubulin splits into two bands: the slower migrating band has been designated the α-subunit, and the faster, the β-subunit. Both monomers, particularly α-tubulin, exhibit anomalous behaviour on SDS gels. The mobility of the α-subunit increases with ionic strength and decreases with pH, whereas that of the β-subunit remains relatively constant (Bryan 1974). Carboxymethylation, carboxyamidomethylation, performic acid oxidation and S-sulphonation cañ each cause major and distinct changes in the electrophoretic mobility of both monomers or of the α-subunit alone (Eipper 1972, 1974).

The reason for this anomalous behaviour is not clear, but can be exploited to provide wide separation of the tubulin monomers. Since the complete sequencing of tubulin subunits from brain has been reported by several groups (Postingl et al. 1981; Valenzuela et al. 1981), the molecular weight of both subunits must now be reduced from the widely cited 55,000 to 50,000 (450–451 and 445 amino acid residues for α- and β-tubulin respectively; Fulton 1982).

Immediately prior to electrophoresis, $(NH_4)_2SO_4$ precipitates are dissolved in extraction buffer, desalted on a small column of Sephadex G25 and acetone-precipitated (Morejohn and Fosket 1982) or concentrated using an Amicon concentration/dialysis cell and PM10 or YM10 membranes, or using Millipore immersible CX ultrafilters. Precipitation methods are the best to employ for small quantities and volumes. Precipitates, pellets of MT's purified by assembly-disassembly cycles, or other concentrated protein samples are dissolved in SDS-sample buffer and heated in a boiling water bath for 1.5 min. Discontinuous electrophoresis (Laemmli 1970) is performed in polyacrylamide slab gels, using a 10–12% resolving gel and 4% stacking gel.

We have obtained excellent separation and sharp protein bands by incorporating 4 M urea and 1 mM DTT into both resolving and stacking gels (Fig. 2). George et al. (1981) cast slab gels using a 4–12% (w/v) acrylamide gradient, with a constant ratio of acrylamide : bis acrylamide of 30 : 0.8, in 6 M urea, 0.1% SDS and 0.375 M Tris-HCl, pH 8.8, with a 0–10% (w/v) sucrose gradient. The stacking gel contained 3% (w/v) acrylamide in 0.1% SDS, 0.0675 M Tris-HCl, pH 6.8.

Fig. 2. α- and β-subunits of tubulin separated by discontinuous SDS-PAGE. The tubulin was prepared from porcine brain homogenates by DEAE-Sephadex chromatography. The *faint band* represents a high molecular weight MAP

When rose tubulin was fractionated by SDS-PAGE, the β-subunit co-migrated with β-tubulin from bovine brain (Morejohn and Fosket 1982). However, the rose α-subunit had a slightly faster mobility than the corresponding α-monomer from brain. Yadav and Filner (1983) reported identical results with *Nicotiana* tubulin, and we have confirmed this consistent discrepancy with tubulin isolated from maize root tips and azuki bean epicotyls (Sabnis and Sek, unpublished).

Reports of tubulin from lower plants also indicate differences, but the results are less in agreement. Kilmartin (1981) reported that yeast tubulin co-migrated with α-tubulin from brain, but showed less correspondence on two-dimensional gels. Baum et al. (1978) reported that putative yeast tubulin in their preparations migrated faster than the α- and β-subunits of brain tubulin (apparent MW of 45,000 and 46,000). Clayton et al. (1979) could only demonstrate one band on autoradiograms of co-polymers of ^{35}S-labelled yeast tubulin and brain tubulin, which co-migrated with the brain α-subunit. White et al. (1983) have shown that tubulin subunits isolated from *Dictyostelium discoideum*, and identified on Western blots by one polyclonal and two monoclonal antibodies against yeast tubulin, do not migrate near the positions of brain tubulin on isoelectric focussing; they focus at a far more basic pH. There are now a number of reports indicating microheterogeneity in tubulin subunits from a number of sources, when sensitive fractionation techniques such as gel isoelectric focussing are employed (Bibring et al. 1976; Gozez and Littauer 1978; Dahl and Weibel 1979; George et al. 1981; Gozez and Sweadner 1981; Denoulet et al. 1982). Caution dictates, therefore, that SDS-PAGE or IEF should only be used to identify putative plant tubulin in conjunction with the assay of one or more of the other characteristic properties of tubulin such as colchicine-binding or re-assembly into MT's.

5 Colchicine-Binding Assay for Tubulin

The discovery that the antimitotic alkaloid, colchicine, interacts directly with the MT subunit was of crucial importance in the original isolation of tubulin (Weisenberg et al. 1968). Micromolar concentrations of applied colchicine are able to deploymerise MT's in animal cells: however, considerably higher concentrations of the drug (up to 10^{-3}M) seem to be required to disassemble plant MT's in situ (reviewed by Hart and Sabnis 1976c). Thus, earlier work suggested that colchicine-binding activity (CBA) could not be applied readily to assay tubulin levels in plant extracts (Haber et al. 1972; Hart and Sabnis 1973, 1976a, b; Heath 1975; Flanagan and Warr 1977). However, with improving methods for the isolation and stabilisation of tubulin from plant sources, and with appropriate controls, CBA may be used as a diagnostic feature in the identification and assay of plant tubulin.

[^{3}H-ring C-methoxy]-colchicine is made up to a suitable concentration and specific activity with unlabelled colchicine. Binding activity in extracts may be assayed by the filter disc method or by gel filtration (Weisenberg et al. 1968; Wilson 1970). Decay of CBA can be rapid in plant preparations (Hart and Sabnis 1976a, b; Okamura 1980; Mizuno et al. 1981) and may be related to the rapid proteolysis of tubulin in plant extracts. Therefore, samples should be assayed as rapidly as possible after preparation, and inhibitors of protease activity should be included at all stages where endogenous proteolytic activity may still be present. Concentrated samples of known protein concentration are incubated for 1 h at 30 °C with ^{3}H-labelled drug. Triplicate 0.1 ml samples are applied to 2.5-cm-diameter discs of Whatman DE 81 paper and the discs washed in a Buchner funnel with 7×50 ml volumes of cold 0.01 M phosphate buffer, pH 6.8, containing 10 mM $MgCl_2$. Alternatively, 1 ml samples of incubation mixture are filtered, at 24 ml h^{-1}, through a 16×1.6 cm column of Sephadex G100, pre-equilibrated with the phosphate buffer. The radioactivity retained on the washed discs or present in column eluate fractions of 1 ml, is dispersed in an appropriate scintillation cocktail and counted (Fig. 3).

Earlier, we have cautioned against the possibility of spurious binding of colchicine to cellular components and have recommended a series of controls that

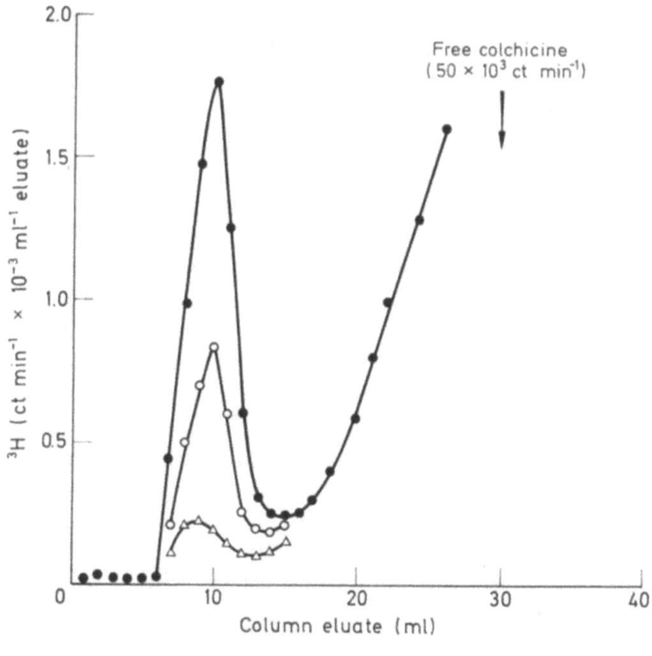

Fig. 3. Elution profile of ^{3}H (ct min^{-1}) after Sephadex G-100 filtration of plant preparations incubated with ^{3}H-colchicine. The 30–45%-saturated ammonium sulphate fraction was resuspended in extraction buffer; ^{3}H-colchicine (1 ml) was incubated with 1.25×10^{-7}M ^{3}H-colchicine at 37 °C for 60 min, and then passed through Sephadex G-100 at 24 ml h^{-1}; column 16×1.6 cm, void volume 9 ml. ●—●, *Heracleum* vascular tissue, 3.6 mg protein ml^{-1}; ▲—▲, same sample with 8.7 mM unlabelled colchicine added before incubation; ○—○, 4-d-old mustard roots, 5 mg protein ml^{-1}

need to be incorporated into CB assays for plant tubulin (Hart and Sabnis 1976b, c). We have advocated the routine use of ^{3}H-lumicolchicine as one of them. Lumicolchicine, the mixture of colchicine isomers derived by UV-irradiation of the drug, mimics many of the cytotoxic actions of colchicine in several lines of mammalian cells, but has been shown not to affect MT's (Wilson et al. 1974). It does not bind to tubulin (Wilson and Freidkin 1967), nor does it seem to affect mitosis (Linskens and Wulf 1953). However, recent work on the sequential formation of lumi-derivatives of colchicine with different effects on seedling growth has undermined the predicted usefulness of lumicolchicine as a control for CBA of tubulin (Sabnis 1981). Complete phototransformation yields a derivative with no detectable biological activity.

6 Immunochemical Methods of Analysis

Tubulin is a highly conserved protein and is generally a poor antigen; nevertheless, both polyclonal and monoclonal antibodies can be raised against the protein (Fuller et al. 1975; Piperno and Luck 1977; Morgan et al. 1978; Gozez and Barnstable 1982). The conserved nature of the protein usually, but not invariably, renders the source of the antigen of little immunological concern. Thus, anti-tubulin antibodies raised against brain tubulin will react in many instances with higher plant tubulin (Lloyd et al. 1979; van der Valk et al. 1980; Wick et al. 1981; De Mey et al. 1982). The antibodies are generally directed against the 6s dimer.

Immunisation protocols are varied and often chosen arbitrarily; hence, they will only be described here in outline, following the procedures described by Brinkley et al. (1980) which we have used successfully. Tubulin, purified by cycles of assembly-disassembly followed by phosphocellulose chromatography, is cross-linked with 5% purified glutaraldehyde at 0 °C overnight with gentle stirring. If precipitation occurs, the cross-linked protein can be collected and washed by repeated centrifugation and resuspension. In the absence of a precipitate, the sample is dialysed against phosphate-buffered saline (PBS). The resuspended or dialysed sample (1–2 mg protein) is vigorously mixed with an equal volume of Freund's adjuvant (complete for initial injections, incomplete for booster injections) preferably using two glass syringes connected with a double Luer-lock fitting. Young adult New Zealand white rabbits are bled to collect preimmune serum for use in controls, and immunised using a multiple-site, multiple-route injection schedule (subcutaneous, subscapular, intravenous, and intramuscular). Antigen doses of 10–20 μg per site are recommended.

Bleeding procedures are described in detail by Brinkley et al. (1980) and Mayer and Walker (1980), and are generally carried out by licensed personnel. After removal of clots, the serum is clarified by centrifugation at 1000 *g* for 10 min, divided into aliquots and stored until required at −20 °C. Repeated freezing and thawing of the sera should be avoided. Antibody titre can be assayed using ELISA (Sect. 6.2).

6.1 Radioimmunoassay

If a radioimmunoassay is planned, the use of unfractionated serum is acceptable. Such assays have been described for the quantitative measurements of tubulin from brain (Morgan et al. 1977; Le Guern et al. 1977), for brain MAP_2 (Nieto et al. 1981), and for putative plant tubulin (Rikin et al. 1982).

Rikin et al. used polyvinylchloride microtitration plates, with 96 conical wells, to provide the solid phase of an indirect assay. Wells are incubated with 15 μg of purified calf brain tubulin in 50 μl of PBS overnight at 4 °C. The wells were washed three times with PBS containing 0.05% Tween 20 and 0.1% gelatin, incubated for 2 h in PBS containing 0.1% gelatin to block unreacted sites, and washed three times with PBS/Tween/gelatin. They were then incubated for 2 h at room temperature with 50 μl of plant extract containing antitubulin serum at a final dilution of 1:50. After three washings, antibody bound to the immobilised tubulin was detected by incubation with 10 μl of ^{125}I-labelled protein A. The wells were washed and counted in a γ-spectrometer. An inhibition curve prepared using soluble brain tubulin, showed that the radioimmunoassay allows quantitative determination of 0.1 to 10 μg tubulin. A standard inhibition curve with brain tubulin was used to estimate the levels of tubulin-like activity in crude plant extracts.

In preference to the indirect RIA, plant tubulin that has been partially purified by procedures described earlier (DEAE-Sephadex chromatography or MT assembly) may be quantitatively assayed by direct RIA. Serial dilutions of the plant protein preparation are immobilised in the wells of a microtitre plate, and incubated with antiserum. After washing, the antigen-antibody complex is detected with ^{125}I-protein A and counted as before. A standard curve is prepared using brain tubulin.

6.2 Enzyme-Linked Immunosorbent Assay (ELISA)

The ELISA procedure is a variant of the RIA method described above, which we have found to be as sensitive, cheaper and much more rapid. Either fixed concentrations (10–100 μg ml^{-1}) or serial dilutions (1–200 μg ml^{-1}) of plant or brain tubulin are prepared in coating buffer (1.59 g Na_2CO_3, 2.39 g $NaHCO_3$, 0.2 g NaN_3, pH 9.6, made up to 1 l) and immobilised on 96-well ELISA plates (50 μl/well) by incubation at room temperature for 2 h. The plates are washed three times with PBS/Tween (8 g NaCl, 0.2 g KH_2PO_4, 2.9 g $Na_2HPO_4 \cdot 12H_2O$, 0.2 g KCl, 0.5 ml Tween 20, 0.2 g NaN_3, pH 7.4, made up to 1 l). Leave plates for 10 min in each wash. Fixed or serial dilutions of anti-tubulin serum, diluted with PBS/Tween are added to the wells (50 μl/well) and incubated for 2 h at room temperature. The plates are washed three times with PBS/Tween.

The second enzyme-linked antibody consists of alkaline phosphatase coupled to goat-antirabbit IgG. This is diluted 1:500 or 1:1000, 50 μl added to each well, and incubated for 1 h at room temperature, followed by three washes with PBS/Tween. The colour reaction is developed using 200 μl/well of p-nitrophenyl phosphate substrate (1 mg ml^{-1}) dissolved in buffer (97 ml diethanolamine, 0.2 g

NaN_3, 100 mg $MgCl_2 \cdot 6H_2O$, 800 ml water; adjust pH to 9.8 with 1 M HCl and make up to 1 l). After 30 min, the reaction is stopped by adding 50 µl of 4 M NaOH to each well. From each well 200 µl is diluted to 1 ml and read in a spectrophotometer at 400 nm, or the ELISA plate may be assayed directly on a Titertek Multiskan or similar automated equipment.

6.3 Antibody Purification on Antigen-Affinity Column

To reduce the chances of spurious interactions, it is preferable to purify the anti-tubulin antibodies using an antigen-CNBr Sepharose affinity column. DEAE-Sephadex and phosphocellulosepurified tubulin is dialysed against borate-saline buffer (BSB; 100 mM H_3BO_3, 25 mM $Na_2B_4O_7$, 75 mM NaCl, pH 8.4) containing 10 mM $MgCl_2$, to remove all GTP and EGTA; amino groups react readily with CNBr-activated resins and compete with the protein. CNBr-Sepharose 4B is also washed with BSB: tubulin is added to the swollen activated Sepharose (2–4 mg protein ml^{-1} gel) and allowed to react at 4 °C for 24 h. Unreacted groups are blocked with ethanolamine, and the column is equilibrated with BSB. It is best to prepare a short wide column which has 10–30 mg tubulin bound (enough to process 50 ml of serum at a time).

The antiserum is dialysed against BSB and clarified by centrifugation. The serum is then passed through the affinity column and the column flushed with BSB till the $O.D._{280}$ of the eluate is less than 0.02. The anti-tubulin IgG is eluted with 4 M $MgCl_2$ in BSB (high salt is prefered to low pH). The yield of anti-tubulin should fall within the range 0.1–1.0 mg anti-tubulin per 10 ml serum. Pool and dialyse antibody-containing fractions against PBS. Divide into aliquote and store at −20 °C.

6.4 Western Blots

Anti-tubulin antibodies can also be employed to identify very low levels of tubulin on SDS-polyacrylamide gels. This may be of particular interest during early steps in the processing of plant extracts. For blots, SDS-gels may be run using separate test samples in individual lanes, or by fractionating a single sample across the width of the gel. Protein bands are transferred across to nitrocellulose sheets using an electroblotting apparatus such as the Biorad Trans-Blot Transfer apparatus, following the recommended procedures and voltage. Efficiency of transfer can be checked by staining a test strip of the nitrocellulose sheet with Amido Black [Naphthalene Black 12B, 0,1% (w/v) in 45% methanol/10% acetic acid] for 15–30 s followed by destaining in water.

For reaction with antibody, the entire nitrocellulose blot, or individual tracks are soaked in a large volume of blocking buffer (3% w/v bovine serum albumin, 10 mM Tris-HCl, pH 7.4, 0.9% w/v NaCl) for 2 h at room temperature in order to saturate reactive sites on the transfer. The transfer sheets are sealed into polythene sacs of minimal required volume and bathed with anti-tubulin diluted to a suitable titre with PBS. After incubation for 2 h, the nitrocellulose strips are removed from the sacs and thoroughly washed with PBS (5 × 100 ml over 1 h). The strips are then resealed into fresh sacs along with ^{125}I-labelled protein A diluted with blocking buffer for 3 h at room temperature. The strips or sheets are thoroughly washed with PBS once again and air-dried at 25 °C. Autoradiograms are prepared by exposing the sheets to pre-flashed X-ray film at −80 °C using an intensifier screen.

Alternatively, the antitubulin may be located on the blots using a peroxidase-linked second antibody. After incubation with anti-tubulin and washing with PBS, the blots are incubated with peroxidase-linked second antibody diluted in blocking buffer (1:500). Following incubation, the blots are washed in blocking buffer, rinsed in PBS, and soaked in a bath of the substrate consisting of 50 μg ml^{-1} of diaminobenzidine-hydrochloride in 0.1 M Tris-HCl, pH 7.6, containing 1.6 μl ml^{-1} of H_2O_2 (6%). The substrate should be prepared immediately before use and protected from light. Location of the first antibody is recognised by brown bands that form in 1–10 min. The reaction is stopped by washing with PBS. The developed blots must be photographed immediately, while still wet, since background develops both with time and drying.

7 Concluding Remarks

Immunochemistry now offers powerful tools for the identification, isolation and characterisation of MT proteins from plants. The possibility of using an antibody-affinity column to isolate and concentrate tubulin from large volumes of plant extract only requires an adequate source of antibody. Hybridoma technology promises not only the specificity of monoclonal antibodies as analytical tools, but the greater quantities of antibodies provided by continuous culture of monoclonal lines or the collection of ascites fluid from tumour-bearing animals injected with hybridoma cells. Equally, immunocytochemistry at the light or electron microscopic level provides a new dimension to the study in situ of the cytoskeletal function of MT's in morphogenesis.

Acknowledgements. I wish to thank Prof. Brian Gunning for the opportunity to learn and develop in his laboratory some of the immunochemical methods described. I am grateful to Drs James Hart and Sue Wick for helpful discussions in the past and to Frank Sek and Ian Skene for excellent technical assistance.

References

Asnes CF, Wilson L (1979) Isolation of bovine brain microtubule protein without glycerol: polymerization kinetics change during purification cycles. Anal Biochem 98:64–73

Baum P, Thorner J, Honig L (1978) Identification of tubulin from the yeast *Saccharomyces cerevisiae*. Proc Natl Acad Sci USA 75:4962–4966

Bergen LG, Borisy GG (1980) Head-to-tail polymerization of microtubules in vitro. Electron microscope analysis of seeded assembly. J Cell Biol 84:141–150

Berkowitz SA, Katagiri J, Binder HK, Williams Jr RC (1977) Separation and characterization of microtubule proteins from calf brain. Biochemistry 16:5610–5617
Bibring T, Baxendall J, Denslow S, Walker B (1976) Heterogeneity of the alpha subunit of tubulin and the variability of tubulin within a single organism. J Cell Biol 69:301–312
Borisy GG, Marcum JM, Olmsted JB, Murphy DB, Johnson KA (1975) Purification of tubulin and associated high molecular weight proteins from porcine brain and characterization of microtubule assembly in vitro. Ann NY Acad Sci 253:107–132
Brinkley BR, Fistel SH, Marcum JH, Pardue RL (1980) Microtubules in cultured cell; indirect immunofluorescent staining with tubulin antibody. Int Rev Cytol 63:59–95
Bryan J (1974) Biochemical properties of microtubules. Fed Proc 33:152–157
Bulinsky JC, Borisy GG (1979) Self-assembly of microtubules in extracts of cultured HeLa cells and the identification of HeLa microtubule-associated proteins. Proc Natl Acad Sci USA 76:293–297
Burgoyne RD, Cumming R (1983) Characterisation of microtubule-associated proteins at the synapse – absence of MAP-2. Eur J Cell Biol 30:154–158
Clayton L, Pogson CI, Gull K (1979) Microtubule proteins in the yeast, *Saccharomyces cerevisiae*. FEBS Lett 106:67–70
Cleveland DW (1982) Treadmilling of tubulin and actin. Cell 28:689–691
Cleveland DW, Hwo S-Y, Kirschner MW (1977) Purification of tau, a microtubule – associated protein that induces assembly of microtubules from purified tubulin. J Mol Biol 116:207–225
Connolly JA, Kalnins VI (1980) The distribution of tau and HMW microtubule – associated proteins in different cell types. Exp Cell Res 127:341–350
Connolly JA, Kalnins VI, Cleveland DW, Kirschner MW (1977) Immunofluorescent staining of cytoplasmic and spindle microtubules in mouse fibroblasts with antibody to tau protein. Proc Natl Acad Sci USA 74:2437–2440
Cote RH, Borisy GG (1981) Head-to-tail polymerization of microtubules in vitro. J Mol Biol 150:577–602
Dahl JL, Weibel VJ (1979) Changes in tubulin heterogeneity during postnatal development of rat brain. Biochem Biophys Res Commun 86:822–828
De Brabander M, Bulinsky JC, Geuns G, De Mey K, Borisy GG (1981) Immunoelectron microscopic localisation of the 210,000-mol wt microtubule – associated protein in culture cells of primates. J Cell Biol 91:438–445
DeMey J, Lambert AM, Bajer AS, Moeremans M, De Brabander M (1982) Visualization of microtubules in interphase and mitotic plant cells of *Haemanthus* endosperm with the immuno-gold staining method. Proc Natl Acad Sci USA 79:1898–1902
Denoulet P, Edde B, Jeantet C, Gros F (1982) Evolution of tubulin heterogeneity during mouse brain development. Biochimie (Paris) 64:165–172
Detrich III HW, Berkowitz SA, Kim H, Williams Jr RC (1976) Binding of glycerol by microtubule protein. Biochem Biophys Res Commun 68:961–968
Detrich III HW, Williams RC (1978) Reversible dissociation of the $\alpha\ \beta$ dimer of tubulin. Biochemistry 17:3900–3907
Eipper BA (1972) Rat brain microtubule protein: purification and determination of covalently bound phosphate and carbohydrate. Proc Natl Acad Sci USA 69:2283–2287
Eipper BA (1974) Properties of rat brain tubulin. J Biol Chem 249:1407–1416
Fine RE (1971) Heterogeneity of tubulin. Nature New Biol 233:283–284
Flanagan D, Warr JR (1977) Colchicine binding of a high-speed supernatant of *Chlamydomonas reinhardi*. FEBS Lett 80:14–18
Frigon RP, Timasheff SN (1975) Magnesium-induced self-associations of calf-brain tubulin I. Stoichiomery, Biochemistry 14:4559–4566
Fuller GM, Brinkley BR, Boughter JM (1975) Immunofluorescence of mitotic spindles by using monospecific antibody against bovine brain tubulin. Science (Wash DC) 187:948–950
Fulton C (1982) Two bar mitzvahs for tubulin. Nature 296:308–309
Fulton C, Simpson PA (1979) Tubulin pools, synthesis and utilization. In: Roberts K, Hyams JS (eds) Microtubules. Academic Press, London, pp 117–174

George HJ, Misra L, Field DJ, Lee JC (1981) Polymorphism of brain tubulin. Biochemistry 20:2402–2409
Gozez I, Barnstable CJ (1982) Monoclonal antibodies that recognise discrete forms of tubulin. Proc Natl Acad Sci USA 79:2579–2583
Gozez I, Littauer UZ (1978) Tubulin microheterogeneity increases with rat brain maturation. Nature 276:411–413
Gozez I, Sweadner KJ (1981) Multiple tubulin forms are expressed by a single neurone. Nature 294:477–480
Gunning BES (1982) The cytokinetic apparatus: its development and spatial regulation. In: Lloyd CW (ed) The cytoskeleton in plant growth and development. Academic Press, New York, pp 229–292
Gunning BES, Hardham AR (1982) Microtubules. Annu Rev Plant Physiol 33:651–698
Haber JE, Peloquin JG, Halvorson HO, Borisy GG (1972) Colcemid inhibition of cell growth and the characterization of a colcemid-binding activity in *Saccharomyces cerevisiae*. J Cell Biol 55:355–367
Heath IB (1975) Colchicine and colcemid binding components of the fungus *Saprolegnia ferax*. Protoplasma 85:177–192
Hart JW, Sabnis DD (1973) Colchicine binding protein from phloem and xylem of a higher plant. Planta 109:147–152
Hart JW, Sabnis DD (1976a) Colchicine binding activity in extracts of higher plants. J Exp Bot 27:1353–1360
Hart JW, Sabnis DD (1976b) Binding of colchicine and lumicolchicine to components in plant extracts. Phytochemistry 15:1897–1901
Hart JW, Sabnis DD (1976c) Colchicine and plant microtubules: a critical evaluation. Curr Adv Plant Sci 26:1095–1104
Hyams JS (1982) Microtubules. In: Lloyd CW (ed) The cytoskeleton in plant growth and development. Academic Press, New York, pp 31–53
Jackson WT (1982) Actomyosin. In: Lloyd CW (ed) The cytoskeleton in plant growth and development. Academic Press, New York, pp 3–29
Jameson L, Frey T, Zeeburg B, Dalldorf F, Caplow M (1980) Inhibition of microtubule assembly by phosphorylation of microtubule-associated proteins. Biochemistry 19:2472–2478
Job D, Fischer EH, Margolis RL (1981) Rapid disassembly of cold-stable microtubules by calmodulin. Proc Natl Acad Sci USA 78:4679–4682
Johnson KA, Borisy GG (1977) Kinetic analysis of microtubule assembly in vitro. J Mol Biol 117:1–31
Kakiuchi S, Sobue K (1983) Control of the cytoskeleton by calmodulin and calmodulin-binding proteins. Trends Biochem Sci 8:59–62
Kennett RH (1981) Hybridomas: a new dimension in biological analysis. In Vitro (Rockville) 17:1036–1050
Kilmartin JV (1981) Purification of yeast tubulin by self assembly in vitro. Biochemistry 20:3629–3633
Laemmli UK (1970) Cleavage of structural proteins during the assembly of the head of bacteriophage T4. Nature 227:680–685
Lazarides E (1982) Intermediate filaments: a chemically heterogeneous, developmentally regulated class of proteins. Annu Rev Biochem 51:219–246
Lee JC, Timasheff SN (1975) The reconstitution of microtubules from purified calf brain tubulin. Biochemistry 14:5183–5187
Lee JC, Tweedy N, Timasheff SN (1978) In vitro reconstitution of calf brain microtubules: effects of macromolecules. Biochemistry 17:2783–2790
Le Guern C, Pradelles P, Dray F, Jeantet C, Gros F (1977) Radioimmunoassay for tubulin detection. FEBS Lett 84:97–100
Linskens HF, Wulf N (1953) Über die Trennung und Mitosewirkung der Lumicolchicine. Naturwissenschaften 40:487–488
Lloyd CW (ed) (1982) The cytoskeleton in plant growth and development. Academic Press, New York

Lloyd CW, Slabas AR, Powell AJ, Macdonald G, Badley RA (1979) Cytoplasmic microtubules of higher plant cells visualised with antitubulin antibodies. Nature 279:239–241

Luck D, Piperno G, Ramanis Z, Huang B (1977) Flagellar mutants of *Chlamydomonas*: studies of radial spoke-defective strains by dikaryon and revertant analysis. Proc Natl Acad Sci USA 74:3456–3460

Luduena RF (1979) Biochemistry of tubulin. In: Roberts K, Hyams JS (eds) Microtubules. Academic Press, New York, pp 65–116

Margolis RL, Wilson L (1978) Opposite end assembly and disassembly of microtubules at steady state in vitro. Cell 13:1–8

Margolis RL, Wilson L (1979) Regulation of the microtubule steady state in vitro by ATP. Cell 18:673–679

Mayer RJ, Walker JH (1980) Immunochemical methods in the biological sciences: enzymes and proteins. Academic Press, New York

Mizuno K, Koyama M, Shibaoka H (1981) Isolation of plant tubulin from azuki bean epicotyls by ethyl-N-phenyl-carbamate – Sepharose affinity chromatography. J Biochem (Tokyo) 89:329–332

Morejohn LC, Fosket DE (1982) Higher plant tubulin identified by selfassembly into microtubules in vitro. Nature 297:426–428

Morgan JL, Rodkey LS, Spooner BS (1977) Quantitation of cytoplasmic tubulin by radioimmunoassay. Science 197:578–580

Morgan JL, Holladay CR, Spooner BS (1978) Species-dependent immunological differences between vertebrate brain tubulins. Proc Natl Acad Sci USA 75:1414–1417

Murphy DB, Borisy GG (1975) Association of high-molecular-weight proteins with microtubules and their role in microtubule assembly in vitro. Proc Natl Acad Sci USA 72:2696–2700

Nieto A, Avila J, Valdivia MM (1981) Comparative measurement by radioimmunoassay of the brain microtubule-associated protein MAP_2. Mol Cell Biochem 37:185–189

Okamura S (1980) Binding of colchicine to a soluble fraction of carrot cells grown in suspension culture. Planta 149:350–354

Olmsted JB, Borisy GG (1975) Ionic and nucleotide requirements for microtubule polymerization in vitro. Biochemistry 14:2996–3005

Parness J, Horowitz SB (1981) Taxol binds to polymerised tubulin in vitro. J Cell Biol 91:479–487

Piperno G, Luck DJL (1977) Microtubular proteins of *Chlamydomonas reinhardtii*: an immunochemical study based on the use of an antibody specific for the B-tubulin subunit. J Biol Chem 252:383–391

Postingl H, Krauhs E, Little M, Kempf T (1981) Complete amino acid sequence of β-tubulin from porcine brain. Proc Natl Acad Sci USA 78:4156–4160

Rikin A, Atsmon D, Gitler C (1982) Extraction and immunochemical assays of a tubulin-like factor in cotton seedlings. Planta 154:402–406

Roberts K, Hyams JS (1979) Microtubules. Academic Press, New York

Rosenbaum JL, Binder LI, Granett S, Dentler WL, Snell W, Sloboda R, Haimo L (1975) Directionality and rate of assembly of chick brain tubulin onto pieces of neurotubules, flagellar axonemes and basal bodies. Ann NY Acad Sci 253:147–177

Rubin RW, Cousins EH (1976) Isolation of a tubulin-like protein from *Phaseolus*. Phytochemistry 15:1837–1839

Runge MS, Detrich III HW, Williams Jr RC (1979) Identification of the major 68.000-dalton protein of microtubule preparations as a 10-nm filament protein and its effects on microtubule assembly in vitro. Biochemistry 18:1689–1698

Sabnis DD (1981) Lumicolchicine as a tool in the study of plant microtubules: some biological effects of sequential products formed during phototransformations of colchicine. J Exp Bot 32:271–278

Sabnis DD, Hart JW (1982) Microtubule proteins and P-proteins. In: Boulter D, Parthier B (eds) Nucleic acids and proteins in plants I. Encyclopedia Plant Physiol, new series, vol 14A. Springer, Berlin Heidelberg New York, pp 401–437

Scheele RB, Borisy GG (1979) In vitro assembly of microtubules. In: Roberts K, Hyams JS (eds) Microtubules. Academic Press, London, pp 175–254
Schiff PB, Horowitz SB (1981) Taxol assembles tubulin in the absence of exogenous guanosine 5 – triphosphate and microtubule-associated proteins. Biochemistry 20:3247–3252
Schiff PB, Fant J, Horowitz SB (1979) Promotion of microtubule assembly in vitro by taxol. Nature 277:665–667
Schleicher M, Iverson DB, Van Eldik LJ, Watterson DM (1982) Calmodulin. In: Lloyd CW (ed) The cytoskeleton in plant growth and development. Academic Press, London, pp 85–106
Sheir-Neiss G, Nardi RV, Gealt MA, Morris WR (1976) Tubulin-like protein from *Aspergillus nidulans*. Biochem Biophys Res Commun 69:285–290
Shelanski ML, Gaskin F, Cantor CR (1973) Microtubule assembly in the absence of added nucleotides. Proc Natl Acad Sci USA 70:765–768
Sherline P, Schiavone K (1978) High molecular weight MAPs are part of the mitotic spindle. J Cell Biol 77:pp R9–R12
Slabas AR, MacDonald G, Lloyd CW (1980) Selective purification of plant proteins which co-polymerise with mammalian microtubules. FEBS Lett 110:77–79
Sloboda RD, Dentler WL, Rosenbaum JL (1976) Microtubule-associated proteins and the stimulation of tubulin assembly in vitro. Biochemistry 15:4497–4505
Snell WJ, Dentler WL, Haimo LT, Binder LI, Rosenbaum JL (1974) Assembly of chick brain tubulin onto isolated basal bodies of *Chlamydomonas reinhardi*. Science 185:357–360
Sternlicht H, Ringel I, Szasz J (1980) The co-polymerization of tubulin and tubulin-colchicine complex in the absence and presence of associated proteins. J Biol Chem 255:9138–9148
Valenzuela P, Quiroga M, Zaldivar J, Rutter WJ, Kirschner MW, Cleveland DW (1981) Nucleotide and corresponding amino acid sequences encoded by α and β tubulin mRNAs. Nature 289:650–655
Vallee R (1980) Structure and phosphorylation of microtubule-associated protein 2 (MAP 2). Proc Natl Acad Sci USA 3206–3210
Vallee RB (1982) A taxol-dependent procedure for the isolation of microtubules and microtubule-associated proteins (MAPs). J Cell Biol 92:435–442
Van der Valk P, Rennie PJ, Connolly JA, Fowke LC (1980) Distribution of cortical microtubules in tobacco protoplasts. An immunofluorescence microscopic and ultrastructural study. Protoplasma 105:27–43
Water RD, Kleinsmith LJ (1976) α- and β-tubulin in yeast. Biochem Biophys Res Commun 70:704–708
Waxman PG, Del Campo AA, Lowe MC, Hamel E (1981) Induction of polymerization of purified tubulin by sulfonate buffers. Eur J Biochem 120:129–136
Webb BC, Wilson L (1980) Cold-stable microtubules from brain. Biochemistry 19:1993–2001
Weingarten M, Suter D, Littman D, Kirschner MW (1974) Properties of the depolymerization products of microtubules from mammalian brain. Biochemistry 13:5529–5537
Weingarten MD, Lockwood AH, Hwo SY, Kirschner MW (1975) A protein factor essential for microtubule assembly. Proc Natl Acad Sci USA 72:1858–1862
Weisenberg RC (1972) Microtubule formation in vitro in solutions containing low calcium concentrations. Science 177:1104–1105
Weisenberg RC, Timasheff SN (1970) Aggregation of microtubule subunit protein. Effects of divalent cations, colchicine and vinblastine. Biochemistry 9:4110–4116
Weisenberg RC, Borisy GG, Taylor EW (1968) The colchicine-binding protein of mammalian brain and its relation to microtubules. Biochemistry 7:4466–4479
White E, Tolbert EM, Katz ER (1983) Identification of tubulin in *Dictyostelium discoideum*: characterization of some unique properties. J Cell Biol 97:1011–1019
Wick SM, Seagull RW, Osborn M, Weber K, Gunning BES (1981) Immunofluorescence microscopy of organized microtubule arrays in structurally stabilized meristematic plant cells. J Cell Biol 89:685–690

Williams Jr RC, Detrich HW (1979) Separation of tubulin from microtubule-associated proteins on phosphocellulose. Accompanying alterations in concentrations of buffer components. Biochemistry 18:2499–2503

Wilson L (1970) Properties of colchicine binding protein from chick embryo brain. Interactions with Vinca alkaloids and podophyllotoxin. Biochemistry 9:4999–5007

Wilson L, Freidkin M (1967) The biochemical events of mitosis. The in vivo and in vitro binding of colchicine in grasshopper embryos and its possible relation to inhibition of mitosis. Biochemistry 6:3126–3135

Wilson L, Bamberg JR, Mizel SB, Gisham LM, Creswell KM (1974) Interaction of drugs with microtubule proteins. Fed Proc 33:158–166

Witman GB, Plummer J, Sander G (1978) *Chlamydomonas* flagellar mutants lacking radial spokes and central tubules: structure, composition and function of specific axonemal components. J Cell Biol 76:729–747

Yadav NS, Filner P (1983) Tubulin from cultured tobacco cells: isolation and identification based on similarities to brain tubulin. Planta 157:46–52

Yelton DE, Scharff MD (1981) Monoclonal antibodies: a powerful new tool in biology and medicine. Annu Rev Biochem 50:657–680

Subject Index

MIX
Papier aus verantwortungsvollen Quellen
Paper from responsible sources
FSC® C105338

If you have any concerns about our products,
you can contact us on
ProductSafety@springernature.com

In case Publisher is established outside the EU,
the EU authorized representative is:
Springer Nature Customer Service Center GmbH
Europaplatz 3, 69115 Heidelberg, Germany

Printed by Libri Plureos GmbH
in Hamburg, Germany